Advances in
Fruit Processing Technologies

Contemporary Food Engineering

Series Editor

Professor Da-Wen Sun, Director

Food Refrigeration & Computerized Food Technology
National University of Ireland, Dublin
(University College Dublin)
Dublin, Ireland
http://www.ucd.ie/sun/

Contemporary Food
Engineering Series
Da-Wen Sun, Series Editor

Advances in Fruit Processing Technologies

Edited by
Sueli Rodrigues
Fabiano Andre Narciso Fernandes

CRC Press
Taylor & Francis Group
Boca Raton   London   New York

CRC Press is an imprint of the
Taylor & Francis Group, an **informa** business

CRC Press
Taylor & Francis Group
6000 Broken Sound Parkway NW, Suite 300
Boca Raton, FL 33487-2742

First issued in paperback 2016

© 2012 by Taylor & Francis Group, LLC
CRC Press is an imprint of Taylor & Francis Group, an Informa business

No claim to original U.S. Government works

Version Date: 20120424

ISBN 13: 978-1-138-19945-3 (pbk)
ISBN 13: 978-1-4398-5152-4 (hbk)

Library of Congress Cataloging-in-Publication Data

Advances in fruit processing technologies / editors, Sueli Rodrigues, Fabiano Andre Narciso Fernandes.
    p. cm. -- (Contemporary food engineering)
    Includes bibliographical references and index.
    ISBN 978-1-4398-5152-4 (hardback)
    1. Fruit--Processing. I. Rodrigues, Sueli. II. Narciso Fernandes, Fabiano Andre.

TP440.A288 2012
664'.8--dc23

2012013482

Visit the Taylor & Francis Web site at
http://www.taylorandfrancis.com

and the CRC Press Web site at
http://www.crcpress.com

Contents

Series Preface

Contemporary Food Engineering

Food engineering is the multidisciplinary field of applied physical sciences combined with the knowledge of product properties. Food engineers provide the technological knowledge transfer essential to the cost-effective production and commercialization of food products and services. In particular, food engineers develop and design processes and equipment to convert raw agricultural materials and ingredients into safe, convenient, and nutritious consumer food products. However, food engineering topics are continuously undergoing changes to meet diverse consumer demands, and the subject is being rapidly developed to reflect market needs.

In the development of food engineering, one of the many challenges is to employ modern tools and knowledge, such as computational materials science and nanotechnology, to develop new products and processes. Simultaneously, improving food quality, safety, and security continues to be a critical issue in food engineering study. New packaging materials and techniques are being developed to provide more protection to foods, and novel preservation technologies are emerging to enhance food security and defense. Additionally, process control and automation regularly appear among the top priorities identified in food engineering. Advanced monitoring and control systems are developed to facilitate automation and flexible food manufacturing. Furthermore, energy saving and minimization of environmental problems continue to be important food engineering issues, and significant progress is being made in waste management, efficient utilization of energy, and reduction of effluents and emissions in food production.

The Contemporary Food Engineering Series, consisting of edited books, attempts to address some of the recent developments in food engineering. The series covers advances in classical unit operations in engineering applied to food manufacturing as well as such topics as progress in the transport and storage of liquid and solid foods; heating, chilling, and freezing of foods; mass transfer in foods; chemical and biochemical aspects of food engineering and the use of kinetic analysis; dehydration, thermal processing, nonthermal processing, extrusion, liquid food concentration, membrane processes, and applications of membranes in food processing; shelf-life and electronic indicators in inventory management; sustainable technologies in food processing; and packaging, cleaning, and sanitation. These books are aimed at professional food scientists, academics researching food engineering problems, and graduate-level students.

The editors of these books are leading engineers and scientists from many parts of the world. All the editors were asked to present their books to address the market's need and pinpoint the cutting-edge technologies in food engineering.

All the contributions have been written by internationally renowned experts who have both academic and professional credentials. All the authors have attempted to provide critical, comprehensive, and readily accessible information on the art and science of a relevant topic in each chapter, with reference lists for further information. Therefore, each book can serve as an essential reference source to students and researchers in universities and research institutions.

Da-Wen Sun
Series Editor

Series Editor

Professor Da-Wen Sun, PhD, is a world authority on food engineering research and education; he is a member of the Royal Irish Academy, which is the highest academic honor in Ireland; he is also a member of Academia Europaea (The Academy of Europe). His main research activities include cooling, drying, and refrigeration processes and systems; quality and safety of food products; bioprocess simulation and optimization; and computer vision technology.

In particular, his innovative studies on vacuum cooling of cooked meat, pizza quality inspection using computer vision, and edible films for shelf-life extension of fruits and vegetables have been widely reported in the national and international media. Results of his work have been published in about 600 papers, including over 250 peer-reviewed journal papers (h-index = 36). He has also edited 13 authoritative books. According to Thomson Scientific's *Essential Science IndicatorsSM* updated as of July 1, 2010, based on data derived over a period of ten years and four months (January 1, 2000–April 30, 2010) from the ISI Web of Science, a total of 2554 scientists are among the top 1% of the most cited scientists in the category of agriculture sciences, and Professor Sun is listed at the top with a ranking of 31.

Dr. Sun received his first class BSc honors and his MSc in mechanical engineering, and his PhD in chemical engineering in China before working at various universities in Europe. He became the first Chinese national to be permanently employed in an Irish university when he was appointed a college lecturer at the National University of Ireland, Dublin (University College Dublin [UCD]), in 1995. He was then continuously promoted in the shortest possible time to the position of senior lecturer, associate professor, and full professor. Dr. Sun is now a professor of food and biosystems engineering and director of the Food Refrigeration and Computerized Food Technology Research Group at UCD.

As a leading educator in food engineering, Dr. Sun has contributed significantly to the field of food engineering. He has guided many PhD students who have made their own contributions to the industry and academia. He has also, on a regular basis, given lectures on the advances in food engineering at international academic institutions and delivered keynote speeches at international conferences. As a recognized authority in food engineering, Dr. Sun has been conferred adjunct/visiting/consulting professorships by over ten top universities in China, including Zhejiang University, Shanghai Jiaotong University, Harbin Institute of Technology, China Agricultural University, South China University of Technology, and Jiangnan University. In recognition of his significant contribution to food engineering worldwide and for his outstanding leadership in the field, the International Commission of Agricultural and

Biosystems Engineering (CIGR) awarded him the CIGR Merit Award in 2000 and again in 2006; the U.K.-based Institution of Mechanical Engineers named him Food Engineer of the Year 2004; in 2008, he was awarded the CIGR Recognition Award in recognition of his distinguished achievements as the top 1% of agricultural engineering scientists around the world; in 2007, he was presented with the only AFST(I) Fellow Award in that year by the Association of Food Scientists and Technologists (India); and in 2010, he was presented with the CIGR Fellow Award (the title of "Fellow" is the highest honor in CIGR and is conferred upon individuals who have made sustained, outstanding contributions worldwide).

Dr. Sun is a fellow of the Institution of Agricultural Engineers and a fellow of Engineers Ireland (the Institution of Engineers of Ireland). He has also received numerous awards for teaching and research excellence, including the President's Research Fellowship, and has received the President's Research Award from UCD on two occasions. He is also the editor in chief of *Food and Bioprocess Technology— An International Journal* (Springer) (2010 Impact Factor = 3.576, ranked at the fourth position among 126 ISI-listed food science and technology journals); series editor of the Contemporary Food Engineering Series (CRC Press/Taylor & Francis Group); former editor of *Journal of Food Engineering* (Elsevier); and an editorial board member of *Journal of Food Engineering* (Elsevier), *Journal of Food Process Engineering* (Blackwell), *Sensing and Instrumentation for Food Quality and Safety* (Springer), and *Czech Journal of Food Sciences*. Dr. Sun is also a chartered engineer.

On May 28, 2010, he was awarded membership to the Royal Irish Academy (RIA), which is the highest honor that can be attained by scholars and scientists working in Ireland. At the 51st CIGR General Assembly held during the CIGR World Congress in Quebec City, Canada, in June 2010, he was elected as incoming president of CIGR and will become CIGR president in 2013–2014. The term of the presidency is six years—two years each for serving as incoming president, president, and past president. On September 20, 2011, he was elected to Academia Europaea (The Academy of Europe), which is functioning as European Academy of Humanities, Letters and Sciences and is one of the most prestigious academies in the world; election to the Academia Europaea represents the highest academic distinction.

Preface

Fruits are major food products and key ingredients in many processed foods. They are a rich source of vital nutrients and constitute an important component of human nutrition. Consumers nowadays are more aware of the importance of healthy foods and require food products with high nutritional value along with high standards of sensory characteristics. Thus, fruit processing has to preserve the nutritional value of the fruit, while also preserving its natural color and flavor. This book reviews new advances in fruit-processing technologies.

Fruits are highly perishable, and about 20%–40% of the fruits produced are wasted from the time of harvesting till they reach the consumers, either in natural form or in processed form. To reduce fruit loss and improve final sensory characteristics of processed fruits, new or improved technologies have been applied to fruit processing. This book reviews new technologies, such as ozone application, ultrasound processing, irradiation application, pulsed electric field, vacuum frying, and high-pressure processing, and improved technologies, such as ultraviolet and membrane processing, enzymatic maceration, freeze concentration, and refrigeration.

The effect of processing on sensory characteristics and nutritional value is addressed in each chapter. New trends in modified atmosphere packaging, effects of processing on aroma, and the use of fruit juices as a vehicle for probiotic microorganisms and prebiotic oligosaccharides as an alternative for dairy products are also covered in this book.

Editors

Sueli Rodrigues is currently a professor of food engineering at the Federal University of Ceará, Fortaleza, Brazil, where she teaches and does research on process and product development. She graduated in chemical engineering from the State University of Campinas, Campinas, São Paulo, Brazil, and received her PhD in chemical engineering in 2003 from the same university.

Dr. Rodrigues has published more than 65 papers in scientific journals. Her research interests include bioprocess, ultrasound, and drying technology, especially with fruit and functional food processing.

Fabiano André Narciso Fernandes is currently a professor of chemical engineering at the Federal University of Ceará, Fortaleza, Brazil, where he teaches and does research on process and product development. He graduated in chemical engineering from the Federal University of São Carlos, São Carlos, São Paulo, Brazil. He received his PhD in chemical engineering in 2002 from the Sate University of Campinas, Campinas, São Paulo, Brazil.

Dr. Fernandes has published more than 90 papers in scientific journals. His research interests include drying technology, ultrasound technology, and the field of reaction engineering.

Contributors

Ingrid Aguiló-Aguayo
Department of Food Technology
University of Lleida
Lleida, Spain

Andréa Cardoso de Aquino
Department of Chemical Engineering
Federal University of Ceará
Ceará, Brazil

Josep Maria Auleda
Department of Agri-Food Engineering
   and Biotechnology
Technical University of Catalonia
Barcelona, Spain

Henriette Monteiro Cordeiro de Azeredo
Embrapa Tropical Agroindustry
Brazilian Agricultural Research
   Corporation
Fortaleza, Brazil

Zoraia de Jesus Barros
Department of Plant, Soil, and Insect
   Sciences
University of Massachusetts
Amherst, Massachusetts

Maria do Socorro Rocha Bastos
Embrapa Tropical Agroindustry
Brazilian Agricultural Research
   Corporation
Fortaleza, Brazil

Elena M. Castell-Perez
Department of Biological and
   Agricultural Engineering
Texas A&M University
College Station, Texas

Edmar Clemente
Department of Chemistry
State University of Maringá
Maringá, Brazil

José Maria Correia da Costa
Department of Food Technology
Federal University of Ceará
Fortaleza, Brazil

Patrick J. Cullen
School of Food Science and
   Environmental Health
Dublin Institute of Technology
Dublin, Ireland

Sunando DasGupta
Department of Chemical Engineering
Indian Institute of Technology
Kharagpur, India

Pedro Elez-Martínez
Department of Food Technology
University of Lleida
Lleida, Spain

Heliofábia Virginia de Vasconcelos Facundo
Department of Food and Experimental
   Nutrition
University of São Paulo
São Paulo, Brazil

Fabiano André Narciso Fernandes
Department of Chemical Engineering
Federal University of Ceará
Fortaleza, Brazil

Deborah dos Santos Garruti
Embrapa Tropical Agroindustry
Brazilian Agricultural Research
    Corporation
Fortaleza, Brazil

Eduard Hernández
Department of Agri-Food Engineering
    and Biotechnology
Technical University of Catalonia
Barcelona, Spain

Tatiana Koutchma
Guelph Food Research Center
Agriculture and Agri-Food Canada
Guelph, Ontario, Canada

Janice Ribeiro Lima
Embrapa Tropical Agroindustry
Brazilian Agricultural Research
    Corporation
Fortaleza, Brazil

Frank Mangan
Department of Plant, Soil, and Insect
    Sciences
University of Massachusetts
Amherst, Massachusetts

Olga Martín-Belloso
Department of Food Technology
University of Lleida
Lleida, Spain

Fátima A. Miller
Center of Biotechnology and Fine
    Chemistry
Biotechnology Higher School
Catholic University of Portugal
Porto, Portugal

Jane de Jesus da Silveira Moreira
Department of Food Technology
Federal University of Sergipe
São Cristóvão, Brazil

Rosana G. Moreira
Department of Biological and
    Agricultural Engineering
Texas A&M University
College Station, Texas

Narendra Narain
Department of Food Technology
Federal University of Sergipe
São Cristóvão, Brazil

Marta Orlowska
Guelph Food Research Center
Agriculture and Agri-Food Canada
Guelph, Ontario, Canada

Mercé Raventós
Department of Agri-Food Engineering
    and Biotechnology
Technical University of Catalonia
Barcelona, Spain

Sueli Rodrigues
Department of Food Technology
Federal University of Ceará
Fortaleza, Brazil

Biswajit Sarkar
University School of Chemical
    Technology
Guru Gobind Singh Indraprastha
    University
New Delhi, India

Cristina L.M. Silva
Center of Biotechnology and Fine
    Chemistry
Biotechnology Higher School
Catholic University of Portugal
Porto, Portugal

Ebenézer de Oliveira Silva
Embrapa Tropical Agroindustry
Brazilian Agricultural Research
    Corporation
Fortaleza, Brazil

Robert Soliva-Fortuny
Department of Food Technology
University of Lleida
Lleida, Spain

Brijesh K. Tiwari
Hollings Faculty
Manchester Food Research Centre
Manchester Metropolitan University
Manchester, United Kingdom

Nédio Jair Wurlitzer
Embrapa Tropical Agroindustry
Brazilian Agricultural Research
    Corporation
Fortaleza, Brazil

1 Ultraviolet Light for Processing Fruits and Fruit Products

Tatiana Koutchma and Marta Orlowska

CONTENTS

1.1 INTRODUCTION

During the last decade, there has been an increase in the production of fresh fruit and fruit products due to the health properties of fruits. Fruit products can be consumed in raw, minimally processed or processed, ready-to-eat/ready-to-drink forms as whole fresh fruits, fresh-cut fruits, and fruits as ingredients, beverages, juices, and jams. The processing of fruits starts after harvesting and consists of four activities: stabilization or preservation, transformation, production of ingredients, and production of fabricated foods. The role of processing technology in each stage implies the control of microbiological, chemical, and biochemical changes, which occur as a result of microbial and enzymatic activities, and oxidation reactions, which can lead to problems of safety, color, flavor, taste, and texture. Processing technologies that do not significantly alter the organoleptic or nutritional qualities of fruits and do not form any undesirable chemical compounds in the product have obvious advantages in modern food production. The interest in so-called minimal processing technologies led to the development of nonthermal or mild heat high-tech methods that have the potential to replace traditional thermal preservation techniques. They result not only in better quality and longer shelf life but also, potentially, in higher nutritional value or products with health benefits. A large number of studies have associated consumption of fruits and their products with decreased risk of development of diseases such as cancer and coronary heart disease (Hansen et al., 2003). This may be due to the presence of health-promoting phytochemicals such as carotenoids, flavonoids, phenolic compounds, and vitamins (Gardner et al., 2000), which have, in some cases, been shown to have disease-preventing properties. In this respect, it is of paramount importance to develop processing methods that preserve not only the safety of fruits but also the sensorial and nutritional quality and bioactivity of the constituents present in fruits and their products.

Ultraviolet (UV) light treatment of foods is a nonthermal physical method of preservation that is free of chemicals and waste effluents, which makes it ecologically friendly. It does not produce by-products. It is safe to use, although precautions must be taken to avoid human exposure to UV light and to evacuate ozone generated by vacuum and far UV wavelengths. Based on recent engineering advances and new scientific data, UV light technology in continuous and pulsed modes (cUV and PL) offers the promise of enhanced microbiological safety of fresh fruits and improved quality of fruit products that have a freshness of flavor, color, texture, and nutritional value closer to those of nontreated products. The discovery of UV inactivation of the chlorine-resistant parasites *Cryptosporidium parvum* and *Giardia* sp. has catalyzed the use of UV light in the water industry (Hijnen et al., 2006). UV has been utilized similarly in the disinfection of air, nonfood contact, and food contact surfaces, and recently was used for treatments of surfaces of solid foods and liquid foods, beverages, and ingredients. Reports are available that indicate that application of UV light can also improve the toxicological safety of foods of plant origin through its ability to reduce levels of toxins such as patulin mycotoxin in fresh apple cider (Dong et al., 2010) and possibly to control browning through its effects on enzymes (Manzocco et al., 2009). Regarding the preservation of organoleptic and nutritional attributes, recent research has shown promising results in the exposure of fruit products to UV irradiation. In addition to higher cost-efficiency, sustainability, and broad antimicrobial effects, UV light not only minimally affects quality attributes but also has beneficial effects on the content of bioactive compounds. It has the potential for obtaining premium quality products that can lead to faster commercialization.

This chapter aims to provide detailed and critical information on the latest applications of continuous and pulsed UV light for processing fresh fruits and fruits products. The fundamental principles and features of UV light generation, propagation, and photochemistry are briefly reviewed, and the control measures to be adopted where UV technology can be utilized to enhance safety during fruit production are analyzed. Particular focus is given to the effects of UV light on the survival of pathogenic and spoilage microorganisms typical to fruits and the environment essential in fruit processing followed by a discussion of recent research into effects of UV light on quality and bioactive compounds.

1.2 BASICS OF UV PROCESSING OF FOODS OF PLANT ORIGIN

1.2.1 UV Light Sources

Light is emitted from gas discharge at wavelengths dependent on its elemental composition and the excitation, ionization, and kinetic energy of those elements. Gas discharges are responsible for the light emitted from UV lamps. UV light transfer phenomenon is defined by the emission characteristics of the UV source along with considering long-term lamp aging and absorbance/scattering of the product. Consequently, the performance of a UV system depends on the correct matching of the UV source parameters to the demands of the UV application. The commercially available UV sources include low- and medium-pressure mercury lamps (LPM and MPM), excimer lamps (EL), pulsed lamps (PL), and light-emitting diodes (LED). LPM and EL are monochromatic sources, whereas emission of MPM and PL is polychromatic. There are no reports on the application of EL in fruit processing, so this UV source is not discussed in this chapter.

1.2.1.1 Mercury Lamps

Mercury vapor UV lamp sources have been successfully used in water treatment for nearly 50 years and are considered as reliable sources for other disinfection treatments that benefit from their performance, low cost, and quality. Typically, three general types of mercury UV lamps are used: low-pressure (LPM), low-pressure high-output (LPHO), and medium-pressure (MPM). These terms are based on the vapor pressure of mercury when the lamps are operating. The effects of mercury vapor pressure on spectra distribution is shown in Figure 1.1. Vapor discharge lamps consist of a UV-transmitting envelope made from a tube of vitreous silica

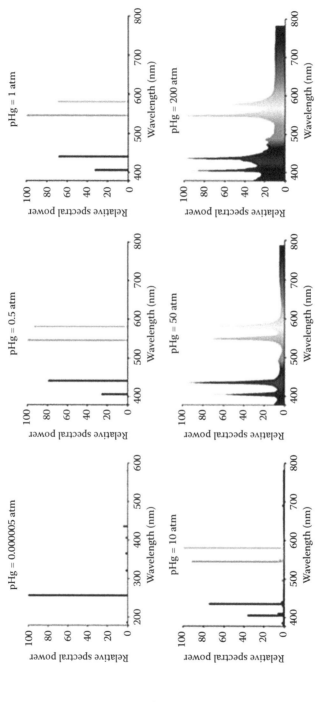

FIGURE 1.1 Effects of vapor pressure of mercury on output spectra distribution.

glass sealed at both ends. The envelope is filled with mercury and an inert gas. Argon, the most common filler, has ionization energy of 15.8 eV, whereas the lowest activated metastable state is at 11.6 eV (Masschelein and Rice, 2002).

LPM lamps are operated at nominal total gas pressures of 10^2–10^3 Pa, which corresponds to the vapor pressure of mercury at a temperature of 40°C. The emission spectrum of LPM is concentrated at the resonance lines at 253.7 and 185 nm. The 253.7 nm line represents around 85% of the total UV intensity emitted and is directly related to the germicidal effect. The wavelength of 253.7 nm is most efficient in terms of germicidal effect since photons are absorbed most by the DNA of microorganisms at this specific wavelength. Light with a wavelength below 230 nm is most effective for the dissociation of chemical compounds. At wavelengths of 185 nm, ozone is produced from oxygen and organic compounds can be oxidized (Voronov, 2007). The photons with the wavelength of 185 nm are responsible for ozone production, and the combination of both wavelengths is a very effective means for photochemical air treatment. The ratio of light at 185 nm to light at 253.7 nm varies from 12% to 34% depending on the operating current, wall temperature, and inert gas. The U.S. FDA regulations approved the use of LPM lamps for juice processing, and they have already been successfully commercialized (U.S. FDA, CFR part 179, 2000).

MPM lamps are operated at a total gas pressure of 104–106 Pa (Masschelein and Rice, 2002). Compared with the LPM lamps, the coolest possible temperature of the MPM is about 400°C, whereas it goes up to 600°C and even 800°C in a stable operation. MPM lamps operate in the potential gradient range of 5–30 W/cm. The emission spectrum of MPM covers wavelengths from about 250 nm to almost 600 nm, which results from a series of emissions in the UV and in the visible range. MPM lamps are not considered to be useful for targeted germicidal treatment; however, their strong UV radiation flux results in high penetration depth. By varying the gas filling, doping, and the quartz material, the spectrum as well as the radiation flux of the UV lamps can be varied and matched to suit specific food processing applications, especially for oxidation or photodegradation.

Recently, LPHO amalgam lamps that contain a mercury amalgam was developed and incorporated into disinfection applications; however, LPM and MPM are the dominant sources for UV disinfection treatment.

1.2.1.2 Environmental Impact

The potential mercury exposure due to lamp sleeve breakage is a health concern. Breakage of lamps can occur when lamps are in operation and during maintenance. The mercury contained within a UV lamp is isolated from exposure by the lamp envelope and surrounding lamp sleeve. For the mercury to be released, both the lamp and lamp sleeve must break. The mercury content in a single UV lamp used for water treatment typically ranges from 0.005 to 0.4 g (5–400 mg). LPM lamps have less mercury (5–50 mg/lamp) compared with LPHO (26–150 mg/lamp) and MPM lamps (200–400 mg/lamp). The EPA established a maximum contaminant level (MCL) for mercury at 0.002 mg/L. The EPA has found mercury to potentially cause kidney damage from short-term exposures at levels above the 0.002 mg/L MCL (EPA, 1995). The concern over the impact of mercury release into the food plant environment stimulated the development and validation of lamps with mercury-free special technologies.

1.2.1.3 Pulsed Lamps

The efficacy of pulsed flash lamps (PL) is potentially greater than continuous sources due to high intensity and broader spectrum. PL technologies are promising due to their instant start, high intensity, and robust packaging with no mercury in the lamp, but more research is needed to establish them for fruit treatment applications. In this technology, alternating current is stored in a capacitor and energy is discharged through a high-speed switch to form a pulse of intense emission of light within about 100 ms. The emission is similar in wavelength composition to the solar light. The UV pulsed devices can deliver high-intensity UV, which can both penetrate opaque liquids better than mercury lamps and provide enhanced treatment rates. Figure 1.2 shows the normalized spectra of these three UV sources—LPM, MPM, and PL. Individual spectra are not comparable on a UV intensity basis but are comparable on a spectral basis with reference to which wavelengths dominate the respective wavelength outputs.

Table 1.1 provides a summary of some of the basic characteristics of common UV sources in commercial use and under development that can be used for comparison purposes. It is evident that no single lamp technology will represent the best source for all food applications. However, situation-specific requirements may dictate a clear advantage for a given process technology. For UV reactors containing LPM or LPHO mercury lamps, UV absorbance and transmittance at 253.7 nm are important design parameters. However, for broadband UV lamps, such as MPM or PL UV lamps, it is important to measure the full scan of absorbance or transmittance in the germicidal region from 200 to 400 nm. Lamps with special technologies such as PL UV and EL are promising due to the different spectral bands or specific wavelengths that they can provide with regard to effects on quality attributes. They also have instant start and robust packaging with no mercury in the lamp. However, more research is needed to establish their suitability for fruit processing applications.

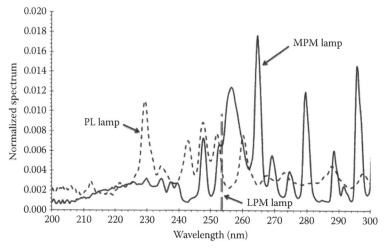

FIGURE 1.2 Comparison of spectrums of continuous (LPM and MPM) lamps and PL UV sources.

TABLE 1.1

Comparison of Efficiency Characteristics of Continuous and Pulsed UV Sources

UV Source	Electrical Efficiency (%)	UV Efficiency (%)	UV Intensity (W/cm²)	Lamp Surface T (°C)	Lifetime, Month	Output Spectrum
LPM	50	38	0.001–0.01	40	18–24	Monochromatic 253.7 nm
Excimer	10–25	10–30	0.05–0.5	Ambient	13	Monochromatic selectable
MPM	15–30	12	12	400–1,000	0.5	Polychromatic 200–400 nm
Flash xenon	45–50	9	600	1,000–10,000	1	Polychromatic 100–1000 nm
Surface discharge	15–20	17	30,000	NA	NA	Polychromatic 200–800 nm

1.2.1.4 Light Emitting Diodes

In recent years, UV-LEDs have been developed with the following advantages: low cost, energy efficiency, long life, easy control of emission, and no production of mercury waste. The wavelength of the commercial UV-LED is around UV-A range (315–400 nm) and enables new applications in existing markets as well as open new areas. An LED is a semiconductor device that emits light when carriers of different polarities (electron and holes) combine generating a photon. The wavelength of the photon depends on the energy difference the carriers overcome in order to combine. An example of a UV-LED system that operates between 210 and 365 nm is the one formed by aluminum nitride (AIN), gallium nitride (GaN), and intermediate alloys. Currently, UV-LEDs are commercially available in research grade and limited quantities and their lifetimes reach the order of 200 h. It is very likely that in the near future, many applications that make use of mercury lamps today will be carried out by UV-LEDs.

1.2.2 UV Light Propagation

UV light emitted from the atoms and ions within the gas discharge of a UV source will propagate away from those atoms and ions. As UV light propagates, it interacts with the materials it encounters through absorption, reflection, refraction, and scattering. Each of these phenomena influences the intensity and wavelength of the UV light reaching the bacteria or chemical compound on the surface or in the liquid.

Absorption (A) of light is the transformation of energy of light photons to other forms of energy as it travels through a substance. *Reflection* (R) is the change in the direction of propagation experienced by light deflected by an interface. *Scattering* is the phenomenon that includes any process that deflects electromagnetic radiation from a straight path through an absorber when photons interact with a particle.

The scattering phenomenon plays an important role in disinfecting food liquids containing particles. Experimental measurements are usually made in terms of *transmittance* of a substance (T) or (UVT), which is defined as the ratio of the transmitted to the incident light irradiance. A convenient way of presenting information about UVT of materials is to give the values of their absorption coefficient at various wavelengths, over a given depth (e.g., 1 cm). Knowing this, the transmittance for any particular depth and the depth of the liquid that will absorb 90% of the energy at 253.7 nm can be calculated. Other important terms to characterize UV light treatments in fruit processing are *fluence rate* and *fluence*. Fluence rate is "the total radiant power incident from all directions onto an infinitesimally small sphere of cross-sectional area dA, divided by dA" (Bolton and Linden, 2003). Fluence is defined as the fluence rate multiplied by the exposure time. The term *UV dose* should be avoided as a synonym of fluence because dose refers in other contexts to absorbed energy, but only a small fraction of all incident UV light is absorbed by microorganisms (Bolton and Linden, 2003). In the case of PL, fluence is determined as energy per pulse multiplied by the number of pulses. The absorbed fluence indicates that radiant energy is available for driving the solution reaction. However, when UV light is absorbed by solution, it is no longer available for inactivating the microorganisms. The remaining interactions, including reflection, refraction, and scattering, change the direction of UV light, but the light is still available for inactivation. The radiant energy delivered to the molecule or microorganism is called the effective or delivered germicidal UV dose. Microbial inactivation depends primarily on the effective dose. The formulas for calculations of the critical UV process parameters are available in the literature (Koutchma et al., 2008).

1.2.3 BASIC PRINCIPLES OF PHOTOCHEMISTRY

Photochemical reactions proceed as a direct result of radiation energy (photons) being introduced to a system. In view of the wavelengths used in most UV-light treatments, the molecules (A) are primarily affected by energy absorption that results in photochemical reactions. In the general case, the process may be viewed as

$$A + h\nu \ \text{®} \ A_+^+ \ \text{®} \ Products \tag{1.1}$$

The first step in this reaction is the absorbance of a photon by a reactant molecule (A), leading to the production of an electronically excited intermediate. The excited state can be for a period of 10^{-10}–10^{-8} s in which the energy of the electrons is increased by the amount of photon energy. Under some conditions, the intermediate may undergo a chemical change to yield products that are relatively stable. For a photochemical reaction to proceed, photons must have sufficient energy to promote the reaction to break or form a bond and photon energy must be absorbed to promote reactions. The bond energies of interest are generally coincident with photon energies in the UV portion of the spectrum. In particular, radiation with wavelength less than approximately 320 nm appears to be sufficiently energetic to promote photochemical reactions in biomolecules. The extent of chemical reaction depends upon the quantum yield and fluence of incident photons. A quantum yield is a ratio of absorbed photons that cause a chemical change to the total absorbed photons. UV light at 253.7 nm has a radiant energy of (472.27 kJ/Einstein) 112.8 kcal /Einstein (one Einstein represents one mole of photons). It is theoretically possible for 253.7 nm light to affect the O–H, C–C, C–H, C–N, H–N, and S–S bonds if it's absorbed.

1.3 UV LIGHT CONTROL MEASURES IN FRUIT PROCESSING FACILITIES

During the manufacturing process, fruits can be exposed to microbiological cross-contamination from surfaces, water, and the air, which can cause their spoilage and raise safety issues. The traditional approach to controlling such contamination has been to target specific sites within the manufacturing environment with cleaning and disinfection regimes. Sanitation, disinfection, and oxidation with UV light is a versatile, environmental-friendly technology, which can be used in fruit processing facilities for the treatment of air, surfaces, and water to reduce microbial contamination in different unit operations of plant foods production that are becoming more and more popular.

1.3.1 AIR TREATMENT

Clean, fresh air is the basis of the industrial production of food products of plant origin. Microorganisms in the air, such as viruses, bacteria, yeasts, and fungi, can contaminate raw materials and intermediate products and spoil finished products during their processing and packaging. LPM sources are used very successfully in these applications, for disinfection in air intake ducting and storerooms and to ensure air of very low germ content in production areas. Short-wave UV radiation at 185 nm produces ozone from the oxygen in the ambient air so that this is activated for the oxidation process. UV oxidation breaks down pollutants in the exhaust air. For providing clean air in sensitive manufacturing food facilities, a combination of filters and UV light has been recommended. Basically, two applications of UV are becoming common. In one, the moving air stream is disinfected in much the same manner as with a water system. In the other application, stationary components of the system such as air conditioning coils, drain pans, and filter surfaces are exposed to help prevent mold and bacteria growth or to disinfect the filter to aid in handling. The UVT in air is higher than that in water, and, therefore, the number of lamps required in a large duct is quite reasonable. Common airborne virus and bacteria are readily deactivated with UV. Fungi (molds and spores) require much higher doses. In the moving air stream, high wattage lamps are used, usually without a quartz sleeve. UV lamp fixtures are placed in such a manner as to completely irradiate surfaces where bacteria and mold might collect and grow. Mathematical modeling software and bioassay testing have been developed to allow efficient design and validation of these systems (Kowalski and Bahnfleth, 2002). Low operating costs and reasonable equipment costs can make UV very cost effective.

1.3.2 WATER TREATMENT

Control of microorganisms in industrial process waters is often necessary to maintain quality of the product or process. The fruit industry is a large-volume consumer of water, and the potential for reuse or recycling of fruit processing water represents an attractive economic benefit to the industry. A combination of UV light and ozone has a powerful oxidizing action to reduce microbial load and the organic content of water to very low levels.

1.3.3 NONFOOD AND FOOD CONTACT SURFACE DISINFECTION

UV light is an economical step toward improved hygiene control measures in the food industry. Mold and biofilms can develop on nonfood surfaces (ceilings, walls, floors) and equipment including tanks and vats, cooling coils, and food contact surfaces of equipment such as cutting equipment and conveyor belts (Kowalski, 2006). In general, standard cleaning and disinfection procedures are adequate to contain these problems, but alternatives are available, including antimicrobial coatings like copper and TiO_2. UV irradiation of food processing equipment and surfaces, cooling coil disinfection systems, whole area UV disinfection, and after-hours irradiation of rooms when personnel are not present are all viable control options for maintaining high levels of sanitation and disinfection in food industry facilities (Kowalski and Dunn, 2002). UV light kills up to 99.9% of total germs on conveyor belts for transporting fruits and vegetables.

1.3.4 PACKAGING

The packaging technologies play an important role in extending the shelf life of fruits. UV light might be applied as pre- or postpackaging technology to reduce the microbial spoilage. As a prepackaging control measure, UV treatment of packaging in fruit filling plants, for example, for lids, cups, sealing, and packaging foils for drinks and beverages, helps to extend the shelf life of fruit stuff. When using cUV and PL as postpackaging treatment for packaged fruits, the considerations about transparency are referred to the packaging materials. For example, materials such as glass, polystyrene, and PET, which allow visible light to penetrate through the container, are not transparent to the UV wavelengths that are essential for microbial inactivation, and therefore, they are not suitable for cUV and PL treatments. On the other hand, polymers such as polyethylene, polypropylene, polybutylene, EVA, nylon, Aclar, and EVOH transmit UV light and hence meet the requirements for PLT very well (Anonymous, 2000). In addition, ink-printed labels or drawings could interfere with the light absorption of the treated item and should be avoided on the surface of packaging materials. Besides the intrinsic transparency of the material, for the success of a UV process it is very critical that the "condition" of the item to be treated is suitable for the penetration of the light. This means that the product surface should be smooth, clear, and without roughness, pores, and grooves, which could "shadow" the microbial cells from the light, causing less complete light diffusion and thus reducing process effectiveness; for the same reason, the item to be treated should be clean and free of contaminating particulates. In addition, items that have a complex geometry could have areas hidden from the light and could require a more accurate design of the treatment chamber in order for the light pulses to reach each point of the product surface.

TABLE 1.2
Application of UV Light Sources as a Control Measure in Fruit Processing

	Reported Applications			
UV Source	Processed Water	Air	Surfaces	Low UVT Juices
LPM	X	X	X	X
MPM	X	X		
Excimer lasers	X		X	X
Pulsed			X	X
LED	X	X	X	

1.3.5 Fresh Fruit and Cut Fruit Surfaces Treatment

CUV and PL treatments result in various levels of inactivation of spoilage and pathogenic microflora on the surface of a wide variety of solid foods. Comprehensive reviews of the literature in this field have been compiled by the U.S. FDA (2000) and by Woodling and Moraru (2005). The variability of the results (a 2- to 8-log reduction was generally reported) is most likely due to the different challenge microorganisms used in various studies, the intensity of the treatment, and the different properties of the treated substrates. Woodling and Moraru (2005) demonstrated that the efficacy of PL is affected by substrate properties such as topography and hydrophobicity, which affect both the distribution of microbial cells on the substrate surface and the interaction between light and the substrate (i.e., reflection and absorption of light). Surface disinfection of fresh and cut fruit products is a basis for longer shelf life. In designing a PL treatment for fruit items, both source (light wavelength, energy density, duration and number of the pulses, interval between pulses) and target (product transparency, color, size, smoothness, and cleanliness of surface) parameters are critical for process optimization, in order to maximize the effectiveness of product microbial inactivation and to minimize product alteration. Such alteration can be mainly determined by an excessive increase in temperature causing thermal damage to fruits and also by an excessive content of UV-C light, which could result in some undesired photochemical damage to fruit itself or to packaging materials. Table 1.2 summarizes current and future applications of cUV and PL available sources in fruit processing for air, surface, water, and low UVT fruit drinks and beverages.

1.4 UV TREATMENT OF WHOLE FRESH FRUITS

1.4.1 Antimicrobial Effect

Traditionally UV-light applications for treatments of whole fruits and vegetables were focused on the disinfection role with the objective to extend the shelf life as naturally occurring microflora may present on the surface of raw produce both of nonpathogenic or spoilage and pathogenic nature (Table 1.3).

TABLE 1.3
Fresh Produce and Typical Microflora Present on the Surface

Commodity	Microflora	Reference
Fruits (in general)	Fungi: *B. cinerea, Aspergillus niger*; Yeasts: *Canidia, Cryptococcus, Fabospora, Kluyveromyces, Pichia, Saccharomyces*, and *Zygosaccharomyces*; Bacteria: *Shigella* spp.	Martin-Belloso et al. (2006)
Carrot	*B. cinerea*	Mercier et al. (1993)
Lettuce	*Enterobacter, Erwinia, Escherichia, Leuconostoc, Pantoea, Pseudomonas, Rahnela, Salmonella, Serratia*, and *Yersinia*	Allende et al. (2006)
Tomato	*B. cinerea*	Charles et al. (2008)
Apple	*E. coli* O157:H7	Martin-Belloso et al. (2006)
Raspberry	*Cyclospora cayetanensis*	Martin-Belloso et al. (2006)
Strawberry	*Campylobacter jejuni* *B. cinerea*	Martin-Belloso et al. (2006), Erkan et al. (2008), Pombo et al. (2011)
Watermelon	*Salmonella* spp., *Shigella* spp.	Martin-Belloso et al. (2006)
Cantaloupe	*Campylobacter jejuni*	Martin-Belloso et al. (2006)
Pineapple	*E. coli* O157:H7, *Salmonella*	Strawn and Danyluk (2010)

During storage, fruits undergo biochemical and physiological changes that can result in loss of nutrients, color changes, and tissue disruption. Along with these undesirable changes, crops become more susceptible to pathogenic decay, which increases the possibility of illness incidences and also causes large economic losses.

Enhanced shelf life of UV-treated fruits can be associated with the germicidal effect on pathogens that may be present on the surface of the crops. However, the UV treatment requires that the whole surface of the object is exposed to the UV light for a time sufficient for any microorganisms present to accumulate a lethal dose. This also means that the topography of the surface determines the efficacy of UV treatment and presents its limitation due to shielding effects. The importance of the fruit positioning during the UV-C exposure of strawberries was reported by Stevens et al. (2005). The authors found that irradiation of the stem ends of the fruits resulted in lower decay during subsequent storage in comparison with the fruits exposing only one or two different sides to UV-C light.

Several studies have shown that UV processing of fresh produce is effective in the reduction of pathogenic bacterial population. For instance, Yaun et al. (2004) inoculated the surface of Red Delicious apples, leaf lettuce, and tomatoes with cultures of *Salmonella* spp. or *Escherichia coli* O157:H7. UV-C (253.7 nm) applied to apples inoculated with *E. coli* O157:H7 resulted in the highest log reduction of approximately 3.3 logs at 240 W/m². Lower log reductions were seen on tomatoes inoculated with *Salmonella* spp. (2.19 logs) and green leaf lettuce inoculated with both *Salmonella* spp. and *E. coli* O157:H7 (2.65 and 2.79 logs, respectively). PL UV light was also applied to reduce the population of pathogenic bacteria on the surface of fruits. For instance, Bialka and Demirci (2007, 2008) exposed blueberries inoculated with *E. coli* O157:H7 and *Salmonella* to the PL (Xenon Corp.) emitting in a range from 100 to 1100 nm for 5, 10, 30, 45, and 60 s. The authors reported reductions between 1.1 and 4.3 $\log_{10}$ CFU/g of *E. coli* O157:H7 and 1.1 and 2.9 $\log_{10}$ CFU/g of *Salmonella*. Due to the high-intensity nature of the PL source, substantial increase in the temperature was observed during fruit processing that could contribute to the microbial reduction. The impact of the PL treatment on the nutrients of treated crops has not been studied yet.

The deterioration of many fresh fruits can be caused by fungi, which give rise to various infections on harvested plant produce. For example, *Monilinia fructicola* is the main cause of brown rot in peaches, apricots, nectarines, and plums. Stevens et al. (1998) revealed that UV treatment can reduce the fungal population on peaches. The surfaces of peaches were inoculated with spores of the *M. fructicola* and then fruits were subjected to the UV light. At UV fluence of $4.8 \, \text{kJ/m}^2$, a decrease in growth of *M. fructicola* by approximately one order of magnitude was observed. Another study performed by Stevens et al. (2005) has shown that UV-C (253.7 nm) treatment at 7.5 and $1.3 \, \text{kJ/m}^2$ resulted in higher resistance to bitter rot (*Colletotrichum gloeosporioides*), brown rot (*M. fructicola*), and green mold (*Penicillium digitatum*) in apples, peaches, and tangerines. González-Aguilar et al. (2001, 2007) demonstrated that exposure to UV-C light in the range of 250–280 nm at $4.93 \, \text{kJ/m}^2$ lowered fungal decay of mango fruits stored for 18 days at 25°C by 60%. Significantly lower incidence of decay was also observed after UV-C treatment in kumquat fruit and bitter orange (*Citrus aurantium*) inoculated with *P. digitatum* (Rodov et al., 1992; Arcas et al., 2000). In the case of papaya fruits inoculated with *Colletotrichum gloeosporioides*, none of the UV-C (253.7 nm) treatments ($0.2–2.4 \, \text{kJ/m}^2$) was effective against anthracnose fungal sporulation (Cia et al., 2007). Another fungus, *Botrytis cinerea*, is the main cause of gray mold rot in many crops. Exposure to the UV-C light with the peak at the wavelength of 253.7 nm reduced the *B. cinerea* growth in carrots (Mercier et al., 1993), tomatoes at UV fluence of $3.7 \, \text{kJ/m}^2$ (Charles et al., 2008), pepper fruits at UV fluence of $7 \, \text{kJ/m}^2$ (Vicente et al., 2005), and strawberries (Erkan et al., 2008; Pombo et al., 2011). Erkan et al. (2008) reported that in UV-treated strawberries at fluence levels of 0.43, 2.15, and $4.30 \, \text{kJ/m}^2$, after 20 days of storage at 10°C the percentage of fungal decay was 49.6, 29.6, and 27.98, respectively, while in control fruits the decay reached 89.98%. Similar observations have been reported by Pombo et al. (2011) who inoculated Strawberries with *B. cinerea* 8 h after UV-C treatment at the dose of $4.1 \, \text{kJ/m}^2$. The reduction of fungal growth was found and can be attributed to the plant defense mechanism against pathogens induced by UV light.

1.4.2 Plant Antimicrobial Defense Mechanism Triggered by UV

Exposure to UV at very low doses over hours or even days triggers a series of biochemical events within the plant tissue. The term *hormesis* has been applied to this type of UV treatment. According to Shama (2007), hormesis involves the use of small doses of potentially harmful agents directed against a living organism or living tissue to elicit a beneficial or protective response. Hormetic UV treatment is distinguished from conventional UV treatment. In conventional treatment, the UV is directed toward microorganisms that are present on the surfaces of an object, whereas in the case of hormetic UV treatment, the object itself is exposed to the incident UV. The purpose of the treatment is to elicit an antimicrobial response in the fruit tissue. Both types of UV treatment employ the same wavelengths; however, for hormetic treatments only low UV doses are applied (Shama and Alderson, 2005). The plant defense mechanism that is triggered by the hormetic UV dose is not yet fully known and understood. Figure 1.3 schematically presents some of the biochemical responses of plant membrane that were recently reported. It was found that UV-C hormetic treatment at UV fluences in the range of $0.4–4.3\,kJ/m^2$ stimulates the activity of several groups of enzymes that play different roles in plant antimicrobial defense actions. This includes (1) enzymes of peroxidases and reductases that are responsible for the oxidative burst and formation of lignin polymers generating structural barriers against invading pathogens; (2) glucanases and chitinases that exhibit lytic activities toward major fungal cell wall components; and (3) l-phenylalanine ammonia lyase (PAL)—involved in biosynthesis of phenolics, which are characterized by antioxidant and antimicrobial activities (Erkan et al., 2008; Pombo et al., 2011).

It was found that the higher accumulation of rishitin in UV-C-treated (253.7 nm, $3.7\,kJ/m^2$) tomato fruits was positively correlated with enhanced resistance against gray mold rot (Charles et al., 2008). In addition, the hormetic UV treatments result in protective effects against microorganisms throughout the entire tissue rather than at its surface only. Stevens et al. (1999) showed that sweet potatoes inoculated with spores of Fusarium solani at a depth of 12 mm below the surface could be successfully protected from infection following hormetic UV treatment. The research attention was also focused on citrus fruits and in fruits where the enhancement of resistance to phytopathogens such as *P. digitatum* has been attributed to accumulation of the phytoalexins scoparone. As example, Ben-Yehoshua et al. (1992) reported that UV illumination of lemon reduced susceptibility to *P. digitatu*, which was directly related to the level of scoparone in the treated fruit.

1.4.3 EFFECTS ON BIOACTIVE COMPOUNDS

The reports related to UV hormesis in fresh produce showed that due to the induction of plant defense mechanisms accumulation of the phytochemicals in the plant cells can occur. Their antimicrobial and antioxidant properties are highly desirable as they can contribute to delaying the onset of ripening and consequently reducing economic losses due to spoilage. Moreover, the formation of bioactive phenolic compounds such as phenolic acids and flavonoids increases the nutritional value of UV-treated commodities. Phenolic acids and flavonoids are characterized by essential health promoting properties such as antiinflammatory, antihistaminic, and antitumor activities.

Several studies reported increase in and better maintenance of phenolics and flavonoid compounds in crops processed with the UV light. The type of the polyphenols as well as their accumulation and better maintenance during storage was highly dependent on the crop commodity and applied UV dose. González-Aguilar et al.

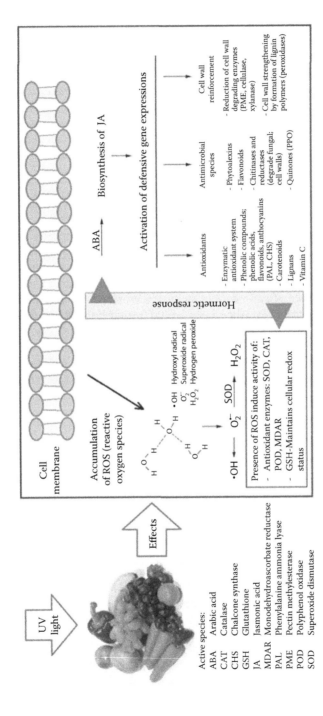

FIGURE 1.3 Schematic representation of biochemical responses in the plant tissue membrane triggered by the hormetic UV treatment. Enzymes that play crucial roles in the plant defensive mechanism are shown.

(2001, 2007) found higher levels of total phenols and polyamine compounds in mangoes irradiated with UV-C at 4.93 kJ/m^2 than in fruits exposed to 2.46 or 9.86 kJ/m^2. In other studies, authors also observed induction of polyamine compounds in peaches after UV-C exposure (González-Aguilar et al., 2004). The accumulation of polyamines in crops might be beneficial in increasing the resistance of fruit tissue to deterioration and chilling injury. In particular, exposure of citrus fruits to UV light was found to be advantageous in terms of the formation of flavonols. For example, Arcas et al. (2000) noted that due to UV-C exposure at 0.72 kJ/m^2, the content of naringin and tangeretin in the peel of *Citrus aurantium* fruits increased by 7% and 55%, respectively. Table 1.4 summarizes results of the recent studies of UV treatments of fruits where UV-related enhancement in the content of the bioactive compounds was observed.

1.4.4 STORAGE OF POST-UV-TREATED FRUITS

The storage conditions, such as temperature or modified atmosphere, can adversely affect the levels of UV-formed phytochemicals. For instance, Vicente et al. (2005) observed increase in the antioxidant capacity in pepper fruits immediately after the UV-C exposure. During subsequent storage at 10°C, the antioxidant capacity of pepper decreased. However, after 18 days of storage, UV-treated fruits showed more antioxidants than control fruit. Allende et al. (2007) studied the effect of the modified atmosphere packaging on the quality of the UV-treated strawberries. The results revealed that strawberries stored under superatmospheric O$_2$ and CO$_2$-enriched concentrations at 2°C showed lower total phenolic contents after 5 days and a vitamin C reduction after 12 days when compared with the fruits that were kept in the air.

1.4.5 FORMATION OF VITAMIN D

Mushrooms are the only plant source of vitamin D$_2$ because they contain a high amount of ergosterol that can be converted to vitamin D$_2$ after exposure to UV irradiation (Mau et al., 1998; Jasinghe and Perera, 2005). All three c-UV bands (UV-A, UV-B, and UV-C) were applied for the postharvest treatment of edible mushrooms. Mau et al. (1998) found UV-B (310 nm) light more effective than UV-C (253.7 nm) in conversion of ergosterol to vitamin D$_2$ in common (*Agaricus bisporus*) mushrooms. It was found that due to exposure for 2 h to UV-B (9.86 kJ/m^2) and UV-C (14.71 kJ/m^2) light, the vitamin D$_2$ content in common mushrooms increased from 2.20 µg/g of dry weight to 12.48 and 7.30 µg/g, respectively. UV-B irradiation also affected the vitamin D$_2$ formation in Shiitake and Straw mushrooms, with the increase rates of 2.15 and 1.86 µg/h, respectively. However, Jasinghe et al. (2006) reported that UV-C exposure (23.0 kJ/m^2) for 2 h resulted in higher yields of vitamin D$_2$ in all treated kinds of mushrooms, Shiitake, Oyster, Abalone, and Button, when compared with the UV-A (25.2 kJ/m^2). It is known that the increase in phenol content might be accompanied by tissue browning. In the case of treatments of Shiitake and Straw mushrooms (Mau et al., 1998; Jiang et al., 2010), the changes in color were not observed. However, Mau et al. (1998) observed that both UV-B and UV-C treatments for 2 h resulted in discoloration of common mushrooms. Therefore, the optimal conditions for UV processing

TABLE 1.4

Examples of UV Treatments of Fruits with the Accumulation of Different Phytochemicals

Commodity	Affected Bioactive Compounds	Number/UV Lamp/ Power Fluence	Reference
Strawberries	Increase in antioxidant capacity and total phenolic content	3/LPM/8 W 2.15 kJ/m^2	Erkan et al. (2008)
Blueberries	Increase in antioxidant capacity, total phenolic and anthocyanins content	15/LPM/8 W, 2.15 and 4.30 kJ/m^2	Wang et al. (2009)
	Increased total phenolic content	1 UV-B fluorescent lamp (305–310 nm) 0.54 kJ/m^2	Eichholz et al. (2011)
Grape berries	Increased resveratrol derivatives content	1/LPM/N/A 0.01 kJ/m^2	Cantos et al. (2000)
		3/UV-B lamp (340 nm)/80 W	Cantos et al. (2000)
		N/A/LPM/510 W	González-Barrio et al. (2009)
Apples	Enhanced anthocyanins content	UV-B lamp (320 nm)	Ubi et al. (2006)
Peaches	Enhanced content of polyamine compounds	N/A/LPM/15 W 8.22 W/m^2	Gonzalez-Aguilar et al. (2004)
Mangoes	Enhanced contents of phenols and polyamine compounds (spermidine, putrescine, spermine)	N/A/LPM/15 W 8.22 W/m^2	González-Aguilar et al. (2001, 2007)
Kumquat	Enhanced phytoalexin scoparone content	LPM 0.2–1.5 kJ/m^2	Rodov et al. (1992)
Orange	Enhanced phytoalexin scoparone content	LPM 1.5–9.0 kJ/m^2	Rodov et al. (1992)
Bitter orange	Enhanced flavonols content (tangeretin)	1/LPM/N/A 0.1 W/m^2	Arcas et al. (2000)
Limon	Increased total phenolic content	6/UV-B lamp (280–400 nm)/N/A 0.052 and 0.077 kJ/m^2	Interdonato et al. (2011)
Pepper fruits	Increased antioxidant capacity	4/LPM/30 W 1, 3, 7 and 14 kJ/m^2	Vicente et al. (2005)
Green tomatoes	Increase in total phenolic content	2/UV-B lamp (311 nm)/N/A 20 and 40 kJ/m^2	Liu et al. (2011)
Onions	Enhanced quercetin content	UV-A (352 nm) 1.84 W/m^2	Higashio et al. (2005)
Shiitake mushrooms	Enhanced vitamin C, total phenolic, and total flavonoids levels	N/A/LPM/20 W 4 kJ/m^2	Jiang et al. (2010)

still need to be determined. As Jasinghe et al. (2006) concluded that the irradiation of 5 g of fresh Shiitake mushrooms for 15 min with UV-A or UV-B is sufficient to obtain the recommended allowances of vitamin D for adults (10 μg/day).

1.4.6 EFFECTS ON GENERAL APPEARANCE

Nutritional value, color, flavor, and texture of fruits are the major factors that indicate product freshness and highly influence the consumer's choice. Deterioration and ripening during storage result in tissue damage, discoloration, and formation of off-flavor. UV technology can be also protective against these symptoms of senescence due to the activation of the plant defensive mechanism by the hormetic UV doses. According to Pombo et al. (2009), delay in the softening of plant produce could be associated with a decrease in the expression of a set of genes involved in cell-wall degradation, during the first hours after UV treatment. It was reported that optimal UV treatment can increase the shelf life of strawberries, apples, peaches, tomatoes, peppers, and broccoli by reducing the respiration rate and weight loss, retaining overall visual quality, delaying the ripening and electrolyte leakage, and maintaining firmness for a longer time, when compared with controls (Lu et al., 1991; Baka et al., 1999; Marquenie et al., 2002; Gonzalez-Aguilar et al., 2004; Lammertyn et al., 2004; Vicente et al., 2005; Costa et al., 2006; Allende et al., 2007; Lemoine et al., 2007; Pombo et al., 2009). In order to increase shelf life, the processing conditions, UV dose (kJ/m^2), and emission spectrums should be optimized for a given commodity of crops. Lammertyn et al. (2004) and Allende et al. (2007) recommended 1.0 kJ/m^2 as optimal fluence for the UV-C processing of strawberries since at higher treatments authors observed browning and dehydration of the sepals. UV-C fluence levels of about 4–5 kJ/m^2 were found to have the most beneficial effect on shelf life and quality of mango fruits (González-Aguilar et al., 2001, 2007) and Shiitake mushrooms (Jiang et al., 2010). Reports are available that application of UV light can protect the color of green commodities. For instance, Costa et al. (2006) and Lemoine et al. (2007) reported that exposure to the UV-C at peak emission of 253.7 nm and at fluence levels of 7–8 kJ/m^2 allowed retaining the highest levels of chlorophyll and hence preserves the green color of broccoli florets. Similarly, UV-B (312 nm) treatment at 8.8 kJ/m^2 delayed the chlorophyll breakdown in the broccoli and lime peel. Moreover, UV treatment resulted in reduced weight loss and shriveling of the lime fruits (Aiamla-or et al., 2009; Srilaong et al., 2011). Aiamla-or et al. (2009) reported that attempts to delay the yellowing of broccoli by UV-A light (342 nm) at 4.5 and 9.0 kJ/m^2 were not effective.

1.5 UV TREATMENT OF FRESH-CUT PRODUCE

Fresh-cut fruits became popular among consumers due to increased preference for minimally processed fresh-like and ready-to-eat products. Mechanical operations of fresh-cut fruits production, such as peeling, slicing, shredding, etc., often result in enzymatic browning, off-flavors, texture breakdown, and lower resistance of fresh-cut produce to microbial spoilage in comparison with the unprocessed commodities (Lemoine et al., 2007) because of presence of natural microflora on the surface of raw commodities as shown in Table 1.3. Therefore, during operations of cutting and shredding, cross-contamination may occur, which might increase the risks of food-borne outbreaks.

To improve hygiene and safety during mechanical processing, sanitizing and dripping treatments are commonly applied. During washing and dipping steps, raw or fresh-cut material is immersed into the tap water containing sanitizing agents (chlorine, sodium hypochlorite) to remove spoilage microorganisms, pesticide residues, and plant debris from product surface (Martin-Belloso et al., 2006). To reduce the usage of sanitizing chemicals, UV light alone or in combination with ozone or another preservative agent was explored as novel processing alternatives. Fonseca and Rushing (2006) examined the effects of UV-C light (1.4–13.7 kJ/m^2 at 253.7 nm) on the quality of fresh-cut watermelon compared with the common sanitizing solutions. Dipping cubes in chlorine (40 µL/L) and ozone (0.4 µL/L) was not effective in reducing microbial populations, and cube quality was lower after these aqueous treatments compared with UV-irradiated cubes or control. In commercial trials, exposure of packaged watermelon cubes to UV-C at 4.1 kJ/m^2 produced more than 1-log reduction in microbial populations by the end of the product's shelf life without affecting juice leakage, color, and overall visual quality. Higher UV doses neither showed differences in microbial populations nor resulted in quality deterioration (13.7 kJ/m^2). Spray applications of hydrogen peroxide (2%) and chlorine (40 µL/L) without subsequent removal of excess water failed to further decrease microbial load of cubes exposed to UV-C light at 4.1 kJ/m^2. It was concluded that when properly used, UV-C light is the only method tested that could be potentially used for sanitizing fresh-cut watermelon. Similarly, exposure of sliced apples to UV-C resulted in higher (~1 log) reduction of *Listeria innocua* ATCC 33090, *E. coli* ATCC 11229, and *Saccharomyces cerevisiae* KE 162 in comparison with apples pretreated with antibrowning and sanitizing agents (1% w/v ascorbic acid—0.1% w/v calcium chloride). The combination of UV-C with antibrowning pretreatment better preserved the color of sliced apples during storage at 5°C for 7 days (Gómez et al., 2010). Other studies have shown that UV-C treatment applied alone was efficient in the reduction of the number of microbiological organisms present on the surface of fresh-cut crops. The examples of successful applications of UV-C light are given in Table 1.5.

Similarly to raw crops, the effectiveness of UV treatment on reduction of microbial deterioration and quality retention was defined by the delivered UV dose and overall characteristics of the surface exposed to the UV light. Allende et al. (2006) found a better preservation of "Red Oak Leaf" lettuce irradiated by UV-C light on both sides of the leaves. As optimal condition for the increasing of the shelf life of "Red Oak Leaf" lettuce, the authors recommended the UV fluence of 2.37 kJ/m^2. Undesirable quality changes occurring at higher fluences included tissue softening and browning. Lamikanra et al. (2005) stressed that the moment of the application of UV light during the fruit processing is an important factor. In their studies, the authors exposed the cantaloupe melon to UV-C at 254 nm during cutting and after cutting of the fruits. Cutting of cantaloupe melon under the UV-C light was as effective as postcut treatment in reduction of yeast, molds, and *Pseudomonas* spp. populations. However, fruit cutting during simultaneous exposure to UV-C resulted

TABLE 1.5

Summary of Studies of the Effect of UV-C Light on Reduction of Microorganisms in Fresh-Cut Produce

Fresh-Cut Commodity	Microbiological Organism	Number/UV Lamp/ Power Fluence	Reference
Watermelon	Mesophilic, psychrophilic, and enterobacteria	15/LPM/36 W 1.6, 2.8, 4.8, 7.2 kJ/m²	Artés-Hernández et al. (2010)
Cantaloupe melon	Yeast, mold, *Pseudomonas* spp., mesophilic aerobes, lactic acid bacteria	1/LPM/N/A 0.0118 kJ/m²	Lamikanra et al. (2005)
Apple	*L. innocua* ATCC 33090; *E. coli* ATCC 11229 and *Saccharomyces cerevisiae* KE 162	2/LPM/15 W 5.6 ± 0.3; 8.4 ± 0.5 and 14.1 ± 0.9 kJ/m²	Gómez et al. (2010)
Pear	*L. innocua* ATCC 33090, *Listeria monocytogenes* ATCC 19114 D, *E. coli* ATCC 11229, and *Zygosaccharomyces bailii* NRRL 7256	2/LPM/15 W 15, 31, 35, 44, 56, 66, 79, and 87 kJ/m²	Schenk et al. (2007)
"Red Oak Leaf" lettuce	*Enterobacter cloacae*, *Enterobacter asburiae*, *Erwinia carotovora* ECC71, *E. coli* RecA_ HB101 and RecA + MC4100, *Escherichia vulneris*, *Escherichia hermannii*, *Leuconostoc carnosum*, *Pantoea agglomerans*, *Pseudomonas fluorescens* Biotype G and A, *Pseudomonas corrugata*, *Pseudomonas putida* C552, *Pseudomonas tolaasii*, *Rahnela aquatilis*, *Salmonella typhimurium*, *Serratia ficaria*, *Serratia plymuthica*, *Serratia liquefaciens*, *Yersinia aldovae*	15/LPM/15 W 1.18, 2.37, 7.11 kJ/m²	Allende et al. (2006)

in improved product quality, that is, reduced rancidity and respiration rate, and also increased firmness retention, when compared with postcut and control samples. Better preservation of fruits processed during the UV exposure can be related to the defense response of the wounded plant enhanced by the UV. Mechanical injury of the plant tissues activates the expression of wound-inducible genes. UV radiation is capable of inducing the expression of plant defense-related proteins that are normally activated during wounding. For example, Lamikanra et al. (2005) reported significant increase in ascorbate peroxidase enzyme activity during storage of cantaloupe melon processed under UV-C light. Peroxidases protect plant cells against oxidation. Higher levels of terpenoids (β-cyclocitral, *cis*- and *trans*-β-ionone, terpinyl acetate, geranylacetone, and dihydroactinidiolide) were found in cantaloupe tissues, which can play important roles as phytoalexins in the disease resistance of a variety of plant families (Lamikanra et al., 2005; Beaulieu, 2007). Significant increase in antioxidative compounds, such as phenolics and flavonoids, was also observed by Alothman et al. (2009) in UV-treated fresh-cut banana, pineapple, and guava fruits. However, decrease in vitamin C was observed in all fruits.

In terms of UV effects on fruits' flavor, Beaulieu (2007) and Lamikanra et al. (2005) reported that fruits processed with UV light preserved their aroma to the same extent as nontreated control samples. Detailed studies of volatile compounds in thin-sliced cantaloupe tissues revealed that UV treatment is not responsible for the chemical transformations to ester bonds, esterase, and lipase decrease. However, Beaulieu (2007) indicated that improper cutting, handling, sanitation treatment, and storage can radically alter the desirable volatile aroma profile in cut cantaloupe and potentially leads to decreased consumer acceptance.

1.6 UV PASTEURIZATION OF FRESH JUICES

Fresh juices are popular beverages in the world's market. They are perceived as wholesome, nutritious, all day beverages. For items such as juices or juice beverages, minimal processing techniques are expected to be used to retain fresh physical, chemical, and nutritional characteristics with extended refrigerated shelf life. The U.S. FDA approval of UV light as an alternative treatment to thermal pasteurization of fresh juice products (U.S. FDA, 2000) led to the growing interest and research in UV technology. Key factors that influence the efficacy of UV treatment of fruit juices include optical properties, design of UV reactors, and UV effects on inactivation of pathogenic and spoilage organisms. There are a number of studies recently published that examined UV light not only as a potential means of alternative pasteurization by studying effects on microflora but also its effects on flavor, color, and nutrient content of fresh juices and nectars (Koutchma, 2009).

1.6.1 UV Absorption of Fresh Juices

Fruit juices are characterized by a diverse range of chemical, physical, and optical properties. Chemical composition, pH, dissolved solids (°Brix), and water activity have to be considered as hurdles that can modify the efficacy of UV microbial inactivation. Optical properties (absorbance and scattering) are the major factors impacting UV light transmission and consequently microbial inactivation. UV absorbance and transmittance at 253.7 nm are important parameters to design UV preservation process using an LPM or LPHO source. In the case of the broadband continuous UV and PL, it is important to measure the spectra of the absorbance or transmittance in the UV germicidal region from 200 to 400 nm. The juices can be transparent if

10% < UVT < 100%, opaque if UVT ~ 0%, or semitransparent if 0 < UVT < 10% for anything in between. In the majority of cases, juices will absorb UV radiation. For example, juices can be considered as a case of semitransparent fluid if they have been clarified (apple, grape, or cranberry juices) or opaque fluids if the juice contains suspended solids (apple cider, orange juice). Juice chemical composition such as vitamin content and concentration of dissolved and suspended solids determines whether the product is transparent, opaque, or semitransparent. The examples of the optical characteristics of clarified fresh juices and opaque juices with particles are shown in Figure 1.4.

The absorption coefficient of three commercial brands of clarified juices such as white grapes, apple, and cranberry falls in the range of 20–26 cm^{-1}, whereas the absorption coefficient of a commercial brand of the orange juice is almost twice higher about 40 cm^{-1} in the range of light path lengths up to 2 mm. As it can be noted in Figure 1.4b, juices with suspended particles did not follow the Beer–Lambert law that is typical behavior for the category of semitransparent juices. The Beer–Lambert law Equation 1.2 is the linear relationship between absorbance (A), concentration of an absorber of electromagnetic radiation (c, mol/L) and extinction coefficient (ε, (L/mol)/cm), or molar absorptivity of the absorbing species, which

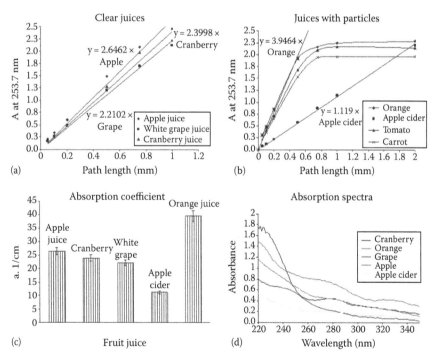

FIGURE 1.4 Absorption characteristics of fruit juices: (a) absorbance at 253.7 nm of clear apple, white grapes, and cranberry juice; (b) absorbance at 253.7 nm of juices with suspended particles orange, tomato, carrot, and apple cider; (c) absorption coefficients; and (d) absorption spectra in UV-C range. (Agriculture and Agri-Food Canada, unpublished data.)

is a measure of the amount of light absorbed per unit concentration absorbance or optical density, and path length of light (d, cm).

$$A = \varepsilon \times c \times d \tag{1.2}$$

This group of juices with suspended solids can be characterized by a nonlinear function of A vs. (ε, c, d) as non-Lambertian liquids. The absorption coefficient of fresh nontreated apple cider that contained suspended particles was approximately $\sim 12\,cm^{-1}$, which is lower than other fruit juices with particles as well as clarified brands. The higher absorbance of the clarified commercial brands can be probably due to contribution of added preservatives and vitamin C. From this prospective, the UV treatment of freshly pressed fruit juices looks more favorable.

1.6.2 Design of UV Systems

A number of continuous flow UV systems were developed and validated for a variety of fruit juices or other fruit beverages ranging from exotic tropical juices and nectars, to the more common apple cider and apple juice. The reactor designs include traditional annular, thin-film, static, and dynamic mixers (Taylor–Coutte UV reactor) and coiled tube devices. Annular type laminar reactors were used for treatment of apple juice and cider (Worobo, 1999) and mango nectar (Guerrero-Beltrán and Barbosa-Cánovas, 2006). The length and gap size can vary depending on the type of treated juice or flow rate. Thin-film reactors are characterized by laminar flow with a parabolic velocity profile. Extensive research of the application of UV light for fresh apple cider by Worobo (1999) yielded a design and production model of a thin-film with 0.8 mm gap "CiderSure" UV reactor, which was approved for safe use to reduce the microbial load of apple cider. UV treatment of orange juice was reported by Tran and Farid (2004) using a vertical single UV lamp thin-film reactor. The thickness of the film was approximately 0.21–0.48 mm. Another commercial thin-film reactor is the PureUV/SurePure reactor that was used for treatment of apple juice, guava-and-pineapple juice, mango nectar, strawberry nectar, and two different orange and tropical juices (Keyser et al., 2008). This reactor is a single-lamp system with a thin fluid film formed between the lamp surface and a surrounding rippled or undulating outer wall. The reactor consisted of inlet, outlet chambers, and a corrugated spiral tube between the chambers. Another type of static mixers is coiled tube UV reactors that are used to increase liquid delivery to UV source by more mixing due to Dean effect (Dean, 1927). Salcor Inc. has promoted a UV reactor in which juice is pumped through the Teflon tubes coiled in a helix, with 12 LPM lamps inside and 12 lamps outside the helix (Anonymous, 1999; Koutchma et al., 2007). The curved flow path can result in a pair of counter-rotating vortices with their axis along the length of the coil. Koutchma et al. (2007) validated the performance of a coiled UV module 420 model (Salcor Inc., Fallbrook, CA) for fresh tropical juice pasteurization. Geveke (2005) processed apple cider with a single lamp UV system surrounded by a coil of UV transparent Chemfluor tubing. Forney et al. (2004) used dynamic mixer Taylor–Coutte design to improve UV inactivation efficiency in apple juice.

1.6.3 INACTIVATION OF PATHOGENIC, NONPATHOGENIC, AND SPOILAGE ORGANISMS

Table 1.6 summarizes results of several reports on inactivation of pathogenic and nonpathogenic bacteria in fruit juices using continuous UV light sources. These data were obtained using static (collimated beam device) and continuous flow UV systems. The approaches to determine UV fluence also differed, so reported results are not directly comparable.

Bobe et al. (2007) studied the presence and concentrations of pathogenic and indicator microorganisms in apple cider processed in Michigan. Neither *E. coli* O157:H7 nor *Salmonella* were detected in any tested cider samples, suggesting a very low frequency of pathogens in apple cider. The persistent and relatively high frequency of generic *E. coli* observed in samples indicated a continued risk of pathogen contamination in apple cider, especially when it is untreated. Basaran et al. (2004) compared log reductions among the *E. coli* strains in the apple cider made of different cultivars. The result failed to show any statistically significant relationship. However, the results of this study indicate that regardless of the apple cultivar used, a minimum 5-log reduction is achieved for all of the strains of *E. coli* O157:H7 tested. Gabriel and Nakano (2009) examined the UV resistance of strains of *E. coli* (K-12 and O157:H7), *Salmonella* (enteritidis and typhimurium), and *Listeria monocytogenes* (AS-1 and M24–1), which were individually suspended in phosphate-buffered saline (PBS) and apple juice prior and exposed to UV radiation (220–300 nm). The AS-1 and M24–1 strains of *L. monocytogenes* were found to be most resistant to UV in PBS (0.28–0.29 min), whereas the AS-1 strain was most resistant in juice (1.26 min). The AS-1 strain of *L. monocytogenes* and *E. coli* O157:H7 were most heat resistant when suspended in PBS (4.41 min) and juice (4.43 min), respectively. Ye et al. (2007) reported that *Yersinia pseudotuberculosis* was less resistant to UV light than *E. coli* K12.

Table 1.7 summarizes results of reported studies in terms of inactivation of spoilage microorganisms in fresh juices. Variations in UV fluence levels can be accounted for due to limitations in dosimetry and fluid absorbance measurements. Molds spores are considered to be very UV resistant, with the resistance higher than that of *B. subtilis* spores, followed by yeasts and lactic bacteria (Warriner et al., 2004, unpublished proprietary data). However, data on UV effectiveness against food-borne pathogenic and spoilage microorganisms of high importance are limited or available in confidential reports and need to be generated. Data generated in the air or water cannot be used for the calculation of UV processing of low UVT food liquids. The results should be considered by juice processors in selecting appropriate surrogate organisms for UV light process lethality validations.

1.6.4 INACTIVATION OF ENZYMES

Enzymatic activity actually depends on the native structure of the protein which, by principle, can be modified following photooxidation promoted by UV and visible light exposure reported to occur via two major routes: (1) direct photooxidation

TABLE 1.6

UV Inactivation of Pathogenic and Nonpathogenic Microorganisms in Fresh Juices

Juice	Type of UV Reactor — Flow Regime	Number/UV Lamp/Power	Gap Size (mm)	Fluence, (mJ/cm^2)	Test Organism	Log (No/N)	Reference
Apple cider	Thin-film laminar	10/LPM	NA	9–61	E. coli O157:H7	3.8	Wright (2000)
Apple cider	Laminar	8/LPM/39 W	0.8	14.32	C. parvum Oocyst	5	Hanes et al. (2002)
Apple cider	Laminar	8/LPM/39 W	0.8	14	E. coli O157:H7 (933, ATCC 43889, and ATCC 43895	5	Basaran et al. (2004)
Apple juice	Petri dish	220–300 nm/15W	D = 5	At 50cm up to 0–33 min	E. coli (K-12 and O157:H7) Salmonella (enteritidis and typhimurium) Listeria monocytogenes (AS-1, M24–1)		Gabriel and Nakano (2009)
Orange juice	Petri dish	4/LPM/30 W		2.19 J/cm^2	E. coli O157:H7	5	Oteiza et al. (2010)
Apple cider	Laminar	8/LPM/ 39 W	0.8	NA	E. coli ATCC 25922	5–6	Worobo (1999)
Apple juice	Thin laminar	8/LPM/ 39 W	0.8	14.5	E. coli K12	3–4	Koutchma et al. (2004)
Apple cider	Turbulent	12/LPM/42 W	5–10	0.75	E. coli K12	<1	Koutchma et al. (2004)
Apple juice	Dean flow	1/LPM/15W	Id 3.6	34 J/mL	E. coli K12 L. innocua	3.4 2.5	Geveke (2005)
Apple juice	Taylor Coutte	4/MPM/0.684	5.5 2	21.7	E. coli 15597	3–5	Forney et al. (2004)
Apple juice	Thin-film laminar	1/LPM/15	5		Yersinia pseudotuberculosis E. coli K 12	1 1	Ye et al. (2007)

TABLE 1.7

UV Inactivation of Spoilage Microorganisms in Fresh Juices

Juice	Flow Regime	Type of UV Reactor Number/UV Lamp/Power	Gap Size (mm)	Fluence (mJ/cm²)	Test Organism	Log (No/N)	Reference
Orange	Thin-film laminar vertical	1/LPM/30 W	0.21–0.48	74	APC	0.53	Tran and Farid (2004)
					Yeasts	0.36	
Apple	Laminar	2/LPM/25 W	NA	45,000	E. coli	1.34	Guerrero-Beltrán and Barbosa-Cánovas (2005)
					APC[a]	4.29	
					Y&M[b]	5.10	
Mango nectar	Laminar	2/LPM/25 W	NA	45,000	APC	2.94	Guerrero-Beltrán and Barbosa-Cánovas (2006)
					Yeasts	2.71	
Model of tropical juices	Turbulent, Dean flow	24/LPM/65 W	ID 10–12	21.5	Yeasts	Up to 6	Koutchma et al. (2007)
Orange					Molds	1.5	
Guava					Molds	1.2	
Carrot					APC	3.2	
Pineapple					Y&M	1.0	
Apple	Turbulent Re > 7500	1–10/LPM/100 W	NA	234	APC	>3.50	Keyser et al. (2008)
					Y&M	>2.99	
Guava-and-pineapple				1,404	APC	3.31	
				468	Y&M	2.23	
Mango nectar				702	APC	0.40	
					Y&M	0.44	
Strawberry nectar				1,404	APC	1.32	
					Y&M	2.45	

[a] aerobic plate count
[b] yeasts and molds

arising from the absorption of radiation by the protein structure or bound chromophore; (2) indirect protein oxidation mediated by singlet oxygen generated by energy transfer by either protein bound or other chromophores (Davies and Truscott, 2001). The effect of UV light on the activity and structure of food enzymes is still a matter of speculation. Limited and controversial information is available in the literature.

Color is a very important quality parameter in fruit juices. It is related to nonenzymatic and enzymatic browning, due to polyphenol oxidase (PPO) activity. The effect of UV light on the inactivation of enzymes related to food quality is diverse. While Noci et al. (2008) reported no effect of UV on apple PPO activity, Manzocco et al. (2009) reported about 80% inactivation of PPO at approximately 1250 mJ/cm^2 of UV fluence. Guerrero-Beltrán and Barbosa-Cánovas (2006) found that after UV treatment of mango nectar at 44,633 mJ/cm^2 PPO reduced activity to 19%. Falguera et al. (2011) irradiated apple juices made from four different varieties (Golden, Starking, Fuji, and King David) for 120 min with a polychromatic mercury lamp of 400 W in a range of 250 and 740 nm with an incident energy of 3.88×10^{-1} E/min. The treatment was effective in the inactivation of PPO after 100 min, while peroxidase was completely destroyed in 15 min in all the four varieties. It should be noted that major absorbance peak of the PPO enzyme matched with the largest peak of the emission spectrum of the lamp.

One important factor in orange juice appearance is the "cloud" formed by pectin. Pectin methylesterase (PME) is an enzyme that tends to de-esterify pectin, and whose inactivation is consequently pursued. Tran and Farid (2004) reported the results of UV treatment of reconstituted orange juice. In addition to the decimal reduction dose for the standard aerobic plate count, effects on shelf life, pH, color, vitamin C, and destruction of PME enzyme were studied. The shelf life of fresh squeezed orange juice was extended to 5 days as a result of limited exposure to UV light of 73.8 mJ/cm^2. No destruction of PME (5%), which is a major cause of cloud loss of juices, was reported whereas the activity of this enzyme was significantly decreased (70%) by mild heat treatment at 70°C for 2 s.

1.6.5 Effects on Essential Vitamins

Vitamins even though they may be present in small amounts in fresh juices, are of concern because some vitamins are considered light sensitive. Water-soluble light-sensitive vitamins include C (ascorbic acid), B12 (cobalamin), B6 (pyridoxine), B2 (riboflavin), and folic acid. Fat-soluble, light-sensitive vitamins include A, K, E (alpha-tocopherol), and carotene. Most studies were conducted on the effects of light on vitamins in the wavelength range of 290–700 nm, which includes both UV and visible light. They have involved exposure to fluorescent lamps, but there are limited data available at 253.7 nm. Since vitamin C is characterized by high UV absorbance within the germicidal wavelength range (peak at approximately of 260 nm) but does not absorb light significantly above 300 nm, the content of vitamin C also affected the magnitude of absorption coefficient. The destruction of vitamin C during exposure to UV light may alter the absorption properties of treated juice. Ye et al. (2007) measured vitamin C content before and after UV treatment. Two brands of packaged apple juice (pasteurized, no preservatives), Sahara Burst and Gordon Food Service, were enriched with Vitamin C. The UV system consisted of four chambers with varied lengths and a single LPM bulb at output power of 25 W at 253.7 nm. Approximately 50% destruction of vitamin C was observed after one complete pass through the system at the slowest flow rate. The effect of vitamin C destruction on the value of the absorption coefficient in apple juice enriched with this vitamin was also measured. After three passes through the UV system at the flow rate of 4 mL/s, the absorption coefficient of apple juice reduced to approximately 20% of initial value. It was concluded that juices enriched with vitamin C require significantly higher doses of UV irradiation for pasteurization purposes. A comparison of vitamin C destruction and inactivation of *E. coli* K12, in commercial apple juice (Motts) exposed to UV at the fluence rate of 1.0 mW/cm², showed that *E. coli* bacteria were more sensitive to UV light exposure with a destruction rate almost of 2.5 times higher compared with samples containing vitamin C. When destruction of vitamin C in apple juice was measured after processing using a commercial multiple lamp UV unit CiderSure1500, it was found that after three consecutive passes through the system at the slowest flow rate of 57 mL/s approximately 50%–60% of initial concentration of vitamin C (25 mg/100 g) remained. Comparison of the destruction of vitamin C in clarified apple juice with absorption coefficient of 15 cm⁻¹ and orange juice of 54 cm⁻¹ after exposing both juices to identical levels of UV fluence of 1.0 mW/cm² in a Petri dish demonstrated that the destruction rate was eight times faster in clarified apple juice due to greater levels of available absorbed energy (Koutchma et al., 2008). Falguera et al. (2011) studied the effect of mercury lamp of 400 W in a range of 250 and 740 nm at incident energy of 3.88×10^{-1} E/min on the content of vitamin C in juices from Golden, Starking, and Fuji. The loss in Golden juice after 120 min of UV irradiation was 5.7%, while in Starking it was 5.6%, and in Fuji 4.0%. In the juice from King David, the loss was 70.0%. This significant difference was attributed to the lack of pigmentation of this juice. In the three first cases, more vitamin C was damaged in the first 60 min than in the second hour, meaning that as pigments were degraded (and the juice color was lighter) its protective effect was less. In the King David juice, the loss after 0 min was 62.4% of the initial content, and after 60 min it was 69.8%.

Vitamin A is another vitamin of great importance in fresh juices because it contributes more than 2% nutritional value to the Recommended Daily Allowance (RDA). After exposure of vitamin A in malate buffer to UV light at the fluence of 200 mJ/cm², approximately 50% of vitamin A initial concentration remained. Orange juice is an essential source of vitamin C and A. One eight fluid ounce (3.69 mL) serving of orange juice contributes approximately 210% of RDA vitamin C and 10% RDA vitamin A to the diet. The destruction of the essential vitamins in orange juice was reported by Anonymous (1999) after treatment in the commercial Salcor UV module (Salcor Co, CA) at a flow rate of 7.5 gpm (28.39 L/min) when total accumulative UV dose was 298.9 mJ/cm². The highest destruction of riboflavin and beta-carotene (~50%) was observed. However, in terms of vitamins C, B6, and A only 16.6%–11% of those vitamins were destroyed after exposure to UV light.

1.6.6 Destruction and Formation of Furan

The presence of furan in processed foods is a concern because furan is listed as "reasonably anticipated to be human carcinogen." Apple juice as a baby food contained furan levels ranging from 2.5 to 8.4 ppb (U.S. FDA, 2004). Furan is formed from carbohydrates, ascorbic acid, fatty acids, and a mixture of the three upon heating (Fan, 2005). Fan and Geveke (2007) investigated whether UV-C induced furan in apple cider and its components and determined furan levels in apple cider exposed to UV-C at doses that would inactivate *E. coli* by 5 log. The results showed that more furan was formed at higher doses (>3.5 J/cm^2) in the freshly prepared cider than in the commercial one. In the commercial fresh apple cider, UVC induced little furan at doses less than 3.5 J/cm^2. When fresh apple ciders were UV treated to achieve the 5-log reduction of *E. coli*, as required by the U.S. FDA, less than 1 ppb furan was found. It was concluded that a significant amount of furan could be accumulated if apple cider was overtreated. Overall, these results suggested that little furan is induced in apple cider if UV-C processing is used for the purpose of apple cider pasteurization. The destruction of d$_4$-furan by UV-C in different solutions and apple cider was also analyzed in this study. There was little destruction of d$_4$-furan at a dose of ~0.9 J/cm^2 when d$_4$-furan in water, glucose, sucrose, ascorbic acid, or apple cider was UV treated, but in fructose solutions, 88% of d$_4$-furan was destroyed. In water, less than 10% of d$_4$-furan was destroyed even at a dose of 9 J/cm^2. In fructose solution, all d$_4$-furan was destroyed at 9 J/cm^2. In a dose response study, it was demonstrated that most d$_4$-furan was degraded even in the low dose (<0.1 J/cm^2). It is possible that the degradation products of fructose may react with furan. The exact specific mechanism requires further study.

1.6.7 DESTRUCTION OF PATULIN

Patulin [4-hydroxy-4H-furo (3, 2-c)-pyran-2-(6H)-one] is a mycotoxin produced by a wide range of molds involved in fruit spoilage, most commonly by *Penicillium expansum*. As with the majority of mycotoxins, patulin is stable and can persist in juice over extended time periods. It is a health concern for both consumers and manufactures. Although several methods for control and elimination of patulin have been proposed, there is no unifying method commercially successful for reducing patulin while keeping produce quality. A few recent studies evaluated the feasibility of UV radiation as a possible commercial alternative for the reduction of patulin in fresh apple juice. UV light is effective to reduce patulin-producing *Penicillium* spores. Dong et al. (2010) used the CiderSure 3500 commercial UV system equipped with 8 LPM lamps for patulin destruction. It was reported that UV exposure of 14.2–99.4 mJ/cm^2 resulted in a significant and nearly linear decrease in patulin levels while producing no quantifiable changes in the chemical composition (i.e., pH, Brix, and total acids) or organoleptic properties of the cider. Even though the products resulting from UV light–induced patulin degradation are yet unknown, UV technology is worth further investigation due to its advantages as potential patulin control measure.

1.7 CONCLUSIONS AND FUTURE TRENDS

UV light technology using continuous and pulsed modes can be a viable nonthermal alternative for fruits and fruit products processing. A number of reports are available on using UV light for eliminating or reducing the levels of undesirable pathogenic, nonpathogenic, and spoilage microorganisms on the surfaces of fresh fruits and fruit products and in juices. In order to achieve the required microbial reduction along with color, texture, and flavor preservation, optimal UV processing conditions and proper UV source has to be found for a given product. Recent studies reported a potential of UV light for enhancement of health promoting compounds such as antioxidants, polyphenols, and flavonoids. Moreover, UV light can be recommended as effective means to control microbial loads in the air, water, nonfood, and food contact surfaces in fruit-processing facilities. A variety of UV sources are commercially available or currently under development that can be applied for specific fruit processing purposes, whereas LPM lamps and xenon PL are currently the dominant sources for UV treatment of fruits since they were approved by the U.S. FDA and Health Canada. A number of UV-light continuous flow systems that included annular laminar and turbulent flow reactors, thin-film devices, and static and dynamic mixers were developed and validated for a variety of fruit juices for pasteurization purposes. The correct UV design can reduce the interference of low UVT and viscosity associated with some juices and therefore improves the UV inactivation efficiency. More work is needed in regard to the design of UV systems capable of delivering sufficient UV doses to all parts of the treated liquid with low UVR such as fruit juices.

Numerous studies cited here have shown the beneficial effects of the UV treatment on the preservation of many fruits, both raw and fresh cut. However, on the basis of the available literature, the mechanism that underlies the hormetic response in fresh produce is still open to debate. In response to the exposure of UV light, plants activate different enzymes peroxidases, reductases, and chitinases, which differ in their chemical structure and absorptive properties in UV-A, UV-B, and UV-C ranges. Therefore, plant response varies depending on applied UV emission spectrum and UV dose. To improve the state of the existing knowledge on UV processing of fresh produce, further studies are necessary that will measure and report conditions and parameters of the UV treatment, such as lamp characteristics, emitted wavelength, and UV fluence levels.

The effect of UV light on the quality of fruits requires further studies. Despite the fact that UV is purely a nonthermal treatment, the possible undesirable effects may include damage to vitamins and proteins, destruction of the antioxidants, changes in color, and formation of off-flavors and aromas depending on UV spectra and applied dose. In addition, the effects of UV light on the potential formation of chemical compounds in foods that may present a health threat should be evaluated to determine whether there is any toxicological or chemical safety concerns associated with products that have undergone UV treatment. Closer examination of UV-light potential to destroy undesirable compounds or pollutants also deserves more attention. Due to low penetration of UV light, the combinations with other postharvest technologies (ozone, ultrasound, modified packaging atmosphere, sanitizing, and antibrowning agents) might be attractive for processors and more efficient. Limited data are available on UV processing combined with other treatments, and further studies are necessary.

LIST OF ABBREVIATIONS

A	Absorption
AlN	Aluminum nitride
EL	Excimer lamp
EPA	U.S. Environmental Protection Agency
EVA	Ethylene vinyl acetate
EVOH	Ethyl vinyl alcohol copolymer
FDA	U.S. Food and Drug Administration
LED	Light emitting diodes
LPHO	Low-pressure high-output lamp
LPM	Low-pressure mercury lamp
MCL	Maximum contaminant level
MPM	Medium pressure mercury lamp
PBS	Phosphate-buffered saline
PL	Pulsed lamp
PME	Pectin methylesterase
PPO	Polyphenol oxidase
R	Reflection
RDA	Recommended daily allowance
T or UVT	Transmittance or transmittance of material in the ultraviolet range
UV	Ultraviolet
UV-A	Ultraviolet light range: 315–400 nm
UV-B	Ultraviolet light range: 280–315 nm
UV-C	Ultraviolet light range: 200–280 nm
cUV	Continuous ultraviolet mode
VUV	Vacuum ultraviolet radiation (100–200 nm)

REFERENCES

Aiamla-or S., Yamauchi N., Takino S., and Shigyo M. 2009. Effect of UV-A and UV-B irradiation on broccoli *Brassica oleracea* L. Italica Group floret yellowing during storage. *Postharvest Biology and Technology*, 54, 177–179.

Allende A., Marín A., Buendía B., Tomás-Barberán F., and Gil M.I. 2007. Impact of combined postharvest treatments UV-C light, gaseous O_3, superatmospheric O_2 and high CO_2 on health promoting compounds and shelf-life of strawberries. *Postharvest Biology and Technology*, 46, 201–211.

Allende A., McEvoy J.L., Luo Y., Artes F., and Wang C.Y. 2006. Effectiveness of two-sided UV-C treatments in inhibiting natural microflora and extending the shelf-life of minimally processed 'Red Oak Leaf' lettuce. *Food Microbiology*, 23, 241–249.

Alothman M., Bhat R., and Karim A.A. 2009. UV radiation-induced changes of antioxidant capacity of fresh-cut tropical fruits. *Innovative Food Science and Emerging Technologies*, 10, 512–516.

Anonymous. 1999. A food additive petition for the use of ultraviolet light in the reduction of microorganisms on juice products. Submitted to FDA regarding CFR 21 179. Glendore Calif. California Day-Fresh Foods Inc., Glendora, CA, pp. 1–117.

Anonymous. 2000. Kinetics of microbial inactivation for alternative food processing technologies. Institute of Food Technologists. *Journal of Food Science.* Supplement http://vm.cfsan.fda.gov/~comm/ift-pref.html (accessed).

Arcas M.C., Botía J.M., Ortuño A.M., and Del Río J.A. 2000. UV irradiation alters the levels of flavonoids involved in the defence mechanism of *Citrus aurantium* fruits against *Penicillium digitatum*. *European Journal of Plant Pathology*, 106, 617–622.

Artés-Hernández F., Robles P.A., Gómez P.A., Tomás-Callejas A., and Artés F. 2010. Low UV-C illumination for keeping overall quality of fresh-cut watermelon. *Postharvest Biology and Technology*, 55, 114–120.

Baka M., Mercier J., Corcuff R., Castaigne F., and Arul J. 1999. Photochemical treatment to improve storability of fresh strawberries. *Journal of Food Science*, 64, 1068–1072.

Basaran N., Quintero-Ramos A., Moake M.M., Churey J.J., and Worobo R.W. 2004. Influence of apple cultivars on inactivation of different strains of *Escherichia coli* O157:H7 in apple cider by UV irradiation. *Applied and Environmental Microbiology*, 70, 6061–6065.

Beaulieu J.C. 2007. Effect of UV irradiation on cut cantaloupe: Terpenoids and esters. *Journal of Food Science*, 724, S272–S281.

Ben-Yehoshua S., Rodov V., Kim J., and Carmeli S. 1992. Preformed and induced antifungal materials of citrus fruit in relation to the enhancement of decay resistance by heat and ultraviolet treatment. *Journal of Agricultural Food Chemistry*, 40, 1217–1221.

Bialka K.L. and Demirci A. 2007. Decontamination of *Escherichia coli* O157:H7 and *Salmonella enterica* on blueberries using ozone and pulsed UV-light. *Journal of Food Science*, 72, M391–M396.

Bialka K.L. and Demirci A. 2008. Efficacy of pulsed UV-light for the decontamination of *Escherichia coli* O157:H7 and *Salmonella spp.* on raspberries and strawberries. *Journal of Food Science*, 73, M201–M207.

Bobe G., Thede D., Eyck T.A.T., and Bourquin L.D. 2007. Microbial levels in Michigan apple cider and their association with manufacturing practices. *Journal of Food Protection*, 70, 1187–1193.

Bolton J.R. and Linden K.G. 2003. Standardization of methods for fluence UV dose determination in bench-scale UV experiments. *Journal of Environmental Engineering*, 129, 209–215.

Cantos E., García-Viguera C., de Pascual-Teresa S., and Tomás-Barberá F.A. 2000. Effect of postharvest ultraviolet irradiation on resveratrol and other phenolics of Cv. Napoleon table grapes. *Journal of Agriculture and Food Chemistry*, 48, 4606–4612.

Charles M.T., Mercier J., Makhlouf J., and Arul J. 2008. Physiological basis of UV-C-induced resistance to *Botrytis cinerea* in tomato fruit I. Role of pre- and post-challenge accumulation of the phytoalexin-rishitin. *Postharvest Biology and Technology*, 47, 10–20.

Cia P., Pascholati S.F., Benato E.A., Camili E.C., and Santos C.A. 2007. Effects of gamma and UV-C irradiation on the postharvest control of papaya anthracnose. *Postharvest Biology and Technology*, 43, 366–373.

Costa L., Vicente A.R., Civello P.M., Chaves A.R., and Martínez G.A. 2006. UV-C treatment delays postharvest senescence in broccoli florets. *Postharvest Biology and Technology*, 39, 204–210.

Davies M.J. and Truscott R.J.W. 2001. Photo-oxydation of proteins and its role in cataractogenesis. *Journal of Photochemistry and Photobiology B: Biology*, 63, 114–125.

Dean W.R. 1927. Note on the motion of fluid in a curved pipe. *Philosophical Magazine Journal of Science*, 4, 208–223.

Dong Q., Manns D.C., Feng G., Yue, T., Churet J.J., and Worobo R.W. 2010. Reduction of patulin in apple cider by UV radiation. *Journal of Food Protection*, 73, 69–74.

Eichholz I., Huyskens-Keil S., Keller A., Ulrich D., Kroh L.W., and Rohn S. 2011. UV-B-induced changes of volatile metabolites and phenolic compounds in blueberries *Vaccinium corymbosum* L. *Food Chemistry*, 126, 60–64.

EPA Office of Water 1995. National primary drinking water regulations contaminant fact sheets inorganic chemicals—Technical version. EPA 811-F-95-002-T, Washington, DC.

Erkan M., Wang S.Y., and Wang C.Y. 2008. Effect of UV treatment on antioxidant capacity, antioxidant enzyme activity and decay in strawberry fruit. *Postharvest Biology and Technology*, 48, 163–171.

Falguera V., Pagán J., and Ibarz A. 2011. Effect of UV irradiation on enzymatic activities and physicochemical properties of apple juices from different varieties. *LWT-Food Science and Technology*, 44, 115–119.

Fan X. 2005. Formation of furan from carbohydrates and ascorbic acid following exposure to ionizing radiation and thermal processing. *Journal of Agricultural and Food Chemistry*, 53, 7826–7831.

Fan X. and Geveke D. 2007. Furan formation in sugar solution and apple cider upon ultraviolet treatment. *Journal of Agricultural and Food Chemistry*, 55, 7816–7821.

Fonseca J.M. and Rushing J.W. 2006. Effect of ultraviolet-C light on quality and microbial population of fresh-cut watermelon. *Postharvest Biology and Technology*, 40, 256–261.

Forney L., Pierson J.A., and Ye Z. 2004. Juice irradiation with Taylor-Coutte flow: UV inactivation of *Escherichia coli*. *Journal of Food Protection*, 67: 2410–2415.

Gabriel A.B. and Nakano H. 2009. Inactivation of *Salmonella*, *E. coli* and *Listeria monocytogenes* in phosphate-buffered saline and apple juice by ultraviolet and heat treatments. *Food Control*, 20, 443–446.

Gardner P.T., White T.A.C., McPhail D.B., and Duthie G.G. 2000. The relative contributions of vitamin C, carotenoids and phenolics to the antioxidant potential of fruit juices. *Food Chemistry*, 68, 471–474.

Geveke D. 2005. UV inactivation of bacteria in apple cider. *Journal of Food Protection*, 68, 1739–1742.

Gómez P.L., Alzamora S.M., Castro M.A., and Salvatori D.M. 2010. Effect of ultraviolet-C light dose on quality of cut-apple: Microorganism, color and compression behavior. *Journal of Food Engineering*, 98, 60–70.

Gonzalez-Aguilar G., Wang C.Y., and Buta G.J. 2004. UV-C irradiation reduces breakdown and chilling injury of peaches during cold storage. *Journal of the Science of Food and Agriculture*, 84, 415–422.

González-Aguilar G.A., Wang C.Y., Buta1 J.G., and Krizek D.T. 2001. Use of UV-C irradiation to prevent decay and maintain postharvest quality of ripe 'Tommy Atkins' mangoes. *International Journal of Food Science and Technology*, 36, 767–773.

González-Aguilar G.A., Zavaleta-Gatica R., and Tiznado-Hernández M.E. 2007. Improving postharvest quality of mango 'Haden' by UV-C treatment. *Postharvest Biology and Technology*, 45, 108–116.

González-Barrio R., Vidal-Guevara M.L., Tomás-Barberán F.A., and Espín J.C. 2009. Preparation of a resveratrol-enriched grape juice based on ultraviolet C-treated berries. *Innovative Food Science and Emerging Technologies*, 10, 374–382.

Guerrero-Beltrán J.A. and Barbosa-Canovas G.V. 2005. Reduction of *Saccharomyces cerevisiae*, *Escherichia coli* and *Listeria innocua* in apple juice by ultraviolet light. *Journal of Food Process Engineering*, 28, 437–452.

Guerrero-Beltrán J.A. and Barbosa-Canovas G.V. 2006. Inactivation of *Saccharomyces cerevisiae* and polyphenoloxidase in mango nectar treated with UV light. *Journal of Food Protection*, 69, 362–368.

Hanes D.E., Orlandi P.A., Burr D.H., Miliotis M.D., Robi M.G., Bier J.W., Jackson G.J., Arrowood M.J., Churey J.J., and Worobo R.W. 2002. Inactivation of *Cryptosporidium parvum* oocysts in fresh apple cider using ultraviolet irradiation. *Applied and Environmental Microbiology*, 68, 4168–4172.

Hansen S.L., Purup S., and Christensen L.P. 2003. Bioactivity of falcarinol and its content in carrots. *Journal of the Science of Food and Agriculture*, 83, 1010–1017.

Higashio H., Hirokane H., Sato F., Tokuda S., and Uragami A. 2005. Effect of UV irradiation after harvest on the content on the flavonoids in vegetables. *Proceedings of the 5th International Postharvest Symposium Acta horticulturae*, 6822, 1007–1012.

Hijnen W.A.M., Beerendonk E.F., and Medema G.J. 2006. Inactivation credit of UV radiation for viruses, bacteria and protozoan oocysts in water: A review. *Water Research*, 40, 3–22.

Interdonato R., Rosa M., Nieva C.B., González J.A., Hilal M., and Prado F.E. 2011. Effects of low UV-B doses on the accumulation of UV-B absorbing compounds and total phenolics and carbohydrate metabolism in the peel of harvested lemons. *Environmental and Experimental Botany*, 70, 204–211.

Jasinghe V.J. and Perera C.O. 2005. Distribution of ergosterol in different tissues of mushrooms and its effect on the conversion of ergosterol to vitamin D2 by UV irradiation. *Food Chemistry*, 92, 541–546.

Jasinghe V.J. and Perera C.O. 2006. Ultraviolet irradiation: The generator of Vitamin D2 in edible mushrooms. *Food Chemistry*, 95, 638–643.

Jiang T., Jahangir M.M., Jiang Z., Lu X., and Ying T. 2010. Influence of UV-C treatment on antioxidant capacity, antioxidant enzyme activity and texture of postharvest shiitake *Lentinus edodes* mushrooms during storage. *Postharvest Biology and Technology*, 56, 209–215.

Keyser M., Müllera I., Cilliersb F.P., Nelb W., and Gouwsa P.A. 2008. UV radiation as a non thermal treatment for the inactivation of microorganisms in fruit juices. *Innovative Food Science and Emerging Technologies*, 9, 348–354.

Koutchma T. 2009. Advances in UV light technology for non-thermal processing of liquid foods. *Food Bioprocess Technology*, 2, 138–155.

Koutchma T., Forney L., and Moraru C. 2008. *Ultraviolet Light in Food Technology: Principles and Applications*, CRS Press, Taylor and Francis, Boca Raton, FL.

Koutchma T., Keller S., Parisi B., and Chirtel S. 2004. Ultraviolet disinfection of juice products in laminar and turbulent flow reactors. *Innovative Food Science and Emerging Technologies*, 5, 179–189.

Koutchma T., Parisi B., and Patazca E. 2007. Validation of UV coiled tube reactor for fresh fruit juices. *Journal of Environmental Science and Engineering*, 6, 319–328.

Kowalski W.J. 2006. *Aerobiological Engineering Handbook: A Guide to Airborne Disease Control Technologies*. New York: McGraw-Hill.

Kowalski W.J. and Bahnfleth W.P. 2002. MERV filter models for aerobiological applications. *Air Media*, Summer Issue, 13–17.

Kowalski W.J. and Dunn C.E. 2002. Current trends in UVGI air and surface disinfection. *Environment Professional*, 86, 4–6.

Lamikanra O., Kueneman D., Ukuku D., and Bett-Garber K.L. 2005. Effect of processing under ultraviolet light on the shelf life of fresh-cut cantaloupe melon. *Journal of Food Science*, 70, C534–C538.

Lammertyn J., De Ketelaere B., Marquenie D., Molenberghs G., and Nicolaï B.M. 2004. Mixed models for multicategorical repeated response: Modelling the time effect of physical treatments on strawberry sepal quality. *Postharvest Biology and Technology*, 30, 195–207.

Lemoine M.L., Civello P.M., Martínez G.A., and Chaves A.R. 2007. Influence of postharvest UV-C treatment on refrigerated storage of minimally processed broccoli *Brassica oleracea* var. Italica. *Journal of the Science of Food and Agriculture*, 87, 1132–1139.

Liu C., Han X., Cai L., Lu X., Ying T., and Jiang Z. 2011. Postharvest UV-B irradiation maintains sensory qualities and enhances antioxidant capacity in tomato fruit during storage. *Postharvest Biology and Technology*, 59, 232–237.

Lu J.Y., Stevens C., Khan V.A., Kabwe M., and Wilson C.L. 1991. The effect of ultraviolet irradiation on shelflife and ripening of peaches and apples. *Journal of Food Quality*, 14, 299–305.

Manzocco L., Quarta B., and Dri A. 2009. Polyphenoloxidase inactivation by light exposure in model systems and apple derivatives. *Innovative Food Science and Emerging Technologies*, 10, 506–511.

Marquenie D., Michiels C.W., Geeraerd A.H., Schenk A., Soontjens C., Van Impe J.F., and Nicolaï B.M. 2002. Using survival analysis to investigate the effect of UV-C and heat treatment on storage rot of strawberry and sweet cherry. *International Journal of Food Microbiology*, 73, 187–196.

Martin-Belloso O., Soliva-Fortuny R., and Oms-Oliu G. 2006. Fresh-Cut Fruits. In: *Handbook of Fruits and Fruit Processing*, Ed. Y.H. Hui, Blackwell Publishing, Oxford, U.K., pp. 129–144.

Masschelein W.J. and Rice R.G. 2002. *Ultraviolet Light in Water and Wastewater Sanitation*, Lewis Publication, Boca Raton, FL.

Mau J.-L., Chen P.-R., and Yang J.-H. 1998. Ultraviolet irradiation increased vitamin D2 content in edible mushrooms. *Journal of Agricultural and Food Chemistry*, 46, 5269–5272.

Mercier J., Arul J., and Julien C. 1993. Effect of UV-C on Phytoalexin accumulation and resistance to *Botrytis cinerea* in stored carrots. *Journal of Phytopathology*, 139, 17–25.

Noci F., Riener J., Walkling-Ribeiro M., Cronin D.A., Morgan D.J., and Lyng J.G. 2008. Ultraviolet irradiation and pulsed electric fields PEF in a hurdle strategy for the preservation of fresh apple juice. *Journal of Food Engineering*, 85, 141–146.

Oteiza J.M., Giannuzzi L., and Zaritzky N. 2010. Ultraviolet treatment of orange juice to inactivate *E. coli* O157:H7 as affected by native microflora. *Food Bioprocess Technology*, 3, 603–614.

Pombo M.A., Dotto M.C., Martínez G.A., and Civello P.M. 2009. UV-C irradiation delays strawberry fruit softening and modifies the expression of genes involved in cell wall degradation. *Postharvest Biology and Technology*, 51, 141–148.

Pombo M.A., Rosli H.G., Martínez G.A., and Civello P.M. 2011. UV-C treatment affects the expression and activity of defense genes in strawberry fruit Fragaria × ananassa, Duch. *Postharvest Biology and Technology*, 59, 94–102.

Rodov V., BenYehoshua S., Kim J.J., Shapiro B., and Ittah Y. 1992. Ultraviolet illumination induces scoparone production in kumquat and orange fruits and improves decay resistance. *Journal of the American Society for Horticultural Science*, 117, 788–791.

Schenk M., Guerrero S., and Maris Alzamora S. 2007. Response of some microorganisms to ultraviolet treatment on fresh-cut pear. *Food and Bioprocess Technology*, 1, 384–392.

Shama G. 2007. A new role for UV? Extensions to the shelf life of plant foods by UV-induced effects. *2007 Ioa-Iuva Joint World Congress*, August 2007, Los Angeles, CA, pp. 27–29.

Shama G. and Alderson P. 2005. UV hormesis in fruits: A concept ripe for commercialisation, *Trends Food Science and Technology*, 16, 128–136.

Srilaong V., Aiamla-or S., Soontornwat A., Shigyo M., and Yamauchi N. 2011. UV-B irradiation retards chlorophyll degradation in lime (*Citrus latifolia* Tan.) fruit. *Postharvest Biology and Technology*, 59, 110–112.

Stevens C., Khan V.A., Lu J.Y., Wilson C.L., Chalutz E., Droby S., Kabwe M.K., Haung Z., Adeyeye O., Pusey L.P., and Tang A.Y.A. 1999. Induced resistance of sweet potato to Fusarium root rot by UV-C hormesis, *Crop Protection*, 18, 463–470.

Stevens C., Khan V.A., Lu J.Y., Wilson C.L., Pusey P.L., Kabwe M.K., Igwegbe E.C.K., Chalutz E., and Droby S. 1998. The germicidal and hormetic effects of UV-C light on reducing brown rot disease and yeast microflora of peaches. *Crop Protection*, 17, 75–84.

Stevens C., Khan V.A., Wilson C.L., Lu J.Y., Chalutz E., and Droby S. 2005. The effect of fruit orientation of postharvest commodities following low dose ultraviolet light-C treatment on host induced resistance to decay. *Crop Protection*, 24, 756–759.

Strawn L.K. and Danyluk M.D. 2010. Fate of *Escherichia coli* O157:H7 and *Salmonella* on fresh and frozen cut pineapples. *Journal of Food Protection*, 73, 418–424.

Tran M.T. and Farid M. 2004. Ultraviolet treatment of orange juice. *Innovative Food Science and Emerging Technologies*, 5, 495–502.

Ubi B.E., Honda C., Bessho H., Kondo S., Wada M., Kobayashi S., and Moriguchi T. 2006. Expression analysis of anthocyanin biosynthetic genes in apple skin: Effect of UV-B and temperature. *Plant Science*, 170, 571–578.

U.S. Food and Drug Administration. 2000. 21 CFR Part 179. Irradiation in the production, processing and handling of food. *Federal Register*, 65, 71056–71058.

U.S. Food and Drug Administration. 2004. Exploratory data on furan in food data. U.S. Food and Drug Administration. http://vm.cfsan.fda.gov/~dms/furandat.html. Accessed February 21, 2006.

Vicente A.R., Pineda C., Lemoine L., Civello P.M., Martinez G.A., and Chaves A.R. 2005. UV-C treatments reduce decay, retain quality and alleviate chilling injury in pepper. *Postharvest Biology and Technology*, 35, 69–78.

Voronov A. 2007. New generation of low pressure mercury lamps for producing ozone. In: *UV and Ozone World Congress, Conference CD Proceedings*, Los Angeles, CA.

Wang C.Y., Chen C.-T., and Wang S.Y. 2009. Changes of flavonoid content and antioxidant capacity in blueberries after illumination with UV-C. *Food Chemistry*, 117, 426–431.

Warriner K., Movahedi S., and Waites W.M. 2004. Laser-based packaging sterilization in aseptic processing. *Improving the Thermal Processing of Foods*, Chapter 14, pp. 277–303.

Woodling S.E. and Moraru C.I. 2005. Influence of surface topography on the effectiveness of pulsed light treatment for the reduction of *Listeria innocua* on stainless steel surfaces. *Journal of Food Science*, 70, 245–351.

Worobo R. 1999. Efficacy of the CiderSure 3500 Ultraviolet light unit in apple cider. In: *CFSAN Apple Cider Food Safety Control Workshop*, July 15–16, 1999.

Wright H.B. 2000. Comparison and validation of UV dose calculations for low- and medium-pressure mercury arc lamps. *Water Environment Research*, 72, 439.

Yaun B.R., Sumner S.S., Eifert J.D., and Marcy J.E. 2004. Inhibition of pathogens on fresh produce by ultraviolet energy. *International Journal of Food Microbiology*, 90, 1–8.

Ye Z., Koutchma T., Parisi B., Larkin J., and Forney L.J. 2007. Ultraviolet inactivation kinetics of *E. coli* and *Y. pseudotuberculosis* in annular reactors. *Journal of Food Science*, 72, E271–E278.

2 High-Pressure Processing

Fabiano A.N. Fernandes

CONTENTS

2.1 INTRODUCTION

The main application of high-pressure processing (HPP) is for the preservation and homogenization of food. The process consists mainly in applying high pressure to food, to inactivate microorganisms and enzymes. Unlike traditional technologies, HPP induces minor changes in the sensorial and nutritional properties of foods (Aparicio et al., 2011).

In recent years, pasteurization of food products using high pressure at room temperature has been accepted by regulatory agencies for commercial use. The quality of such products displays better nutritional retention; fresh flavor; improved color, texture, and taste; and extended shelf life compared with thermally pasteurized products.

Applications of high pressure include processing of fruits, vegetables, meats, seafood, dairy, and egg products. High pressure has also been used successfully to pasteurize foods with high acid content, extending their shelf lives considerably (Welti-Chanes et al., 2005).

Pressurized products, such as Japanese jams, jellies, and sauces (Rastogi et al., 2007) and avocado sauce (guacamole) were first marketed to consumers in the early 1990s in the United States. In the early 2000, an increase in the number of high-pressure units was observed. Since then, the number of high-pressure units has been growing around the world.

2.2 MECHANISM

Pressure is applied isostatically (equally and instantaneously) and transmitted to all parts of food. In combination with mild heat, it inactivates pathogenic and spoilage vegetative microorganisms and quality deteriorative enzymes, thus extending the shelf life and preserving the original properties of the food (Mertens and Knorr, 1992).

During pressurization, compression heating induces an increase in temperature both in the food to be treated and in the pressure-transmitting fluid, which surrounds the food. The temperature change depends not only on the pressure (P) and on the temperature of the process (T) but also on the thermodynamic properties (thermal expansion coefficient, specific volume, and heat capacity) of each substance involved:

$$\frac{dT}{dP} = \frac{\alpha \cdot T \cdot v}{C_P}, \tag{2.1}$$

where

C_P is the heat capacity
v is the specific volume
α is the thermal expansion coefficient

2.3 EQUIPMENT

High-pressure equipment offers several unique characteristics based on different operation parameters, such as range of operating pressure; the systems used to pressurize, heat, and cool down the process; the volume of the chamber; and the layout of the system, among others.

The typical pressure range of high-pressure equipment is around 600 MPa, but some equipment are capable of achieving up to 1400 MPa. The temperature may change from 20°C to 150°C in just seconds. Equipment layouts, horizontal or vertical, are based on the type of product processed and the space available in the industrial facility. Vessel capacity may range from 35 to 320 L in industrial-scale equipment. The horizontal layout offers some advantages: the traceability of the product during processing, for example, is controlled because the inlet and outlet are located on different sides of the equipment, making the installation of equipment easier and cheaper than with the vertical layout; the equipment is not very high and does not require additional equipment to raise the inlet and outlet, making the process of loading and unloading products easier (Hernando-Sáinz et al., 2008).

2.3.1 Cost

The main cost involved in HPP is the equipment and its installation. The average cost of HPP ranges from U.S.$0.05 to U.S.$0.5 per liter or per kilogram depending on processing conditions (Rastogi et al., 2007; Hernando-Sáinz et al., 2008). The cost of the process in 2011 makes the technology suitable only for products of premium quality.

Combining preservation factors with pressure can reduce the level of pressure required for processing, thus reducing the cost of the product.

2.4   MICROBIAL INACTIVATION

In current commercial applications, HPP is an essentially "nonthermal" pasteurization process, in which a food is subjected to pressures up to 600 MPa and initial temperatures lower than 40°C for 1–15 min, depending on the product application (Barbosa-Cánovas and Juliano, 2008).

HPP has been successful in commercial processes against pathogenic bacteria such as *Listeria monocytogenes*, *Escherichia coli*, *Salmonella*, *Vibrio*, yeasts, molds, and deteriorative microorganisms (Balasubramaniam et al., 2008).

High pressure in combination with subzero temperatures has successfully inactivated some bacteria, such as *L. monocytogenes*, using pressures up to 200 MPa (for 100 min) in combination with temperature ranging from −18°C to −10°C (Ritz et al., 2008). Such a process is known as nonthermal high pressure, which refers to the use of elevated pressure (up to 680 MPa) at room temperature (20°C). Although important microbial inactivation has been achieved under these conditions, some microorganisms are resistant to being killed with only pressure and require the addition of a mild thermal treatment for inactivation. Bacteria have certain cell layers that offer resistance to preservation factors; for example, Gram-positive bacteria have a strong cell wall that wraps around the inner cellular content, protecting it from the exterior. However, to maintain good product quality, HPP should be carried out at low temperature.

E. coli and *Listeria innocua* were tested in kiwifruit and pineapple juice at 0°C and −10°C using 300 MPa of pressure. Inactivation was achieved after 5 min of treatment, with enhancement of treatment in pineapple juice when pressure was applied in a pulsed manner (Buzrul et al., 2008).

Pressure in the range of 250–300 MPa in combination with dissolved CO_2 was able to inactivate between 7 and 8 log of *E. coli* and *Staphylococcus aureus* (Wang et al., 2010). Pressure in this range in combination with CO_2 can reduce processing cost, making the process more accessible and feasible for a higher number of products.

Pressure in the range of 450–500 MPa in combination with some natural antimicrobials has been reported to inactivate *Salmonella* cells up to 8 log. Using pressure alone or nisin alone (antimicrobial) did not allow inactivation of *Salmonella* cells (Lee and Kaletunc, 2010).

Additives such as citric acid, adipic acid, C8-sugar ester, C10-sugar ester, tannin, nisin, wasabi extract, e-polylysine, and protamine have enhanced the results of microbial inactivation when each additive was applied simultaneously with high pressure (Ogihara et al., 2009).

The treatment medium is a factor that should be taken into account during HPP, because the inactivation depends heavily on the composition of the medium.

The more complex a medium is, the more difficult the inactivation is (Narisawa et al., 2008; Escriu and Mor-Mur, 2009).

No sublethal injury after high pressure treatment was reported in *L. innocua, E. coli, S. aureus, Yersinia enterocolitica, Salmonella typhimurium,* and *L. monocytogenes* by using a selective environment such as the addition of salt, SDS, or low pH, specific selective medium for the microorganism or propidium iodide staining (Wuytack et al., 2002, 2003; Diels et al., 2005; Briñez et al., 2006a,b, 2007; Roig-Sagués et al., 2009).

Several microorganisms and enzymes were inactivated by applying high pressure for more than 15 min at room temperature, which may elevate the cost of the process. In these cases, it is important to analyze the process from an economical point of view to determine whether a particular product justifies the use of a longer process.

The combination of pressure and temperature has been reported to inactivate *Salmonella* and *E. coli* cells. Application of pressure alone did not inactivate these bacteria; however, the pressure (500 MPa) plus temperature (45°C) for 2 min allowed inactivation of both microorganisms up to 5 log (Neetoo and Chen, 2010).

The inactivation kinetics of *Lactobacillus plantarum* in mandarin juice presented a synergic effect between pressure and temperature. High-pressure treatment achieved a 2.4 log cycles after 120 MPa at 30°C for 10 s and a 6.1 log cycles after 400 MPa at 45°C for 1 min (Carreño et al., 2011).

2.5 SPORE INACTIVATION

Spores have proven to be resistant to high-pressure treatment. Regardless of treatment intensity, spore inactivation has not been observed with pressure alone. Most high-pressure processed foods are chilled or are highly acidic. Low-acid foods (pH > 4.5) require the use of heat in combination with high pressure. Some degree of spore inactivation is usually obtained applying pressures above 500 MPa and temperatures above 60°C (Wuytack and Michiels, 2001; Reddy et al., 2006; Ju et al., 2008; Robertson et al., 2008; Chaves-López et al., 2009; Ramaswamy et al., 2010).

Applying high pressure in a pulsed manner: two pulses of very high pressure (690–1700 MPa) in combination with high temperature (60°C–90°C) can inactivate spores in low-acid products, such as vegetables, seafood, eggs, milk, meat, and pasta (Meyer et al., 2000), and can also be applied to fruit and fruit juices.

Spore inactivation at a lower pressure range has been achieved using a process called high-pressure sterilization (HPS) or pressure assisted thermal sterilization (PATS). This process applies pressure ranging from 600 to 800 MPa and initial chamber temperature ranging from 60°C to 90°C for short processing times (5 min or less) (Matser et al., 2004; Juliano et al., 2009).

Six tasks need to be performed during the HPS process: (a) sample vacuum packaging and product lodging, (b) preheating to target temperature, (c) product equilibration to initial temperature, (d) product temperature increase to pressurization temperature by compression heating, (e) product temperature decrease during decompression, and (f) product cooling to ambient temperature (Barbosa-Cánovas and Juliano, 2008).

A critical stage of HPS is compression heating, when the product is pressurized up to room temperature. In this stage, an increase in the food's temperature and compression fluid occurs (Ting et al., 2002). For example, when the initial temperature of water is around 90°C and the pressure is 100 MPa, an increase of 5.3°C is observed because of compression heating. Higher temperature increase was observed for some foods such as olive oil and tomato sauces (Matser et al., 2004; Wilson et al., 2008). Although this temperature increase may enhance microbial and spore inactivation, it will also have some undesired impact on the quality of the product.

2.6 VIRUS INACTIVATION

Inactivation of viruses using HPP is achieved more easily because viruses are round shaped and formed only by protein, making their protection limited. HPP has been successfully tested for Hepatitis A virus, poliovirus, aichivirus, coxsackievirus B5, coxsackievirus A9, feline calicivirus, murine norovirus, and rotavirus (Kovac et al., 2010).

Given their tissue structure, viruses are easily inactivated at moderate pressures under refrigerated conditions or temperatures above 30°C. However, the presence of acids, NaCl, or sucrose in the medium may have a protective effect on viruses against HPP (Kingsley and Chen, 2008).

HPP has been effective in inactivating the norovirus MNV-1. This inactivation was enhanced at temperatures below 20°C, especially at 5°C, which gave the greatest degree of inactivation, reducing the virus titer by 5.5 log. Higher temperatures resulted in lower inactivation (<1.5 log). As pressure and treatment times increased, virus inactivation also increased. The inactivation of norovirus was characterized by a rapid initial inactivation of the viral count followed by tailing caused by diminishing inactivation rate (Kingsley et al., 2007).

2.7 ENZYME INACTIVATION

Enzyme inactivation in HPP is challenging as in other technologies. Some enzymes may be inactivated, while others may be activated during processing. Also, inactivation occurs in some products under specific conditions, whereas in the same products under different conditions enzyme activation is quite possible. Pressure alone is usually not enough to inactivate enzymes (Hendrickx et al., 1998).

The effect of high pressure on enzymes is related to conformational or structural changes in the protein, changes in enzyme–substrate interaction, and changes in the cell membrane (Ludikhuyze et al., 2002).

Polyphenoloxidase (PPO) is usually related to enzymatic browning in fruits, and the effects of HPP has been studied for several fruits, such as apple, apricot, avocado, guava, grape, pear, plum, strawberry, and tomato (Ludikhuyze et al., 2002). Results from these studies showed that enzymatic browning in fruits is affected more by phenol concentration in the medium than by PPO concentration. High pressure generates cell permeabilization and, as a result, browning is generated when the physical stress produced by pressure is enough to alter the cell (Dörnenburg and Knorr, 1998).

Pectin methylesterase (PME) is a good example of enzyme inactivation/activation. PME is an enzyme of either plant or microbial origin that catalyzes the demethoxylation of pectin, a major plant cell wall polysaccharide contributing to tissue integrity and rigidity. Studies on PME in tomato juice showed that enzyme activation was evident in the tomato juice at pressures ranging from 300 to 500 MPa, whereas the lowest activity of PME was observed at 100 MPa. Despite the activity of PME under high pressure, improvement in all quality properties at the highest pressure tested was observed by Aparicio et al. (2011).

High-pressure treatments require less time to reach the same PME residual level compared with thermal treatment. Results for watermelon juice and white grape juice showed that high-pressure treatment required between 5 and 6 min to reach a PME residual level of 75%, while traditional thermal treatment requires about 20 min to reach the same residual level (Guiavarc'h et al., 2005; Zhang et al., 2011).

The application of high pressure decreases the degree of methyl-esterification in fruits and vegetables. The decrease observed for the high-pressure process is higher than the decrease observed for thermally treated fruits. The decrease in the degree of methyl-esterification is attributed to the stimulation of chemical demethoxylation under pressure (De Roeck et al., 2008, 2009).

2.8 HOMOGENIZATION

Homogenization is applied in the food industry to decrease the particle size of different food products. New homogenization equipment has been developed and is able to reach pressure values of up to 400 MPa, whereas conventional homogenization is carried out until pressure up to 40 MPa.

High-pressure homogenization (HPH) or dynamic high pressure (DHP) is a continuous process with treatment time lasting several seconds, which makes it more suitable for liquid food processing (Popper and Knorr, 1990). The benefits of HPH include increasing viscosity, reducing phase separation, and improving texture and color uniformity (Thakur et al., 1995).

A setback of this technology is that considerable temperature increase is also produced (estimated as 1.5°C–2.5°C/10 MPa), depending on the fluid viscosity (Popper and Knorr, 1990). This temperature increase may have some undesired impact on the quality of the product.

During HPH, homogenization is obtained by forcing a product through a small orifice between the homogenizing valve and the valve seat. HPH causes changes in the particle structure by disaggregation of cell clusters and disruption of cell structures. Different mechanisms, including turbulence, shear, and cavitation, have been proposed to cause disruption of cell walls during HPH (Stang et al., 2001). As a result, HPH treatment reduces the particle size, converting part of the sedimentable pulp into colloidal pulp. Transmittance is also significantly reduced leading to a more opaque juice, which may be caused by greater stabilization of the juice cloud.

Homogenized tomato pulp (at 2100 MPa) contained small cell clusters, which showed signs of tearing. No cell clusters were found in tomato pulp homogenized at 4750 MPa. The pulp homogenized at 4750 MPa did not present cell clusters, and the pulp consisted of a limited number of single intact cells and broken cell material. No intact cells were found in pulp homogenized above 4750 MPa. In this case, the cell material seemed to be uniformly distributed inside the pulp (Colle et al., 2010).

As a general tendency, as the homogenization pressure increases, the percentage of small particles increases. In addition, the increase in pressure results in a narrower particle size distribution (PSD).

HPH reduces the length of fibers and tends to fibrillate the fibers' end. Under the proper conditions, homogenization will sufficiently fibrillate the fibers to allow absorption of great amounts of liquid in the product. Overprocessing may reduce the length of the fibrous materials to a point where they are no longer able to form an effective network (Gallaher et al., 1999).

2.9   EFFECT OF HIGH PRESSURE ON SENSORY CHARACTERISTICS

High pressure is considered as an excellent option for processing fruits because of the minimal changes in the sensory characteristics with its use. Preservation of the sensory characteristics is partially explained by the low effect of HPP on the covalent bonds of low molecular weight, such as color and flavor compounds (Oey et al., 2008a; De Roeck et al., 2010).

2.9.1   COLOR

The color of many fruit products such as jams, fruit juices, and purée is generally preserved during HPP at ambient temperature (Ludikhuyze and Hendrickx, 2002; Guerrero-Beltrán and Barbosa-Cánovas, 2004; Guerrero-Beltrán et al., 2005, 2006). Color preservation indicates a limited effect of HPP on the pigments (e.g., chlorophyll, carotenoids, anthocyanins, etc.) responsible for the color of fruits. Chlorophyll pigments are very susceptible to degradation when high pressure is combined with moderate to high temperatures (Weemaes et al., 1999; Oey et al., 2008a).

The effect of HPP on the color of fruit products depends on the processing conditions. High pressure usually does not present any effect on color degradation at room or moderate temperature (less than 40°C).

Lycopene is considered pressure stable in tomato products. Anthocyanins are stable during pressurization but are not during storage, probably due to incomplete enzyme inactivation, substrate specificity, or presence of ascorbic acid in the product (Oey et al., 2008a).

Changes in carotenoids have not been observed with high pressure, as in mandarin juice. Color parameters a and b significantly decreased possibly due to a decrease in pulp particle size that affected juice opaqueness.

Discoloration in apple purée was attributed to destruction of pigment (Landl et al., 2010). HPP increased insignificantly the L^* values indicating a lightening of the purée surface. This phenomenon was also observed in tomato-based products (Sánchez-Moreno et al., 2006) and might be caused by cell disruption during high-pressure treatment. Color parameters a^* and b^* remained constant during HPP of apple purée (Landl et al., 2010). Ahmed et al. (2005) observed that color quality parameters remained almost constant after high-pressure treatment of mango pulp, indicating pigment stability.

Pomegranate juice was subjected to HPP at a wide range of pressures (up to 900 MPa) and temperatures (−20°C–90°C). Results showed that the juice retained its initial intense red color in the product and that the concentration of anthocyanins was similar to that in the fresh product (Ferrari et al., 2010).

The effect of high pressure on the color of the watermelon juice showed significant color change with a ΔE higher than 3.0. Treatment at 600 MPa was more effective in preserving the color of watermelon juice. Treatments at 600 and 900 MPa significantly decreased the browning degree in watermelon juice (Zhang et al., 2011). The results reported by Zhang et al. (2011) corroborates with data reported by Castellari et al. (2000) for grape juice.

During storage, discoloration of pressurized samples occurred, which was indicated by the reduction in the lightness variable L^* and the increase in the color parameter a^*. No statistically significant differences were observed when pressure was increased from 300 to 600 MPa. ΔE^* values for apple purée were 16.4 and 17.7 for pressure of 400 and 600 MPa, respectively, which was much higher than ΔE^* for mild pasteurization (5.3). Similar results were obtained by Guerrero-Beltrán et al. (2006) who treated standardized mango purée at 552 MPa for 5 min.

2.9.2 FLAVOR

High pressure does not significantly alter the components related to flavor. Some changes have been reported in hexanal, as observed in strawberries and tomato juice. Additionally, new compounds were found in strawberries after HPP, enhancing the characteristic flavor of the fruit.

The effect of HPP on flavor is still a research field that has to be explored since few reports are available in the open literature.

2.9.3 TEXTURE

Texture is a major quality indicator of fruits and is significantly changed during thermal sterilization, mainly due to beta-eliminative depolymerization of the pectic polysaccharides of the cell wall (Sila et al., 2006). Depolymerization of pectic polysaccharides leads to pectin solubilization and, consequently, to decreased intercellular adhesion, resulting in tissue softening.

Depolymerization of pectic polysaccharides is strongly dependent on the pH, the degree of pectin methyl-esterification, and the presence of ions (Keijbets and Pilnik, 1974; Sajjaanantakul et al., 1989, 1993).

Texture improvement can be achieved by making fruit tissue less vulnerable to the beta-elimination reaction occurring during heating, or in the case of HPP, compression heating. This can be achieved by lowering the degree of esterification of the pectin, stimulating the activity of the enzyme PME. Nonmethoxylated carboxyl groups can be cross-linked by calcium ions present in the tissue or added artificially to the fruit enhancing the intercellular adhesion (Sila et al., 2005; Van Buggenhout et al., 2009). The treatment of foodstuffs with calcium solution may improve the texture significantly during HPP (De Roeck et al., 2010).

2.10 EFFECT OF HIGH PRESSURE ON FOOD CELL STRUCTURE

High pressure can modify the microstructure of fruits, especially when the fruit contains entrapped air molecules, causing changes in the tissue such as softening (Rastogi et al., 2007). Changes in the fruit tissue structure usually result from damage in the cell structure, deforming the cell walls and releasing cell serum into the medium. Such changes in cell increased the permeability of the membrane, allowing the interchange of water and metabolites in the microstructure of the product (Oey et al., 2008b).

The combined effect and manipulation of pressure and temperature to activate or inactivate the pectinases makes it possible to create a certain texture in the fruit product, softening or increasing its firmness. PME and polygalacturonase (PG) activity in orange, strawberry, tomato, banana, and plum can be modified to achieve final texture requirements.

The application of high pressure may change the rheological properties of fresh and processed fruit products, such as purées, juices, and nectars. Changes on rheological properties have been reported for apple, pear, orange, pineapple, tomato, and mango (Oey et al., 2008b; Sila et al., 2008).

HPP tends to increase the viscosity of fruit purée. The increase in viscosity is attributed to an increase in the linearity of cell walls and volumes of particles due to the permeabilization of cell walls (Sánchez-Moreno et al., 2003; Oey et al., 2008b).

The viscosity in mango pulp increased after moderate HP treatments (100–200 MPa for 15 or 30 min at 20°C), decreasing after higher pressures were applied (400 MPa) (Ahmed et al., 2005). The increase in viscosity in apple purée was observed only at high pressures (400 MPa), showing significantly different responses to high pressure (Landl et al., 2010). Tomato purée processed under high pressure (400 MPa/15 min/25°C) showed higher consistency compared with conventionally thermally treated purées (Sánchez-Moreno et al., 2006). The same trend was observed in navel orange juice treated at 600 MPa for 4 min at 40°C (Polydera et al., 2005).

2.11 EFFECT OF HIGH PRESSURE ON NUTRITIONAL VALUE

Oxygen concentration plays an important role in vitamin C degradation during HPP. When oxygen concentration is low, the degradation of this vitamin is limited. In addition, when sugar concentration is high, the degradation of this vitamin is limited, because of the protective effect against pressure. The combination of high pressure with temperature above 60°C increases the degradation of ascorbic acid significantly.

A decrease of 30% in Vitamin C was observed in the case of tomato purée. The combination of high pressure (100–400 MPa) with higher temperatures (>60°C) showed a higher decrease in the content of Vitamin C after treatment (Sánchez-Moreno et al., 2003, 2006).

Vitamin B is stable when high pressure is used at room temperature. Little information is available regarding the behavior of fat-soluble vitamins (Oey et al., 2008b).

Water-soluble polysaccharides decrease with increasing pressure, probably due to structural rearrangement. No difference in the alkali-soluble polysaccharides and cellulose were observed after HPP (Yang et al., 2009).

Total lycopene content did not change during HPP. Isomerization of lycopene did not occur in some fruit pulp and purées, as in the case of tomato pulp. The application of high pressure with pasteurization also did not affect the total lycopene content in tomato pulp and in watermelon juice (Pérez-Conesa et al., 2009; Colle et al., 2010; Zhang et al., 2011). Total lycopene concentration in watermelon juice did not change with increasing pressure (Zhang et al., 2011), while total lycopene has been reported to retain its maximum content at 400 MPa and display a significant loss above 500 MPa (Qiu et al., 2006).

Trans-lycopene concentration of watermelon juice decreased with decreasing PME residual levels. Concentration of trans-lycopene of the watermelon juice subjected to the high-pressure treatment was significantly higher than that subjected to thermal treatments (Zhang et al., 2011).

2.12 CONCLUDING REMARKS

One of the main challenges in researching nonthermal technologies, such as high-pressure processing, is to establish shorter processing times when compared with thermal processing. HPP is a promising nonthermal technology because it has little impact on the quality parameters of fruits and fruit products.

Studies still have to be carried out regarding the sterilization of low-acid foods with combined elevated pressure (600–800 MPa), moderate temperatures (30°C–90°C), and short processing times (less than 5 min). The combination of HPP with other chemical compounds has to be investigated further aiming at higher log reduction in spore count.

REFERENCES

Ahmed, J., Ramaswamy, H.S., and Hiremath, N. 2005. The effect of high pressure treatment on rheological characteristics and colour of mango pulp. *International Journal of Food Science and Technology* 40: 885–895.

Aparicio, C., Otero, L., Sanz, P.D., and Guignon, B. 2011. Specific volume and compressibility measurements of tomato paste at moderately high pressure as a function of temperature. *Journal of Food Engineering* 103: 251–257.

Balasubramaniam, V.M., Farkas, D., and Turek, E.J. 2008. Preserving foods through high-pressure processing. *Journal of Food Science* 62: 32–38.

Barbosa-Cánovas, G.V. and Juliano, P. 2008. Food sterilization by combining high pressure and thermal energy. In *Food Engineering Integrated Approaches*, eds. G.F. Gutiérrez-López, G.V. Barbosa-Cánovas, and J. Welti-Chanes. New York: Springer.

Briñez, W.J., Roig-Sagués, A.X., Hernández, M.M., and Guamis, B. 2006a. Inactivation of *Listeria innocua* in milk and orange juice by ultrahigh-pressure homogenization. *Journal of Food Protection* 69: 86–92.

Briñez, W.J., Roig-Sagués, A.X., Hernández, M.M., and Guamis, B. 2007. Inactivation of *Staphylococcus* spp. strains in whole milk and orange juice using ultra high pressure homogenization at inlet temperatures of 6 and 20°C. *Food Control* 18: 1282–1288.

Briñez, W., Roig-Sagués, A.X., and Herrero, M. 2006b. Inactivation by ultrahigh-pressure homogenization of *Escherichia coli* strains inoculated into orange juice. *Journal of Food Protection* 69: 984–989.

Buzrul, S., Alpas, H., Largeteau, A., and Demazeau, G. 2008. Inactivation of *Escherichia coli* and *Listeria innocua* in kiwifruit and pineapple juices by high hydrostatic pressure. *International Journal of Food Microbiology* 124: 275–278.

Carreño, J.M., Gurrea, M.C., Sampedro, F., and Carbonell, J.V. 2011. Effect of high hydrostatic pressure and high-pressure homogenisation on *Lactobacillus plantarum* inactivation kinetics and quality parameters of mandarin juice. *European Food Research Technology* 232: 265–274.

Castellari, M., Matricardi, L., Arfelli, G., Carpi, G., and Galassi, S. 2000. Effects of high hydrostatic pressure processing and of glucose oxidase-catalase addition on the colour stability and sensorial score of grape juice. *Food Science and Technology International* 6: 17–23.

Chaves-López, C., Lanciotti, R., Serio, A., Paparella, A., Guerzoni, E., and Suzzi, G. 2009. Effect of high pressure homogenization applied individually or in combination with other mild physical or chemical stress on *Bacillus cereus* or *Bacillus subtilis* spore viability. *Food Control* 20: 691–695.

Colle, I., Van Buggenhout, S., Van Loey, A., and Hendrickx, M. 2010. High pressure homogenization followed by thermal processing of tomato pulp: Influence on microstructure and lycopene in vitro bioaccessibility. *Food Research International* 43: 2193–2200.

De Roeck, A., Duvetter, T., Fraeye, I., Van der Plancken, I., Sila, D.N., Van Loey, A., and Hendrickx, M. 2009. Effect of high-pressure/high-temperature processing on chemical pectin conversions in relation to fruit and vegetable texture. *Food Chemistry* 115: 207–213.

De Roeck, A., Mols, J., Duvetter, T., Van Loey, A., and Hendrickx, M. 2010. Carrot texture degradation kinetics and pectin changes during thermal versus high-pressure/high-temperature processing: A comparative study. *Food Chemistry* 120: 1104–1112.

De Roeck, A., Sila, D.N., Duvetter, T., Van Loey, A., and Hendrickx, M. 2008. Effect of high pressure/high temperature processing on cell wall pectic substances in relation to firmness of carrot tissue. *Food Chemistry* 107: 1225–1235.

Diels, A.M.J., De Taeye, J., and Michiels, C.W. 2005. Sensitisation of *Escherichia coli* to antibacterial peptides and enzymes by high pressure homogenization. *International Journal of Food Microbiology* 105: 165–175.

Dörnenburg, H. and Knorr, D. 1998. Monitoring the impact of high pressure processing on the biosynthesis of plant metabolites using plant cell cultures. *Trends in Food Science and Technology* 9: 355–361.

Escriu, R. and Mor-Mur, M. 2009. Role of quantity and quality of fat in meat models inoculated with *Listeria innocua* or *Salmonella typhimurium* treated by high hydrostatic pressure and refrigerated stored. *Food Microbiology* 26: 834–840.

Ferrari, G., Maresca, P., and Ciccarone, R. 2010. The application of high hydrostatic pressure for the stabilization of functional foods: Pomegranate juice. *Journal of Food Engineering* 100: 245–253.

Gallaher, D.M., Gallaher, K.L., and Gallaher, D.E. 1999. Method for improving the texture of tomato paste products. Patent number 5965190.

Guerrero-Beltrán, J.A. and Barbosa-Cánovas, G.V. 2004. High hydrostatic pressure processing of peach puree with and without antibrowning agents. *Journal of Food Processing and Preservation* 28: 69–85.

Guerrero-Beltrán, J.A., Barbosa-Cánovas, G.V., Moraga-Ballesteros, G., Moraga-Ballesteros, M.J., and Swanson, B.G. 2006. Effect of pH and ascorbic acid on high hydrostatic pressure-processed mango puree. *Journal of Food Processing and Preservation* 30: 582–596.

Guerrero-Beltrán, J.A., Swanson, B.G., and Barbosa-Cánovas, G.V. 2005. Shelf life of HHP-processed peach puree with antibrowning agents. *Journal of Food Quality* 28: 479–491.

Guiavarc'h, Y., Segovia, O., Hendrickx, M., and Van Loey, A. 2005. Purification, characterization, thermal and high-pressure inactivation of a pectin methylesterase from white grapefruit (*Citrus paradisi*). *Innovative Food Science and Emerging Technologies* 6: 363–371.

Hendrickx, M., Ludikhuyze, L., Van den Broeck, I., and Weemaes, C. 1998. Effects of high pressure on enzyme related to food quality. *Trends in Food Science and Technology* 9: 197–203.

Hernando-Sáinz, A., Tárrago-Mingo, S., and Purroy-Balda, F. 2008. Advances in design for successful commercial high pressure food processing. *Food Australia* 60: 154.

Ju, X.R., Gao, Y.L., Yao, M.L., and Qian, Y. 2008. Response of *Bacillus cereus* spores to high hydrostatic pressure and moderate heat. *LWT Food Science and Technology* 41: 2104–2112.

Juliano, P., Knoerzer, K., and Barbosa-Cánovas, G.V. 2009. High-pressure thermal processes: Thermal and fluid dynamic modeling principles. In: *Mathematical Modeling of Selected Food Processes*, ed. R. Simpson. Boca Raton, FL: CRC Press.

Keijbets, M.J. and Pilnik, W. 1974. Beta-elimination of pectin in presence of anions and cations. *Carbohydrate Research* 33: 359–362.

Kingsley, D.H. and Chen, H. 2008. Aqueous matrix compositions and pH influence feline calicivirus inactivation by high pressure processing. *Journal of Food Protection* 71: 1598–1603.

Kingsley, D.H., Holliman, D.R., Calci, K.R., Chen, H., and Flick, G.J. 2007. Inactivation of a norovirus by high-pressure processing. *Applied and Environmental Microbiology* 73: 581–585.

Kovac, K., Diez-Valcarce, M., and Hernández, M. 2010. High hydrostatic pressure as emergent technology for the elimination of foodborne viruses. *Trends in Food Science and Technology* 21: 558–568.

Landl, A., Abadias, M., Sárraga, C., Viñas, I., and Picouet, P.A. 2010. Effect of high pressure processing on the quality of acidified Granny Smith apple purée product. *Innovative Food Science and Emerging Technologies* 11: 557–564.

Lee, L. and Kaletunc, G. 2010. Inactivation of *Salmonella* Enteritidis strains by combination of high hydrostatic pressure and nisin. *International Journal of Food Microbiology* 140: 49–56.

Ludikhuyze, L. and Hendrickx, M.E.G. 2002. Effects of high pressure on chemical reactions related to food quality. In: *Ultra High Pressure Treatments of Food*, eds. M.E.G. Hendrickx and D.W. Knorr. New York: Kluwer Academic, pp. 167–188.

Ludikhuyze, L., Van Loey, A., Indrawati, S.D., and Hendrickx, M.E.G. 2002. Effects of high pressure on enzymes related to food quality. In: *Ultra High Pressure Treatments of Foods*, eds. M.E.G. Hendrickx and D. Knorr. New York: Kluwer Academic, pp. 115–166.

Matser, A.M., Krebbers, B., van den Berg, R.W., and Bartels, P.V. 2004. Advantages of high pressure sterilization on quality of food products. *Trends in Food Science and Technology* 15: 79–85.

Mertens, B. and Knorr, D. 1992. Developments of nonthermal processes for food preservation. *Food Technology* 46: 124–133.

Meyer, R.S., Cooper, K.L., Knorr, D., and Lelieveld, H.L.M. 2000. High pressure sterilization of foods. *Food Technology* 54: 67–72.

Narisawa, N., Furukawa, S., Kawarai, T., Ohishi, K., Kanda, S., Kimjima, K., Negishi, S., Ogihara, H., and Yamasaki, M. 2008. Effect of skimmed milk and its fractions on the inactivation of *Escherichia coli* K12 by high hydrostatic pressure. *International Journal of Food Microbiology* 124: 103–107.

Neetoo, H. and Chen, H. 2010. Inactivation of *Salmonella* and *Escherichia coli* O157:H7 on artificially contaminated alfalfa sees using high hydrostatic pressure. *Food Microbiology* 27: 332–338.

Oey, I., Lille, M., and Van Loey, A. 2008a. Effect of high-pressure processing on colour, texture and flavor of fruit based food products: A review. *Trends in Food Science and Technology* 19: 320–328.

Oey, I., Van der Plancken, I., Van Loey, A., and Hendrickx, M. 2008b. Does high pressure processing influence nutritional aspects of plant based food systems? *Trends in Food Science and Technology* 19: 300–308.

Ogihara, H., Yatuzuka, M., Horie, N., Furukawa, S., and Yamasaki, M. 2009. Synergistic effect of high hydrostatic pressure treatment and food additives on the inactivation of *Salmonella* enteritidis. *Food Control* 20: 963–966.

Pérez-Conesa, D., García-Alonso, J., García-Valverde, V., Iniesta, M.D., Jacob, K., Sánchez-Siles, L.M., Ros, G., and Periago, M.J. 2009. Changes in bioactive compounds and antioxidant activity during homogenization and thermal processing of tomato puree. *Innovative Food Science and Emerging Technologies* 10: 179–188.

Polydera, A.C., Stoforos, N.G., and Taoukis, P.S. 2005. Effect of high hydrostatic pressure treatment on post processing antioxidant activity of fresh Navel orange juice. *Food Chemistry* 91: 495–503.

Popper, L. and Knorr, D. 1990. Applications of high-pressure homogenization for food preservation. *Food Technology* 44: 84–89.

Qiu, W.F., Jiang, H.H., Wang, H.F., and Gao, Y.L. 2006. Effect of high hydrostatic pressure on lycopene stability. *Food Chemistry* 97: 516–523.

Ramaswamy, H.S., Shao, Y., and Zhu, S. 2010. High-pressure destruction kinetics of *Clostridium sporogenes* ATCC11437 spores in milk at elevated quasi-isothermal conditions. *Journal of Food Engineering* 96: 249–257.

Rastogi, N.K., Raghavarao, K.S.M.S., Balasubramaniam, V.M., Niranjan, K., and Knorr, D. 2007. Opportunities and challenges in high pressure processing of foods. *Critical Reviews in Food Science* 47: 69–112.

Reddy, N.R., Tetzloff, R.C., Solomon, H.M., and Larkin, J.W. 2006. Inactivation of *Clostridium botulinum* nonproteolytic type B spores by high pressure processing at moderate to elevated high temperatures. *Innovative Food Science and Emerging Technologies* 7: 169–175.

Ritz, M., Jugiau, F., Federighi, M., Chapleau, N., and De Lamballerie, M. 2008. Effects of high pressure, subzero temperature, and pH on survival of *Listeria monocytogenes* in buffer and Smoked Salmon. *Journal of Food Protection* 71: 1612–1618.

Robertson, R.E., Carroll, T., and Pearce, L.E. 2008. Bacillus spore inactivation differences after combined mild temperature and high pressure processing using two pressurizing fluids. *Journal of Food Protection* 71: 1186–1192.

Roig-Sagués, X., Velázquez, R.M., Montealegre-Agramont, P., López-Pedemonte, T.J., Briñez-Zambrano, W.J., Guamis-López, B., and Hernández-Herrero, M.M. 2009. Fat content increases the lethality of ultra-high-pressure homogenization on *Listeria monocytogenes* in milk. *Journal of Dairy Science* 92: 5396–5402.

Sajjaanantakul, T., Van Buren, J.P., and Downing, D.L. 1989. Effect of methyl-ester content on heat degradation of chelator-soluble carrot pectin. *Journal of Food Science* 54: 1272–1277.

Sajjaanantakul, T., Van Buren, J.P., and Downing, D.L. 1993. Effect of cations on heat degradation of chelator-soluble carrot pectin. *Carbohydrate Polymers* 20: 207–214.

Sánchez-Moreno, C., Cano, M.P., De Ancos, B., Plaza, L., Olmedilla, B., Granado, F., and Martín, A. 2003. Effect of Orange juice intake on vitamin C concentrations and biomarkers of antioxidant status in humans. *American Journal of Clinical Nutrition* 78: 454–460.

Sánchez-Moreno, C., Plaza, L., De Ancos, B., and Cano, M.P. 2006. Impact of high-pressure and traditional thermal processing of tomato puree on carotenoids, vitamin C and antioxidant activity. *Journal of the Science of Food and Agriculture* 86: 171–179.

Sila, D.N., Duvetter, T., De Roeck, A., Verlent, I., Smout, C., Moates, G.K., Hills, B.P., Waldron, K.K., Hendrickx, M., and Van Loey, A. 2008. Texture changes of processed fruits and vegetables: Potential use of high pressure processing. *Trends in Food Science and Technology* 19: 309–319.

Sila, D.N., Smout, C., Elliot, F., Van Loey, A., and Hendrickx, M. 2006. Non-enzymatic depolymerization of carrot pectin: Toward a better understanding of carrot texture during thermal processing. *Journal of Food Science* 71: E1–E9.

Sila, D.N., Smout, C., Vu, T.S., Van Loey, A., and Hendrickx, M. 2005. Influence of pretreatment conditions on the texture and cell wall components of carrots during thermal processing. *Journal of Food Science* 70: E85–E91.

Stang, M., Schuchmann, H., and Schubert, H. 2001. Emulsification in high-pressure Homogenizers. *Chemical Engineering and Technology* 24: 151–157.

Thakur, B.R., Singh, R.K., and Handa, A.K. 1995. Effect of homogenization pressure on consistency of tomato Juice. *Journal of Food Quality* 18: 389–396.

Ting, E., Balasubramaniam, V.M., and Raghubeer, E. 2002. Determining thermal effects in high-pressure processing. *Food Technology* 56: 31–35.

Van Buggenhout, S., Sila, D.N., Duvetter, T., Van Loey, A., and Hendrickx, M. 2009. Pectins in processed fruits and vegetables: Part III—Texture engineering. *Comprehensive Reviews in Food Science and Food Safety* 8: 105–117.

Wang, L., Pan, J., Xie, H., Yang, Y., and Lin, C. 2010. Inactivation of *Staphylococcus aureus* and *Escherichia coli* by the synergistic action of high hydrostatic pressure and dissolved CO_2. *International Journal of Food Microbiology* 144: 118–125.

Weemaes, C., Ooms, V., Indrawati, Ludikhuyze, L., Van den Broeck, I., and Van Loey, A. 1999. Pressure-temperature degradation of green color in broccoli juice. *Journal of Food Science* 64: 504–508.

Welti-Chanes, J., López-Malo, A., and Palou, E. 2005. Fundamentals and applications of high pressure processing of foods. In *Novel Food Processing Technologies*, eds. G.V. Barbosa-Cánovas, M.S. Tapia, and M.P. Cano. New York: CRC Press.

Wilson, D.R., Lukasz, D., Stringer, S., Moezelaar, R., and Brocklehurst, T.F. 2008. High pressure in combination with elevated temperature as a method for the sterilization of food. *Trends in Food Science and Technology* 19: 289–299.

Wuytack, E.Y., Diels, A.M.J., and Michiels, C.W. 2002. Bacterial inactivation by high-pressure homogenization and high hydrostatic pressure. *International Journal of Food Microbiology* 77: 205–212.

Wuytack, E.Y., Duong Thi Phuong, L., Aertsen, A., Reyns, K.M.F., Marquenie, D., De Ketelaere, B., Masschalck, B., Van Opstal, I., Diels, A.M.J., and Michiels, C.W. 2003. Comparison of sublethal injury induced in *Salmonella enterica* serovar Typhimurium by heat and by different nonthermal treatments. *Journal of Food Protection* 66: 31–37.

Wuytack, E.Y. and Michiels, C.W. 2001. A study on the effects of high pressure and heat on *Bacillus subtilits* spores at low pH. *International Journal of Food Microbiology* 64: 333–341.

Yang, B., Jiang, Y., Wang, R., Zhao, M., and Sun, J. 2009. Ultra-high pressure treatment effects on polysaccharides and lignins of longan fruit pericarp. *Food Chemistry* 112: 428–431.

Zhang, C., Trierweiler, B., Li, W., Butz, P., Xu, Y., Rüfer, C.E., Ma, Y., and Zhao, X. 2011. Comparison of thermal, ultraviolet-c, and high pressure treatments on quality parameters of watermelon juice. *Food Chemistry* 126: 254–260.

3 Ultrasound Applications in Fruit Processing

Fabiano A.N. Fernandes and Sueli Rodrigues

CONTENTS

3.1 INTRODUCTION

Nonthermal technologies are preservation treatments that are effective at ambient or sublethal temperatures, thereby minimizing negative thermal effects on food nutritional and quality parameters. Novel nonthermal processes that can ensure product safety and maintain the desired nutritional and sensory characteristics have been of great interest in the food industry.

The use of ultrasound in the food industry is increasing, and in recent years much research has been carried out to understand and determine the efficiency of ultrasound application in several areas of the food industry.

Ultrasound has been used in drying, washing, removal of pesticides, sanitization, freezing, extraction, and in quality assessment. In this chapter, ultrasound applications are described and its advantages and problems are discussed.

3.2 MINIMAL PROCESSING

3.2.1 WASHING

The starting point to prepare fruits for sale is to cut and wash them. Continuous washing equipment has been used for this purpose. This washing equipment comprises mostly tanks in which the products come into intimate contact with a flow of wash liquid. The wash liquid is mostly water with added substances for destroying the bacterial count of the product.

Vegetal products are "dirty" by their nature, since they originate from the countryside normally rich in characteristic microbe flora, in addition to soil and foreign bodies of various kinds. In addition to a series of microorganisms innocuous to human health, the fruits can also contain harmful pathogens. As such, washing is carried out not only to eliminate visible dirt but also to reduce the total pathogen count. Mirco (2005) has described a continuous apparatus to wash and sanitize vegetable products with good results.

A few studies have been carried out with ultrasound-assisted washing of fruits and vegetables, but these studies have shown the great potential of the use of ultrasound in the removal of dirt and especially pesticides. Removal of dirt has been proved very efficient when ultrasound is applied and has been used for years in the removal of grease and oxides from metals and minerals. Removal of grease and oxides is a harder task than the removal of dirt from fruits, and as such the use of ultrasound is also efficient in washing the surface of fruits.

Ultrasound has been tested and showed to improve pesticide removal from some fruits (Table 3.1). The ultrasonic process has been shown to be much faster and efficient than the conventional liquid–solid extraction technique for the removal of α-cypermethrin and dimethoate.

Washing of oranges with soap in an ultrasonic bath decreased the amount of thiabendazole (pesticide) by 50% in the orange peel. The removal of imazalil (pesticide) from oranges was most effective applying ultrasonic wave in a water bath. The effectiveness of ultrasonic treatment in removing residues of imazalil may result from two effects: ultrasound facilitates diffusion of imazalil out of the peel pores or imazalil decomposes by treatment with ultrasound (Kruve et al., 2007).

TABLE 3.1

Removal of Pesticides from Fruits

Compound	Fruit	Time (min)	Efficiency (%)	Reference
Dimethoate	Olive	20	90	Peña et al. (2006)
α-cypermethrin	Olive	20	99	Peña et al. (2006)
Thiabendazole	Orange		50	Kruve et al. (2007)

In related studies on pesticide removal, some vegetables such as cabbage and green pepper were subjected to pesticide washing by ultrasound application. The application of ultrasound for 10 min with strong alkaline-electrolyzed water reduced methamidophos and dimethoate concentrations in cabbage by 33.8% and 31.6%, respectively, and in green pepper by 17.0% and 28.7%, respectively. Continuously changing electrolyzed water further decreased the methamidophos and dimethoate concentrations in leafy cabbage by 79.3% and 89.3%, respectively (Lin et al., 2006).

The potential use of ultrasound on removal of pesticides is evidenced by the development of several analytical methods to quantify pesticides and herbicides in fruits. Most methods require the immersion of the fruit or vegetable in an ultrasonic bath to remove the pesticide or herbicide from the sample tissue, which is the basis for washing of fruit. Analytical methods have been developed to identify and quantify the presence of abamectin (Hernández-Borges et al., 2007), carbaryl (Pia et al., 2008), carbedazim (Pia et al., 2008), dimethoate (Pia et al., 2008), fenitrothion (Cao et al., 2005), glyphosate (Granby et al., 2003), imidacloprid (Cao et al., 2005; Liu et al., 2005; Pia et al., 2008), monocrotophos (Pia et al., 2008), organophosphate compounds (Xiang et al., 2005; Ramos et al., 2008), parathion (Cao et al., 2005), pentachloronitrobenzene (PCNB) and its metabolites (Li et al., 2009), simazine (Pia et al., 2008), and triazine (Rodríguez-Gonzalo et al., 2009).

Washing and removal of pesticides can be enhanced by the application of ultrasound. Actual technology allows removal of pesticides from the fruit peel or skin, while little is known on the applicability of the method in the removal of pesticide from the inner part of the fruit.

To summarize, the removal of pesticides may be attained by the immersion of the fruit (whole or in pieces) in an ultrasonic bath for a period of 20–30 min. Increase in removal efficiency can be attained by constant renewal of the washing fluid inside the bath and by the use of strong alkaline-electrolyzed water.

3.2.2 Sanitization

The U.S. Food and Drug Administration (FDA) require processors to achieve a 5 log reduction in the numbers of the most resistant pathogens in their finished products. Power ultrasound has shown promise as an alternative technology to thermal treatment for food processing or even associated with thermal treatment (thermosonication) and has been identified as a potential technology to meet the FDA requirement.

Pathogenic organisms of most concern in minimum processed food products are *Escherichia coli*, *Listeria monocytogenes*, *Shigella*, *Salmonella*, and hepatitis A virus.

These organisms cause diseases, which were linked to the consumption of contaminated fresh fruits (Davis et al., 1988; Farber et al., 1990; O'Mahony et al., 1990; CDC, 1997a,b). Ultrasound application was shown to reduce the microorganism count for *E. coli, Listeria innocua, L. monocytogenes, Saccharomyces cerevisiae, Salmonella, Shigella boydii,* and *Streptococcus mutans.*

The lethal effect of ultrasound has been attributed to the cavitation phenomenon, due to the formation of bubbles or cavities in liquids. The collapse of the bubbles produces intense localized changes in pressure and temperature, causing shear-induced breakdown of cell walls, disruption and thinning of cell membranes, and DNA damage via free radical production (Manvell, 1997). Ultrasound causes cytolytic effects; however, the cell disruption rate depends upon the type of microorganism and the experimental conditions.

Several Gram-negative bacteria have a higher susceptibility toward ultrasound compared with Gram-positive bacteria. Gram-positive bacteria are less sensitive to ultrasound as they usually have a thicker and a more tightly adherent layer of peptidoglycans than Gram-negative bacteria. Ananta et al. (2005) showed that the major action site of ultrasound in inducing lethal effects on Gram-positive and Gram-negative bacteria was not necessarily the cytoplasmic membrane and cell death could occur without any severe damage of membranes. In the absence of thermal effect, cell death seemed to result from nonmembrane-related degradation. For example, *L. rhamnosus* was killed by ultrasound application even though the integrity of the cytoplasmic membrane of these cells was not seriously affected by ultrasound.

Sporulated microorganisms appear to be more resistant to ultrasound than vegetative ones, and fungi are more resistant in general than vegetative bacteria.

Microbial inactivation by ultrasound depends on many factors that are critical to the outcome of the process. These factors are design parameters (ultrasonic power and wave amplitude, temperature, volume of the food to be processed), product parameters (composition and physical properties of the food), and microbial characteristics (Guerrero et al., 2005). It is also important to notice that there exists heterogeneity within the population with respect to their capacity of resisting the deteriorating impact of ultrasound.

Ultrasonic power and the sonication time showed significant positive effects on the inactivation of cells. Ultrasonic irradiation may result in great damage to cell microstructure, and this damage on structural integrity can be the direct reason for the death of cells (Zhou et al., 2007).

The value of decimal reduction times (D_R) decreases with the increase in the ultrasonic power. A higher decrease is observed for low power density and a lower decrease is observed for higher powers (Figure 3.1). The effect of ultrasound may not improve the value of decimal reduction times if high ultrasonic power is applied.

Several factors, including vessel geometry, composition of the medium, surface strength, viscosity, or volume of the sample, can have an influence on the lethal effects of ultrasound. The influence of these factors on the lethality of the ultrasonic treatment is due to their effect on the acoustic field. Vessel geometry can affect the location of maximum and minimum intensity zones of the generated stationary acoustic field. Factors such as the composition of the medium, surface tension, or viscosity affect the absorption of the acoustic energy and its availability for the product lethal effects.

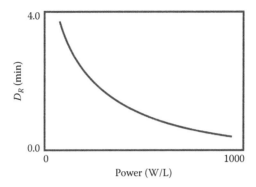

FIGURE 3.1 Relationship between D_R and power density.

Ultrasound alone is not very effective in killing microorganisms in food since long treatment times (>10 min) are required. The high levels needed could adversely modify nutritional and sensory properties of the food. This limitation has suggested that ultrasound could be more effective when used in combination with other techniques such as heat (thermosonication) and pressure (manosonication) for inactivation of pathogenic and spoilage microorganisms. In addition, the application of ultrasound in combination with natural and synthetic antimicrobials has demonstrated to enhance its inhibitory effect on a variety of microorganisms.

A study carried out by Seymour et al. (2002) showed that the frequency of ultrasound treatment (25, 40, and 70 kHz) had no significant effect on decontamination efficiency. The use of hot water (55°C) in combination with heat and ultrasound contributed to an increase of lethality (Scouten and Beuchat, 2002). However, above 60°C, the rate of death by ultrasound was not significantly different from that in the thermal treatment.

The combined effect of low-frequency ultrasound (20 kHz) with temperature (from 45°C to 55°C) on the survival of a strain of *Saccharomyces cerevisiae* suspended in water was studied by Ciccolini et al. (1997). Application of ultrasound at nonlethal temperature did not display a deactivating action, but at a high temperature the synergy between ultrasound and temperature was observed and confirmed. The ultrasonic waves were not able to destroy the yeast's cells but were able to damage the cells, increasing their sensitivity to heat. The optimal ultrasonic power for a maximal deactivating effect was 100 W. Thermosonication treatment was considerably more effective for *E. coli* ($D_{45°C} = 1.74$ min, $D_{50°C} = 0.89$ min) and for *Staphylococcus epidermidis* ($D_{45°C} = 2.08$ min, $D_{50°C} = 0.94$ min) (Czank et al., 2010).

E. coli O157:H7 were inactivated by ultrasound application in salt solutions (Na_2PO_4, NaCl, $NaNO_3$, NH_4Cl, $CaCl_2$, and $AlCl_3$) at concentrations ranging from 0% to 5% (w/v). Increases in sonication treatment time, intensity, and temperature led to increased lethality of *E. coli* O157:H7. At 40°C, the ultrasonic treatment led to total inactivation for cells suspended in 0.5% NH_4Cl, 1% $CaCl_2$, 2% NaCl, 5% Na_2PO_4, and all concentrations of $AlCl_3$ after 10 min. Complete inactivation also occurred at 0°C for 5% $AlCl_3$, which was shown to be the most effective salt (Stanley et al., 2004).

Inactivation of *E. coli* O157:H7 was attributed to cavitation-induced shear forces, reaction of cavitation-generated hydrogen peroxide with microbial cell wall constituents, and electrostatic interactions of dissociated salts with cell membranes (Stanley et al., 2004). Extensive damage for ultrasound-treated *E. coli* K12 cells, including cell perforation, was observed. Perforation is a unique phenomenon found on ultrasound-treated cells that could be caused by liquid jets generated by cavitation.

Due to its complexity and the protective nature of *Lactobacillus sporogenes*, the inactivation effect of ultrasound on this microorganism was much smaller compared with irradiation. However, the combination of ultrasound and γ-irradiation reduced the irradiation dose to achieve the same microbial lethality values for *Lactobacillus sporogenes* as γ-irradiation alone (Ye et al., 2009).

3.2.2.1 Antimicrobials

Ultrasound was highly effective in inducing damage on the lipopolysaccharide layer of the outer membrane of Gram-negative bacteria. One practical consideration, which can be drawn, based on this observation is the possibility to combine ultrasound processing with antimicrobials, whose bactericidal efficacy is reduced by the presence of an intact outer membrane.

Chlorine (in the form of hypochlorite or dioxide, or directly in the form of gas) is the most used substances to wash fruits. Chlorine has a relatively low cost and is simple to use. However, the use of gaseous chlorine is harmful to the process operators and there is the possibility of formation of chloramines, which remain as residues on the washed product and are potentially toxic for the final consumer.

Reductions in *Salmonella typhimurium* obtained by cleaning with ultrasound with water and ultrasound with chlorinated water were 1.5 and 2.7 logs, respectively (Seymour et al., 2002).

The presence of vanillin or citral increases the bactericidal effect of sonication and reduces the inactivation times. When both antimicrobials are added together and ultrasound is applied, death time means between 1.6 and 2.6 min were obtained for *L. monocytogenes* in orange juice (processed at 45°C with high-intensity ultrasound—600 W, 20 kHz) (Ferrante et al., 2007).

Chitosan, a natural mucopolysaccharide, has been considered a very interesting biopolymer for its antimicrobial and functional properties and low toxicity. Chitosan inhibits the growth of various fungi by mechanisms that differ distinctly from that of the current synthetic fungicides. Chitosan makes polyelectrolyte complexes with the acidic and basic groups of the cell surface disordering it and rendering the cell wall more viscoelastic. In addition, chitosan increases membrane permeability of cells due to its polycationic nature.

Chitosan, at a 1000 ppm, can reduce the population of *S. cerevisiae* in approximately 1 log cfu/g after 30 min of exposure. The combination of chitosan, at a 1000 ppm with sonication, led to a 3 log cfu/g reduction in viable cells. Exposing the cells to chitosan for 90 min did not increase the inactivation as compared with incubation for 30 min (Guerrero et al., 2005).

The combination of ultrasound and ozone application is also possible. The application of ozone in association with ultrasound, however, is less effective than the application of chlorine unless effective delivery of ozone is achieved (Kim et al., 1999).

The total count of *Pseudomonas fluorescens* reduced to 1.4 log cfu/g in a treatment chamber flushed with gaseous ozone while sonicating, which was 2× higher than washing with water only (Kim et al., 1999).

3.2.2.2 Sanitization of Whole Fruits or Fruit Pieces

Sanitization of fresh fruit (whole or in pieces) is required to reduce the risk of microbial spoilage and contamination by pathogenic microorganisms. The cleaning action of cavitation caused by ultrasound application appears to remove cells attached to the surface of fresh produce, rendering the pathogens more susceptible to the sanitizer.

No significant differences, in decay index, were observed in strawberries subjected to ultrasound when comparing the control and 25 or 28 kHz ultrasound treatment. Treatment with 40 or 59 kHz ultrasound, however, significantly inhibited the increase in decay index. The decay index in 40 and 59 kHz ultrasound treatments was 43.7% and 29.0%, respectively, lower than that in the control on the 8th day of storage. Ultrasound treatment with 40 kHz was more effective in reducing strawberry decay due to microbial spoilage (Cao et al., 2010).

3.2.2.3 Sanitization of Juices and Purees

The application of ultrasound can be of great interest in liquid food processing to sanitize and extend the shelf life of fruit juices and purees. Ultrasound has proved to reduce the concentration of microorganisms in fruits juices either alone or in combination with known biocides. The use of ultrasound was shown to reduce the need of biocides reducing the cost and safety of sanitization processes.

The inactivation of microorganisms by ultrasound application usually follows a first-order kinetics throughout the sonication time. The use of high-intensity ultrasound treatment in a continuous flow mode, with major application in industrial needs, generates inactivation curves similar to those obtained under batch processes.

Titratable acidity, pH, and soluble solids content are not affected by ultrasound sanitization treatment. Minor changes in color and turbidity for ultrasound-treated samples are observed.

Because of the acidity, the primary spoilage of raw fruits are yeasts and moulds, predominantly molds and *Saccharomyces* spp. This fact in addition to the great resistance of these species to ultrasonic treatment suggested *S. cerevisiae* as one of the target microorganisms for fruit juice treatments.

Inactivation tests were performed with *E. coli* K12 at 40°C, 45°C, 50°C, 55°C, and 60°C with and without ultrasound. The inactivation tests showed that sonication increased *E. coli* K12 cell destruction by 5.3 log, 5.0 log, and 0.1 log cycles at 40°C, 50°C, and 60°C, respectively, in apple cider. The additional destruction due to sonication was more pronounced at sublethal temperatures (Ugarte-Romero et al., 2006). Also in apple cider, a 5 log reduction was achieved for *E. coli* O157:H7 and *L. monocytogenes* with the use of copper ion water (1 ppm) in combination with sodium hypochlorite (100 ppm) followed by sonication at 44 kHz (Rodgers and Ryser, 2004).

In peptone water and apple juice, UV-C radiation provoked higher *E. coli* ATCC 35218 inactivation than ultrasound treatment. The effect of UV-C light in orange juice was enhanced by the combination with ultrasound (Char et al., 2010).

Ultrasonic treatment (10 min) of *Salmonella* in peptone water produced up to 4.0 log reduction in viable counts (Lee et al., 1989).

Ultrasonic power had only a slight effect on the inactivation rate of *Lactobacillus sporogenes* in tomato puree. The effect of ultrasonic processing time had no statistically significant effect on the inactivation rate. The combination of ultrasound and irradiation showed that the required irradiation dose decreased linearly with the increase in ultrasonic power intensity. The optimum process parameters for the inactivation of *Lactobacillus sporogenes* in tomato paste were ultrasonic power of 120 W, ultrasonic processing time of 25 min, and irradiation dose of 6.5 kGy. The ultrasound treatment prior to irradiation for the inactivation of *Lactobacillus sporogenes* in tomato paste was unsuitable for reducing the irradiation dose (Ye et al., 2009).

3.2.2.4 Modeling

The relationship between decimal reduction time (D_R) and the power density (P_D) can be modeled using a logarithmic equation.

$$D_R = a - b \cdot \ln(P_D), \tag{3.1}$$

where
  a and b are the equation parameters
  D_R is the decimal reduction time (min)
  P_D is the ultrasound power density (W/m^3)

To integrate all the variables (T, t, and P_D) in a single model, an equation was built considering the logarithmic relationship between the kinetic constants (D_R and z) and the volume of treatment (Cabeza et al., 2010):

$$\ln\left(\frac{N}{N_0}\right) = \frac{1}{a - b \cdot \ln(P_D)} \times t \times \exp\left(\frac{T - T_{ref}}{c - d \cdot \ln(P_D)}\right), \tag{3.2}$$

where
  a, b, c, and d are parameters of the equation
  N is the number of microorganisms
  N_0 is the initial number of microorganism
  P_D is the power density (W/m^3)
  t is the time (min)
  T is the temperature (K)
  T_{ref} is the reference temperature (K)

3.2.3 BLANCHING

Blanching has an important role in the inactivation of enzymes and microorganisms and in the stabilization of color. However, the degree of thermal treatment during blanching can also have adverse effects on the quality of dried and frozen fruits, such as modifications to the cellular structure and leaching of vitamins and minerals into the blanching medium (Katsaboxakis, 1984).

When fruits are blanched, an initial brightening of the color is observed, because of the removal of gases and air around the surface of the fruit, and the expulsion of air from extracellular space (Priestley, 1979). With further heating, an undesirable fading of color begins. To reduce the adverse effects of blanching, thermosonication blanching can be employed.

Thermosonication treatment can be used to minimize color changes, since blanching times can be reduced (Cruz et al., 2007). The thermosonication treatment may inactivate peroxidase (POD) at less severe blanching conditions, consequently retaining vitamin C and antioxidants content at higher levels. Higher retention of vitamin C was observed in thermosonication treatment (about 94%) as compared with heat blanching, which reduced vitamin C content to 29% (Cruz et al., 2008).

3.2.4   INACTIVATION OF ENZYMES

Ultrasound alone or in combination with heat or pressure or both heat and pressure is effective against various food enzymes pertinent to the fruit juice industry such as lipoxygenase, POD, and polyphenol oxidase (PPO), as well as heat-resistant lipase and protease. The effectiveness of ultrasound for control of enzymatic activity is strongly influenced by intrinsic and extrinsic factors such as enzyme concentration, temperature, pH, and composition of the medium. Ultrasound application is more effective in inhibiting enzyme activity when combined with other processes, such as high pressure or heat.

The application of ultrasound could be useful to reduce the activity of enzymes that promote enzymatic browning in fruits. HO and HO_2 free radicals, formed during cavitation of bubbles, may attack enzymes in the cavitation field and destroy hydrogen bonds and hydrophobic interactions between the enzyme subunits, initiating their dissociation that leads to the loss of the catalytic activity of the enzyme (Tarun et al., 2006).

Ultrasonic inactivation mechanisms may be specific to the enzyme under investigation and may depend on amino acid composition and the conformational structure of the enzyme. Manothermosonication inactivates POD by splitting its prosthetic heme group (López and Burgos, 1995). Lipoxygenase appears to be inactivated by a free radical–mediated mechanism (López and Burgos, 1995) and by denaturation of proteins (Mason, 1998).

Cloud loss or clarification of juice samples mainly occurs due to enzymatic precipitation of cloud in citrus juices caused by a series of events initiated by pectin methylesterase (PME). Sequential cleavage of the methyl esters at C6 of galacturonic acid residues in pectin produces free acid groups. When of sufficient size, such blocks on adjacent pectin molecules can be cross-linked by divalent cations, leading to protein precipitation (Cameron et al., 1998).

PME activity significantly decreases as a function of sonication time and ultrasonic power density (W/mL). A 62% inactivation of PME was observed at 1.05 W/mL for 10 min (Tiwari et al., 2009a). Reduction of PME activity in orange juice and in lemon juice resulted in an improved cloud stability due to mechanical damage of the PME protein structure.

Thermosonication (20 kHz, 50°C–75°C) enhanced the inactivation rates of polygalacturonase (PG) and PME in tomato juice (Raviyan et al., 2005; Terefe et al., 2009).

The inactivation rate of PME was increased sixfold and the inactivation rate of PG by fourfold in the temperature range 60°C–75°C, with the highest increase corresponding to the lowest temperature (50°C). Thermosonication inactivation of PME was described by first-order kinetics.

Thermosonication reduced the activity of lemon pectinesterase by 83% (50°C, 63 min of processing) while ultrasound alone (25°C) reduced the activity of the enzyme by 30% (Kuldiloke et al., 2007).

Surface color is one of the most important quality attributes in fruits because consumers usually judge the quality of fresh-cut fruits based on their appearance. Shelf life of fresh-cut fruits is restricted by physiological injury resulting from essential processing operations, including peeling, coring, and cutting. These processes cause browning on the cut surface, which limits the development and commercialization of fresh-cut fruits.

PPO and POD are the enzymes involved in the browning process. Browning occurs almost instantly when the cell structure is destroyed, and the enzyme and substrate are mixed. PPO catalyzes the hydroxylation of monophenols (monophenolase) and oxidation of o-diphenols to o-quinones (diphenolase), which subsequently polymerize to yield undesirable brown pigments in the presence of oxygen (Espín et al., 1998). POD is relevant to enzymatic browning since diphenols may function as reducing substrate in the enzyme reaction and could promote darkening in fruit products during processing and preservation (Chisari et al., 2007).

Ultrasound may allow ascorbic acid or other antioxidant to act inside the cells disrupted by ultrasonic treatment, hence enhancing antioxidant action and inhibiting enzyme activity. Low monophenolase activity was reported when ultrasound (40 kHz) was applied in combination with ascorbic acid (1%) in fresh-cut apples. The activity of monophenolase when ultrasound (40 kHz) was applied in combination with ascorbic acid (1%) was lower than when ultrasound or ascorbic acid were used alone (Jang and Moon, 2011).

Ultrasound treatment has stimulated PPO activity. Higher diphenolase activities were reported for fresh-cut apples treated by ultrasound application as compared to that of other inhibiting processes. However, when ultrasound (40 kHz) was applied in combination with ascorbic acid (1%), it completely inhibited the diphenolase activity on the treatment day and showed low activity during the storage period—12 days (Jang and Moon, 2011).

Ultrasound application significantly lowered POD activity in fresh-cut apples, but the POD activity showed a steady increase during storage time—12 days. The combination of ultrasound application with ascorbic acid (1%) reduced the POD activity in fresh-cut apples during the storage period—12 days (Jang and Moon, 2011).

3.2.5 Preservation of Enzyme Activity

The use of free radical scavengers can reduce the inactivation of enzymes in fruits because these scavengers reduce the amount of HO and HO_2 free radicals in the processing media. Astragalin, silybin, naringin, hesperidin, quercetin, kaempferol, and morin could be used to protect enzyme activity (Tarun et al., 2006).

Ultrasound may also increase the activity of some enzymes. Ultrasonic preirradiation of lipases in aqueous buffer and organic solvents enhanced the activity of the enzyme and the activities of the hydrolytic (esterase) and transesterification (Shah and Gupta, 2008).

3.3   FRUIT TISSUE STRUCTURE

Ultrasonic waves cause a rapid series of alternative compressions and expansions, in a similar way to a sponge when it is squeezed and released repeatedly (sponge effect). The sponge effect results in the creation of microscopic channels in the fruit tissue (Tarleton, 1992; Tarleton and Wakeman, 1998; Fuente-Blanco et al., 2006). Studies with ultrasound and ultrasound-assisted osmotic dehydration have shown that different fruits respond differently to the application of these treatments (Fernandes et al., 2008a, 2009; Rodrigues et al., 2009a,b).

Microscopic image analysis of fresh fruits shows typical thin-walled cells with normal morphology, with few or no visible intercellular spaces (Figure 3.2).

The microscopic image analysis of fresh sapota presents two types of cells: parenchyma cells and dense cells (Figure 3.3). The parenchyma cells (white cells

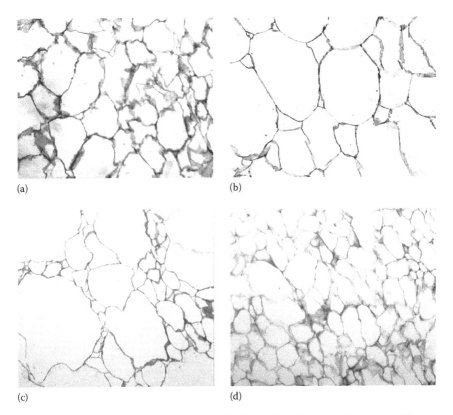

(a)   (b)

(c)   (d)

FIGURE 3.2   Micrographs of fresh fruits: (a) pineapple; (b) melon; (c) papaya; (d) malay apple.

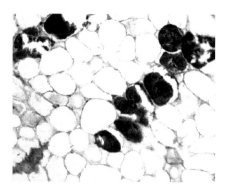

FIGURE 3.3 Micrographs of fresh sapotas.

in Figure 3.3) comprise mostly round-shaped cells. Only few parenchyma cells are slightly distorted. The tissue presents few small cell interspaces, mostly near the junction between three cells. Dense cells (black cells in Figure 3.3) are found near one another and contain high concentration of phenolic compounds.

Few changes are detected in fruit tissue structure during the first 10–20 min under ultrasound application (Figure 3.4). Melon tissue presented two distinct regions: one region where the cells became bloated because of water gain and a

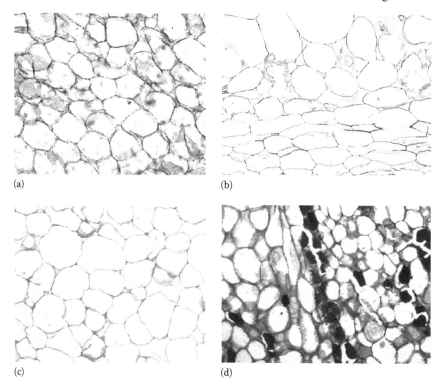

FIGURE 3.4 Micrographs of fruit tissue subjected to 20 min of ultrasound: (a) pineapple; (b) melon; (c) papaya; (d) sapota.

second region where the cells became much smaller and needle shaped. These needle-shaped cells were not observed in the fresh fruit and were formed during ultrasonic treatment. In sapota, the cells became severely distorted in some regions and the dense cells presented a higher degree of collapse. The cells became elongated forming channels similar to the microscopic channels observed in melons. In papaya, the tissue showed a high degree of disruption of cells, creating several large cell interspaces.

After 30 min of ultrasonic treatment, the tissue cells become distorted and microscopic channels begin to be formed (Figure 3.5). In some regions, the junctions between adjacent cells are present and the intercellular spaces are reduced while in other regions disruption of the cells is observed causing an increase in intercellular spaces. After this period, the microscopic channels may become broader and some new smaller microscopic channels can appear. In papaya, microscopic voids were mostly formed by disruption of contiguous cells, which produced large cell interspaces.

Formation of microscopic channels in pineapples was mostly caused because of loss of cellular adhesion (disruption of contiguous cells), which produced large cell interspaces (Fernandes et al., 2009). The formation mechanism of microscopic channels in pineapples differed from the mechanism observed in melons (Fernandes et al., 2008a), where microscopic channels were formed by flattening and elongation of cells.

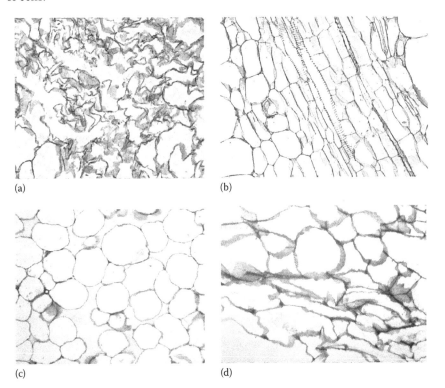

(a)   (b)

(c)   (d)

FIGURE 3.5 Micrographs of fruit tissue subjected to 30 min of ultrasound: (a) pineapple; (b) melon; (c) papaya; (d) malay apple.

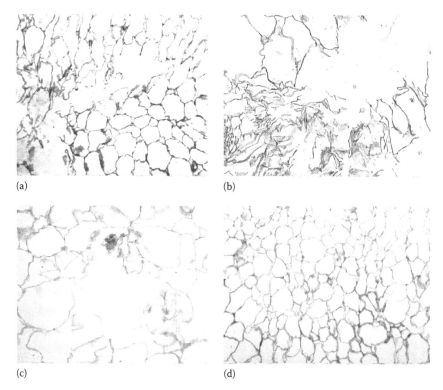

(a)   (b)

(c)   (d)

FIGURE 3.6   Micrographs of fruit tissue subjected to 30 min of ultrasound-assisted osmotic dehydration (osmotic solution = 35% w/w): (a) pineapple, (b) melon, (c) papaya, and (d) malay apple.

Significant differences are observed when an ultrasound-assisted osmotic dehydration is employed. The use of an osmotic solution results in the formation of larger microchannels and rupture of cells. After a 30 min immersion in an osmotic solution with sugar content of 35 °Bx, the cells become more distorted (Figure 3.6), and the formation of microscopic channels is usually accompanied by breakdown (rupture) of cell walls.

Loss of adhesion of the cells is observed in some regions of the tissue causing an increase in intercellular spaces that may be caused by the solubilization of chelator-soluble pectin of the middle lamella. Chelator-soluble pectin is the substance that most contributes to cell adhesion and firmness and can be solubilized in the early stages of osmotic dehydration.

Longer microscopic channels are also formed by breakdown of dense cells. Most dense cells have their inner phenolic content fragmented. The breakdown of cells creates large regions where no cell membrane was observed (Figure 3.6c). The breakdown of dense cells is more intense than when distilled water is used as the liquid medium. The higher osmotic pressure gradient may be responsible for the enhancement of cell breakdown.

A greater change is observed when high sugar content osmotic solutions (70 °Bx) are applied. The use of high sugar content osmotic solutions produces microchannels

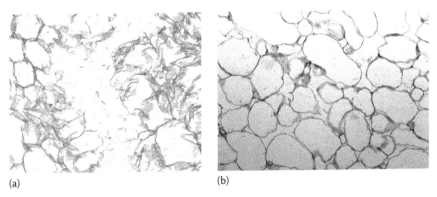

(a)　　　　　　　　　　　　　　　　　　　(b)

FIGURE 3.7　Micrographs of fruit tissue subjected to 10 min of ultrasound-assisted osmotic dehydration (osmotic solution = 70% w/w): (a) pineapple and (b) sapota.

in the first 10 min of processing and is also accompanied by a significant degree of cell rupture (Figure 3.7). Cells may become severely distorted and solubilization of pectin can be visually noted by the decreasing cell wall strength.

3.4　PHYSICOCHEMICAL CHARACTERISTICS

The rheology of plant-derived food products is a key parameter in determining consumer acceptance of juices and purees, because it influences visual appearance, mouthfeel, and flavor release of the product.

The viscosity profile of food products is influenced by insoluble solids such as pectin, hemicelluloses, cellulose, proteins, and lignin. These insoluble components exist in a continuous aqueous matrix of soluble pectin, organic acids, sugars, and salts. In most applications, these naturally present viscosity-modifying compounds are not sufficient to provide the optimum viscosity for such products, and the viscosity of purees must be increased by the use of thickening agents, such as starches and gums.

The use of ultrasound in the range of 20–300 kHz can cause the disruption of cells in fruits and vegetables, which impacts the viscosity of juices and purees (Bates et al., 2005).

3.5　NUTRITIONAL VALUE

The nutritional value of the food that is subjected to ultrasound may suffer some changes. Most of the changes induced by ultrasound are related to the formation of •OH radicals during cavitation of the liquid bubbles. The •OH radicals react mainly with ascorbic acid, vitamins, and antioxidants, thus reducing the nutritional value of fruits. Minimization of nutritional loss may be attained by adding small amounts of scavengers in the liquid medium.

Ascorbic acid is well known for its preferential reaction with •OH radicals (Ueda et al., 1996; Sprinz et al., 1998). The addition of ascorbic acid in the sonication medium can be expected to react with most of the •OH radicals generated by acoustic cavitation, which may minimize the reaction between •OH radicals and food compounds. Cao et al. (2010) observed that the addition of 0.1 mol/L ascorbic acid almost eliminated the formation of hydrogen peroxide.

Addition of ethanol to the sonication medium also reduces the formation of •OH radicals. The addition of 0.1 mol/L ethanol to the sonication medium reduces •OH formation in about 50% compared to a 95% reduction in the presence of ascorbic acid (Ashokkumar et al., 2008). The addition of alcohols, such as ethanol, is known to decrease the cavitation bubble temperatures (Rae et al., 2005).

3.5.1 VITAMIN C

Increasing processing time and ultrasound density (W/mL) decrease the ascorbic acid content in juices and whole fruits; however, the degradation rate obtained for sonicated juice is lower than the degradation rate observed for thermally processed orange juice (Polydera et al., 2003).

For orange juice, the highest degradation of ascorbic acid (5% loss) was obtained at an ultrasound density of 0.8 W/mL and 10 min of processing time (Tiwari et al., 2009). Minimal effect on vitamin C was also reported for orange juice during storage at 10°C. The shelf life of sonicated juice was significantly higher than that of thermally processed juice because of the lower degradation rate found for sonicated juices (Tiwari et al., 2008b,c).

A decrease in vitamin C content in strawberry occurs after harvest. The decay rate was lower when the fruit was subjected to ultrasound at 40 and 59 kHz, increasing shelf life. No significant difference in vitamin C content between control and 25 or 28 kHz ultrasound-treated fruit was observed. The better quality parameters of strawberries when subjected to 40 and 59 kHz ultrasound is due to inhibited decay incidence and microbial population of strawberries, thus decreasing the loss of quality parameters (Cao et al., 2010).

3.5.2 ANTIOXIDANTS AND ANTHOCYANINS

Anthocyanins are plant pigments responsible for the orange, red, and blue colors of several fruits. They are known for their pharmacological properties and are employed for therapeutic purposes (Wang et al., 1997; Duthie et al., 2000). The protective effects of fruits against diseases have been attributed to several antioxidants present in fruits and vegetables (Hertog et al., 1995; Clifford, 2000; Rentzsch et al., 2007; Zafra-Stone et al., 2007).

The stability of anthocyanins is influenced by pH, light, oxygen, enzymes, ascorbic acid, and high temperature during processing and storage (Jackman et al., 1987; Wang and Xu, 2007). The stability of anthocyanins is also influenced by the interaction of anthocyanin with ascorbic acid (Markakis et al., 1957; Shrikhan and Francis, 1974; Skrede et al., 1992).

Degradation of anthocyanins and antioxidants may happen during ultrasound application because of cavitation, which may form free radicals that attack anthocyanins and antioxidants. The anthocyanin degradation may also be due to the presence of other organic acid or ascorbic acid and can be related to oxidation reactions, promoted by the interaction of free radicals formed during sonication (Portenlänger and Heusinger, 1992). Hydroxyl radicals produced by cavitation may degrade anthocyanins by opening of rings and formation of chalcone mainly due to the temperature rise that occurs during sonication (Sadilova et al., 2007). Pyrolysis within cavitation bubbles or nuclei is likely to be the major reaction path for the degradation of polar compounds (Sivasankar et al., 2007).

The degradation of ascorbic acid or anthocyanins may be caused by either thermolysis or combustion occurring inside the bubble or by reaction with hydroxyl radicals, leading to the formation of oxidation products occurring on the bubble surface. Ascorbic acid is more degraded by ultrasound application than anthocyanins. Petrier et al. (1998, 2007) reported that when two compounds with differing physicochemical properties are subjected to sonication, the more volatile compounds degrade first.

The level of anthocyanin and antioxidant degradation due to ultrasound application is relatively low, and ultrasound processing technology may be considered as a potential alternative technique when retention of nutritional quality is a priority. Degradation of anthocyanin from blackberry juice and from strawberry juice was lower than 5% even when high amplitude levels of ultrasonic waves were applied (Tiwari et al., 2009b,c).

Low degradation levels were also observed for blueberry juice subjected to ultrasound during 4–30 min. The content of anthocyanins in juices sonicated up to 32 min did not alter the amount of cyanidin 3-glucoside and cyanidin 3-(6-malonyl) glucoside (anthocyanins). Phenolic compounds were also not degraded by ultrasound, since no changes in concentration were observed for ellagitannins like lambertianin C and sanguiin H-6 (Wong et al., 2010).

3.5.3 TOTAL SOLUBLE SOLIDS

The amount of total soluble solids decreases with ultrasound application, if the fruit is immersed in distilled water or in a hypotonic solution. In general, the loss of soluble solids increases with decreasing ultrasonic frequency. The loss of total soluble solids may be as high as 50% if immersed in distilled water for more than 30 min.

Total soluble solids loss in whole fruits during storage may be inhibited by ultrasound application due to reduction in microbial count (Cao et al., 2010).

3.5.4 MINERALS

Application of ultrasound decreases the amount of minerals in fruit samples. Loss of minerals is observed after 10–20 min of sonication time and is dependent on the mineral and on the fruit. Sample size was an important factor for mineral loss. Smaller samples increase the removal of minerals (Nascentes et al., 2001). Losses up to 60% may be observed for calcium; 75% for magnesium, zinc, and manganese; and 85% for iron (Nascentes et al., 2001).

Sonication times over 30 min have decreased the loss of minerals. Different species can be formed during sonication, and it is possible that a longer sonication time provokes the recombination of different structures in the liquid, forming new species and reducing ions in the liquid phase (Carvalho et al., 1995).

3.6 SENSORY CHARACTERISTICS

3.6.1 COLOR

Few studies have addressed the effect of ultrasound application in the color of foods subjected to some kind of ultrasonic treatment. The results obtained by several researchers did not provide any special trend.

Sonication of orange juice was studied by Gómez-López et al. (2010), and they reported slight changes in the color of orange juice subjected to sonication. Juice samples became slightly lighter, greener, and more yellow. However, no statistical difference was observed on panel-tested color of orange juice, even at extreme treatment conditions (Gómez-López et al., 2010). Valero et al. (2007) reported that ultrasound treatment had no effect on browning pigment production in batch conditions. Ultrasound processing of juices showed minimal effect on the degradation of key quality parameters such as color and ascorbic acid in orange juice during storage at 10°C (Tiwari et al., 2009b). This positive effect of ultrasound is assumed to be due to the effective removal of occluded oxygen from the juice (Knorr et al., 2004).

Studies with blueberry juice showed that no significant differences were found for different sonication times for color parameters. All parameters remained constant for all times, indicating the absence of overall color degradation between untreated and treated samples (Wong et al., 2010).

3.6.2 FLAVOR AND AROMA

Few studies have addressed the effect of ultrasound application in aroma and flavor of foods subjected to some kind of ultrasonic treatment. Overall, no significant change was observed in sonicated fruits and juices. Some residual bitterness was detected in some juices.

Orange juice was not affected by sonication. No statistical difference was observed on panel-tested aroma and flavor of orange juice, even at extreme treatment conditions (Gómez-López et al., 2010). The same result was found for blueberry juice. No significant differences were detected for sonication times ranging from 4 to 30 min. An off-flavor, described as a "cooked flavor" by some panelists, was reported for blueberry juice sonication for more than 8 min. The off-flavor may be caused by degradation of aromatic compounds or production of aromatic compounds that may have occurred during sonication, presenting a low sensory perception threshold (Wong et al., 2010).

3.6.3 TEXTURE

Texture is a critical quality attribute in the consumer acceptability of fruits. Conservation of fruit texture is usually related to the conservation of the fruit tissue structure. Changes in the tissue structure will affect the texture of the final product.

The tissue structure of fruits subjected to ultrasound does not change significantly. As such, the texture of the product after sonication also does not change significantly. Changes are expected for fruits subjected to ultrasound-assisted osmotic dehydration, which may induce severe breakdown of cells.

Strawberry is a soft fruit that suffers a rapid loss of firmness during storage, which contributes greatly to its short postharvest life and susceptibility to fungal contamination. Ultrasonic treatment with 40 kHz has inhibited fruit softening and maintained the fruit firmness during storage. Firmness on the 8th day of storage in strawberries treated with 40 kHz ultrasound was 17.5% higher than that in non-treated fruits (Cao et al., 2010). Inhibition of softening in strawberries is linked with the inhibition of PG and PME activity, which are associated with strawberry softening (Nogata et al., 1993; Vicente et al., 2005).

3.7 PROCESSING

The use of ultrasound within the food industry has been growing and in recent years, many applications of ultrasonic energy have been investigated concerning the use of ultrasound to directly affect the process or product.

3.7.1 PRETREATMENT FOR DRYING PROCESSES

Drying can be applied to reduce postharvest loss of fruits and also to produce dehydrated fruit used in foodstuffs formulations, providing an extension of shelf life, lighter weight for transportation, and less space for storage. Many fruits are used in dehydrated foodstuffs (sauces, food mixes, grain mixes, and others) and benefit from drying to extend their shelf life.

Conventional air-drying is energy intensive and consequently cost intensive because it is a simultaneous heat and mass transfer process accompanied by phase change. A pretreatment, such as ultrasonic pretreatment, can be used to reduce the initial water content or to modify the fruit tissue structure in such a way that air-drying becomes faster.

The rapid alternative compressions and expansions caused by ultrasound application facilitate moisture removal. In addition, ultrasound produces cavitations that help in removing strongly attached moisture. Deformation of porous solid materials, such as fruits, caused by ultrasonic waves is responsible for the creation of microscopic channels that reduce the diffusion boundary layer and increase the convective mass transfer in the fruit.

The ultrasonic pretreatment prior to drying consists in the immersion of the fruit in distilled water to which ultrasound is applied. The application of ultrasound should be carried out for at least 10 min, and after 20–30 min the changes in the effective water diffusivity becomes slight (Fernandes and Rodrigues, 2007a; Rodrigues and Fernandes, 2007; Fernandes et al., 2008b,c; Oliveira et al., 2010).

The advantage of using ultrasound is that the process can be carried out at ambient temperature. No heating is required and consequently the probability of degradation is reduced (Mason, 1998). The water to sample ratio should be maintained low. Water to sample ratios from 3:1 to 4:1 (weight basis) are preferred because the area of the required vessel (or ultrasonic bath) can be small. Higher water to sample ratios require larger vessels and a more powerful ultrasound to generate an ultrasonic intensity above 4000 W/m^2, increasing power consumption.

At the end of the ultrasonic pretreatment, little change is observed in the moisture content, and the effect of the ultrasonic pretreatment on drying is mainly observed during the air-drying stage where a significant increase in the effective water diffusivity is observed.

Melon and pineapple lose water to the liquid medium during the pretreatment when distilled water is used as the medium solution (Fernandes and Rodrigues, 2007b; Fernandes et al., 2008b). Papaya, sapota, genipap, and banana show an increase in water gain (Table 3.2). Melon and pineapple presented high moisture content (90% for melon and 83% for pineapple) and were significantly affected by the sponge effect of ultrasonic waves. The continuous squeeze and release of the tissue of the fruit, on a microscopic level, expelled water to the liquid medium, despite the osmotic pressure gradient that tends to transfer water to the fruit. For the fruit that gained water, the water gain is not significant and, except for genipaps, it is lower than 10%. Genipaps presented a water gain of 14.9% (Fernandes and Rodrigues, 2008b).

Sugar gain is negative for all fruits subjected to ultrasound in distilled water, indicating that the fruit loses soluble solids to the liquid medium. This result is expected because of the concentration gradient of soluble solids between the fruit and the liquid medium that favors the mass transfer of soluble solids from the fruit to the liquid medium. The loss of soluble solids for each fruit is also different, ranging from 1.5% to 52.9%. Fruits with high initial moisture content lose more soluble solids to the liquid medium. This behavior is explained by the diffusion of soluble solids in water. Fruits with high initial moisture content may ease the diffusion of soluble solids toward the liquid medium.

The loss of soluble solids is also influenced by the effects of ultrasound on the tissue structure of the fruit. Microscopic channels ease the diffusion of water and soluble solids in and out of the fruit. The number, length, and width of the microscopic channels affect the mass transfer between the fruit and the liquid medium.

TABLE 3.2

Water Loss and Sugar Gain of Fruits Subjected for 30 min to Ultrasound Pretreatment

Fruit	Sugar Gain (%)	Water Loss (%)	Reference
Banana (*Musa* spp.)	−21.3	−7.2	Fernandes and Rodrigues (2007a)
Genipap (*G. americana* L.)	−8.2	−14.9	Fernandes and Rodrigues (2008b)
Malay apple (*Syzygium malaccense* L.)	−17.0	−14.7	Oliveira et al. (2010)
Melon (*Cucumis melo* L.)	−52.2	+5.5	Fernandes and Rodrigues (2007b)
Papaya (*Carica papaya* L.)	−1.5	−5.7	Fernandes et al. (2008c)
Pineapple (*Ananas comosus* L.)	−18.9	+3.1	Fernandes et al. (2008b)
Pinha (*Annona squamosa* L.)	−52.9	+18.7	Fernandes and Rodrigues (2008b)
Sapota (*Achras sapota* L.)	−3.8	−6.6	Fernandes and Rodrigues (2008a)

TABLE 3.3

Effective Water Diffusivity of Fruits Submitted to Ultrasound Pretreatment

Fruit	Effective Diffusivity of the Fresh Fruit (m^2/s)	Effective Diffusivity of the Fruit Treated in Ultrasound for 30 min (m^2/s)	Reference
Banana (*Musa* spp.)	1.28×10^{-9}	1.41×10^{-9}	Fernandes and Rodrigues (2007a)
Genipap (*G. americana* L.)	1.41×10^{-8}	1.56×10^{-8}	Fernandes and Rodrigues (2008b)
Malay apple (*Syzygium malaccense* L.)	6.52×10^{-11}	2.97×10^{-10}	Oliveira et al. (2010)
Melon (*Cucumis melo* L.)	5.00×10^{-9}	6.97×10^{-9}	Fernandes and Rodrigues (2007b)
Papaya (*Carica papaya* L.)	5.90×10^{-9}	5.76×10^{-9}	Fernandes et al. (2008c)
Pineapple (*Ananas comosus* L.)	4.97×10^{-10}	4.89×10^{-10}	Fernandes et al. (2008b)
Pinha (*Annona squamosa* L.)	8.87×10^{-10}	4.41×10^{-9}	Fernandes and Rodrigues (2008b)
Sapota (*Achras sapota* L.)	7.31×10^{-9}	7.12×10^{-9}	Fernandes and Rodrigues (2008a)

An increase in the number, length, and width of the microscopic channels increases mass transfer and reduces drying time. Bananas (*Musa* sp. L.) increased the effective water diffusivity by 10.1%, similar to genipap (*Genipa americana* L.) that presented an increase by 10.6% (Table 3.3).

The increase in the effective water diffusivity results in a faster drying time (Table 3.4). The total processing time observed for sonicated melons to reach a final moisture content of $0.5\,g_{water}/g_{dry\ solids}$ was 200 min lower (24.6% lower) than the total processing time for melons *in natura* (Table 3.4). The use of ultrasound with pineapples, however, showed an increase in total processing time; thus, this pretreatment is not suitable for pineapples.

The ultrasonic process is cost effective for several fruits because of the lower energy consumption during the air-drying step. The ultrasonic process will not be cost effective when the effective water diffusivity in the fresh fruit and in the fruit treated with ultrasound is similar.

3.7.1.1 Modeling

No phenomenological mathematical models that describe the changes in effective water diffusivity as a function of the time spent in ultrasound are described in the literature, and further studies on modeling the ultrasonic process still need to be carried out to develop a model that considers the changes on the fruit tissue structure and their consequences. A phenomenological mathematical model of the ultrasonic pretreatment should consider the mass transfer coefficient and also the changes in the mass transfer coefficient caused by the formation of microscopic channels.

TABLE 3.4

Total Processing Time for Drying of Fruits Subjected to Ultrasonic Pretreatment

Fruit	Final Moisture Content (g/g_dry solids)	Pretreatment Time (min)	Air-Drying Time (min)	Total Time (min)	Reference
Banana	0.05	—	839	839	Fernandes and Rodrigues (2007a)
Banana	0.05	10	793	803	Fernandes and Rodrigues (2007a)
Banana	0.05	20	733	753	Fernandes and Rodrigues (2007a)
Banana	0.05	30	761	791	Fernandes and Rodrigues (2007a)
Pineapple	1.25	—	473	473	Fernandes et al. (2008b)
Pineapple	1.25	10	463	473	Fernandes et al. (2008b)
Pineapple	1.25	20	493	513	Fernandes et al. (2008b)
Pineapple	1.25	30	520	550	Fernandes et al. (2008b)

Empirical correlations can be obtained from experimental data. The effective water diffusivity for a given fruit can be obtained by a polynomial equation based on pretreatment time:

$$D = a + b \cdot t_U + c \cdot t_U^2 + d \cdot t_U^3 \qquad (3.3)$$

where

D is the effective water diffusivity (m²/s)

t_U is the time spent under ultrasonic pretreatment (min)

3.7.2 OSMOTIC DEHYDRATION

Ultrasound application can be carried out using distilled water or an osmotic solution as the liquid medium. The choice of liquid medium has a great influence on the process, leading to different behaviors regarding water loss and soluble solids gain. A higher water loss is observed when an osmotic solution is used as the liquid medium when compared with the ultrasound pretreatment using distilled water.

Table 3.5 shows the results for ultrasound-assisted osmotic dehydration for several fruits. In general, water loss increases with increasing concentration of the solution. Soluble solids gain also increases with increasing concentration of the solution. The gain of sugars occurs because of the sugar concentration gradient (osmotic pressure) between the fruit and the osmotic solution, which favor the mass transfer of sugar from the osmotic solution to the fruit and the mass transfer of water from the fruit to the osmotic solution.

TABLE 3.5

Sugar Gain and Water Loss for Fruit Subjected to Ultrasound-Assisted Osmotic Dehydration for 30 min

Fruit	Sugar Content (°Bx)	Sugar Gain (%)	Water Loss (%)	Reference
Malay apple	25	26.4	−14.5	Oliveira et al. (2010)
Papaya	35	27.9	3.2	Fernandes et al. (2008c)
Pineapple	35	13.5	8.3	Fernandes et al. (2008b)
Pineapple	70	34.1	14.1	Fernandes et al. (2008b)
Sapota	35	−8.1	−2.4	Fernandes and Rodrigues (2008)
Sapota	70	10.0	13.2	Fernandes and Rodrigues (2008)

Pineapple lost between 5.1% and 8.3% of its initial water content when immersed in an osmotic solution of 35 °Bx and lost between 9.8% and 14.2% of its initial water content when immersed in an osmotic solution of 70 °Bx. Pineapple gained soluble solids during ultrasound-assisted osmotic dehydration. An increase by 13.5% of sugar in pineapple was observed when the fruit was immersed in an osmotic solution of 35 °Bx and an increase by 34.1% was observed when immersed in an osmotic solution of 70 °Bx (Fernandes et al., 2008b). Similar behavior is observed for papaya (Fernandes et al., 2008c). Water loss and weight gain for apple were significantly higher in samples treated under ultrasound application than in samples subjected to osmosis statically or under mechanical agitation (Cárcel et al., 2007).

Low water loss during ultrasound-assisted osmotic dehydration is related to the starch content of the fruit. High concentration of starch favors water uptake, a behavior that has been observed for sapota (16% ± 5% of starch/dry basis) and for vegetables with high starch content such as potatoes (Witrowa-Rajchert and Lewicki, 2006). After 10 min under ultrasound-assisted osmotic dehydration, sapota lost 12.1% and 11.2% of its soluble solid to the osmotic solution, respectively, for osmotic solutions of 35 and 70 °Bx. The pretreatment carried out with an osmotic solution of 35 °Bx showed a slight reduction in sugar loss after 30 min, whereas when the pretreatment was carried out with an osmotic solution of 70 °Bx, the fruit showed a sugar gain of 10.0% after 30 min under ultrasound-assisted osmotic dehydration (Fernandes and Rodrigues, 2008a).

Sapota presented an expressive water loss when the fruit was immersed in an osmotic solution of 70 °Bx, a condition that resulted in the highest water loss after 30 min subjected to ultrasound application (13.2%). However, the ultrasound-assisted osmotic dehydration did not show significant difference at lower soluble solids content (35 °Bx) when compared with the pretreatment carried out using distilled water as the liquid medium (Fernandes and Rodrigues, 2008a).

The effect of ultrasound-assisted osmotic dehydration on the effective water diffusivity also differed among fruits (Table 3.6). The increase observed in most fruits is caused by the breakdown of cells observed when an osmotic solution is used, creating an easier path for water to diffuse toward the surface of the fruit. When a low degree of cell breakdown is observed in the osmotic treatment, the effective

TABLE 3.6
Water Effective Diffusivity for Fruit Subjected to Ultrasound-Assisted Osmotic Dehydration for 30 min

Fruit	Sugar Content of Osmotic Solution (°Bx)	Effective Diffusivity of the Fresh Fruit (m²/s)	Effective Diffusivity of the Fruit Treated in Ultrasound for 30 min (m²/s)	Reference
Malay apple	25	6.52×10^{-11}	3.90×10^{-10}	Oliveira et al. (2010)
Papaya	35	5.00×10^{-9}	7.32×10^{-9}	Fernandes et al. (2008c)
Pineapple	35	5.90×10^{-9}	4.80×10^{-10}	Fernandes et al. (2008b)
Pineapple	70	5.90×10^{-9}	7.10×10^{-9}	Fernandes et al. (2008b)
Sapota	35	7.31×10^{-9}	4.86×10^{-9}	Fernandes and Rodrigues (2008)
Sapota	70	7.31×10^{-9}	5.72×10^{-9}	Fernandes and Rodrigues (2008)

water diffusivity is lower than the water diffusivity of the fresh fruit. This behavior was observed for sapota and pineapple subjected to ultrasonic waves immersed in an osmotic solution of 35 °Bx. In this case, the sugar incorporated by the fruit creates an extra resistance for water to diffuse toward the fruit surface, which is not compensated by the microscopic channels that might have become saturated with sucrose molecules.

Simal et al. (1998) found that the water diffusivity in apples subjected to ultrasound application was higher than the water diffusivity of the fresh fruit. The values reported by Simal et al. (1998) for apples were similar to the values reported for pineapple and Malay apple (Oliveira et al., 2010). Ultrasonic treatments increased the water diffusivity in apple by 117%.

3.7.3 Air-Drying Processes

Ultrasound is an attractive means of drying heat-sensitive foods because they can be dried more rapidly and at lower temperatures than in the conventional hot-air driers. Air-borne power ultrasound introduces pressure variations at gas/liquid interfaces and therefore increases the evaporation rate of moisture (Garcia-Perez et al., 2007).

Ultrasound-assisted air-drying is carried out in a container consisting of an aluminum vibrating cylinder driven by a piezoelectric composite transducer generating a high-intensity ultrasonic field in its inside. The driving transducer consists of an extensional piezoelectric sandwich element together with a mechanical amplifier. The ensemble has to be resonant at the frequency of the selected vibration mode of the chamber. This extensional transducer is screwed to the central part of the cylindrical chamber, which corresponds to the point of maximum displacement.

Garcia-Perez et al. (2006, 2007) have studied the influence of ultrasound-assisted air-drying on the mass transfer process during convective air drying. Their results show that the drying rate increased with increasing power. A linear relationship was found between the average effective moisture diffusivity or the mass transfer coefficient and the electric power applied to the transducer. No influence on the drying rate was observed for low ultrasound intensity.

The acoustic energy was shown to decrease as air velocity gets higher due to the disruption of the acoustic field by air flow. However, for highly porous products, like lemon peel, the influence of power ultrasound was observed in the whole air velocity range, even at high air velocity levels.

3.7.4 FREEZING

Ultrasound has applications in controlling crystallization processes because sonication can enhance both nucleation rate and crystal growth rate by producing nucleation sites.

The presence of large ice crystals within the frozen tissue may result in mechanical damage and drip loss, thus reducing the frozen fruit quality. The size and location of ice crystals are closely related to the rate of freezing (Martino et al., 1998; Ngapo et al., 1999; Delgado and Sun, 2001). Rapid freezing produces small intracellular ice, while slow freezing produces large ice crystals. Thus, the rate of freezing is recognized as critical in the final quality of the frozen product.

Rapid and even seeding of ice crystals occurs under sonication, leading to shorter time between the initiation of crystallization and the complete formation of ice, reducing cell damage. The cavitation bubbles act as nuclei for crystal growth (Mason, 1998).

Fruit tissues have semirigid cellular structure and exhibit less resistance to the expansion of ice crystals in volume. As such, fruit tissue is prone to being subjected to irreversible freezing damage. Four contributory processes are proposed for the damages observed in fruit tissue: chill damage, solute-concentration damage, dehydration damage, and mechanical damage from ice crystals (Reid, 1994). Chilling damage is a result of exposing the fruit to low temperature. Solute-concentration damage is caused by the increase in the concentration of solutes in the unfrozen liquid with the formation of ice crystals. Dehydration damage results from the increase in solute concentration in the unfrozen liquid and the osmotic transfer of water from the cell to the unfrozen liquid. Mechanical damage happens because of the formation of hard ice crystals. These damages result in loss of function in the cell membrane, disruption of metabolic systems, protein denaturation, permanent transfer of water to the extracellular environment, enzyme inactivation, and extensive cell rupture.

The preservation of cellular structure is a function of the freezing rate. Faster freezing rate produces numerous small ice crystals, resulting in less damage to cellular structure. Slow freezing rate leads to large ice crystals, which may cause the breakdown of cells.

High-power ultrasound carried out for a long period accelerates the freezing process. However, the process may be affected by the heat produced by ultrasound application. As such, optimum ultrasonic parameters and operation conditions need to the carefully chosen to optimize the formation of small ice crystals.

Few articles have addressed ultrasound-assisted freeze-drying. The few available studies report that vegetables frozen using this technology retain the integrity of the cells and that the quality of the frozen vegetable had superior quality than the vegetable frozen by traditional freezing (Sun and Li, 2003).

3.7.5   EXTRACTION PROCESSES

Extraction is a mass transfer operation, which depends on the nature of the solute and solvent, the selectivity of the solvent, and the level of convection in the medium. The Soxhlet method involves washing of the solid mass with a solvent that has high solubility and selectivity for the solute. The Soxhlet extraction mechanism is mainly diffusion, and the procedure does not involve application of any shear stress to the biomass. The results show that relying simply on diffusion through the cell membrane is a slow process and results in low yield of extracted product.

Most extraction methods (cold methods) involve simultaneous extraction and partitioning by mixing of the solute or fruit samples in a mixture of solvents. The mixture forms two phases after completion of extraction. These methods also do not involve application of shear stress to the sample; thus, the predominant mechanism is also diffusion.

To enhance the efficiency of the cold methods, it is important to disrupt the cell walls to release products into the solvent mixture. In this case, the extent of extraction should be independent of the solvent properties. For this process, sonication could provide the means to disrupt the cell walls of the fruit sample.

Ultrasonic extraction involves sonication of the sample in the solvent. Ultrasound causes cavitation of the liquid, which are nucleation, growth, and transient impulsive collapse of microscopic bubbles in the liquid. The chemical effect of cavitation is the generation of highly reactive radicals due to dissociation of the entrapped vapor molecules in the cavitation bubble at the extreme conditions reached inside the bubble at the moment of transient collapse. Furthermore, the implosion of the cavitation bubbles inside and outside a biological sample may contribute to rupture of cell walls (Mason et al., 1996). The association of the cold methods with ultrasound extraction contributes to increasing the amount of product extracted from a fruit sample or fruit powder. Histological studies showed that the application of ultrasound increased the number of disrupted cells, thus contributing to a higher yield of extracted product (Ranjan et al., 2010).

Compared with conventional solvent extraction, the use of ultrasound is more effective in extracting compounds at lower temperatures. Compared with traditional Soxhlet extraction method, ultrasonic-assisted extraction reduces extraction time, extraction temperature, and solvent consumption (Paniwnyk et al., 2009).

Ultrasound-assisted extraction depends mainly on processing time and ultrasound power. Extraction efficiency usually increases with increasing processing time and ultrasound power. The effect of ultrasound frequency (25–100 kHz) on the extraction efficiency is negligible.

The main advantage of ultrasound-assisted extraction is that this process is less dependent on the extraction solvent than other extraction methods. However, some solvents, such as chloroform and dichloromethane, may contribute to weakening the fruit cell wall, thus also leading to a more intense extraction of products from the fruit cells.

Ethanol is usually a less effective solvent for the extraction than methanol under conventional procedures. However, under ultrasonic-assisted extraction, ethanol showed product yields similar to or greater than methanol. Thus, ethanol is also a viable solvent for ultrasound-assisted extraction.

Ultrasonic extraction has been shown to be effective up to a solvent volume of 125 L, indicating the potential for the use of ultrasound on a commercial scale.

3.7.5.1 Extraction of Phenolics and Antioxidants

The operating conditions of ultrasound extraction may promote, in some level, the degradation of certain anthocyanins due to the ultrasonic frequency used because these compounds are highly sensitive. Anthocyanins may degrade even during the short ultrasonic treatment (Santos et al., 2010).

Processing time showed to be an important variable to extract phenolics and antioxidants from Campbell Early grape, while the concentration of ethanol and temperature showed less significant effect (Ghafoor et al., 2009). The optimal conditions for ultrasound-assisted extraction of bioactive compounds from grape seed were 53% w/v ethanol, 56°C, and 29 min for the maximum total phenolic compounds (5.44 mg GAE/100 mL); 53% w/v ethanol, 60.6°C, and 30.6 min for the maximum antioxidant activity (12.31 mg/mL); and 52% w/v ethanol, 55°C, and 29.5 min for the maximum total anthocyanins (2.28 mg/mL).

Long processing times were also required for the extraction of phenolics from coconut powder. The effects of toasting time, toasting temperature, and extraction time were evaluated by Rodrigues and Pinto (2008). The extraction time strongly affected the phenolic content in the extracts. Long extraction periods (>50 min) at low toasting temperatures (100°C) with long toasting periods (60 min) maximized the extraction of phenolics from coconut powder.

A combined short ultrasound-assisted extraction followed by conventional solvent extraction under agitation may maximize phenolic extraction. Santos et al. (2010) have applied this combined technique to extract ascorbic acid and anthocyanins from jabuticaba skins. Ultrasound was applied for 10 min followed by extraction with ethanol, resulting in the maximum extraction of ascorbic acid and anthocyanins.

The extraction of phenolic acids from citrus peel increased with sonication time, especially at low temperature (up to 30°C). Extraction of caffeic, p-coumaric, ferulic, sinapic, and p-hydroxybenzoic acid increased with increasing extraction time. Ultrasonic temperature showed a positive effect on the extraction of phenolics. From 15°C to 40°C, the contents of caffeic, p-coumaric, ferulic, sinapic, and p-hydroxybenzoic acid increased by 545%, 357%, 405%, 346%, and 74%, respectively, after 10 min of processing time (Ma et al., 2008a).

Temperature showed to have a higher effect on the extraction yield for phenolic compounds than processing time. An increase in temperature favors the extraction process by accelerating both the solubility of the solute and the diffusion coefficient. However, higher temperatures may also induce degradation of some phenolic compounds. Degradation of caffeic acid occurs at temperature above 40°C, as it does for sinapic, p-coumaric, and ferulic acids but at a lower degradation rate. Cinnamic acids and benzoic acids are more temperature stable (Ma et al., 2008a).

The degradation of phenolics is assumed to depend on the number and type of substituents in their aromatic ring. The greater number of hydroxylic-type substituents and the smaller number of methoxylic-type substituents contributed to higher thermal degradation of phenolics during sonication (Liazid et al., 2007).

As for all ultrasound-assisted extraction processes, ultrasonic power has a positive effect on the extraction of phenolic acids from fruits. Increasing the ultrasonic power increases the content of caffeic, ferulic, sinapic, and vanillic acid in the extract. Most phenolic compounds show stability when higher ultrasonic power is applied (Ma et al., 2008b).

3.7.5.2 Extraction of Flavonoids

Unsuccessful recovery of flavonoids from citrus by-products is mainly caused by the absence of an effective extraction procedure. Studies showed that ultrasound extraction can be successfully used to extract several flavonoids from citrus peels: hesperidin, neohesperidin, diosmin, nobiletin, and tangeritin.

Aqueous extraction of lime, orange, and tangerine peels was studied (Londoño-Londoño et al., 2010). The optimized process presented high yield (40.25 ± 12.09 mg of flavonoid fraction/g peel), and the total phenolic content in flavonoid fractions obtained from different sources were 74.80 ± 1.90, 66.36 ± 0.75, and 58.68 ± 4.01 mg GAE/g, for lime, orange, and tangerine, respectively. Orange peel contained hesperidin, neohesperidin, diosmin, nobiletin, and tangeritin, being the most complex source. Tangerine peel was the simplest source and contained only hesperidin and neohesperidin.

3.7.5.3 Extraction of Polysaccharides and Carbohydrates

Polysaccharides and carbohydrates can be extracted from several fruits by short application of ultrasound (<20 min). Extraction of polysaccharides may decrease with the extension of ultrasonic time when high ultrasonic power is used, which is possibly caused by degradation of polysaccharides by ultrasonic wave (Yang et al., 2008; Fu et al., 2010).

The molecular weight of the polysaccharides extracted applying ultrasound may be lower than the molecular weight of the polysaccharides extracted by hot water. This result suggests that ultrasonic waves may degrade high molecular weight polysaccharides forming low molecular weight polysaccharides (Mislovicova et al., 2000; Zhou and Ma, 2006).

The yield of polysaccharides extracted from *Zizyphus jujuba* cv. *jinsixiaozao* was 20.2% above that of classical extraction, and the purity increased by a factor of 1.2. Optimum extraction conditions were temperature in the range of 45°C–53°C, processing time of 20 min, and water/solid mass ratio of 20:1 (Li et al., 2007).

Extraction of polysaccharides from longan fruit was higher when low power density was applied in the extraction process. Application of low power density yielded approximately 30% more polysaccharides than those with high power density, which may have degraded the extracted polysaccharides. Temperature and processing time did not significantly affect the extraction of polysaccharides in longan fruit. These polysaccharides increased DPPH radical scavenging activity with the extension of ultrasonic time at low ultrasonic power but decreased at high ultrasonic power. The optimal conditions to obtain the highest recovery and strongest DPPH radical scavenging activity of polysaccharides were 1200 W/L, 22 min, and 60°C. Experiments were carried out in a 40 kHz equipment using a 4:100 fruit to water mass ratio (Yang et al., 2008).

The effect of temperature on the recovery of polysaccharides was studied by Fu et al. (2010). Higher extraction of polysaccharides was obtained at 75°C. Lower temperatures showed lower yield in polysaccharides, as did higher temperatures (95°C). High temperatures, however, induced the breakdown of the polysaccharides and consequently lower yield.

3.7.5.4 Extraction of Aromas

Ultrasonic-assisted extraction in combination with vacuum distillation have provided extracts with higher flavoring strength due to the increased concentration of desirable oxygenated compounds (from 5 to 8 times) compared with hydrodistillation (Da Porto and Decorti, 2009).

REFERENCES

Ananta, E., Voigt, D., Zenker, M., Heinz, V., and Knorr, D. 2005. Cellular injuries upon exposure of *Escherichia coli* and *Lactobacillus rhamnosus* to high-intensity ultrasound. *Journal of Applied Microbiology* 99: 271–278.

Ashokkumar, M., Sunartio, D., Kentish, S., Mawson, R., Simons, L., Vilkhu, K., and Versteeg, C. 2008. Modification of food ingredients by ultrasound to improve functionality: A preliminary study on a model system. *Innovative Food Science and Emerging Technologies* 9: 155–160.

Bates, D.M., Bagnall, W.A., and Bridges, M.W. 2005. Method of treatment of vegetable matter with ultrasonic energy. European Patent EP1562446.

Cabeza, M.C., Cárcel, J.A., Ordóñez, J.A., de la Hoz, L., Garcia, M.L., and Benedito, J. 2010. Relationships among selected variables affecting the resistance of *Salmonella enterica*, serovar Enteritidis to thermosonication. *Journal of Food Engineering* 98: 71–75.

Cameron, R.G., Baker, R.A., and Grohmann, K. 1998. Multiple forms of pectinmethylesterase from citrus peel and their effects on juice cloud stability. *Journal of Food Science* 63: 253–256.

Cao, S., Hu, Z., Pang, B., Wang, H., Xie, H., and Wu, F. 2010. Effect of ultrasound treatment on fruit decay and quality maintenance in strawberry after harvest. *Food Control* 21: 529–532.

Cao, H., Yue, Y., Hua, R., Tang, F., Zhang, R., Fan, W., Chen, H., and Hua, Y. 2005. HPTLC determination of imadacloprid, fenitrothion and parathion in Chinese cabbage. *Journal of Planar Chromatography—Modern TLC* 18: 151–154.

Cárcel, J.A., Benedito, J., Rosselló, C., and Mulet, A. 2007. Influence of ultrasound intensity on mass transfer in apple immersed in a sucrose solution. *Journal of Food Engineering* 78: 472–479.

Carvalho, L.R.F., Souza, S.R., Martinis, B.S., and Korn, M. 1995. Monitoring of the ultrasonic irradiation effect on the extraction of airborne particulate matter by ion chromatography. *Analytica Chimica Acta* 317: 171–179.

CDC. 1997a. Hepatitis A associated with consumption of frozen strawberries. *Morbidity and Mortality Weekly Report* 46: 288–295.

CDC. 1997b. Outbreaks of *Escherichia coli* 01 57:H7 infection associated with eating alfalfa sprouts. *Morbidity and Mortality Weekly Report* 46: 741–744.

Char, C.D., Mitilinaki, E., Guerrero, S.N., and Alzamora, S.M. 2010. Use of high-intensity ultrasound and UV-C light to inactivate some microorganisms in fruit juices. *Food and Bioprocess Technology* 3: 797–803.

Chisari, M., Barbagallo, R.N., and Spagna, G. 2007. Characterization of polyphenol oxidase and peroxidase and influence on browning of cold stored strawberry fruit. *Journal of Agricultural and Food Chemistry* 55: 3469–3476.

Ciccolini, L., Taillandier, P., Wilhem, A.M., Delmas, H., and Strehaiano, P. 1997. Low frequency thermo-ultrasonication of *Saccharomyces cerevisiae* suspensions: Effect of temperature and of ultrasonic power. *Chemical Engineering Journal* 65: 145–149.

Clifford, M.N. 2000. Anthocyanins nature, occurrence and dietary burden. *Journal of the Science of Food and Agriculture* 80: 1063–1072.

Cruz, R.M.S., Vieira, M.C., and Silva, C.L.M. 2007. Modelling kinetics of watercress (*Nasturtium officinale*) colour changes due to heat and thermosonication treatments. *Innovative Food Science & Emerging Technologies* 8: 244–252.

Cruz, R.M.S., Vieira, M.C., and Silva, C.L.M. 2008. Effect of heat and thermosonication treatments on watercress (*Nasturtium officinale*) vitamin C degradation kinetics. *Innovative Food Science & Emerging Technologies* 9: 483–488.

Czank, C., Simmer, K., and Hartmann, P.E. 2010. Simultaneous pasteurization and homogenization of human milk by combining heat and ultrasound: Effect on milk quality. *Journal of Dairy Research* 77: 183–189.

Da Porto, C. and Decorti, D. 2009. Ultrasound-assisted extraction coupled with under vacuum distillation of flavour compounds from spearmint (carvone-rich) plants: Comparison with conventional hydrodistillation. *Ultrasonics Sonochemistry* 16: 795–799.

Davis, H., Taylor, J.P., Perdue, J.N., Stelma, Jr., G.N., Humphrey, Jr., J.M., Rowntree, R., and Greene, K.D. 1988. A shigellosis outbreak traced to commercially distributed shredded lettuce. *American Journal of Epidemiology* 128: 1312–1321.

Delgado, A.E. and Sun, D.W. 2001. Heat and mass transfer models for predicting freezing process-A review. *Journal of Food Engineering* 47: 157–174.

Duthie, G.G., Duthie, S.J., and Kyle, J.A.M. 2000. Plant polyphenols in cancer and heart disease: Implications as nutritional antioxidants. *Nutrition Research Reviews* 13: 79–106.

Espín, J.C., García-Ruiz, P.A., Tudela, J., Varón, R., and García-Cánovas, F. 1998. Monophenolase and diphenolase reaction mechanisms of apple and pear polyphenol oxidases. *Journal of Agricultural and Food Chemistry* 46: 2968–2975.

Farber, J.M., Carter, A.O., Varughese, P.V., Ashton, F.E., and Ewan, E.P. 1990. Listeriosis traced to the consumption of alfalfa tablets and soft cheese. *New England Journal of Medicine* 332: 338.

Fernandes, F.A.N., Gallão, M.I., and Rodrigues, S. 2008a. Effect of osmotic dehydration and ultrasound pre-treatment on cell structure: Melon dehydration. *LWT—Food Science and Technology* 41: 604–610.

Fernandes, F.A.N., Gallão, M.I., and Rodrigues, S. 2009. Effect of osmosis and ultrasound on pineapple cell tissue structure during dehydration. *Journal of Food Engineering* 90: 186–190.

Fernandes, F.A.N., Linhares, Jr., F.E., and Rodrigues, S. 2008b. Ultrasound as pre-treatment for drying of pineapple. *Ultrasonics Sonochemistry* 15: 1049–1054.

Fernandes, F.A.N., Oliveira, F.I.P., and Rodrigues, S. 2008c. Use of ultrasound for dehydration of papayas. *Food and Bioprocess Technology* 1: 339–345.

Fernandes, F.A.N. and Rodrigues, S. 2007a. Use of ultrasound as pre-treatment for drying of fruits: Dehydration of banana. *Journal of Food Engineering* 82: 261–267.

Fernandes, F.A.N. and Rodrigues, S. 2007b. Use of ultrasound as pretreatment for dehydration of melons. *Drying Technology* 25: 1791–1796.

Fernandes, F.A.N. and Rodrigues, S. 2008a. Dehydration of sapota (*Achras sapota* L.) using ultrasound as pre-treatment. *Drying Technology* 26: 1232–1237.

Fernandes, F.A.N. and Rodrigues, S. 2008b. Application of ultrasound and ultrasound-assisted osmotic dehydration in drying of fruits. *Drying Technology* 26: 1509–1516.

Ferrante, S., Guerrero, S., and Alzamora, S.M. 2007. Combined use of ultrasound and natural antimicrobials to inactivate *Listeria monocytogenes* in orange juice. *Journal of Food Protection* 70: 1850–1856.

Fu, L., Chen, H., Dong, P., Zhang, X., and Zhang, A. 2010. Effects of ultrasonic treatment on the physicochemical properties and DPPH radical scavenging activity of polysaccharides from mushroom *Inonotus obliquus*. *Journal of Food Science* 75: C322–C327.

Fuente-Blanco, S., Sarabia, E.R.F., Acosta-Aparicio, V.M., Blanco-Blanco, A., and Gallego-Juárez, J.A. 2006. Food drying process by power ultrasound. *Ultrasonics Sonochemistry* 44: e523–e527.

Garcia-Perez, J.V., Carcel, J.A., Benedito, J., and Mulet, A. 2007. Power ultrasound mass transfer enhancement in food drying. *Food and Bioproducts Processing* 85: 247–254.

Garcia-Perez, J.V., Carcel, J.A., Fuente-Blanco, S., and Riera-Franco de Sarabia, E. 2006. Ultrasonic drying of foodstuff in a fluidized bed: Parametric study. *Ultrasonics* 44: e539–e543.

Ghafoor, K., Choi, Y.H., Jeon, J.Y., and Jo, H. 2009. Optimization of ultrasound-assisted extraction of phenolic compounds, antioxidants, and anthocyanins from grape (*Vitis vinifera*) seeds. *Journal of Agriculture and Food Chemistry* 57: 4988–4994.

Gómez-López, V.M., Orsolani, L., Martínez-Yépez, A., and Tapia, M.S. 2010. Microbiological and sensory quality of sonicated calcium-added orange juice. *LWT—Food Science and Technology* 43: 808–813.

Granby, K., Johannesen, S., and Vahl, M. 2003. Analysis of glyphosate residues in cereals using liquid chromatography-mass spectrometry (LC-MS/MS). *Food Additives and Contaminants* 20: 692–698.

Guerrero, S., Tognon, M., and Alzamora, S.M. 2005. Response of *Saccharomyces cerevisiae* to the combined action of ultrasound and low weight chitosan. *Food Control* 16: 131–139.

Hernández-Borges, J., Ravelo-Pérez, L.M., Hernández-Suárez, E.M., Carnero, A., and Rodríguez-Delgado, M.A. 2007. Analysis of abamectin residues in avocados by high-performance liquid chromatography with fluorescence detection. *Journal of Chromatography A* 1165: 52–57.

Hertog, M.G.L., Kromhout, D., Aravanis, C., Blackburn, H., Buzina, R., and Fidanza, F. 1995. Flavonoid intake and long-term risk of coronary heart disease and cancer in the seven countries study. *Archives of Internal Medicine* 155: 381–386.

Jackman, R.L., Yada, R.Y., Tung, M.A., and Speers, R.A. 1987. Anthocyanins as food colorants—A review. *Journal of Food Biochemistry* 11: 201–247.

Jang, J.H. and Moon, K.D. 2011. Inhibition of polyphenol oxidase and peroxidase activities on fresh-cut apple by simultaneous treatment of ultrasound and ascorbic acid. *Food Chemistry* 124: 444–449.

Katsaboxakis, K.Z. 1984. The influence of the degree of blanching on the quality of frozen vegetables. In: *Thermal Processing and Quality of Foods*, eds. P. Zeuthen, J.C. Cheftel, C. Eriksson, M. Jul, H. Leniger, P. Linko, G. Varela, and G. Vos, pp. 559–578. New York: Elsevier Applied Science Publishers.

Kim, J.G., Yousef, A.E., and Chism, G.W. 1999. Use of ozone to inactivate microorganisms on lettuce. *Journal of Food Safety* 19: 17–34.

Knorr, D., Zenker, M., Heinz, V., and Lee, D.U. 2004. Applications and potential of ultrasonics in food processing. *Trends in Food Science and Technology* 15: 261–266.

Kruve, A., Lamos, A., Jekaterina, K., and Herodes, K. 2007. Pesticide residues in commercially available oranges and evaluation of potential washing methods. *Proceedings of the Estonian Academy of Sciences: Chemistry* 56: 134–141.

Kuldiloke, J., Eshtiaghi, M., Zenker, M., and Knorr, D. 2007. Inactivation of lemon pectinesterase by thermosonication. *International Journal of Food Engineering* 3: art. 3.

Lee, B.H., Kermasha, S., and Baker, B.E. 1989. Thermal, ultrasonic and ultraviolet inactivation of *Salmonella* in thin films of aqueous media and chocolate. *Food Microbiology* 6: 143–152.

Li, J.W., Ding, S.D., and Ding, X.L. 2007. Optimization of the ultrasonically assisted extraction of polysaccharides from *Zizyphus jujuba cv. jinsixiaozao*. *Journal of Food Engineering* 80: 176–183.

Li, J., Dong, F., Liu, X., Zheng, Y., Yao, J., and Zhang, C. 2009. Determination of pentachloronitrobenzene and its metabolites in ginseng by matrix solid-phase dispersion and GC-MS-MS. *Chromatographia* 69: 1113–1117.

Liazid, A., Palma, M., Brigui, J., and Barroso, C.G. 2007. Investigation on phenolic compounds stability during microwave-assisted extraction. *Journal of Chromatography A* 1140: 29–34.

Lin, C.S., Tsai, P.J., Wu, C., Yeh, J.Y., and Saalia, F.K. 2006. Evaluation of electrolysed water as an agent for reducing methamidophos and dimethoate concentrations in vegetables. *International Journal of Food Science and Technology* 41: 1009–1104.

Liu, H., Song, J., Zhang, S., Qu, L., Zhao, Y., Wu, Y., and Liu, H. 2005. Analysis of residues of imidacloprid in tobacco by high-performance liquid chromatography with liquid-liquid partition cleanup. *Pest Management Science* 61: 511–514.

Londoñõ-Londoño, J., Lima, V.R., Lara, O., Gil, A., Pasa, T.B.C., Arango, G.J. and Pineda, J.R.R. 2010. Clean recovery of antioxidant flavonoids from citrus peel: Optimization on aqueous ultrasound-assisted extraction method. *Food Chemistry* 119: 81–87.

López, P. and Burgos, J. 1995. Lipoxygenase inactivation by manothermosonication: Effects of sonication physical parameters, pH, KCl, sugars, glycerol, and enzyme concentration. *Journal of Agricultural and Food Chemistry* 43: 620–625.

Ma, Y.Q., Ye, X.Q., Fang, Z.X., Chen, J.C., Xu, G.H., and Liu, D.H. 2008a. Phenolic compounds and antioxidant activity of extracts from ultrasonic treatment of satsuma mandarin (*Citrus unshiu* Marc.) peels. *Journal of Agriculture and Food Chemistry* 56: 5682–5690.

Ma, Y.Q., Ye, X.Q., Hao, Y.B., Xu, G.N., Xu, G.H., and Liu, D.H. 2008b. Ultrasound-assisted extraction of hesperidin from penggan (*Citrus reticulata*) peel. *Ultrasonics Sonochemistry* 15: 227–232.

Manvell, C. 1997. Minimal processing of food. *Food Science and Technology Today* 11: 107–111.

Markakis, P., Livingston, G.E., and Fellers, C.R. 1957. Quantitative aspects of strawberry pigment degradation. *Food Research* 22: 117–130.

Martino, M.N., Otero, L., Sanz, P.D., and Zaritzky, N.E. 1998. Size and location of ice crystals in pork frozen by high-pressure-assisted freezing as compared to classical methods. *Meat Science* 50: 303–313.

Mason, T.J. 1998. Power ultrasound in food processing—The way forward. In: *Ultrasound in Food Processing*, eds. M.J.W. Povey and T.J. Mason, pp. 104–124. London, U.K.: Thomson Science.

Mason, T.J., Paniwnyk, L., and Lorimer, J.P. 1996. The use of ultrasound in food technology. *Ultrasonics Sonochemistry* 3: S253–S256.

Mirco, G.C. 2005. Process for continuously washing fruit and vegetables by ultrasound, and relative apparatus. World Intellectual Property Organization Patent WO205104880.

Mislovicova, D., Masarova, J., Bendzalova, K., Soltes, L., and Machova, E. 2000. Sonication of the chitin-glucan, preparation of water-soluble fractions and characterization by HPLC. *Ultrasonics Sonochemistry* 7: 63–68.

Nascentes, C.C., Korn, M., and Arruda, M.A.Z. 2001. A fast ultrasound-assisted extraction of Ca, Mg, Mn and Zn from vegetables. *Microchemical Journal* 69: 37–43.

Ngapo, T.M., Babare, I.H., Reynolds, J., and Mawson, R.F. 1999. Freezing rate and frozen storage effects on the ultrastructure of samples of pork. *Meat Science* 53: 159–168.

Nogata, Y., Ohta, H., and Voragen, A.G.J. 1993. Polygalacturonase in strawberry fruit. *Phytochemistry* 34: 617–620.

O'Mahony, M., Cowden, J., Smyth, B., Lynch, D., Hall, M., Rowe, B., Teare, E.L., Tettmar, R.E., Rampling, A.M., Coles, M., Gilbert, R.J., Kingcott, E., and Bartlett, C.L.R. 1990. An outbreak of *Salmonella* saint-paul infection associated with bean sprouts. *Epidemiology and Infection* 104: 229–235.

Oliveira, F.I.P., Gallao, M.I., Rodrigues, S., and Fernandes, F.A.N. 2010. Dehydration of Malay apple (*Syzygium malaccense* L.) using ultrasound as pre-treatment. *Food and Bioprocess Technology*. DOI: 10.1007/s11947-010-0351-3.

Paniwnyk, L., Cai, H., Albu, S., Mason, T.J., and Cole, R. 2009. The enhancement and scale up of the extraction of anti-oxidants from *Rosemarinus officinalis* using ultrasound. *Ultrasonics Sonochemistry* 16: 287–292.

Peña, A., Ruano, F., and Mingorance, M.D. 2006. Ultrasound-assisted extraction of pesticides from olive branches: A multifactorial approach to method development. *Analytical and Bioanalytical Chemistry* 385: 918–925.

Petrier, C., Combet, E., and Mason, T. 2007. Oxygen-induced concurrent ultrasonic degradation of volatile and non-volatile aromatic compounds. *Ultrasonics Sonochemistry* 14: 117–121.

Petrier, C., Jiang, Y., and Lamy, M.F. 1998. Ultrasound and environment: Sonochemical destruction of chloroaromatic derivatives. *Environmental Science and Technology* 32: 1316–1318.

Pia, J., Xia, X.X., and Liang, J. 2008. Analysis of pesticide multi-residues in leafy vegetables by ultrasonic solvent extraction and liquid chromatography-tandem mass spectrometry. *Ultrasound Sonochemistry* 15: 25–32.

Polydera, A.C., Stoforos, N.G., and Taoukis, P.S. 2003. Comparative shelf life study and vitamin C loss kinetics in pasteurised and high pressure processed reconstituted orange juice. *Journal of Food Engineering* 60: 21–29.

Portenländer, G. and Heusinger, H. 1992. Chemical reactions induced by ultrasound and γ-rays in aqueous solutions of l-ascorbic acid. *Carbohydrate Research* 232: 291–301.

Priestley, R.J. 1979. *Effects of Heating on Foodstuffs*. London, U.K.: Applied Science Publishers.

Rae, J., Ashokkumar, M., Eulaerts, O., von Sonntag, C., Reisse, J., and Grieser, F. 2005. Estimation of ultrasound induced cavitation bubble temperatures in aqueous solutions. *Ultrasonics Sonochemistry* 12: 325–329.

Ramos, J.J., Rial-Otero, R., Ramos, L., and Capelo, J.L. 2008. Ultrasonic-assisted matrix solid-phase dispersion as an improved methodology for the determination of pesticides in fruits. *Journal of Chromatography A* 1212: 145–149.

Ranjan, A., Patil, C., and Moholkar, S.V. 2010. Mechanistic assessment of microalgal lipid extraction. *Industrial and Engineering Chemical Research* 49: 2979–2985.

Raviyan, P., Zhang, Z., and Feng, H. 2005. Ultrasonication for tomato pectinmethylesterase inactivation: Effect of cavitation intensity and temperature on inactivation. *Journal of Food Engineering* 70: 189–196.

Reid, D.S. 1994. Basic physical phenomena in the freezing and thawing of plant and animal tissues. In: *Frozen Food Technology* (2nd edn.), ed. L. Mallett, pp. 1–19. Glasgow, U.K.: Blackie Academic and Professional.

Rentzsch, M., Schwarz, M., and Winterhalter, P. 2007. Pyranoanthocyanins an overview on structures, occurrence, and pathways of formation. *Trends in Food Science and Technology* 18: 526–534.

Rodgers, S.L. and Ryser, E.T. 2004. Reduction of microbial pathogens during apple cider production using sodium hypochlorite, copper ion, and sonication. *Journal of Food Protection* 67: 766–771.

Rodrigues, S. and Fernandes, F.A.N. 2007. Use of ultrasound as pretreatment for dehydration of melons. *Drying Technology* 25: 1791–1796.

Rodrigues, S., Gomes, M.C.F., Gallão, M.I., and Fernandes, F.A.N. 2009a. Effect of ultrasound-assisted osmotic dehydration on cell structure of sapotas. *Journal of the Science of Food and Agriculture* 89: 665–670.

Rodrigues, S., Oliveira, F.I.P., Gallão, M.I., and Fernandes, F.A.N. 2009b. Effect of immersion time in osmosis and ultrasound on papaya cell structure during dehydration. *Drying Technology* 27: 220–225.

Rodrigues, S. and Pinto, G.A.S. 2008. Ultrasound extraction of phenolic compounds from coconut (*Cocos nucifera*) shell powder. *Journal of Food Engineering* 80: 869–872.

Rodríguez-Gonzalo, E., Carabias-Martínez, R., Cruz, E.M., Domínguez-Álvarez, J., and Hernández-Méndez, J. 2009. Ultrasonic solvent extraction and nonaqueous CE for the determination of herbicide residues in potatoes. *Journal of Separation Science* 32: 575–584.

Sadilova, E., Carle, R., and Stintzing, F.C. 2007. Thermal degradation of anthocyanins and its impact on color and in vitro antioxidant capacity. *Molecular Nutrition and Food Research* 51: 1461–1471.

Santos, D.T., Veggi, P.C., and Meireles, M.A.A. 2010. Extraction of antioxidant compounds from Jabuticaba (*Myrciaria cauliflora*) skins: Yield, composition and economical evaluation. *Journal of Food Engineering* 101: 23–31.

Scouten, A.J. and Beuchat, L.R. 2002. Combined effects of chemical, heat and ultrasound treatments to kill *Salmonella* and *Escherichia coli* O157:H7 on alfalfa seeds. *Journal of Applied Microbiology* 92: 668–674.

Seymour, I.J., Burfoot, D., Smith, R.L., Cox, L.A., and Lockwood, A. 2002. Ultrasound decontamination of minimally processed fruits and vegetables. *International Journal of Food Science and Technology* 37: 547–557.

Shah, S. and Gupta, M.N. 2008. The effect of ultrasonic pre-treatment on the catalytic activity of lipases in aqueous and non-aqueous media. *Chemistry Central Journal* 2: art. 1.

Shrikhan, A.J. and Francis, F.J. 1974. Effect of flavonols on ascorbicacid and anthocyanin stability in model systems. *Journal of Food Science* 39: 904–906.

Simal, S., Benedito, J., Sánchez, E.S., and Roselló, C. 1998. Use of ultrasound to increase mass transport rate during osmotic dehydration. *Journal of Food Engineering* 36: 323–336.

Sivasankar, T., Paunikar, A.W., and Moholkar, V.S. 2007. Mechanistic approach to enhancement of the yield of a sonochemical reaction. *AIChE Journal* 53: 1132–1143.

Skrede, G., Wrolstad, R.E., Lea, P., and Enersen, G. 1992. Color stability of strawberry and blackcurrant syrups. *Journal of Food Science* 57: 172–177.

Sprinz, H., Beckert, D., and Brede, O. 1998. Reactions of OH radicals with ascorbic acid: A pulse radiolysis and Fourier transform ESR study. *Journal of Radioanalytical and Nuclear Chemistry* 232: 39–41.

Stanley, K.D., Golden, D.A., Williams, R.C., and Weiss, J. 2004. Inactivation of *Escherichia coli* O157:H7 by high-intensity ultrasonication in the presence of salts. *Foodborne Pathogens and Disease* 1: 267–280.

Sun, D.W. and Li, B. 2003. Microstructural change of potato tissues frozen by ultrasound-assisted immersion freezing. *Journal of Food Engineering* 57: 337–345.

Tarleton, E.S. 1992. The role of field-assisted techniques in solid/liquid separation. *Filtration Separation* 3: 246–253.

Tarleton, E.S. and Wakeman, R.J. 1998. Ultrasonically assisted separation process. In: *Ultrasounds in Food Processing*, eds. M.J.W. Povey and T.J. Mason, pp. 193–218. Glasgow, U.K.: Blackie Academic and Professional.

Tarun, E.I., Kurchenko, V.P. and Metelitsa, D.I. 2006. Flavonoids as effective protectors of urease from ultrasound inactivation in solutions. *Bioorganism Khimica* 32: 391–398.

Terefe, N.S., Gamage, M., Vilkhu, K., Simons, L., Mawson, R., and Versteeg, C. 2009. The kinetics of inactivation of pectin methylesterase and polygalacturonase in tomato juice by thermosonication. *Food Chemistry* 117: 20–27.

Tiwari, B.K., Muthukumarappan, K., O'Donnell, C.P., and Cullen, P.J. 2008. Modelling colour degradation of orange juice by ozone treatment using response surface methodology. *Journal of Food Engineering* 88: 553–560.

Tiwari, B.K., Muthukumarappan, K., O'Donnell, C.P., and Cullen, P.J. 2009a. Inactivation kinetics of pectin methylesterase and cloud retention in sonicated orange juice. *Innovative Food Science and Emerging Technologies* 10: 166–171.

Tiwari, B.K., O'Donnell, C.P., Muthukumarappan, K., and Cullen, P.J. 2009b. Effect of sonication on orange juice quality parameters during storage. *International Journal of Food Science and Technology* 44: 586–595.

Tiwari, B.K., O'Donnell, C.P., Muthukumarappan, K., and Cullen, P.J. 2009c. Ascorbic acid degradation kinetics of sonicated orange juice during storage and comparison with thermally pasteurised juice. *LWT—Food Science and Technology* 42: 700–704.

Ueda, J.I., Saito, N., Shimazu, Y., and Ozawa, T. 1996. A comparison of scavenging abilities of antioxidants against hydroxyl radicals. *Archives of Biochemistry and Biophysics* 333: 337–384.

Ugarte-Romero, E., Feng, H., Martin, S.E., Cadwallader, K.R., and Robinson, S.J. 2006. Inactivation of *Escherichia coli* with power ultrasound in apple cider. *Journal of Food Science* 71: E102–E108.

Valero, M., Recrosio, N., Saura, D., Muñoz, N., Martí, N., and Lizama, V. 2007. Effects of ultrasonic treatments in orange juice processing, *Journal of Food Engineering* 80: 509–516.

Vicente, A.R., Costa, M.L., Martinez, G.A., Chaves, A.R., and Civello, P.M. 2005. Effect of heat treatment on cell wall degradation and softening in strawberry fruit. *Postharvest Biology and Technology* 38: 213–222.

Wang, H., Cao, G., and Prior, R.L. 1997. Oxygen radical absorbing capacity of anthocyanins. *Journal of Agriculture and Food Chemistry* 45: 304–309.

Wang, W. and Xu, S. 2007. Degradation kinetics of anthocyanins in blackberry juice and concentrate. *Journal of Food Engineering* 82: 271–275.

Witrowa-Rajchert, D. and Lewicki, P.P. 2006. Rehydration properties of dried plant tissues. *International Journal of Food Science and Technology* 41: 1040–1046.

Wong, E., Vaillant, F., and Pérez, A. 2010. Osmosonication of blackberry juice: Impact on selected pathogens, spoilage microorganisms, and main quality parameters. *Journal of Food Science* 75: 468–474.

Xiang, Z.X., Zhao, W.J., and Gao, S.L. 2005. Determination of 11 kinds of organophosphate pesticide residues in *Flos lonicerae* by solid-phase extraction. *Journal of China Pharmaceutical University* 36: 334–337.

Yang, B., Zhao, M., Shi, J., Yang, N., and Jiang, Y. 2008. Effect of ultrasonic treatment on the recovery and DPPH radical scavenging activity of polysaccharides from longan fruit pericarp. *Food Chemistry* 106: 685–690.

Ye, S.Y., Qiu, Y.X., Song, X.L., and Luo, S.C. 2009. Optimization of process parameters for the inactivation of *Lactobacillus sporogenes* in tomato paste with ultrasound and ^{60}Co-γ irradiation using response surface methodology. *Radiation Physics and Chemistry* 78: 227–233.

Zafra-Stone, S., Yasmin, T., Bagchi, M., Chatterjee, A., Vinson, J.A., and Bagchi, D. 2007. Berry anthocyanins as novel antioxidants in human health and disease prevention. *Molecular Nutrition and Food Research* 51: 675–683.

Zhou, L.Z., Li, B., Li, L., and Zhang, X.M. 2007. Inactivation of yeast cells by ultrasound disinfection and related influences. *Journal of South China University of Technology* 35: 121–125.

Zhou, C. and Ma, H. 2006. Ultrasonic degradation of polysaccharide from a red algae (*Porphyra yezoensis*). *Journal of Agriculture and Food Chemistry* 54: 2223–2228.

4 Membrane Applications in Fruit Processing Technologies

Sunando DasGupta and Biswajit Sarkar

CONTENTS

4.1 INTRODUCTION

Raw fruit juice is turbid, viscous, and dark in color. Apart from lower molecular weight components like sugar, acid, salt, flavor, and aroma compounds, it contains significant amounts of macromolecules (100–1000 ppm) such as polysaccharides (pectins, cellulose, hemicellulose, and starch), haze-forming components (suspended solids (SSs), colloidal particles, proteins, and polyphenol), etc. Hence, the juice needs to be clarified before its commercial use. Clarification involves the removal of such macromolecules. An enzyme treatment of raw juice is usually carried out with enzymes (pectinase and amylase) to reduce the pectic substances and starch content. This enzyme treatment reduces the cloudiness and makes the filtration process easier by lowering its viscosity. The enzyme treatment is followed by the addition of fining agents such as gelatin, bentonite, etc., which are used to enhance the settling of formed flocs. Then, SSs, colloidal particles, proteins, etc., are removed by conventional filtration. Filter aids such as diatomaceous earth or kieselguhr are used to facilitate the filtration process. These conventional methods to clarify fruit juice are labor intensive, time-consuming, and batch operated. The use of additives (fining agents and filter aids) may leave a bitter taste in the juice. Moreover, the solids obtained after filtration, which contain enzymes, filter aids, and fining agents, cannot be reused and cause pollution problems due to their disposal. In this regard, energy-efficient membrane processes, particularly microfiltration (MF) and ultrafiltration (UF), represent a valid alternative for the clarification of additive-free high-quality fruit juices with natural fresh taste. The objective of UF/MF of fruit juice is to retain high molecular weight pectin and its derivative, proteins, colloids, etc., and to allow low molecular weight solute-like sugars, acid, salt, etc., to permeate through the membrane together with water. These processes are athermal and involve no phase change or addition of chemicals.

The production of concentrated fruit juices is of interest at the industrial level since it reduces the storage volume, thus reducing the package, transportation, and storage cost, and also facilitates the preservation. The concentration of fruit juices is usually carried out in a multistage vacuum evaporator. However, high-energy consumption, off-flavor formation, color changes, and reduction of nutritional value due to thermal effects are the main disadvantages of traditional evaporation processes. Alternative techniques to thermal evaporation to obtain stable concentrated juice with its original color, aroma, nutritional value, and structural characteristics involve freeze concentration (cryoconcentration) and membrane processes. The cryoconcentration technique requires significant energy consumption. Moreover, in this process, the achievable concentration (50 °Bx) is lower than the values obtained by evaporation (60–65 °Bx). Reverse osmosis (RO) seems to overcome some of the drawbacks associated with thermal evaporation with less energy consumption; however, it is generally used as a preconcentration technique allowing concentration values of about 25–30 °Bx due to high osmotic pressure limitations. Recently, direct osmosis concentration (DOC), osmotic distillation (OD), and membrane distillation (MD) have been proposed as attractive membrane processes allowing very high concentrations (above 65 °Bx) to be reached under atmospheric pressure and ambient temperatures. Complete utilization of the potential of these techniques may be achieved by the integration of the different membrane processes. Considerable research has been carried out for the clarification and concentration of a variety of fruit juices including apple, mosambi, pineapple, grapefruit, kiwi, passion fruit, etc.

This chapter provides a review of the recent advances in membrane processes for clarification and concentration of fruit juices, including the application of RO, MD, OD, and integrated membrane processes.

4.2 CLARIFICATION OF FRUIT JUICE BY UF AND MF

4.2.1 MEMBRANE

The basic objective of a membrane is to act as a thin selective barrier that allows the passage of one or more components of a mixture while retaining the other components in the presence of appropriate driving force. According to Hwang and Kammermeyer (1975), a membrane is defined as "a region of discontinuity interposed between two phases." Lakshminarayanaiah (1984) defined membrane as a "phase that acts as a barrier to prevent mass movement but allows restricted and/or regulated passage of one or more species through it." Membranes can be classified based on their nature, structure, separation of mechanism, and applications. Most of the membranes in practical applications are organic polymers. The widely used polymers for synthesis of membranes are cellulose acetate, aromatic polyamides and polyimides, polysulfones, cellulose nitrate, polyethersulfones, polycarbonate, polypropylene, polytetrafluoroethylene, polyvinylidene fluoride, etc. The polymer should have suitable properties for the specific applications. For example, cost-effective hydrophilic cellulose acetate polymer offers minimum fouling, a wide range of pore size formation in the membrane, ease of manufacture, and low protein-binding tendency, but it is highly biodegradable and poorly resistant to chlorine. Hydrophobic polyethersulfone offers a wide range of pH and temperature tolerance, is easy to fabricate, and has good chlorine and alkaline resistance but is highly fouled by any protein-containing streams. Polytetrafluoroethylene offers resistance toward harsh chemicals and etching agents, but it is degraded by sterilizing gamma rays that are commonly used in many biomedical applications. In general, the important properties of any good membrane are (1) high permeability and high selectivity; (2) resistance to adsorption and fouling; (3) ease of fabrication in a wide variety of configuration; (4) stability and compatibility with the condition of application; (5) good mechanical strength, etc. Inorganic membranes are also used in several applications. Some of the disadvantages associated with the polymeric membrane can be overcome by using the inorganic membrane due to its distinct properties. Inorganic membranes are of two types—metallic and ceramic. The higher permeate flux coupled with long shelf life, backflushing capability, a wide range of operating conditions (temperature, pressure, pH), and low replacement cost make the inorganic membrane attractive despite the higher initial cost and large pumping capacity.

4.2.2 MEMBRANE FOULING AND ITS CONTROL

One of the major limitations of the membrane-based system in many applications is the decline in permeate flux. This flux decline is due to two distinct phenomena, namely, concentration polarization and fouling of the membrane. Accumulation of solute particles over the membrane surface leads to concentration polarization (Blatt et al., 1970). This results in enhancement of the solute concentration at the membrane–solution interface. It is also to be noted that concentration polarization is reversible in nature and can be removed by backflushing, application of high cross-flow velocities, or cleaning protocol. On the other hand, fouling of the membrane is an irreversible phenomenon. The nature and extent of membrane fouling strongly depend on the physicochemical nature of the membrane and solute. Surface morphology, solute–solute, and solute–membrane interactions are important in understanding of fouling phenomena. The mechanisms involved are adsorption of solute particles over the pore mouth or inside the pore, making the pore completely or partially blocked. Pore blockage increases the membrane resistance, while cake formation over the membrane surface offers resistance to permeate flow. Thus, pore blocking and cake formation are the two mechanisms of membrane fouling. This effect is permanent and membrane permeability cannot be recovered to its earlier value completely even after washing of the membrane. Therefore, this is also termed as irreversible fouling. Several studies have been undertaken to reduce the concentration polarization and associated fouling to maintain high permeate flux and to retrieve the native membrane rejection characteristics. These include (1) feed pretreatment, (2) membrane material, (3) use of turbulence promoter, (4) pulsatile flow, (5) gas sparging, (6) external forces, (7) backflushing, and (8) turbulent flow.

4.2.2.1 Feed Pretreatment

The performance of any membrane-based fruit juice clarification strongly depends on the presence of polysaccharides such as pectin and starch in it. Pectin makes the clarification process difficult as it forms a gel-type layer over the membrane surface in the presence of sugar and acid while starch causes haze in juice. Therefore, permeate flux and yield decrease significantly. In order to degrade the polysaccharides, an enzymatic treatment of raw juice is generally carried out with enzymes such as pectinases and amylases. Pectinases hydrolyze pectin and cause the pectin–protein complexes to flocculate. Amylases decompose the starch molecules that may cause cloudiness during storage (Kilara and Van Buren, 1989). This reduces the viscosity and fouling, resulting in higher permeate flux and lower energy consumption. Alvarez et al. (1998a) reported that enzymatic depectinization of apple juice resulted in higher permeate flux of clarified juice with lower turbidity, viscosity, and total pectin content during clarification of apple juice using a commercial zirconium oxide UF membrane (Table 4.1).

4.2.2.2 Membrane Material

The fouling can be reduced by using hydrophilic polymer as membrane material or by applying a surface modification to make the surface more hydrophilic. Macrosolutes, which contain hydrophobic parts, are adsorbed easily on hydrophobic membrane. The adsorption layer on hydrophobic surface offers higher resistance to permeate flow and is difficult to wash away compared with that on hydrophilic surface.

TABLE 4.1
Characteristics of Apple Juices Treated with Different Amounts of Pectinex 3XL before UF

Parameter	Concentration of Pectinex 3XL (FDU/g Pectin)				
	0	100	200	300	400
Turbidity (NTU)	440	97.0	52.3	27.9	26.0
Viscosity $\times 10^3$ Pa s (at T = 50°C)	3.58	1.044	0.977	0.965	0.97
Reduction of total pectin (%)	0	43.0	63.0	80.0	80.0
Steady-state permeate flux (L/m² h) at T = 50°C and P = 400 kPa	38	55	75	125	125

Sources: Alvarez, S. et al., Colloids Surf. A Physicochem. Eng. Aspects, 138, 377, 1998a; Alvarez, S. et al., Sep. Purif. Technol., 14, 209, 1998b.

The surface roughness has a significant effect on fouling (Fane and Kim, 1988). Fane and Kim observed that with increase in surface roughness, loss of permeate flux increases.

4.2.2.3 Turbulent Flow

The system performance in terms of flux can be improved either by reducing concentration polarization by increasing the mass transfer from the membrane surface to the bulk or by reducing fouling by increasing the wall shear rate. This is mostly done either by increasing cross-flow velocity or by changing of the channel geometry.

4.2.2.4 Use of Turbulence Promoter

Many hydraulic approaches, either increasing the fluid velocity or creating the turbulent behavior, have been developed for cross-flow UF to suppress the increase in concentration polarization and progressive fouling along the flow path. Metal grill, static rods, spiral wound (Geraldes et al., 2002), and disc- and doughnut-shaped inserts (Howell et al., 1993) are some of the various types of turbulence promoters used. Due to obstruction in flow path, turbulent promoter creates localized turbulence in its neighborhood on the membrane surface, and the generated turbulent eddies enhance mixing on the membrane surface. Therefore, concentration polarization and fouling decrease, leading to an enhanced permeate flux. During cross-flow UF of simulated fruit juice (a mixture of pectin and sucrose), by incorporation of cylindrical promoters placed perpendicular to the flow path, permeate flux improvement of about 65% as compared with the no-promoter condition was achieved (Pal et al., 2008).

4.2.2.5 Pulsatile Flow

Unsteady flows and oscillation flows can be generated by pulsations into the feed or permeate channels. Pulsations reduce the formation of a fouling layer of rejected particles on the membrane by exerting a "scouring effect." (Su et al., 1993). A three-fold increase in permeate flux was observed using periodically spaced, doughnut-shaped baffles in UF tubes together with pulsed flows with an oscillation frequency up to 2.5 Hz (Finnigan and Howell, 1989). During MF of apple juice, using a 1 Hz pulsed flow and a ceramic membrane, flux improvements of about 45% and 140% have been reported (Gupta et al., 1992, 1993). Amar et al. (1990) reported that using the superimposition of pulsation on the inlet flow, the permeate flux was increased from 200 to 250 L/m^2 h for 200 min, with a ceramic membrane (Figure 4.1) during apple juice clarification. Significant improvement in permeate flux is also observed in the case of intermittent operation. This may be explained by the fact that during shut off of UF system, the sudden release of pressure on the deposited layer results in part of the fouling materials rebounding and disassociating from the membrane. On restart-up operation, the cross-flow wave flushes the loosened fouling materials away from the membrane surface resulting in an enhancement of permeate flux. Chiang and Yu (1987) reported that during UF of passion fruit juice with intermittent on–off operation, permeate flux increased about 40% compared with that without intermittent on–off operation. Unsteady flows can also be obtained using the intermittent jet of the feed in the channel. This causes the feed velocity to abruptly change, resulting

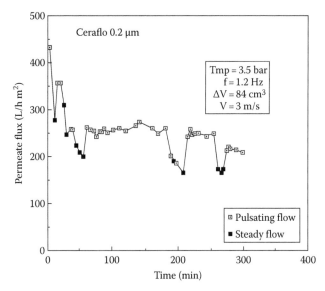

FIGURE 4.1 Effect of pulsating inlet flow on fouling of membrane (apple juice treated with pectinase); Tmp, transmembrane pressure; f, frequency of pulsations; ΔV, displaced volume; V, inlet velocity. (Reprinted from Amar, R.B. et al., *J. Food Sci.*, 55, 1620, 1990. With permission.)

in the formation of large vortices (Arroyo and Fonade, 1993). Kim and Chang (1991) used periodic back-pulsing at a frequency of $0.67\,min^{-1}$ during filtration of a mixed hemoglobin (62.5 kDa) and dextran (10 kDa) solution through a 30 kDa molecular weight cut-off membrane and obtained an almost threefold enhancement in permeate flux. Ding et al. (1991) found a twofold increase in the permeate flux of plasma when being filtered through a MF membrane by periodically squeezing the tubing conveying the retentate, thereby pulsating the flow and pressure.

4.2.2.6 Backflushing, Pulsing, and Shocking

The life of membrane can be increased by periodic backwashing (BW). The permeate itself is often used as the backwash fluid. Periodic BW is carried out by pumping permeate back into to the feed chamber to remove the particles deposited in pores and also on the membrane surface. The backwash pressure is generally higher than the normal operating inlet pressure. Sometimes, backflushing is done by pressurized air instead of permeate. Most backflushing is done with permeate for 1–5 s at a frequency of 1–10 times per minutes at a pressure of 0.1–1 MPa. Amar et al. (1990) investigated the effect of BW for 2–5 s every 2–5 min with compressed nitrogen gas (backwash pressure almost twice the transmembrane pressure [TMP]) pectinase using a ceramic MF membrane. They observed that the permeate flux increased about 166% with BW for pectinase-treated apple juice when compared with permeate flux without backwashing (Figure 4.2).

Su et al. (1993) reported that the improvement of permeate flux ranged between 100–110 and 70–80 L/m² h for vacuum-filtered apple juice and pre-vacuum-filtered juices, respectively, for most of the experimental time using nitrogen backwash

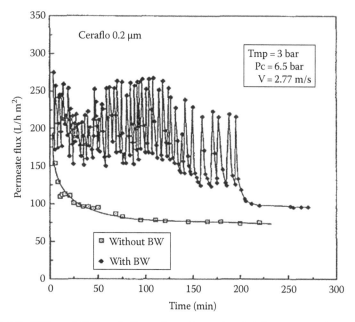

FIGURE 4.2 Effect of BW on fouling of membrane (apple juice treated with pectinase); Tmp, transmembrane pressure; V, inlet velocity; Pc, backwash pressure. (Reprinted with permission from Amar, R.B. et al., *J. Food Sci.*, 55, 1620, 1990.)

every 15 min. In both cases, backwash improved the permeate flux about 11 L/m² h. Rapid BW seems to be more effective, and it is commonly termed as "back pulsing" or "backshocking." The backpulses are of short duration (0.1 s or less) and are operated continuously or periodically.

4.2.2.7 Gas Sparging

The technique of gas–liquid two-phase cross-flow filtration, via the injection of gas bubbles into the feed stream, was found to be effective in enhancing the performance of membrane process. The introduction of gas slugs into the liquid stream can increase turbulence on membrane surface and restrict the growth of concentration boundary layer, leading to enhancement in the flux of the filtration process. It has been observed that the enhancement of gas sparging, in terms of the increase in permeate flux, is dependent on membrane modules. As high as a threefold flux increase was found when ultrafiltering dextran solutions in tubular membranes (Cui and Wright, 1994), whereas flux enhancement by gas sparging for membrane modules such as flat sheet (Li et al., 1998), hollow fiber (Bellara et al., 1996), and spiral wound (Cui et al., 1996) membranes was less pronounced, with an observed flux increase from 7% to 60%. Bellara et al. (1996) reported flux enhancements of 20%–50% for dextran and 10%–60% for albumin using gas–liquid two-phase crossflow in hollow fiber membranes. Youravong et al. (2010) investigated the effect of gas sparging on permeate flux, fouling, and quality of clarified wine during clarification of pineapple wine using a tubular ceramic MF membrane. They found that the permeate flux increased up to 138%. However, further increase in the gas sparging rate did not improve permeate flux compared with that without gas sparging. They also observed the negative effect of gas sparging, which caused a loss of alcohol content in the wine.

4.2.2.8   External Force

The use of additional forces such as electrical, sonic, and magnetic fields to improve the performance of filtration has gained increasing attention in recent years.

4.2.2.8.1   Ultrasonic Field

The passage of ultrasound waves through a suspension can cause many phenomena, including particle dispersion, cavitation, viscosity reduction, and changes in particle surface properties. Cavitation, which is observed as the rapid formation and collapse of gaseous microbubbles at the membrane surface, causes trapped particulates at the pore entry regions to be loosened. The cross-flow stream is then able to carry the particle away from the membrane surface, resulting in a decrease in concentration polarization (Kobayashi et al., 1999; Lamminen et al., 2004). This leads to an increase in permeate flux. However, ultrasound is not effective at removing the fouling material trapped inside pores (Kokugan et al., 1995). There are several factors that influence the effectiveness of the ultrasound treatment such as frequency, power intensity, etc. Lower ultrasound frequencies are found to have higher particle-removing efficiencies than higher frequencies (Kobayashi et al., 1999; Lamminen et al., 2004). In general, an increase in power intensity will result in an increase in the sonochemical effects. With increasing power intensity to the system, both the number of cavitation bubbles formed and the size of the cavitating zone increase. In addition, the hydrodynamic turbulence increases with increased power intensity (Lamminen et al., 2004). The higher the power intensity during ultrasonic irradiation, the better the membrane cleaning and the greater the flux obtained. However, application of high-power intensity sometimes causes membrane erosion though different membrane materials have different durability in ultrasonic treatment. A continuous use of ultrasonic waves from the start of filtration has been found very effective in many investigations (Tarleton and Wakeman, 1990; Matsumoto et al., 1996). However, the continuous use of ultrasound is undesirable in terms of energy consumption. The use of an intermittent ultrasonic field was found effective in terms of both cost and flux enhancement (Matsumoto et al., 1996; Muthukumaran et al., 2004). Sabri et al. (1997) reported that flux enhancements were up to 400% higher with intermittent ultrasound at a frequency of 45 kHz than in the reference cases during filtration of wastewater effluents from pulp and paper mills and brackish water.

4.2.3   Effects of Different Process Parameters on Permeate Flux

4.2.3.1   Effects of Flow Rate on Permeate Flux

The feed flow rate is an important parameter for the performance of the fruit juice clarification process. The effects of cross-flow rate on the permeate flux are observed during clarification of various fruit juices such as kiwifruit, carrot, apple, etc. (Vladisavljevic et al., 2003; Cassano et al., 2004). It is observed that flux increases

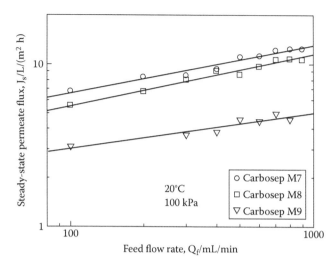

FIGURE 4.3   UF of apple juice. Effect of feed flow rate on steady-state permeate flux for all three membranes. (Reprinted from Vladisavljevic et al., *J. Food Sci.*, 60, 241, 2003. With permission.)

with cross-flow velocity. This can be explained by the fact that at a higher flow rate, development of concentration polarization and subsequent gel-type layer formation on the membrane surface are restricted due to shearing force imposed by cross-flow rate. Clarification of depectinized apple juice using ceramic tubular membranes with 300 kDa (M7), 50 kDa (M8), and 30 kDa (M9) molecular weight cut-off was investigated by Vladisavljevic et al. (2003). They observed that fouling resistance decreased with feed flow rate, and higher permeate flux was obtained in the case of a less-resistive M9 membrane (Figure 4.3).

4.2.3.2   Effects of Transmembrane Pressure on Permeate Flux

Cassano et al. (2003) have described the variation of steady-state permeate flux with applied TMP during UF of carrot juice, as illustrated in Figure 4.4. For small pressures, the solvent flux is proportional to the applied pressure. At low pressure range, significant increase in permeate flux is observed with pressure due to enhanced driving force, and a weak pressure dependence is observed at higher values of operating pressure. This is due to the direct consequence of the gel-type layer formation over the membrane surface. The enhanced driving force due to increase in pressure is compensated by the increasing gel layer thickness. Hence, the permeate flux increases up to a limiting value, and on further pressure increase, no significant increase in flux is observed.

4.2.3.3   Effects of Operating Temperature on Permeate Flux

In general, a higher operating temperature leads to a higher permeate flux, keeping other operating conditions unchanged. The effect of temperature on the permeate flux can be observed in Figure 4.5 during clarification of apple juice. Higher flux at higher temperature is observed due to decrease in juice viscosity and increase in mass transfer coefficient according to the film model (Fane and Fell, 1987). According to

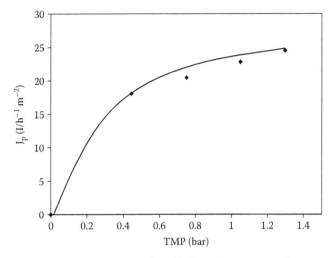

FIGURE 4.4 UF of carrot juice. Effect of the TMP on the permeate flux, operating conditions: Temperature = 23.5°C; Velocity = 0.14 m/s. (Reprinted from Cassano, A. et al., *J. Food Eng.*, 57, 153, 2003. With permission.)

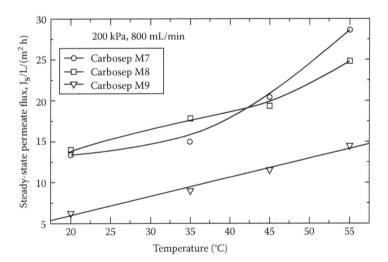

FIGURE 4.5 UF of apple juice. Effect of temperature on steady-state permeate flux for all three membranes. (Reprinted from Vladisavljevic et al., *J. Food Sci.*, 60, 241, 2003. With permission.)

Figure 4.5, an increase in temperature from 20°C to 55°C enhanced the permeation flux irrespective of the membrane molecular weight cut-off. It is in dissimilarity to what Jiraratananon and Chanachai (1996) observed in UF of passion fruit juice. They observed the decrease in permeate flux by an increase in temperature from 40°C to 50°C due to the effect of increased bulk viscosity and gel-type layer formation on the membrane surface caused by cross-linking between deposited entangled pectin and starch molecules, resulting in an increase in the resistance to flow.

4.2.4 MEMBRANE MODULES

Membrane clarification of juices has been studied using membrane modules of different configurations, such as plate-and-frame (Sheu et al., 1987), hollow fiber (Constenla and Lozano, 1996), tubular (Alvarez et al., 1996), spiral wound (Wu et al., 1990; Ghosh et al., 2000), and dead-end or stirred batch cells (Riedl et al., 1998; Sulaiman et al., 1998). The plate-and-frame module consists of a number of flat sheet membranes, a support plate, and a channel spacer in a sandwich arrangement. The membranes can be sealed to the plates using gaskets. The feed channel typically ranges from 0.03 to 0.1 cm in height. The feed is distributed at one end of the module, and the retentate is collected at the opposite end. Hollow fiber module consists of bundles of fiber (50–10,000) with a dense skin layer that gives the membrane permselectivity and a more open matrix that gives the membrane its structural integrity. Fiber diameter varies from about 0.2 to 2.5 mm with the fiber wall thickness around 0.2 mm. These modules are generally operated with the feed flow into the fiber, with the permeate moving radially outward through the fiber walls. The tubular module is similar in design to the hollow fiber module but employs much larger diameter tubes (typically 0.3–2.5 cm). The feed flows through the bore of the tubes while the permeate flows radially outward across the membrane and the support tube and is collected from the permeate outlet ports. Spiral wound modules consist of envelops of flat sheet membranes with their active sides facing each other, separated by thin, mesh-like spacer material. These envelops are spirally wound around a central perforated tube to form several narrow slits for fluid flow. The thin spacers are used between the envelops to establish the desired feed channel height. The open end of the envelops are sealed around the perforated tube. The feed is fed at one end of the module and flows along the length of the module while the retentate is collected from the end of the tube. The permeate flows through the membrane into the permeate channel and spirals toward the perforated center tube. Advantages and disadvantages of various modules are summarized in Table 4.2.

TABLE 4.2
Comparison of Different Membrane Modules

Module	Packing Density (m²/m³)	Energy Cost (Pumping)	Channel Spacing (cm)	Particulate Plugging	Ease of Cleaning	Self-Support	Holdup Volume
Flat plate	300	Moderate	0.03–0.25	Moderate	Good	Not self-supporting	Low
Tubular	60	High	1.0–2.5	Low	Excellent	Not self-supporting	High
Hollow fiber	1200	Low	0.02–0.25	High	Fair	Self-supporting	Low
Spiral wound	600	Low		Very high	Moderate	Self-supporting	Low

Jiraratananon et al. (1997) investigated the UF of pineapple juice using a tubular membrane. The module housing was stainless steel with an equivalent diameter of 350 mm. There were 19 tubular membrane channels, each channel was 4 mm in diameter, 500 nm long, and the effective area was 0.12 m². They used the MF and UF membranes of pore sizes 0.1 and 0.01 μm, respectively. The obtained permeate flux and rejection of macromolecules were 6.37×10^3 m³/m² h and 84%–87%, respectively, at a pressure of 300 kPa and a cross-flow velocity of 2.0 m/s. de Barros et al. (2003) studied the cross-flow UF of depectinized pineapple juice tubular ceramic membrane (0.01 μm) and polysulfone hollow fiber membrane (100 ka). They observed that the permeated flux obtained in tubular ceramic membrane is higher than that obtained in polysulfone hollow fiber membrane. Higher permeate flux in tubular ceramic membrane can be explained by the fact that the turbulent regime flow increases the rate of solute diffusion from the membrane surface to the bulk. This leads to less compact cake, causing a higher permeate flux compared with the flow obtained due to cake formation and the laminar flow in hollow fiber membrane. Sarkar et al. (2008a) studied the UF of enzyme-treated mosambi juice in a flat sheet rectangular channel 6.5 mm in height under a laminar flow regime using 50,000 (MWCO) polyethersulfone membrane. The permeate flux was about 10.5 L/m² h at a pressure of 500 kPa and a cross-flow velocity of 0.12 m/s.

4.2.5 Electric Field–Assisted Clarification of Fruit Juice

4.2.5.1 Basic Principle

To overcome problems with membrane fouling and concentration polarization, an external electric field can be applied across the membrane during cross-flow membrane filtration of a solution containing a charged solute. In this process, the electrical field acts as an additional driving force to the TMP. External electric field from a regulated DC power supply is applied across the membrane surface in the form of continuous or pulse mode. This process utilizes an electrophoretic force that drags the charged solutes away from the membrane surface, thus limiting the solute accumulation on the membrane surface (Figure 4.6). The concentration polarization layer and deposited layer over the membrane surface are thereby reduced, resulting in an increase in permeate flux.

4.2.5.2 Apparatus Design

Two different modules are widely used for electric field–assisted UF such as flat sheet (Sarkar et al., 2008c) and tubular (Yukawa et al., 1983). An electric field can be applied across the membrane with one electrode on either side of the membrane (Figure 4.7a), or the electric field may be applied between the membrane and another electrode. Tubular module consists of one electrode that is centrally located within the tubular membrane (Figure 4.7b).

4.2.5.3 Use of Continuous Electric Field

The application of continuous electric field can be found in several areas of industrial applications, for example, separation and fractionation of protein solution, separation of biopolymer, in the treatment of mineral and biological slurry, gelatin solution,

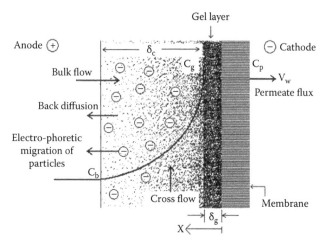

FIGURE 4.6 Schematic of electro-UF process showing the concentration boundary layer and gel layer over the membrane surface. Concentration of solute in bulk, membrane surface, and permeate are C_b, C_g, and C_p, respectively. δ_c and δ_g are the thickness of concentration boundary layer and gel layer, respectively. (Reprinted from Sarkar, B. et al., *J. Membr. Sci.*, 311, 112, 2008c. With permission.)

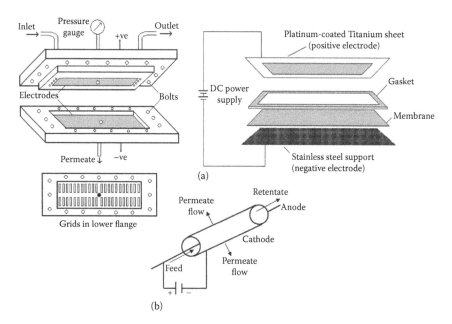

FIGURE 4.7 Schematic of electro-UF: (a) electric field is applied across a flat sheet membrane, and (b) electric field is applied to the electrode within the tubular membrane.

oil-in water emulsion, etc. Flux improvements in the range of 2–10 times have been reported for BSA (Radovich and Sparks, 1980; Radovich et al., 1985; Wakeman, 1998; Zumbusch et al., 1998; Oussedik et al., 2000), mineral and biological slurry (Weigert et al., 1999), bentonite and algal cells (Moulic et al., 1976), biopolymer (Hofmann, 2003; Park, 2006), and gelatin (Yukawa et al., 1983; Guizard et al., 1989). The application of electric field was studied in detail during the UF of citrus fruit juice (Sarkar et al., 2008a,b,c, 2009, 2010). Fruit juice contains low-molecular-weight solutes like sugar, acid, salt, flavor, and aroma compounds and high-molecular-weight solutes, mainly pectin, protein, microorganisms, etc. During membrane clarification, the smaller components permeate through membrane while the larger species, mainly pectin, are retained. In fruit juice, the typical concentration of pectin substances is up to 1.0% (Thakur et al., 1997). Pectin, a complex polysaccharide, has a tendency to form gel in the presence of sugar and acid. Pectin gel is a cross-linked polymer molecule network in a liquid medium (Thakur et al., 1997; Lee et al., 2007). During clarification of fruit juice, pectin forms a gel-type layer over the membrane surface, which offers extra resistance to the permeate flow apart from membrane resistance (Rai et al., 2006a). Pectin being negatively charged at its natural pH (about 3.0), the application of external electric field during cross-flow membrane clarification of fruit juice offers great promise in enhancing permeate flux. Thus, the application of an appropriate external electric field may significantly reduce the gel-type layer formation over the membrane surface caused by charged pectin molecules and consequently may increase the permeate flux (throughput) of the membrane system during clarification. Sarkar et al. (2008a) found flux improvements of up to 34% at 400 V/m for mosambi juice.

4.2.5.4 Use of Pulsed Electric Field

Despite the good filtration performance in electric field–assisted membrane filtration, the process is not yet used in large-scale production. The major problems associated with the use of a continuous electric field during UF include changes in feed properties due to electrode reaction (Bowen and Subuni, 1991), requirement of high energy, rise of temperature, etc. The method cannot be effectively used for feeds of high electrical conductivity and heat sensitivity. The conductivity of the feed solution is critical. A low conductivity is preferred in this process, since the energy consumption increases with increasing conductivity (Weigert et al., 1999; Bargeman et al., 2002). A high conductivity also increases the amount of heat and electrolyte gases produced at the electrodes. Weigert et al. (1999) showed that if the conductivity of a cristobalit solution exceeded 2 mS/cm, the advantage of using electro- MF vanished, since the energy requirements of the electric field became too high. These problems can be circumvented by the use of pulsed electric field instead of a continuous one. Some references of the use of pulsed electric fields are available in the literature (Wakeman and Tarleton, 1987; Bowen et al., 1989; Bowen and Subuni, 1992; Robinson et al., 1993). Sarkar et al. (2008b) investigated the effect of pulsed electric field during clarification of mosambi (*Citrus Sinensis* (L.) *Osbeck*) juice, using a 30 kDa molecular weight cut-off membrane in cross-flow UF mode under laminar flow regime (Figure 4.8). They observed that at a pulse ratio of 3:1 (3 s on and 1 s off), the flux increased up to 39% at 500 V/m for mosambi juice compared with zero electric field. Moreover, pulsed electric field required 22% less energy per unit volume

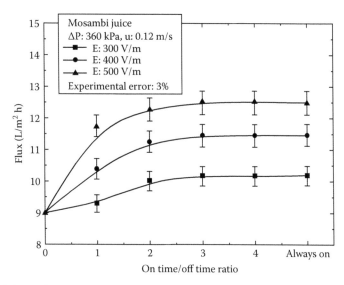

FIGURE 4.8 Variation in steady-state average permeate flux with pulse ratio for various electric fields during clarification of mosambi juice. The solid lines are only guides for the reader. (Reprinted from Sarkar, B. et al., *Sep. Purif. Technol.*, 63, 582, 2008b. With permission.)

TABLE 4.3
Analytical Measurements on Mosambi (*Citrus sinensis (L.) Osbeck*) Fruit Juice

Juice	Pectin (kg/m³)	TSS (°Brix)	pH	Density × 10⁻³ (kg/m³)	Viscosity × 10³ (Pa s)	Conduc- tivity (m S/cm)	Zeta Potential × 10³ (V)	Color (A440)	Clarity % (T660)
Mosambi juice	2.5	8.5	4.10	1.04	2.45	4.0	−21.0	1.25	29.8
Mosambi juice (Enzyme treated)	1.6	8.5	4.15	1.03	1.05	4.0	−21.8	1.1	34
UF clarified juice at E = 0 V/m	Nil	8.4	3.98	1.03	0.87	4.0	—	0.21	98.5
UF clarified juice at E = 500 V/m	Nil	8.4	4.4	1.03	0.87	3.9	—	0.22	98
UF clarified juice E = 500 V/m with pulse ratio 3:1	Nil	8.4	4.6	1.03	0.87	3.9	—	0.22	98

of permeate compared with continuous mode to produce almost the same magnitude of permeate flux. As shown in Table 4.3, pectin content in fresh mosambi juice was completely removed by both UF and electro-UF (continuous and pulse mode) without change in total soluble solid (TSS) content and the resulting clarified juice had lower viscosity and negligible turbidity.

4.2.6 MODELS FOR PREDICTION OF PERMEATE FLUX

4.2.6.1 Analysis of Steady-State Permeate Flux

Many models for cross-flow membrane filtration of fruit juice have been reported in the literature. These models are resistance-in-series model and gel layer model.

4.2.6.2 Resistance-in-Series Model

According to this model, during membrane filtration, flux declines due to various resistances to solvent flow such as membrane resistance (R_m), CP resistance (R_{cp}), cake or gel layer resistance (R_d), and pore blocking resistance (R_p). Therefore, total resistance during membrane filtration can be expressed as

$$R_t = R_m + R_{cp} + R_d + R_p. \tag{4.1}$$

The overall flux decline is represented by the following phenomenological equation:

$$V_w = \frac{\Delta P}{\mu(R_m + R_{cp} + R_d + R_p)}. \tag{4.2}$$

4.2.6.3 Membrane Resistance

Membrane resistance is caused by the membrane itself:

$$R_m = \frac{1}{\mu L_p} = \frac{\Delta P}{\mu V_0}, \tag{4.3}$$

where

μ is the viscosity of the permeate

L_p, ΔP, and V_0 are the membrane permeability, TMP, and pure water permeate flux, respectively

Membrane permeability can be obtained from the slope of the graph of experimentally measured pure water flux vs. TMP. The extent of resistance depends on the membrane thickness and various morphological features such as the tortuosity, porosity, and pore size distribution. For a membrane with pores assumed to be cylindrical capillaries of uniform radius perpendicular to the face of the membrane, the resistance can also be calculated using Hagen–Poiseuille equation:

$$R_m = \frac{8d_m}{n_p \pi r_p^4}, \tag{4.4}$$

where

n_p is the number of pores per unit membrane area

d_m is the membrane thickness

r_p is the radius of the membrane pores

This equation indicates that membrane resistance increases with increasing the membrane thickness, and decreases with increasing the pore size and pore density.

4.2.6.4 Concentration Polarization (C_P) Resistance

During filtration, solutes are transported to the membrane surface by permeation drag. Concentration polarization develops near the membrane surface, where the solute concentration is much higher than the bulk concentration. This exerts a resistance toward mass transfer, that is, the C_P resistance. Due to C_P, the accumulated solute and particle concentrations become so high that the deposition of cake or gel layer can be formed near the membrane surface, which exerts the deposition resistance. This resistance can be controlled by the changing process parameters or by rinsing water.

4.2.6.5 Deposition Resistance

During fruit juice clarification, the macromolecules such as pectin, hemicellulose, cellulose, proteins, etc., are retained by the membrane. These in the presence of sugar and acid form a gel-type layer over the membrane surface, which causes membrane fouling and reduces the rate of filtration. Using classical cake filtration theory, the gel layer resistance (R_d) can be expressed in terms of gel layer thickness as

$$R_d = \alpha(1 - \varepsilon_g)\rho_g L, \tag{4.5}$$

where α is a specific gel resistance that is obtained from Kozeny–Carman equation as

$$\alpha = 180 \frac{(1 - \varepsilon_g)}{\varepsilon_g^3 d_p^2 \rho_g}, \tag{4.6}$$

where ε_g, ρ_g, and d_p are the porosity, density, and diameter of the gel forming particles, respectively. The typical range of specific gel resistance varies from 1.56 to 3.5×10^{16} m/kg (Sarkar et al., 2009).

4.2.6.6 Pore-Blocking Resistance

Membrane fouling occurs by accumulation of solute particles on the membrane surface (known as concentration polarization) and adsorption/clogging inside the pore resulting in pore blocking. Hermans and Bredee (1936) have proposed four different kinds of pore-blocking laws: complete blocking, intermediate blocking, standard blocking, and cake layer formation (Figure 4.9).

Hermia (1982) has revised all blocking laws and reformulated the four laws in a common frame for constant pressure dead-end filtration of power-law non-Newtonian fluids. Complete blocking model assumes that particles arrive at the membrane and seal the membrane pores such that the particles are not superimposed upon the other. The blocked surface area is proportional to the permeate volume. This type of fouling

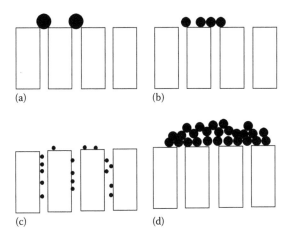

FIGURE 4.9 Mechanism of membrane fouling: (a) complete pore blocking; (b) partial pore blocking; (c) intermediate pore blocking; (d) cake filtration.

occurs when the size of the solute particles is greater than the size of the membrane pores. Pore blocking takes place on the membrane surface. In intermediate blocking model, the number of blocked pores or surface is also assumed to be proportional to the permeate volume, but it is less restrictive in such a way that not every particle necessarily blocks the pores and particles may settle on other particles. This type of fouling occurs when the size of the solute particles is similar to the size of the membrane pores. In standard blocking model, the size of the solute is much smaller than the size of the membrane pores; thus, the solutes can enter the pores. Pore blocking occurs inside the membrane pores and thus reduces the pore volume. The decrease in pore volume is also proportional to the permeate volume. The cake filtration model is used to explain the case of large solute particles, which cannot enter most pores, and hence, deposition takes place on the membrane surface and on the previously deposited solute particle layer. Jiraratananon et al. (1997) calculated various resistances using resistance-in-series model during filtration of pineapple juice by dynamic membrane and by UF membrane (Table 4.4). In their studies, they estimated polarized layer resistance by cleaning membrane with distilled water after each experiment at high cross-flow velocity and low pressure. They estimated deposition resistance using experimental flux data. However, they neglected the pore-blocking resistance.

4.2.6.7 Gel Layer Model

Gel layer model is developed from the pressure independent and mass transfer controlled film theory. During UF, solutes are convected by the TMP gradient toward the membrane surface. The lower-molecular-weight solutes permeate through the membrane, whereas higher-molecular-weight solutes are retained, forming a gel layer over the membrane surface. Since the bulk concentration of higher-molecular-weight solutes is much less than that of the gel layer concentration, a concentration boundary layer forms from the bulk of the solution up to the gel layer. This prompts a diffusion of gel-forming solutes from the gel layer toward the bulk of the solution due to concentration gradient. At steady state, the rate of convective movement of solute particles

TABLE 4.4

Resistances Obtained from Filtration of Pineapple Juice Using Dynamic Membrane and UF Membrane for 2 h

	Resistances (10^{12}/m)		
	R_m	R_p	R_t
UF membrane	6.74	20.38	6.88
Dynamic membrane	8.99	21.97	57.12

Source: Reprinted from Jiraratananon, R. et al., *J. Membr. Sci.*, 129, 135, 1997. With permission.

Operating conditions: 25°C, 300 kPa, cross-flow velocity 2.0 m/s.

toward the membrane surface is equal to the rate of migration of solute particles away from the membrane surface to the bulk due to back diffusion:

$$V_w C + D \frac{dC}{dX} = V_w C_p, \tag{4.7}$$

where

$V_w C$ is the convective transport of the solute toward the membrane

$D(dC/dX)$ is the diffusive back transport of solute from the membrane surface to the bulk

X is the distance from the membrane

$V_w C_p$ is the permeation of solute through the membrane

If the retention is not 100%, there will be a solute permeation through the membrane of $V_w C_p$. By integrating Equation 4.10 under the boundary conditions where $C = C_g$ at $X = \delta_g$ and $C = C_b$ at $X = \delta_g + \delta_c$, the following expression is obtained:

$$V_w = k \ln \frac{C_g - C_p}{C_b - C_p}, \tag{4.8}$$

where

V_w is the permeate flux

C_g is the concentration of solute at the membrane surface

C_p is the solute concentration in the permeate

C_b is the bulk concentration

δ_g is the gel layer thickness

δ_c is the thickness of the concentration boundary layer over the gel layer

k is the mass transfer coefficient equal to D/δ_c

When the solute is completely retained by the membrane, the solvent flow through the membrane increases with TMP until a critical concentration is reached (C_g). On further increase in pressure, the solute concentration at the membrane surface does not increase so the gel layer may become thicker; thus, C_g is independent of operating conditions. This implies that the gel layer resistance (R_g) to solvent flow increases so that the gel layer becomes the limiting factor in determining permeate flux. The total resistance can then be represented by two resistances in series, that is, the gel layer resistance and the membrane resistance. For the gel layer region, flux can be written as

$$V_w = k \ln \frac{C_g}{C_b}. \tag{4.9}$$

The permeate flux can be controlled by the rate at which solute is transferred from membrane surface to the bulk. Since the values of C_g and C_b are fixed by the physicochemical properties of the feed, flux can be increased by improving the mass transfer coefficient. The mass transfer coefficient can be estimated for laminar flow using the Leveque relationship and for turbulent flow using the Dittus–Boelter relationship.

For laminar flow and rectangular flow channel, Re < 1800

$$Sh = 1.86 \left(Re\, Sc\, \frac{de}{L} \right)^{0.33}. \tag{4.10}$$

For laminar flow and tubular flow channel, Re < 1800

$$Sh = 1.62 \left(Re\, Sc\, \frac{d}{L} \right)^{0.33}. \tag{4.11}$$

For laminar flow and cross-flow cell, Re < 1800

$$Sh = 1.47 \left(Re\, Sc\, \frac{h}{R} \right)^{0.33}. \tag{4.12}$$

For turbulent flow and rectangular flow channel, Re > 4000

$$Sh = 0.023\, Re^{0.8}\, Sc^{0.33}. \tag{4.13}$$

These relations were originally developed for heat and mass transfer in nonporous channel. Sarkar et al. (2008a, 2010) developed the following Sherwood number relations for a laminar and turbulent flow regime for porous media including the effect of suction from the first principles as

$$Sh = 2.1 \left(Re\, Sc\, \frac{de}{L} \right)^{0.33} \quad (Re < 1800) \tag{4.14}$$

$$Sh = 0.32\,Re^{0.58}\left(Sc\frac{de}{L}\right)^{0.33} \quad (Re > 4000) \tag{4.15}$$

where

Sh (= k/D) is the Sherwood number, k being the mass transfer coefficient and D
  being the diffusivity of the solute
Re is the Reynolds number
Sc is the Schmidt number
de (4× half of channel height) is the equivalent diameter of the flow channel
L is the length of the channel

4.2.6.8 Analysis of Transient Permeate Flux

Various flux-decline mechanisms are proposed in the literature for fruit juice clarification, for example, gel or cake layer controlled flux decline, and various steps for
blocking the membrane pores, such as complete pore blocking, standard pore blocking, intermediate pore blocking, etc.

4.2.6.9 Gel Layer Model

The rate of gel layer formation on the membrane surface is given by a solute mass
balance in the concentration boundary layer (Figure 4.6) as

$$\rho_g\frac{dL}{dt} = V_w(C - C_p) + D\frac{dC}{dX}, \tag{4.16}$$

where

ρ_g is the density of the gel layer
L is the gel layer thickness
t is the filtration time
V_w is the permeate flux
C is the solute concentration

The concentration of the gel layer is considered to be constant. Therefore, the relevant boundary conditions are

$$C = C_b \quad \text{at } X = \delta_g + \delta_c \tag{4.17}$$

$$C = C_g \quad \text{at } X = \delta_g. \tag{4.18}$$

Equation 4.16 can be solved using the boundary conditions, to obtain the growth of
the gel layer thickness for fully retentive membrane as

$$\rho_g\frac{dL}{dt} = \left[\frac{C_g - C_b\exp(V_w/k)}{1 - \exp(V_w/k)}\right]V_w. \tag{4.19}$$

Equation 4.19 can be solved with the following initial condition, $L = 0$ at $t = 0$, with knowledge of mass transfer coefficient. Thus, the variation of the gel layer thickness and permeate flux with respect to the time of operation can be obtained.

4.2.6.10 Pore-Blocking Model

Four different kinds of pore-blocking laws are proposed by Hermans and Bredee (1936). These are complete blocking, intermediate blocking, standard blocking, and cake layer formation. Hermia (1982) has revised all blocking laws and reformulated the four laws in a common frame for constant pressure dead-end filtration of power-law non-Newtonian fluids (Table 4.5).

The flux decline data are usually described by only one of these mechanisms (Bowen and Subuni, 1991; Hwang et al., 2007) or sometimes more than one mechanism in succession (Ho and Zydney, 2000; Yuan et al., 2002). In most of the studies, it is proposed that the initial rapid flux decline arises from the pore blockage by the physical deposition of the solutes on the membrane surface followed by the cake formation over the initially blocked regions of the membrane for long-term slow flux decline. Field et al. (1995) first introduced the cross-flow removal mechanism into Harmia's approach and derived the governing equations of various blocking laws for cross-flow filtration as

$$-\frac{dV_w}{dt}(V_w^{n-2}) = k_0(V_w - V_w^*), \tag{4.20}$$

where V_w^* represents the limit value of the permeate flux obtained in steady-state conditions. For cake filtration $n = 0$, for complete pore blocking $n = 2$, whilst for the intermediate mechanism $n = 1$. The model parameters, k_0 and n, are phenomenological coefficient and general index, respectively, both depending on fouling mechanism. These parameters are generally estimated using flux decline data through nonlinear regression optimization technique. Rai et al. (2006b) studied the cross-flow UF of depectinized mosambi juice using various membranes. They observed that complete or partial pore blocking occurred during the early stage of the filtration and the growth of the cake type of layer over the membrane surface dictated the long-term

TABLE 4.5
Model Equations for Various Pore-Blocking Models

Blocking Filtration Laws	N	Model Equations
Complete pore-blocking model	2	$V_w = V_0 \exp(-k_c t)$, where, $K_c = k_A V_0$
Intermediate blocking model	1	$\frac{1}{V_w} = \frac{1}{V_0} + k_i t$, where $k_i = k_A$
Standard blocking model	1.5	$\frac{1}{V_w^{0.5}} = \frac{1}{V_0^{0.5}} + k_s t$, where $k_s = 2k_A \frac{A}{A_0} V_0^{0.5}$
Cake filtration model	0	$\frac{1}{V_w^2} = \frac{1}{V_0^2} + k_d t$, where $k_d = \frac{2R_g k_D}{V_0 R_m}$

flux decline. In their analysis, they include one mechanism at a time for the entire duration of filtration. de Barros et al. (2003) investigated the fouling mechanism of cross-flow UF of depectinized pineapple juice, using the modified equations of Field et al. (1995). Their analysis revealed that filtration using hollow fiber is controlled by cake filtration and complete pore-blocking mechanism controls in ceramic tubular membrane. Similar analysis was carried out by Todiso et al. (1996) for UF of orange juice. Analysis of their results revealed that the separation process is controlled by cake filtration at lower Reynolds number (5000) and complete pore blocking was the prevailing flux decline mechanism at higher Reynolds number (>7000). During the UF of blood orange juice, Cassano et al. (2007) observed that standard pore blocking (pore constriction) governed the flux decline at low Reynolds number (5800), while complete pore blocking was the prevailing flux decline mechanism at higher Reynolds number(>7000). Mondal and De (2009) presented a generalized formulation for steady-state continuous filtration, considering sequential occurrences of complete pore blocking and cake filtration. They analyzed flux decline in terms of various nondimensional numbers. They validated their proposed model using the literature data of de Barros et al. (2003) for UF of pineapple juice.

4.3 CONCENTRATION OF FRUIT JUICE

4.3.1 REVERSE OSMOSIS

4.3.1.1 Performance of RO Process

Reverse osmosis is a process in which pressure is applied to reverse the osmotic flow of water (from solution side to the water side) across the semipermeable membrane. The potential of RO membrane was first identified in the desalination of seawater in the 1960s. The application of RO for the concentration of fruit juice has been an area of active research to the fruit-processing industry for about 35 years. RO has several advantages over the conventional concentration techniques such as low operating temperature, lower energy consumption, lower capital equipment cost, increase in flavor and aroma retention, minimum thermal damage, etc. However, due to high osmotic pressure limitation, RO processes on fruit juice seem more economical in the sugar concentration range, up to 20–30 °Bx, which is less than the concentrated juice produced by evaporation techniques (>42 °Bx). RO processes can be used as a preconcentration step to provide a feed for evaporation. Combination of RO and evaporation is found to be advantageous due to low energy consumption and high production capacity (Sheu and Wiley, 1983). It is possible to produce highly concentrated fruit juice (42–60 °Bx) using a combination of high- and low-retention RO membranes (Cheryan, 1998).

The application of RO for the concentration of fruit juice was first investigated by Merson and Morgan (1968). In their studies with orange and apple juice using cellulose acetate RO membrane, they found that oil-soluble aromas were retained while water-soluble aromas passed through the membrane. Several reports on the concentration of various fruit juices such as apple, orange, lemon, pear, grape fruit, kiwi, pineapple, passion fruit, etc., using RO are widely available in the literature. The performance of any RO process depends on the two major factors, namely, retention of juice constituents (aroma and flavor compounds) and permeate flux. Feed juice consists of various flavor compounds in terms of molecular weights, polarities, Taft numbers, solubilities, and volatilities. Each compound would give different retention characteristics under any given set of operating conditions. Most of the studies have focused on the effects of various operating conditions (feed concentration, TMP, cross-flow velocity, and temperature), different types of membranes, and membrane configurations on the retention of juice constituents and permeate flux.

4.3.1.2 Effects of Different Process Parameters on Permeate Flux and Retention

4.3.1.2.1 Effect of Membrane Configuration

Membrane configuration has a significant effect on the retention of flavor compounds during apple juice concentration (Chou et al. 1991). Permeation of n-hexanal is found to be about 11.5% in the plate-and-frame configuration compared with 0.5% in the spiral wound. The higher permeation is probably due to higher membrane packing density and membrane area of the spiral wound (a total membrane surface area of $3.90\,m^2$) when compared with the plate and frame (a total membrane surface area of $0.36\,m^2$). A lower concentration of n-hexanal in the permeate is observed in the spiral wound system due to increased retention and increased flavor compound absorption due to greater membrane surface.

4.3.1.2.2 Effect of Membrane Type

Types of membranes are found to have significant effect on both permeate flux and solute retention. Chua et al. (1988) studied permeate flux and flavor retention by CA and polyamide (PA) membranes. Polyamide was shown to be advantageous in terms of both permeate flux and flavor retention than cellulose acetate membrane. Sheu and Wiley (1983) studied different cellulose acetate (CA) and high resistance (HR) RO membranes to concentrate apple juice. They reported that during concentration of apple juice from 10 to 20 °Bx, at a pressure of 0.45 MPa and at a temperature of 20°C, CA-865 membrane results in a permeate flux of $26.9\,L/m^2\,h$ compared with the HR-95 membrane with a permeate flux of $15.9\,L/m^2\,h$. Permeate flux about $15.0\,L/m^2\,h$ seems to be economically feasible in the RO process. The HR membrane generally shows better apple flavor volatile retention than the CA membranes for a fixed operating pressure. About 88% and 16.9% retention of apple flavor volatiles are noticed using HR-95 and CA-865 membranes, respectively. Hence, the application of CA membranes with low retention of flavor volatiles results in significant quality loss leading to less suitability for apple juice concentration. Characteristics of the permeate collected from different membrane processes are shown in Table 4.6. Lower retention of flavor volatiles using cellulose acetate membranes is also observed by Matsuura et al. (1974). Chou et al. (1991) investigated the retention of apple juice volatiles using polyether-urea and polyamide thin film composite RO membranes and reported that the apple juice concentrated using polyamide membrane would result in a more intense flavor than juice concentrated using the polyether-urea membrane. Moreover, polyamide membranes have longer shelf life than cellulose acetate.

TABLE 4.6
Characteristics of Permeate Collected from Different Membrane Processes

Membrane Type	Operating Pressure (Bar)	Water Removal (%)	% Soluble Solids	pH	Titrable Acidity (%)	Total Flavor Volatiles (%)	Total Apple Volatiles (%)	Ethanol (%)
CA-885[a]	35	75.3	3.6	4.52	0.051	16.4	11.3	20.1
	40	73.7	3.3	4.53	0.042	20.0	13.6	24.8
	45	72.3	2.7	4.55	0.039	21.3	13.3	26.7
CA-990[a]	35[b]	54.4[b]	0.2[b]	4.63[b]	0.012[b]	31.3[b]	19.1[b]	47.2[b]
	40	65.1	0.4	4.6	0.017	21.4	13.8	28.8
	45	65.1	0.2	4.61	0.015	24.2	15.8	30.3
HR-95[b]	35	53.7	<0.1	5.91	<0.002	71.4	62.0	73.8
	40	54.0	<0.1	5.92	<0.002	85.2	87.7	75.0
	45	53.5	<0.1	5.89	<0.002	83.8	85.0	77.2
	40 (45°C)	53.2	<0.1	5.82	<0.002			
HR-98[b]	35	53.8	<0.1	5.45	<0.002	68.9	60.9	71.3
	40	53.5	<0.1	5.47	<0.002	86.0	85.6	74.0
	45	53.6	<0.1	5.3	<0.002	86.0	87.0	75.2
	40 (45°C)	53.2	<0.1	5.46	<0.002			

Source: Reprinted from Sheu, M.J. and Wiley, R.C., *J. Food Sci.*, 48, 422, 1983. With permission.
[a] Concentrated to 25 °Bx.
[b] Concentrated to 20 °Bx.

4.3.1.2.3 Effect of Feed Concentration

Permeate flux is a strong function of feed concentration during the RO process. The concentration polarization increases with feed concentration, resulting in a higher osmotic pressure near the membrane–solution interface during RO process and, thereby, a decrease in the available driving force (i.e., the TMP). This leads to a decrease in permeate flux.

4.3.1.2.4 Effect of Temperature

Temperature has a significant effect on both permeation rate and recovery of aroma and flavor compounds. Alvarez et al. (1997) have observed the temperature dependence on the permeate flux during apple juice concentration by RO using polyamide tubular membrane (Figure 4.10). An increase in operating temperature leads to an increase in the permeation rate of aroma and flavor compounds while their retention decreases with increase in temperature. The effect of temperature on the permeation rate is due to its effect on membrane permeability, viscosity, and diffusivity. At higher temperature, membrane permeability is higher, the viscosity of the solution decreases, and diffusivity increases. Flux enhancement of about 85% has been gained by an increase in temperature of 20°C. Chou et al. (1991) have investigated the effect of operating temperature on the recovery and permeation rates of different types of apple flavor compounds during RO of apple juice spiral wound polyamide membrane.

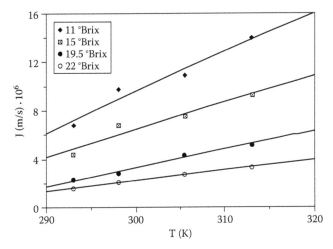

FIGURE 4.10 RO of apple juice. Influence of temperature on permeate flux at constant concentration. Operating conditions: 2.2 m/s, 550 kPa. (Reprinted from Alvarez, V. et al., *J. Membr. Sci.*, 127, 25, 1997. With permission.)

They have found that with an increase in temperature from 20°C to 40°C, recoveries are decreased from 41.7% to 29.2% for hexanol and from 56.6% to 50.8% for ethyl-2-methylbutyrate while permeation rates are increased from 0.2% to 0.7% for hexanol and from 0.15% to 0.25% for ethyl-2-methylbutyrate. According to Sheu and Wiley (1983), the concentration of aroma-stripped apple juice by RO at higher temperature is feasible with the advantages of high permeate flux and high recovery of nonvolatile components. They have reported that a twofold increase in permeate flux with an increase in the retention of soluble solids and sugars is obtained with an increase in temperature of 25°C (Figure 4.11).

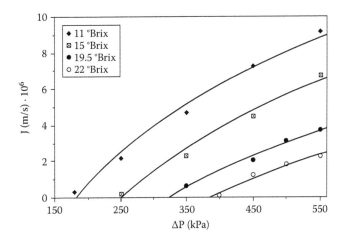

FIGURE 4.11 RO of apple juice. Influence of transmembrane pressure on permeate flux at constant concentration. Operating conditions: 298 K, 2.2 m/s. (Reprinted from Alvarez, V. et al., *J. Membr. Sci.*, 127, 25, 1997. With permission.)

4.3.1.2.5 Effect of Pressure

A significant enhancement in permeate flux is observed with an increase in TMP. Figures 4.12 and 4.13 show that an increase in pressure not only increases the permeate flux but also increases the maximum concentration that can be obtained, the limit being 19, 24.5, and 27.5 °Bx at 0.35, 0.45, and 0.55 MPa, respectively (Alvarez et al., 1997). Chou et al. (1991) have observed that with an increase in pressure, there is a significant enhancement in percent permeation of flavor compounds with higher molecular weight and more negative Taft number. However, no significant effect is observed in the case of flavor compounds with lower molecular weight and less negative Taft number. Recovery of flavor compounds in the retentate is also found to be

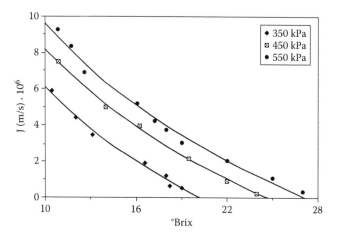

FIGURE 4.12 RO of apple juice. Permeate flux variation as a function of concentration at different transmembrane pressure. Operating conditions: 298 K, 2.2 m/s. (Reprinted from Alvarez, V. et al., *J. Membr. Sci.*, 127, 25, 1997. With permission.)

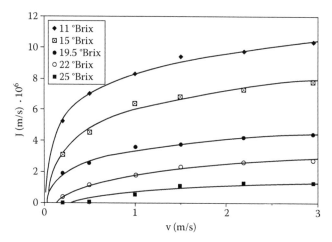

FIGURE 4.13 RO of apple juice. Influence of tangential velocity on permeate flux at constant concentration degree. Operating conditions: 298 K, 550 kPa. (Reprinted from Alvarez, V. et al., *J. Membr. Sci.*, 127, 25, 1997. With permission.)

pressure dependent in their studies. An increase in pressure results in an increase in the permeate flux. Therefore, operational time decreases and consequently smaller flavor compound losses from the retentate, which is attributed to volatilization and adhesion to the membrane.

4.3.1.2.6 Effect of Cross-Flow Velocity

During concentration by RO, cross-flow velocity has a positive effect on permeate flux. It is observed that permeate flux increases with an increase in cross-flow velocity for a fixed feed concentration (Alvarez et al., 1997). Figure 4.13 shows that the effect of cross-flow velocity is more significant at low velocities while this effect is reduced for values higher than 1.5–2 m/s. Increasing cross-flow velocity increases the process efficiency through reduction of either filtration area or operational time but cannot change the maximum limit of juice concentration.

4.3.1.3 Models for Prediction of Permeate Flux

Several studies have focused on modeling of the RO process. The two most important models for the prediction of permeate flux and retention of aroma compounds during concentration of fruit juice by RO are, namely, (1) solution–diffusion model combined with film model (Lonslade, 1972) (2) preferential sorption–capillary flow model (Kimura and Sourirajan, 1967).

4.3.1.3.1 Solution–Diffusion Model

Solution–diffusion model assumes that sorption of both the solvent and the solute occurs at the upstream surface of the membrane in accordance with phase equilibrium consideration, followed by diffusion through a nonporous and homogeneous diffusive barrier under chemical potential gradient in an uncoupled manner. According to this model, the permeate flux is given by

$$V_w = L_p[\Delta P - \Delta \pi], \tag{4.21}$$

where
    V_w is the permeate flux
    ΔP is the TMP
    L_p is the permeability coefficient of water
    $\Delta \pi$ is the osmotic pressure difference between the retentate and the permeate sides of the membrane, defined as

$$\Delta \pi = \pi(c_m) - \pi(c_p), \tag{4.22}$$

where c_m and c_p are the solute concentration on the membrane at the feed side and permeate side, respectively. At steady state, the concentration polarization phenomenon can be written by the film theory as

$$V_w = k \ln \frac{c_m - c_p}{c_o - c_p}, \tag{4.23}$$

where k is the mass transfer coefficient defined as D/d, D being the solute diffusivity and d the boundary layer thickness. For a fully retentive RO membrane, c_p, in Equations 4.22 and 4.23, can be neglected to obtain:

$$V_w = L_p \left[\Delta P - \pi(c_m) \right] \tag{4.24}$$

$$V_w = k \ln \frac{c_m}{c_o}. \tag{4.25}$$

Fruit juice contains sugar, salt, acid, and a large number of aroma and flavor compounds. For the fruit juice containing n solutes, the process can be explained with the following n + 1 equation system:

$$V_w = L_p \left[\Delta P - \pi(c_{mi}) \right] \tag{4.26}$$

$$V_w = k_i \ln \frac{c_{mi}}{c_{oi}}, \tag{4.27}$$

where the subscript "i" denotes the solute present in the juice. The mass transfer coefficient, k_i, can be estimated using Sherwood number relationship obtained from heat and mass transfer analogies for flow through a nonporous geometry. The permeability coefficient of water (L_p) is obtained from the slope of the experimental curve of permeate water flux vs. TMP (using distilled water as feed). In fruit juices, the major contributors to osmotic pressure are sugars (hexoses, disaccharides) and organic acids (Merson and Morgan, 1968). Various relationships are available in the literature for the calculation of osmotic pressure. The simplest method is the van't Hoff equation:

$$\pi = RT \frac{c_{mi}}{M_n}. \tag{4.28}$$

Thussen (1970) presented an empirical equation for the osmotic pressure of fruit juices:

$$\text{Osmotic pressure (MPa)} = \frac{13.375C}{(1-C)}, \tag{4.29}$$

where C is the weight fraction of dissolved solids. According to Matsuura et al. (1974), for the fruit juice solution, the plot of π (osmotic pressure) vs. π/X_c is a straight line. The following correlation is proposed:

$$\frac{\pi}{X_c} = a\pi + b, \tag{4.30}$$

where a and b are the constant characteristics of the fruit juice solution and the relationship is valid up to a carbon weight fraction (X_c) of about 12×10^{-2}. Values of a, b, and X_c are given in Table 4.7 for the calculation of osmotic pressure of various juices.

TABLE 4.7

Data of a, b, and X_c for Calculation of Osmotic Pressure $\pi(psi)/X_c = a\pi + b$

Name of Juice	a	b	$X_c \times 10^2$
Apple, pineapple, orange, grape fruit, and grape	3.94	3560	—
Lime juice	3.31	3997	3.51
Lemon juice	2.59	4442	3.21
Prune juice	3.31	4217	7.56
Carrot juice	4.93	3088	3.87
Tomato juice	8.95	4187	2.59

Source: Reprinted from Matsuura, T. et al., *J. Food Sci.*, 39, 704, 1974. With permission.

4.3.1.3.2 Physicochemical Properties of Fruit Juice

Constenla et al. (1989) have proposed empirical relationships based on their experimental data for the estimation of viscosity and density of apple juice as a function of concentration and temperature as

$$\ln\left(\frac{\mu}{\mu_w}\right) = \frac{AX}{(100-BX)}, \tag{4.31}$$

where $A = -0.258 + 817.11/T$ and $B = 1.891 - 3.021 \times 10^{-3} T$

$$\rho = 0.8278 + 0.34708 \exp(0.01X) - 5.479 \times 10^{-4} T, \tag{4.32}$$

where

  μ and μ_w are the apple juice and water viscosities (kg/m s), respectively
  ρ is the apple juice density (kg/m³)
  X is the apple juice concentration in °Brix
  A and B are constants
  T is the absolute temperature

During the study of apple juice concentration by RO, Alvarez et al. (1997) have used the following expression reported by Gladdon and Dole (1953) to estimate the effect of concentration on the diffusion coefficient of the solute:

$$D_n = D_{on}\left(\frac{\mu}{\mu_w}\right)^{0.45}, \tag{4.33}$$

where D_n and D_{on} are the diffusion coefficients (m²/s) for the solute, in the mixture and in the dilute water, respectively. From the knowledge of L_p, transmembrane pressure, cross-flow velocity, physicochemical properties of fluid (density, viscosity, diffusivity), dimension of flow geometry (equivalent diameter, length), and osmotic pressure, the permeate flux can be estimated.

4.3.1.3.3 Preferential Sorption-Capillary Flow Model

The preferential sorption-capillary flow model assumes that the transport of solute and solvent occurs through the pores in the membrane permselective layer. According to this mechanism, RO is governed both by the porous structure of the membrane surface and preferential sorption at the membrane–solution interface under the given experimental conditions. Preferential sorption at the membrane–solution interface is a function of the solute–solvent–membrane interactions. These interactions arise in general from polar, steric, nonpolar, and ionic character of each one of the three components in the RO system. The polar parameter gives measures of the solute's acidity or basicity. The nonpolar parameter quantifies the extent of hydrophobic interactions between the nonpolar part of the membrane and solute. These two factors would influence the composition of the solution at the membrane–solution interface. It is observed that for the nonionized solute molecule with not more than three straight chain carbon atoms in its structure and not associated with a polar functional group, the separation in RO is governed by polar and steric effects only. These effects can be expressed quantitatively in terms of appropriate free energy and steric parameters, respectively. This concept has been successfully used for the quantification of permeate flux and retention of apple flavor compounds during the RO separation process (Alvarez et al. 1998a). According to this model, the solvent flux can be expressed by Equation 4.21 and the solute flux can be expressed by the following equation:

$$V_s = \frac{D_{AM}K}{l_m}(c_m - c_p), \tag{4.34}$$

where $D_{AM}/K\, l_m$ is the solute transport parameter, which is shown to be constant for a particular membrane–solute system; D_{AM} the diffusivity of the solute in the membrane; K the distribution ratio of the solute between the aqueous solution and the membrane; l_m the membrane thickness. The solute concentration on the membrane surface at the feed side can be written by the film theory as

$$V_s + V_w = k \ln\left[\frac{c_m - c_p}{c_o - c_p}\right] \tag{4.35}$$

and in terms of mole fraction

$$V_s + V_w = kc_1 \ln\left[\frac{x_m - x_p}{x_o - x_p}\right], \tag{4.36}$$

where
  c_1 is the molar concentration (mol/m^3) of the bulk solution
  x_o, x_m, x_p are the mole fraction of the solute in the bulk solution, on the membrane
    at the feed side and permeate side, respectively

The mole fraction of the solute in the permeate can be calculated as

$$x_p = \frac{V_s}{V_s + V_w}. \qquad (4.37)$$

Equation 4.35 can be written as

$$V_s = \frac{D_{AM}K}{l_m} c_1 \left(\frac{1-x_p}{x_p} \right)(x_m - x_p). \qquad (4.38)$$

Concentration polarization can be expressed by the film theory as

$$\frac{c_m - c_p}{c_0 - c_p} = \exp\left(\frac{v}{k} \right). \qquad (4.39)$$

Assuming constant molar concentration, c_1 permeate flux (v) can be written as

$$v = \frac{V_s + V_w}{c_1}. \qquad (4.40)$$

Therefore, from Equations 4.39 and 4.40, solvent flux (V_w) can be written as

$$V_w = c_1 k(1 - x_p)\ln \frac{x_m - x_p}{x_0 - x_p}. \qquad (4.41)$$

The solute transport parameter (D_{AM}/Kl_m) for different membrane materials can be written in terms of the polar and steric Taft numbers and the Small's number of the solute as (Alvarez et al., 1998b)

$$\ln\left(\frac{D_{AM}}{Kl_m} \right) = \ln c^* + \rho^* \sigma^* + \delta^* Es^* + w^* s^*, \qquad (4.42)$$

where
  c^* is a constant depending on the porous structure of the membrane
  σ^* is the polar Taft number
  Es^* is the steric Taft number
  s^* is the Small's number of the solute in the bulk solution
  ρ^*, δ^*, and w^* are the coefficients associated with each number

The coefficient δ^* also depends on the porous structure of the membrane surface and the chemical nature of both solute and membrane material. The Small's number gives a measure of the nonpolar character of the compound. It is only considered when the molecular structure of the solute contains a straight chain involving more than three carbon atoms not associated with a polar functional group. The Taft number gives measure of the electron withdrawing capacity of the substituent group in a polar molecule. Basicity of the molecule increases with increasing the negativity of the Taft number, resulting in an increase in solute rejection by the membrane. The quantities $\rho^*\sigma^*$, w^*s^*, and δ^*Es^* in Equation 4.42 represent the contributions of polar effect, non-polar effect, and steric effect to the solute transport parameter. The solute transport parameter can also be expressed in terms of appropriate free energy and steric parameters (Matsuura and Sourirajan, 1973):

$$\ln\left(\frac{D_{AM}}{Kl_m}\right) = \ln c^* + \left(-\frac{\Delta\Delta G}{RT}\right) + \delta^*Es^* + w^*s^*, \tag{4.43}$$

where the term $(-\Delta\Delta G/RT)$ is the polar free energy parameter for the solute. The quantity $(\Delta\Delta G)$ can be obtained from the following relation:

$$\Delta\Delta G = \Delta G_1 - \Delta G_2, \tag{4.44}$$

where

  ΔG denotes the free energy of the solute–solvent interaction
  The subscripts 1 and 2 refer to the membrane solution interface and the bulk phase, respectively

RO separations of completely ionized inorganic solutes are governed by electrostatic interactions. Hence, the steric and nonpolar terms are neglected in Equation 4.43 for the calculation of the term (D_{AM}/Kl_m). However, for nonionized polar organic solutes, the term (D_{AM}/Kl_m) is usually estimated by estimating the corresponding value for a completely ionized inorganic solute taken as a reference such as NaCl (Matsuura et al., 1974).

  For the known values of membrane permeability, operating conditions (feed concentration, TMP, and cross-flow velocity), composition and physicochemical properties of feed, experimental values of permeation rate, and solute rejection, it is possible to calculate the flux of solute, solvent, and membrane surface concentration using Equations 4.38 through 4.44. Assuming solute transport parameter to be constant, the solute rejection can be predicted.

  Alvarez et al. (1997) used the solution–diffusion model for the prediction of permeate flux during concentration of apple juice. Variations of permeate flux with cross-flow velocity for different feed concentrations are represented in Figure 4.11. The solid lines represent the values of permeate flux predicted by the model and the points the experimental values.

4.3.2 DIRECT OSMOSIS CONCENTRATION

4.3.2.1 Basic Principle

DOC is one of the promising membrane processes capable of producing concentrated fruit juice with original flavor and color characteristics of the fruit operated at low temperature and low pressure. In this process, an aqueous solution of an osmotic agent is used to develop high osmotic pressure than that of the fruit juice. As no significant hydraulic pressure is used in this process, water would be transported from fruit juice to the osmotic agent solution due to osmotic pressure gradient across a semipermeable membrane. The transfer of water occurs from the feed side to osmotic agent side till the osmotic pressures on both the sides become equal. An osmotic agent is usually a highly water-soluble solid, hygroscopic, nontoxic, inert toward the flavor, odor, and color of the foodstuff, and is not allowed to pass through the membrane. The most commonly used osmotic agents are sodium chloride, sucrose or glycerol, cane molasses or corn syrup, etc. Generally, higher concentration and lower molecular weight of solute would result in higher osmotic pressure. For example, the osmotic pressure of a 74 °Bx high fructose corn syrup is about 27 MPa, which is higher than 9 MPa for 42 °Bx pulpy orange juice. Direct osmosis offers several advantages such as low temperature and low-pressure process, low energy consumption, higher retention of the TSS in the product, low membrane replacement cost, and achievement of higher concentration (45–60 °Bx) without product deterioration. Disadvantages are low permeation rate and large investment cost.

4.3.2.2 Performance of the DOC Process

The concept of the DOC process was first applied in the concentrating of grape juice using a flat sheet cellulose acetate membrane (Popper et al., 1966). In Popper et al.'s study of concentration of grape juice from 16 to 60 °Bx, an average osmotic flux of about 2.5 L/m² h was obtained. Osmotic flux during this process is affected by a number of parameters including feed concentration, temperature, osmotic medium concentration, feed flow rate, etc.

4.3.2.2.1 Effect of Feed Concentration

Osmotic flux generally decreases with increasing the juice concentration. Petrotos et al. (1998) studied the effect of tomato juice concentration on the osmotic flux using brine sodium chloride as the osmotic medium. They found a linear relationship between osmotic flux and juice concentration with a negative slope. The decrease in osmotic flux with increase in juice concentration is attributed to the reduction of osmotic pressure gradient across the membrane due to higher juice osmotic pressure at high concentration and the reduction of overall mass transfer coefficient due to increasing juice viscosity at higher juice concentration. This study suggests that for a thick and viscous juice, direct concentration process is unfeasible.

4.3.2.2.2 Effect of Temperature

The performance of DOC process can be increased by increasing the processing temperature. An increase in temperature lowers the viscosity and diffusivity of the fluid

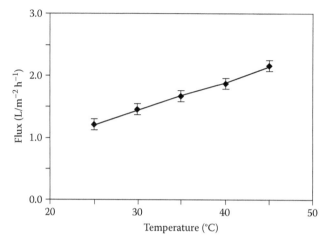

FIGURE 4.14 Effect of feed temperature on transmembrane flux using sucrose (40%, w/w)–sodium chloride (12%, w/w) solution during the concentration of pineapple juice. (Reprinted from Ravindra Babu, B. et al., *J. Membr. Sci.*, 280, 185, 2006. With permission.)

involved in the process resulting in an enhancement of osmotic flux. Petrotos et al. (1998) have found osmotic flux improvements up to 64% (from 1.48 to 2.43 kg/m² h) during the concentration of tomato juice when temperature increases from 26°C to 60°C. Similar positive effects of temperature on the direct osmotic flux during concentration of fruit juice are also reported by Beaudry and Lampi (1990) and Wrolstad et al. (1993). Ravindra Babu et al. (2006) reported that about 78% increase in transmembrane flux was observed with an increase in temperature from 25°C to 45°C during concentration of pineapple juice by direct osmosis using an aqueous solution of sucrose (40%, w/w)–sodium chloride (12%, w/w) combination as an osmotic agent (Figure 4.14).

4.3.2.3 Effect of Nature and Concentration of Osmotic Medium

The nature and concentration of osmotic medium has a significant role in the performance of direct osmotic process. The performance largely depends on the water diffusion from the membrane active layer through the membrane-backing material and osmotic medium polarization layer (external polarization), which in turn depends on the physical properties of osmotic medium. Low viscous osmotic solution has dual effects: (1) it reduces the mass transfer resistance of the external polarization layer and (2) it increases the solute diffusivity, which in turn reduces the resistivity of the backing material, thus lowering the internal polarization layer. Both lead to higher osmotic flux. The effects of various osmotic medium concentrations (brine solution) on the direct osmotic flux are investigated during concentration of tomato juice (Petrotos et al., 1998). The observed enhancement in osmotic flux is shown about 92% for the membrane AFC99 with thickness of 400μm, when brine concentration increases from 5.65% to 22.24%. They have also observed a sharp reduction in overall mass transfer coefficient (ratio of the osmotic flux to the difference of osmotic pressure between brine and tomato juice) with increasing brine concentration. According to Petrotos et al. (1998) the reduction of overall mass transfer coefficient is due to changes in the physical properties of the brine solution, mainly viscosity (increases with increasing concentration), affecting the resistance of the brine side film coefficient and thus the diffusion through the membrane backing material. Ravindra Babu et al. (2006) used a mixed osmotic agent—a mixture of sucrose 40% (w/w) and sodium chloride 12% (w/w)—for the concentration of pineapple juice by direct osmosis. The pineapple juice was concentrated up to a TSS content of 60 °Bx at ambient temperature with higher fluxes and lower salt migration (0.58%). The sucrose–sodium chloride combination was able to overcome the drawback of sucrose (low flux) and sodium chloride (salt migration) as osmotic agents during the direct osmosis process when they were applied alone.

4.3.2.4 Effect of Hydrodynamic Conditions

The osmotic flux generally increases with an increase in the flow rate of the juice to be concentrated. The effect of juice flow rate on the direct osmotic flux during concentration of tomato juice with 23.25% ± 0.15% average brine concentration as an osmotic medium is studied by Petrotos et al. (1998). About 32% increase in osmotic flux is noticed with increase in tomato juice flow rate from 109 to 502 L/h. This osmotic flux increase can be attributed to the flow rate, which results in higher shearing force affecting the resistance to mass transfer of the polarized layer close to the membrane surface, thus increasing the value of osmotic flux. Ravindra Babu et al. (2006) observed that at a fixed Reynolds number of pineapple juice feed side, the transmembrane flux was found to increase about 32.6% (Figure 4.15) with increase in Reynolds number of mixed osmotic agent solution (sucrose, 40% (w/w) and sodium chloride, 12% (w/w)) from 0.2 to 1.6 (from 25 to 100 mL/min). They also obtained 64% enhancement of transmembrane flux with increase in Reynolds number of pineapple juice (50 °Bx) from 0.2 to 0.85 (from 25 to 100 mL/min) at a fixed Reynolds number of the same osmotic agent of 1.6 (100 mL/min).

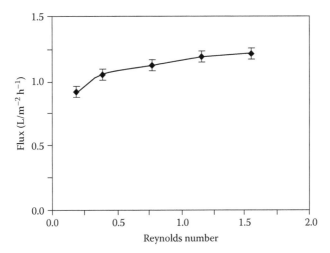

FIGURE 4.15 Effect of osmotic agent side hydrodynamic condition on transmembrane flux during the concentration of pineapple juice. (Reprinted from Ravindra Babu, B. et al., *J. Membr. Sci.*, 280, 185, 2006. With permission.)

TABLE 4.8

The Effect of Membrane Thickness on the Performance of the Direct Osmosis Process

Overall Membrane Thickness (μm)	Thickness of Membrane Active Layer (μm)	Thickness of Membrane Backing Material Layer (μm)	Osmotic Flux (kg/m² h)
400	140	260	3.1
500	140	360	1.48
600	140	460	0.64

Source: From Petrotos, K.B. et al., *J. Membr. Sci.*, 150(1), 99, 1998.

4.3.2.5 Effect of Membrane Thickness

Petrotos et al. (1998) have investigated the concentration of tomato juice by direct osmosis, using commercial thin film composite RO membrane with uniform active layer thickness and varying membrane-backing material thickness. Membrane thickness is found to have a significant role in the formation of overall mass transfer coefficient. A sharper reduction of overall mass transfer coefficient with thinner membrane is noticed. The thinner membrane has a greater contribution of the osmotic medium film coefficient in the formation of the overall mass transfer coefficient due to reduction in membrane resistance. An exponential decreasing trend of the flux is observed with increasing the overall membrane thickness (Table 4.8). Similar effect of the thickness of membrane backing material on the osmotic flux is also observed by Loeb et al. (1997).

4.3.2.5.1 Apparatus Design

Different types of membrane modules with specific design are constructed for direct osmosis process. Flat sheet and tubular configurations are most common. A simple schematic of DOC process is shown in Figure 4.16 (Herron et al., 1994).

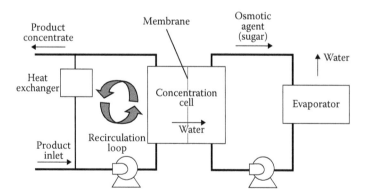

FIGURE 4.16 Simplified DOC flow sheet. (Reprinted from Herron, J.R. et al., Medina, Osmotic concentration apparatus and method for direct osmotic concentration of fruit juice, U.S. Patent 5,281,430, January 25, 1994. With permission.)

In this process, juices of high solid contents can be concentrated with minimum fouling because the solids are not convected toward the membrane surface as it operates under low pressure. The fruit juices can be fed to the module in a continuous, multistage process. At each stage, the fruit juice is recycled through a heat exchanger (if refrigeration is desired) and then through the DOC modules. The juice is kept in inert atmosphere to increase the flavor and aroma retention. The osmotic agent solution is fed either once counter currently through the entire process or is recycled continuously through an evaporator. If there is a large quantity of very concentrated solution that is diluted at the plant, and if this solution could be used as the osmotic agent solution, the dilution can be performed within the juice concentration modules. This would result in substantial cost savings for the juice concentration process (Girard and Fukumoto, 2000b). Generally, the osmotic agent solution is diluted and reconcentrated using evaporation. Therefore, instead of juice, the osmotic agent solution is exposed to the high temperature, allowing the improvement of concentrate quality. In continuous process, it is possible to obtain a steady-state osmotic flux. However, during the concentration process, osmotic flux gradually declines with time. This is due to increase in osmotic pressure of juice as it is concentrated with time. Hence, osmotic pressure differences between the osmotic agent solution and the juice decrease, resulting in flux decline. During the experiment, the flux of a depectinized raspberry juice at 30°C declined from 1.37 to 0.3 L/m² h, while the osmotic agent solution decreased from 69 to 60 °Bx (Beaudry and Lampi, 1990). A flat osmotic apparatus with a special configuration to promote turbulence was developed by Petrotos and Lazarides (2001). They reported that with this apparatus an average flux of 4.5 kg/m² h was obtained during concentration of tomato juice from 5 to 16 °Bx at room temperature and low pressure. It has also been reported that tubular module offers some advantages such as turbulent flow, less pressure drop, larger section of the membrane without fouling, and concentration polarization in concentration of juice of high solid content, leading to an increase in osmotic flux (Jiao et al., 2004).

4.3.3 MEMBRANE DISTILLATION

4.3.3.1 Basic Principle

MD is a separation process in which a microporous hydrophobic membrane separates two aqueous solutions, at different temperatures. This process is based on the formation of a vapor on the side of the warmer solution–membrane interface, to the transport of vapor phase through the microporous membrane, and to its condensation at the cold membrane solution interface. The driving force for MD is the partial pressure difference induced by the temperature gradient between the two solution–membrane interfaces. Heat is also transported within the vapor across the membrane (Figure 4.17). In this process, the membrane acts as a physical support for the vapor–liquid interface. A liquid–vapor interface is formed at the pore mouths, where liquid and vapor are in equilibrium; inside the membrane pores only a gaseous phase is present through which vapors are transported as long as a partial pressure difference is maintained. This process takes place at atmospheric pressure and at a temperature much lower than its

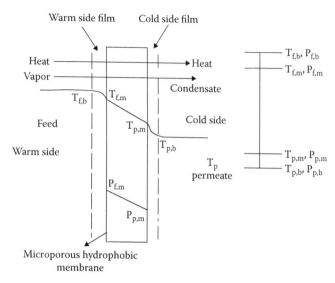

FIGURE 4.17 Schematic of the heat and mass transport profile in membrane distillation. $T_{f,b}$, $T_{f,m}$: temperature at the bulk and membrane of the feed side, respectively; $T_{p,b}$, $T_{p,m}$: temperature at the bulk and membrane of the permeate side, respectively; $P_{f,m}$, $P_{p,m}$: vapor pressure of the membrane surface at the feed side and permeate side, respectively.

boiling point, with net flux always from warm solution to cold solution. The MD includes the following characteristics: (1) the membrane should be porous, (2) the membrane should not be wetted by the process liquids, (3) no capillary condensation should take place inside the pores of the membrane, (4) the membrane must not alter the vapor–liquid equilibrium of the different components in the process liquids, (5) at least one side of the membrane should be in indirect contact with the process liquid, (6) for each component, the driving force of this membrane operation is a partial pressure gradient in the vapor phase. There are various types of MDs depending on the methods employed to impose a vapor pressure difference across the membrane to drive flux (Figure 4.18). The permeate side of the membrane may consist of a condensing fluid in direct contact with the membrane (direct contact MD), a condensing surface separated from the membrane by an air gap (air gap MD), a sweeping gas (sweeping gas MD), or a vacuum (vacuum MD). The type of MD employed is dependent upon the permeate composition, flux, and volatility. The direct contact MD configuration is suitable for applications such as desalination or the concentration of aqueous solutions (orange juice), in which water is the major permeate component. For the separation of volatile organic or a dissolved gas from an aqueous solution, either sweeping gas MD or vacuum MD should be used. Air gap MD, which is the most versatile MD configuration, can be applied to almost any application. The advantages of MD compared with other traditional technologies of concentration of juices are (1) the evaporation surface can be made to various membrane configurations, (2) capable of producing high quality concentrates, (3) low operating temperatures (28°C–48°C), an opportunity to achieve high contents of dry

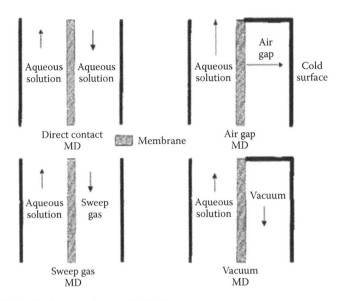

FIGURE 4.18   Various membrane distillation processes.

substances (60%–70%), (4) low energy consumption, and (5) corrosion or fouling are less. Because the process can take place at normal pressure and at lower temperature, MD could be used to concentrate to high osmotic pressures aqueous solutions of solutes sensitive to high temperatures. Therefore, MD is frequently used for concentrating fruit juices such as apple and orange with better flavor and color (Drioli et al., 1992; Calabro et al., 1994).

4.3.3.2   Apparatus Design

A large variety of membrane modules have been used in MD process. This includes plate-and-frame, spiral wound and hollow fiber, etc. (Carlsson, 1983; Andersson et al., 1985; Calabro et al., 1994; Lagan et al., 2000). A well-designed membrane module should provide high rates of heat and mass transfer between the bulk solution and the solution–membrane interface. Gore and Associates (United States), The Swedish Development Co., and Enka AG (Germany) started testing their own membranes in MD units for commercialization. Gore and Associates (United States) developed a spiral wound module for "Gore-Tex Membrane Distillation." The Swedish Development Co. used a plate-and-frame module for "SU Membrane Distillation," and Enka AG (Germany), developed a hollow fiber module for "Transmembrane Distillation" (Lawson and Lloyd, 1997). For the concentration of orange juice, Drioli et al. (1992) investigated MD process, using a commercial plate PVDF membrane (Millipore Corp.) with a nominal pore size of 0.22 μm and a laminated hydrophobic microporous membrane (G0712) with a pore size of 0.2 μm (Gelman Science Tech. Ltd). They found that the flux of PVDF membrane was remarkably higher than that of G0712 membrane. Calabro et al. (1994) reported that single strength orange juice (10.5 °Bx) could be concentrated to 31.5 °Bx in an MD laboratory plant (Figure 4.19) using a plate poly-vinylidene fluoride (PVDV) hydrophobic membrane

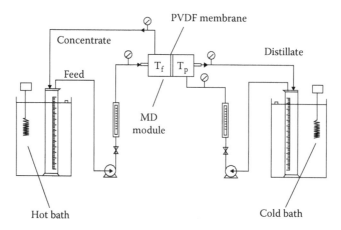

FIGURE 4.19 Schematic of membrane distillation laboratory plant. (Reprinted from Calabro, V. et al., *Ind. Eng. Chem. Res.*, 33, 1803, 1994. With permission.)

with a nominal pore diameter of 0.22 μm, porosity of 75%, thickness of 140 μm, and effective area 20.42 cm^2 (Millipore Corp.). Nene et al. (2002) concentrated a clarified sugar cane juice (20 °Bx) by MD process using a plate-and-frame module (polypropylene membrane of area 75 cm^2, nominal pore size of 0.2 μm). An overall flux of 10 kg/m^2 h was obtained in this study.

Curcio et al. (2000) studied the concentration of single strength apple juice up to 64 °Bx by MD using a polypropylene hollow fiber membrane module (ENKA Microdyn MD-020-2N-CP) with a membrane area of 0.1 m^2 and nominal pore diameter of 0.45 μm. In their study, they obtained an overall permeate flux of 1–1.5 kg/m^2 h.

4.3.3.3 Effect of Different Process Parameters on Permeate Flux and Retention

4.3.3.3.1 Effect of UF Pretreatment

Calabro et al. (1994) investigated the concentration of orange juice in an integrated UF and MD system to compare the performance of the juice concentration in the presence or in the absence of a pretreatment. They observed that UF of orange juice using polypropylene hollow fibers (ENKA Inc.) with an area of 0.1 m^2 and 0.2 μm pore size resulted an increase in permeate flux in MD without flux decay over that observed for unultrafiltered juice (Figure 4.20). This was attributed to a decrease in viscosity at the concentration boundary layer near the membrane surface due to removal of pulp and pectin content.

4.3.3.3.2 Effect of Concentration of Feed Juice

Studies on MD of sugar solution and mandarin orange juice show that MD flux is significantly affected by the feed concentration (Kimura et al., 1987). Permeate flux gradually decays with an increase in feed concentration. This flux decay can be attributed to the reduction of the driving force due to the decrease in the vapor pressure of the feed solution and to the exponential increase in the viscosity of

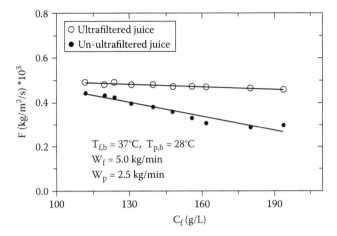

FIGURE 4.20 Permeate mass flux vs. feed concentration with and without UF pretreatment. W_f, $T_{f,b}$ and W_p, $T_{p,b}$ are the corresponding mass flow rate and temperature of feed and permeate, respectively. (Reprinted from Calabro, V. et al., *Ind. Eng. Chem. Res.*, 33, 1803, 1994. With permission.)

the juice solution with concentration. A comparison between MD and RO was reported by Kimura et al. (1987). They observed that during concentration of mandarin orange juice (Figure 4.21), higher permeate fluxes are achieved in the MD process than in RO, at higher concentration ratios (ratio between actual and initial feed concentrations).

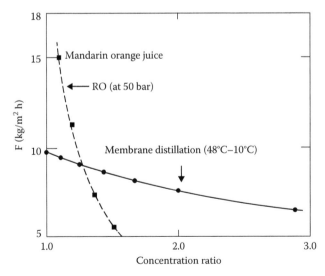

FIGURE 4.21 Comparison of fluxes of membrane distillation and reverse osmosis for concentration of mandarin orange juice. (Reprinted from Kimura, S. et al., *J. Membr. Sci.*, 33(3), 285, 1987. With permission.)

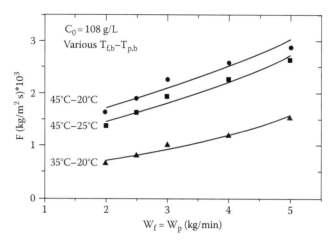

FIGURE 4.22 Permeate mass flux vs. feed and permeate flow rates at different temperature conditions; W_f and W_p are the mass flow rate of feed and permeate, respectively. (Reprinted from Calabro, V. et al., *Ind. Eng. Chem. Res.*, 33, 1803, 1994. With permission.)

4.3.3.3.3   Effect of Flow Rate

Generally, the transmembrane flux increases with the cross-flow rate both in the feed and distillate streams in the MD process (Figure 4.22). This flux increase can be attributed to the increasing cross-flow rate, resulting in stronger shear stress, which causes a lower accumulation of particulates such as pectin and cellulose on the active membrane surface and thus reduces the polarization effects and membrane fouling resulting in higher flux. Moreover, a lower cross-flow velocity causes hindering the heat transfer from the bulk of the solution to and from the membrane surface, leading to a more severe temperature polarization.

The viscosity of the juice at high concentration induces high polarization phenomena near the membrane surface. Temperature polarization is more important than concentration polarization and is observed mainly on the feed side. Therefore, the transmembrane flux was more sensitive to variation in the feed flow rate. Flux rates were found to be dependent upon the temperature polarization coefficient (TPC) rather than the concentration polarization coefficient (CPC) (Lagan et al., 2000). The temperature polarization coefficient represents the thermal driving force reduction on the membrane, and the CPC represents the reduction of concentration driving force on the membrane. It was observed that to obtain the same increase in the TPC it is necessary to increase the distillate flow rate by 190%, whereas only 50% in the feed stream was necessary.

4.3.3.3.4   Effect of Operating Temperature

The effect of temperature difference between feed juice and the cooling water on MD flux was studied by several investigators. It is generally observed that the MD flux gradually increases with an increase in the temperature difference between feed juice and the cooling water. Gunko et al. (2006) studied the permeation of apple juice by direct contact MD process using vinylidene fluoride hydrophobic MF membrane

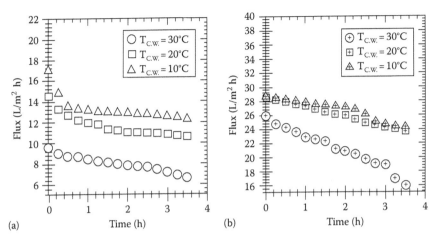

FIGURE 4.23 Permeate fluxes at flow rate of 0.8 m/s, different cold water temperatures with constant hot juice temperature: (a) $T_{h.j.}$ = 50°C and (b) $T_{h.j.}$ = 70°C. (Reprinted from Gunko, S. et al., *Desalination*, 190, 117, 2006. With permission.)

(MFFK-3 type, NPO Polymersyntes, Russian Federation, area 0.049 m², nominal mean pore size 0.45 μm and a porosity 80%–85%) at three constant temperatures of juice (50°C, 60°C, and 70°C) with the three cooling water temperatures (10°C, 20°C, and 30°C). They observed that the temperature between hot juice and cooling water is the controlling factor of the capacity of the process. Their experimental results revealed that at 50°C, the flux increased about 80% with decreasing the cooling water temperature, $T_{c.w.}$, from 30°C to 10°C, while this increase was only 10% at 70°C (Figure 4.23). For the same difference in temperatures between hot and cold surfaces of a membrane $\Delta T = T_{h.j.} - T_{c.w.}$, the difference in vapor pressure is found to be larger when the hot juice temperature is increased, resulting in the performance of the process more dependent on the temperature of hot juice.

Similar observations were also made by Calabro et al. (1994) in the concentration of orange juice by MD. They reported that permeate flux gradually increased with an increase in the feed juice temperature for each flow rate at fixed inlet temperature of the cooling water. The increase in feed temperature increased all the transport parameters, with an improvement in the driving force (Drioli et al., 1992).

4.3.3.4 Effect of Membrane Properties

Because of the potential of MD, various membranes have been and are being used by several investigators. Most of the membranes are composed of organic polymers, such as polyvinyldifluoride (PVDF), polytetrafluoroethylene (PTFE), and polypropylene (PP). The size of membrane micropores can range between 0.2 and 1.0 μm with porosity from 60% to 80% and the overall thickness from 80 to 250 μm. The composite MD membrane has been successfully developed and utilized by several investigators (Xu et al., 2004; Peng et al., 2005). These membranes consist of a hydrophobic top layer, made of PTFE or PVDF, covering a hydrophilic layer made of cellulose acetate, polysulfone, or cellulose nitrate (Cheng and Wiersma, 1982). Membrane characteristics (thermal conductivity, thickness, void fraction, tortuosity) and operating conditions in the two liquid phases greatly influence the performance of the MD process. In general, the thinner the membrane and the larger the surface porosity and pore size of the membrane, the higher the permeate flux. Gostoli and Sarti (1987) have shown that for pure water, molar flux in direct contact MD across a relatively thick membrane is inversely proportional to membrane thickness (δ); however, flux becomes independent of membrane thickness at small thicknesses (i.e., when $\delta \ll k_m/U$ where, k_m is the thermal conductivity of the membrane and $1/U = (1/h_f. + 1//h_p)$, the combined feed and permeate boundary layer resistance). This may be a result of the increased conductive heat loss associated with thinner membranes. During MD of tap water, Gryta (2007) observed that the presence of low porosity layer (thickness below 1 μm) caused the reduction of air permeability to 36.8%, whereas the MD permeate flux was decreased only by 15%. Inorganic membranes are also used in MD for their chemical, thermal, and mechanical stability compared with organic membranes.

4.3.3.5   Effect of MD on the Retention of Fruit Juice Components

During the concentration of apple juice, the use of PVDF membrane was found to have almost full retention of orange juice compounds such as soluble solids, sugars, and organic acids, with satisfactory color and flavor contents (Drioli et al., 1992). During the concentration of black-currant fruit juice (enzyme treated, prefiltered by MF, and preconcentrated by RO) by hollow fiber MD process using hydrophobic polypropylene membrane, Kozak et al. (2009) reported that there were significant improvements in TSS, density, total acid content (TAC), and anthocyanin content in the concentrate (Table 4.9).

4.3.4   OSMOTIC DISTILLATION

4.3.4.1   Basic Principle

OD is a membrane-based separation technique with great potential for the production of high-quality concentrated fruit and vegetable juices such as orange juice, tomato juice (Durham and Nguyen, 1994), apple and grape juices (Sheng et al., 1991), passion fruit juice (Vaillant et al., 2001), and other fruit juices (Girard et al., 2000a; Petrotos et al., 2001). This process is also known as osmotic evaporation, osmotic

---

TABLE 4.9

Determination of TSS, Density, TAC, and AC in Samples of Black Currant Juice Concentrated by Membrane distillation at $\Delta T = 15°C$

Sample	TSS (°Bx)	Density (g/cm³)	TAC (%)	Anthocyanin Content C (g/L)
Raw juice	15.0	1.049	3.65	1.201
MD feed	22.0	1.059	5.12	1.868
MD product	58.2	1.292	16.00	3.805

Source: Reprinted from Kozak, A. et al., *Desalination*, 241, 309, 2009. With permission.

---

concentration by membrane, membrane evaporation, isothermal MD, or gas membrane extraction. OD is usually operated under atmospheric pressure and at ambient temperature, thus avoiding thermal and mechanical degradation with the preservation of the aroma and flavor contents of the fresh juice. The OD process uses a microporous hydrophobic membrane to separate the liquid feed (juice) by a hypertonic solution (generally concentrated brine) flowing downstream a membrane. The water vapor pressure gradient across the hydrophobic membrane is maintained by lowering the vapor pressure on the downstream side relative to that on the upstream side by using a concentrated brine stripper such as salt solution of $MgSO_4$, $CaCl_2$, $KHPO_4$, etc. The hydrophobic nature of the membrane prevents infiltration of aqueous solutions to the pores, resulting in air entrapped within the membrane pores forming the gas membrane. The difference in solute concentration, and so in water activity between the two sides of the membrane, induces at the vapor–liquid interfaces a vapor pressure difference causing a diffusion of water vapor across the pores (Gostoli, 1999). The mass transfer mechanism occurs in three steps: evaporation at the dilute vapor–liquid interface, diffusion of vapor through the membrane pores from higher-vapor pressure side to the lower one, and condensation of the vapor at the membrane-brine interface. OD is very close to the mechanism of MD, although there are some significant differences. Both the processes involve sustaining a water vapor pressure gradient, that is, chemical potential gradient across the membrane pores, in order to get the thermodynamic force causing the diffusion process. In the case of MD, a temperature difference induces a corresponding vapor pressure gradient across the membrane pores, whereas, in the case of OD, this gradient is due to the difference in the composition of the bulk liquid phases adjoining the membrane at both sides. The major problems associated with RO is the osmotic pressure limitation which restricts the concentration of juice up to 25–30 °Bx, while in MD some loss of volatile aroma compounds and thermal degradation due to the heat requirement for the feed stream in order to maintain the water vapor pressure gradient. These problems can be circumvented by the use of OD with the achievable concentration levels close to the values presently obtained by evaporation.

4.3.4.2 Apparatus Design

The OD processes are carried out in batch as well as continuous mode of operation for the concentration of fruit juices. In both modes of operations, the most important design parameters include (1) plant capacity in terms of volume of feed to be concentrated, (2) concentration of solute in feed and product, (3) water vapor pressure–concentration relationship for both feed and strip solution, and (4) membrane characteristics (hydrophobicity, thickness, pore size, porosity, tortuosity, etc.). Sheng et al. (1991) concentrated various fruit juices (orange, apple, and grape) in the range of 65–70 °Bx by OD process in a specially designed Syrinx plate and frame configuration SR-72 membrane module fitted with PTFE membrane (area: $0.7\,m^2$; thickness: $100\,\mu m$, pore size: $0.2\,\mu m$) with 3 mm spacing between the membranes under batch mode of operation. An OD laboratory bench plant supplied by Hoechst-Celanese Corporation (Wiesbaden, Germany) equipped with a Liqui-Cell Extra-Flow 2.5 × 8 in. membrane contactor (Hoechst-Celanese Corporation, Wiesbaden, Germany) (Table 4.10) was used to concentrate

TABLE 4.10

Data Sheet of Liquid-Cell Extra-Flow 2.5 × 8 in. Membrane Contactor

Fiber Characteristics

Fiber type	Celgard microporous polypropylene hollow fiber
External diameter	300 μm
Internal diameter	220 μm
Wall thickness	0.03 mm
Porosity	40%
Length	0.16 m

Cartridge Operating Limits

Max. transmembrane differential pressure	4.2 kg/cm² (60 psi)
Max. operating temperature range	40°C (104°F)

Cartridge Characteristics

Cartridge dimensions (D × L)	8 × 28 cm (2.5 × 8 in.)
Effective surface area	1.4 m² (15.2 ft²)
Effective area/volume	29.3 cm²/cm³
Fiber potting material	Polyethylene

the clarified fruit juices such as citrus and carrot (Cassano et al., 2003), kiwifruit (Cassano et al., 2004), and cactus pear (Cassano et al., 2007). The clarified juice was recirculated through the shell side of the membrane module, while the stripping solution, 60 w/w% calcium chloride dehydrate solution, was recirculated in a tube side with countercurrent mode. The clarified juice was concentrated by this process up to a total soluble content of about 60 °Bx at an average throughput of about 1 kg/m² h. The schematic of OD laboratory plant is shown in Figure 4.24.

4.3.4.3 Effect of Different Process Parameters on Permeate Flux and Retention

4.3.4.3.1 Effect of Pretreatment

It was found that the pretreatment of fresh juice (apple, kiwifruit, passion, carrot, etc.) resulted in an increase in the efficiency of OD process (Vaillant et al., 2001; Cassano et al., 2003, 2004). The pretreatment of juice was usually carried out using enzyme, UF, MF, etc., and sometimes a combination of both. Pretreatment of grape juice by UF was shown to result in an improved flux during subsequent concentration of permeate by OD. The flux increase was attributed to a reduction in the viscosity of the concentrated juice–membrane boundary layer due to removal of pectin, protein, etc. (Bailey et al., 2000). The enzymatic treatment followed by the UF of apple juice was shown to decrease the levels of pectin and proteins significantly, thus reducing the deposition on the hydrophobic surface and thereby membrane fouling. Such a deposition improved membrane wetting and could eventually result in an undesired convective flow of liquid through the membrane in the OD process.

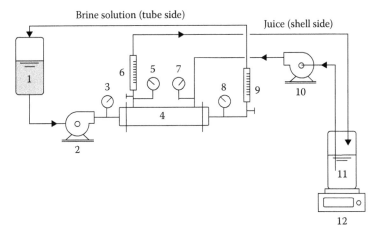

FIGURE 4.24 Schematic of OD laboratory plant. (1) Extracting solution tank, (2) brine pump, (3,5,7,8) manometer, (4) OD membrane module, (6,9) flow meter, (10) feed pump, (11) feed tank, (12) digital balance. (Reprinted from Cassano, A. et al., *J. Food Eng.*, 57, 153, 2003. With permission.)

4.3.4.3.2 Effect of Solute Content

Cassano et al. (2004) performed OD experiments during the concentration of kiwifruit juice to observe the effect of the brine concentration on the evaporation fluxes (Figure 4.25). They observed that the OD flux was significantly affected by the salt content of the brine as a 50% mass fraction reduction of $CaCl_2\,2H_2O$ (from 60.0 w/w% ($a_w = 0.28$) to 30.0 w/w% ($a_w = 0.80$)) leads to a 70% vapor flux decline in OD flux from 1.16 to 0.35 kg/m² h. Similar trends were also observed by Courel et al. (2000) in the concentration of sucrose solution by OD. They observed that decreasing the salt content in the brine of about 30% resulted in 64% flux decline from 10.3 to 3.7 kg/m² h.

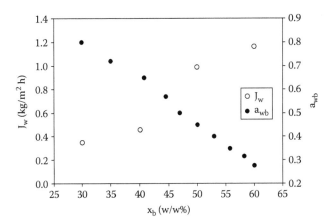

FIGURE 4.25 Evaporation flux (J_w) and water activity of the brine (a_{wb}) vs. brine solute content (x_b). Operating conditions: T = 25°C; juice flow rate = 29.8 L/h; brine flow rate = 37.8 L/h. (Reprinted from Cassano, A. et al., *Food Res. Int.*, 37(2), 139, 2004. With permission.)

This result may be explained by the strong dependence of the water activity of the brine on salt content. This activity effect was more pronounced than the flux increase that could be expected from improvement of transport properties: both density and viscosity of the brine decreased with dilution, while the diffusion coefficient increased. This reduction of mass transfer resistance in the salt solution rendered into a vapor flux improvement, which was in fact masked by the activity effect.

Courel et al. (2000) used OD for the concentration of sucrose solution using a flat sheet membrane. They observed that with increasing sucrose content from 0 to 65 w/w%, the OD flux decreased from 10.3 to 1.1 kg/m² h (Figure 4.26). Unlike the salt effect, the water activity of the sugar solution decreased only by 13.2% in the range of sucrose concentration considered, and the flux decay was not attributed to a water activity effect. The viscosity of the sugar solution increased exponentially with the solute content while the diffusion coefficient strongly decreased. The presence of sugar at higher concentration resulted in vapor flux decay by reducing the transport properties of the solutions and specifically by increasing viscosity. During the concentration of cactus pear juice from 11 to 61.4 °Bx by OD, Cassano et al. (2007) reported that at low TSS concentration of the feed, the flux decline was more attributable to the dilution of the stripper while at higher TSS concentration, it was mainly due to exponential increase in juice viscosity with juice concentration. Similar observations are also reported in the literature during concentration of passion fruit and kiwifruit juice by OD (Vaillant et al., 2001; Cassano et al., 2004).

Vaillant et al. (2001) observed similar trends in the concentration of microfiltered passion fruit juice on an industrial scale by OD. A pilot plant equipped with a module containing 10.2 m² of polypropylene hollow fibers was used to concentrate passion fruit juice up to TSS content higher than 60 °Bx at 30°C. An average evaporation flux of almost 0.75 kg/h was obtained with water, 0.65 kg/h when juice was concentrated to 40 °Bx, and 0.50 kg/h when it reached 60 °Bx. OD can also be carried out in a two-stage process to concentrate juice from 14 to 60 °Bx with a constant evaporation flux of around 0.62 kg/h. Two-stage process showed almost 20% saving in membrane's surface area.

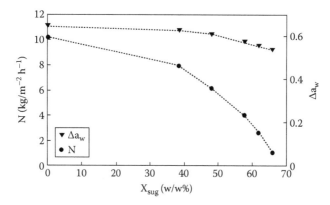

FIGURE 4.26 Effect of sucrose content of the dilute solution on mass transfer: vapor flux (N) and water activity difference between sucrose solution and brine (Δa_w) vs. sucrose content (X_{sug}). (Reprinted from Courel, M. et al., *J. Membr. Sci.*, 170, 281, 2000. With permission.)

4.3.4.3.3 Effect of Velocity

At high concentration level (>60 °Bx), evaporation rate was found to be strongly dependent on tangential velocity (Vaillant et al., 2001). With decreasing the tangential velocity of juice from 0.24 to 0.09 m/s, flux declined by almost 20%. This can be explained by the fact that at higher concentration, viscosity increases exponentially with severe concentration polarization near the membrane surface, which is highly sensitive to tangential velocity and, consequently, to increase flux with increase in tangential velocity. The circulation velocity of brine also has a significant effect on the improvement of OD flux. Courel et al. (2000) observed that with increasing circulation velocity of brine from 0.2 to 2.2 m/s, flux increased more than twofold. At higher velocity (>1.7 m/s), improvement was not significant. This flux increase can be attributed to the strong shear stress along the condensation side of the membrane imposed by the circulation velocity of brine. At higher circulation velocity, the effect of concentration polarization became negligible resulting in no further significant increase in flux.

4.3.4.3.4 Effect of Water Vapor Pressure Difference between Juice and Stripping Solution

The effect of operating conditions (juice flow rate, juice concentration, and temperature) on the OD flux during the concentration of apple, orange, and grape juice through a PTFE membrane (0.7 m²) with pore size of 0.2 μm and an overall thickness of 100 μm was studied by Sheng et al. (1991) under batch mode operation. Their results revealed that OD flux decreased with the increase in juice concentration, and it was shown to depend strongly on the osmotic pressure difference ($\Delta\pi$) between the feed juice and the stripping solution. When decreasing $\Delta\pi$ from 416 atm (low juice concentration) to 280 atm (high juice concentration), that is, about 33%, a fivefold decrease in OD flux was observed. The effect of the water vapor pressure differences between juice and brine solution (ΔP) on the OD flux was also observed by Cassano et al. (2004) in the concentration of kiwifruit juice (Figure 4.27).

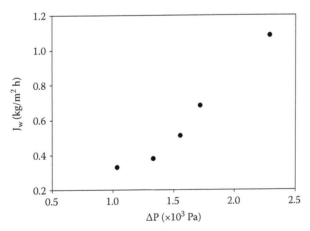

FIGURE 4.27 Effect of water vapor pressure difference (ΔP) between juice and brine on the evaporation flux (J_w). (Reprinted from Cassano, A. et al., *Food Res. Int.*, 37(2), 139, 2004. With permission.)

4.3.4.3.5   Effect of OD on the Retention of Fruit Juice Components

Barbe et al. (1998) investigated the influence of different types of hydrophobic micro-porous membranes on the retention of organic volatile flavor/fragrance components during OD of Gordo grape juice and Valencia orange juice. Their study revealed that membranes with relatively large pore sizes at the surface were shown to be associated with higher organic volatiles retention per unit water removal than those with smaller surface openings. A simple model based on differences in their resistances to organic volatiles transport was proposed. Accordingly, pores with larger diameters at the membrane surface allowed greater intrusion of the liquid feed and brine streams with enhanced stagnation relative to that which occurs in membranes with smaller pore entrances. This resulted in an increase in the thickness and hence resistance of the boundary layer at the pore entrance. Shaw et al. (2001) analyzed the retention of flavor compounds during concentration of orange and passion fruit juices by OD distillation process in a pilot-scale osmotic evaporator containing $10.3\,m^2$ of polypropylene hol-low fibers. Both juices were concentrated threefold to 33.5 and 43.5 °Bx, respectively. Their analysis revealed that OD processing involved a loss of about 32% of volatile components in orange and about 39% in passion fruit juices. Alves and Coelhoso (2006) studied the potential of OD and MD processes for the retention of aroma compounds using model solutions containing citral and ethyl butyrate, two relevant aroma compounds of the orange juice aroma profile. A higher retention per amount of water removal was reported with the OD process. Retentions of citral and ethyl butyr-ate were about 12% and 14%, respectively, for the OD process, while 42% and 48%, respectively, for the MD process. During the concentration of UF clarified cactus pear juice (11 °Bx) by a laboratory bench plant hollow fiber module, Cassano et al. (2007) studied the potential of OD process by characterizing the concentrate in terms of TSS, total antioxidant activity (TAA), ascorbic acid, citric acid, and glutamic acid. They found the reduction of ascorbic acid and citric acid in the final retentate of the OD process of about 3.5% and 5%, respectively, with respect to UF clarified juice while the TAA and glutamic acid content in concentrate remained almost unaltered inde-pendently by the achieved level of the concentration (61 °Bx). According to Flath et al. (1967), the compounds associated with the characteristic delicious apple-like aroma are ethyl 2-methylbutyrate, 1-hexanal, and trans-2-hexanal of which trans-2-hexanal was the most abundant volatile in the juice. The loss of trans-2-hexanal in apple juice was accelerated by increased temperature and operating time of the concentration process applied. Membrane-based concentration processes allowed to preserve trans-2-hexanal considerably, while a significant loss of trans-2-hexanal was observed in the concentration of apple juice by thermal evaporation technique. A higher retention of trans-2-hexanal was observed by Onsekizoglu et al. (2010) with the osmotic evapora-tion process (about 48%–52%) compared with MD process (about 46%–47%).

4.4   INTEGRATED MEMBRANE PROCESSES

Integrated membrane separation processes are gradually becoming consolidated sys-tems for the industrial production of highly concentrated fruit juice. Concentration of fruit juice with high solid and polysaccharide content by RO or OD results in a lower permeate flux due to an increase in the viscosity of concentration boundary layer near the membrane surface. Enzymatic treatment of fruit juice followed by membrane clarification by UF or MF treatment has been shown to decrease the levels of solid contents, polysaccharides, and proteins significantly, leading to a decrease in viscosity and turbidity of the clarified juice. If the concentration processes such as RO or OD are combined with enzymatic treatment, UF, MF, etc., the viscosity in the concentration boundary layer decreases resulting in an increase in permeate flux. Hence, studies have mainly focused on increasing the effectiveness of the concentration process in terms of permeate flux, retention of color, aroma compounds, and TAA by integration of other membrane processes, namely, UF and MF. Therefore, the benefits in terms of low energy consumption, preservation of aroma, nutritional value, and organoleptic properties of juice, quality improvement of final product, and increase in permeate flux could be achieved using integrated membrane process.

The combination of UF and RO as an integrated membrane process for the concentration of fruit juice was disclosed by Lawhon and Lusas (1987) in their patent. At the beginning, UF was used to clarify the juice. The permeate containing almost all the flavor and aroma characteristics was concentrated by RO more than 42 °Bx. The retentate containing all the SSs, pectins, and the spoilage microorganisms was subjected to heat treatment to inactivate the spoilage microorganisms and to improve the stability of the finished product when mixed with RO concentrate. The reconstituted juice was found to have a quality close to that of fresh juice. Walker (1990) illustrated the concept of two-stage RO system after MF or UF to produce concentrated juice (Figure 4.28). UF or MF was first used to clarify (to separate out the pulp from the serum) orange juice. The clarified juice was concentrated by RO (three units in series) with high-retention polyamide hollow fiber membranes (98.5% salt retention). The retentate leaving the final high-retention RO unit was further concentrated by low-retention RO membranes (two units in series, first with 93% and second with 97% salt retention). The finished product of 54 °Bx was obtained by blending the retentate from the second low-rejection unit having a sugar concentration of 63 °Bx with the pasteurized UF retentate.

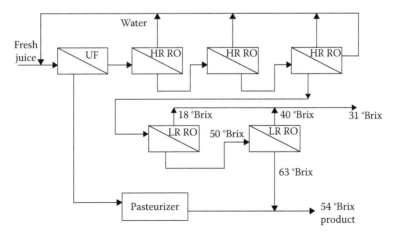

FIGURE 4.28 Schematic of the combined high-retention (HR)-low-retention (LR) RO process for juice concentration.

Yu and Chiang (1986) used the combination of UF and evaporation for concentrating passion fruit juice. Raw juice was clarified by UF after the pretreatment with pectinase, centrifugation, and pasteurization. UF-clarified passion fruit juice showed higher evaporation rate than juice with solid due to improved heat transfer. The final concentrated product was obtained after blending the evaporated concentrate with the UF retentate. The reconstituted juice and the fresh juice were not significantly different in terms of the desired components although the evaporation caused 20% flavor loss. Johnson (1993) studied a combined membrane-evaporation process for concentrating orange juice. He used UF membrane to separate SSs from the raw juice. The UF-clarified juice was then concentrated by a conventional evaporator. The UF retentate may be pasteurized by a heat exchanger. This combined membrane-evaporation technique can produce concentrated juice more than 80 °Bx, which may be used to formulate new fruit beverage juice blends. The combination of enzyme membrane reactor (EMR), RO, and pervaporation (PV) was successfully performed in laboratory and pilot plant unit for the production of apple juice and apple juice aroma concentrates (Alvarez et al., 2000). In the proposed scheme (Figure 4.29), the raw apple juice was clarified in an enzyme membrane reactor followed by the preconcentration of the clarified juice up to 25 °Bx in RO with the permeate flow of 75–110 L/h. Then the RO retentate was fed to the pervaporation unit to recover and concentrate the aroma compounds followed by a final evaporation step to concentrate apple juice up to 72 °Bx. More than 90% rejection of aroma

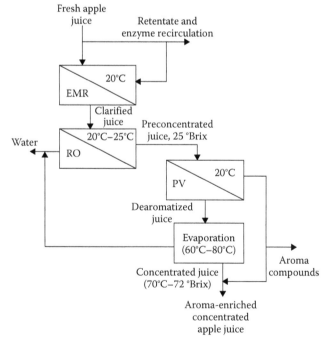

FIGURE 4.29 Integrated membrane process for the production of clarified apple juice concentrate and apple juice aroma. EMR, enzymatic membrane reactor; RO, reverse osmosis; PV, pervaporation. (From Alvarez, S. et al., *J. Food Eng.*, 46, 109, 2000.)

compounds was achieved for most of the compounds considered. They also reported that for the production of clarified apple juice, the use of membrane reactor could be handy in terms of lower enzyme cost, fouling, and cleaning.

An integrated RO–NF membrane system for concentrated fruit juice was proposed by Nabetani (1996). The fresh juice (10 °Bx) is initially concentrated to 30 °Bx with RO membranes and then the RO retentate is concentrated up to 45 °Bx using NF membranes. The NF permeate is recycled to the feed of RO unit. This system is shown to be suitable for concentrating various fruit juices, with advantages of not only retaining the fresh juice flavor but also of energy savings of 12.5% and 20% when compared with evaporation and freeze concentration, respectively. An integrated membrane process for producing blood orange juice was proposed by Galaverna et al. (2008). The process was based on the initial clarification of fresh blood orange juice by UF. Then the clarified orange juice was concentrated to a final value of 60 °Bx in two different configurations such as UF–RO–OD or UF–OD. The process, UF–RO–OD, in which RO, used as a preconcentration technique (up to 25–30 °Bx), followed by OD, up to a final concentration of about 60 °Bx and the process, UF–OD, in which the RO treatment was omitted. Results revealed that the value of TAA in the final product obtained by UF–OD was not significantly different from that obtained by UF–RO–OD. The final products had a brilliant red color with high antioxidant activity, large amounts of natural bioactive components, and aroma, characteristics that were significantly lost during conventional thermal evaporation. For the production of concentrated citrus and carrot juice with high nutrition value, a similar type of integrated membrane process scheme was used (Cassano et al., 2003) shown in Figure 4.30. The clarification of raw juice was first carried out by UF. Then the UF permeate was preconcentrated by RO; a further concentration of RO retentate was performed by OD. This juice was finally concentrated with 60–63 °Bx with high antioxidant activity.

Cassano et al. (2004) proposed an integrated membrane process by a sequence of UF and OD for the production of concentrated kiwifruit juice of high quality and high nutrition value. The raw juice after enzyme treatment was initially clarified by UF. Then the clarified juice was submitted to the concentration step by OD.

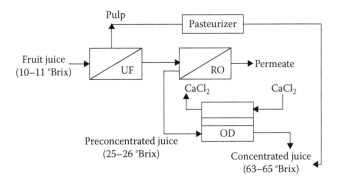

FIGURE 4.30 Integrated membrane processes for the clarification and concentration of citrus and carrot juices. (Reprinted from Cassano, A. et al., *J. Food Eng.*, 57, 153, 2003. With permission.)

TABLE 4.11

Analytical Measurement of Fresh Kiwifruit Juice and Samples
from UF and OD Treatments

Sample	TSS (°Bx)	pH	SS (w/w%)	Turbidity (NTU)	Ascorbic[a] Acid (mg/100 g)	TAA[a] (mM trolox)
Fresh juice	12.5	3.58	5.16	299.5	69.6	16.0
UF permeate	12.1	3.60	0	0	63.3	15.3
UF retentate	13.5	3.58	51.5	1336.7	62.8	15.6
OD retentate	65.8	3.40	0	0	69.6	14.1

Source: Reprinted from Cassano, A. et al., *Food Res. Int.*, 37(2), 139, 2004. With permission.
[a] Values referred to 12.5 °Bx. With permission.

The concentrated juice with a final concentration of TSS of 65.8 °Bx was achieved at an average throughput of about 1 kg/m² h. However, a little decrease in TAA was observed during the UF-OD operation, whereas the contribution of ascorbic acid to the TAA remained unchanged. The fresh kiwifruit juice and the samples obtained from UF and OD operation were characterized in terms of TSS, pH, SS, turbidity, ascorbic acid, and TAA, which are shown in Table 4.11.

Cassano et al. (2007) also studied the potentiality of a membrane-based process for the clarification and the concentration of the cactus pear fruit juice. The fresh juice, with a TSS content of about 11 °Bx, was initially clarified by a UF step, on laboratory scale, according to the batch concentration procedure at a TMP of 2.15 bar, an axial retentate flow rate of 18 L/h, and a temperature of 25°C ± 2°C. The clarified juice was then concentrated by OD up to TSS content of 61 °Bx at 28°C. An initial evaporation flux of 1.16 kg/m² h was obtained using a calcium chloride dehydrate solution at 60 w/w% as stripper. During the concentration by OD, the TAA of the juice remained almost constant for the various levels of the TSS achieved.

4.5 CONCLUSION

The evolution of the use of membrane applications in fruit processing has been reviewed. The process of clarification and concentration of fruit juice using a variety of membrane processes including types of membranes used, membrane fouling and its control, membrane modules, and existing models are discussed in detail. Different flux enhancement techniques, for example, electric field, pulsatile flows, etc., are introduced with applicative examples. The use of reverses osmosis (RO), direct osmosis concentration (DOC), membrane distillation (MD), and osmotic distillation (OD) for the concentration of fruit juice using different membrane processes are described along with the underlying physical principles for each of them; design features of each process and the effect of different process parameters on permeate flux and retention are discussed. Finally, the concepts of integrated membrane separation processes are introduced for the industrial production of highly concentrated fruit juice. The applications of membrane processes for fruit juice processing are legion.

REFERENCES

Alvarez, V., S. Alvarez, F.A. Riera, R. Alvarez, Permeate flux prediction in apple juice concentration by reverse osmosis, *J. Membr. Sci.*, 127 (1997) 25–34.

Alvarez, S., R. Alvarez, F.A. Riera, J. Coca, Influence of depectinization on apple juice ultrafiltration, *Colloids Surf. A Physicochem. Eng. Aspects*, 138 (1998a) 377.

Alvarez, V., L.J. Andres, F.A. Riera, R. Alvarez, Microfiltration of apple juice using inorganic membranes: Process optimization and juice stability, *Can. J. Chem. Eng.*, 74 (1996) 156.

Alvarez, S., F.A. Riera, R. Alvarez, J. Coca, Permeation of apple aroma compounds in reverse osmosis, *Sep. Purif. Technol.*, 14(1–3) (1998b) 209–220.

Alvarez, S., F.A. Riera, R. Alvarez, J. Coca, F.P. Cuperus, S. Bouwer, et al., A new integrated membrane process for producing clarified apple juice and apple juice aroma concentrate, *J. Food Eng.*, 46 (2000) 109.

Alves, V.D., I.M. Coelhoso, Orange juice concentration by osmotic evaporation and membrane distillation: A comparative study, *J. Food Eng.*, 74 (2006) 125.

Amar, R.B., B.B. Gupta, M.Y. Jaffrin, Apple juice clarification using mineral membranes: Fouling control by backwashing and pulsating flow, *J. Food Sci.*, 55 (1990) 1620.

Andersson, S.L., N. Kjellander, B. Rodesjo, Design and field tests of a new membrane distillation desalination process, *Desalination*, 56 (1985) 345.

Arroyo, G., C. Fonade, Use of intermittent jets to enhance flux in cross flow filtration, *J. Membr. Sci.*, 80 (1993) 117–129.

Bailey, A.F.G., A.M. Barbe, P.A. Hogan, R.A. Johnson, J. Sheng, The effect of ultrafiltration on the subsequent concentration of grape juice by osmotic distillation, *J. Membr. Sci.*, 164 (2000) 195.

Barbe, A.M., J.P. Bartley, A.L. Jacobs, R.A. Johnson, Retention of volatile organic flavour/fragrance components in the concentration of liquid foods by osmotic distillation, *J. Membr Sci.*, 145(1) (1998) 67–75.

Bargeman, G., J. Houwing, I. Recio, G.H. Koops, H.C. Van der Horst, Electromembrane filtration for the selective isolation of bio-active peptides from an αs2-casein hydrolysate, *Biotechnol. Bioeng.*, 80 (2002) 599.

Beaudry, E.G., K.A. Lampi, Membrane technology for direct osmosis concentration of fruit juices, *Food Technol.*, 44(6) (1990) 121.

Bellara, S.R., Z.F. Cui, D.S. Pepper, Gas sparging to enhance permeate flux in ultrafiltration using hollow fibre membranes, *J. Membr. Sci.*, 121 (1996) 175.

Blatt, W.F., A. Dravid, A.S. Michaels, L. Nelson, Solute polarization and cake formation in membrane ultrafiltration: Causes, consequences and control techniques, in: J.E. Flinn (Ed.), *Membrane Science and Technology*, Plenum Press, New York, 1970.

Bowen, W.R., R.S. Kingdon, H.A.M. Subuni, Electrically enhanced separation processes: The basis of in situ intermittent electrolytic membrane cleaning (IIEMC) and in situ electrolytic membrane restoration (IEMR), *J. Membr. Sci.*, 40 (1989) 219.

Bowen, W.R., H.A.M. Subuni, Pulsed electrokinetic cleaning of cellulose nitrate microfiltration membrane, *Ind. Eng. Res.*, 31 (1992) 515.

Bowen, W.R., H.A.M. Subuni, Electrically enhanced membrane filtration at low cross-flow velocities, *Ind. Eng. Res.*, 30 (1991) 1573–1579.

Calabro, V., B. Jiao, E. Drioli, Theoretical and experimental study on membrane distillation in the concentration of orange juice, *Ind. Eng. Chem. Res.*, 33 (1994) 1803.

Carlsson, L., The new generation in sea water desalination: SU membrane distillation system, *Desalination*, 45 (1983) 221.

Cassano, A., E. Drioli, G. Galaverna, R. Marchelli, G. Di Silvestro, P. Cagnasso, Clarification and concentration of citrus and carrot juices by integrated membrane processes, *J. Food Eng.*, 57 (2003) 153.

Cassano, A., B. Jiao, E. Drioli, Production of concentrated kiwifruit juice by integrated membrane processes, *Food Res. Int.*, 37(2) (2004) 139.

Cassano, A., C. Conidi, R. Timpone, M. D'Avella, E. Drioli, A membrane-based process for the clarification and the concentration of the cactus pear juice, *J. Food Eng.*, 80 (2007) 914–921.

Cheng, D.Y., S.J. Wiersma, Composite membrane for a membrane distillation system, U.S. Patent 4,316,772 (1982).

Cheryan, M., *Ultrafiltration and Microfiltration Handbook,* Technomic Publishing Company, Lancaster, PA, 1998, p. 83.

Chiang, B.H., Z.R. Yu, Fouling and flux restoration of ultrafiltration of passion fruit juice, *J. Food Sci.*, 52 (1987) 369.

Chou, F., R.C. Wiley, D.V. Schlimme, Reverse osmosis and flavor retention in apple juice concentration, *J. Food Sci.*, 56 (1991) 484.

Chua, H.T., M.A. Rao, T.E. Acree, D.H. Cunningham, Reverse osmosis concentration of apple juice: Flux and flavor retention by cellulose acetate and polyamide membranes, *J. Food Process Eng.*, 9 (1988) 231.

Constenla, D.T., J.E. Lozano, Predicting stationary permeate flux in the ultrafiltration of apple juice, *Lebensmittel-Wissenschaftund Technologie*, 29 (1996) 587.

Constenla, D.T., J.E. Lozano, G.H. Crapiste, Thermophysical properties of clarified apple juice as a function of concentration and temperature, *J. Food Sci.*, 54 (1989) 663.

Courel, M., M. Dornier, J.M. Herry, G.M. Rios, M. Reynes, Effect of operating conditions on water transport during the concentration of sucrose solutions by osmotic distillation, *J. Membr. Sci.*, 170 (2000) 281.

Cui, Z.F., Experimental investigation on enhancement of cross-flow ultrafiltration with air sparg-ing, in: R. Patterson (Ed.), *Proceedings of the 3rd international conference on Effective Membrance Processes-New Perspectives*, Mech. Eng. Pub. Ltd., London, U.K., May 12–14, 1993, p. 237.

Cui, Z.F., K.I.T. Wright, Gas-liquid two-phase cross-flow ultrafiltration of dextrans and BSA solution, *J. Membr. Sci.*, 90 (1994) 183.

Cui, Z.F., K.I.T. Wright, D.H. Glass, Experimental evidence on the mechanism of enhancement of ultrafiltration with gas sparging, in: *Proc. 1996 IChemE Research Event*, Leeds, U.K., 1996, p. 304.

Curcio, E., G. Barbieri, E. Drioli, Operazioni di distillazione a membrana nella concentrazione dei succhi di frutta, *Industrie delle Bevande*, XXIX(April) (2000) 113–121.

de Barros, S.T.D., C.M.G. Andrade, E.S. Mendes, L. Peres, Study of fouling mechanism in pineapple juice clarification by ultrafiltration, *J. Membr. Sci.*, 215 (2003) 213.

Ding, L., J.M. Laurent, M.Y. Jaffrin, Dynamic filtration of blood: A new concept for enhancing plasma filtration, *Hemapheresis*, 14 (1991) 365.

Drioli, E., B. Jiao, V. Calabro, The preliminary study on the concentration of orange juice by membrane distillation, in: *Proceedings of VII International Citrus Congress*, Acireale, Italy, March 8–13, 1992.

Durham, R.J., M.H. Nguyen, Hydrophobic membrane evaluation and cleaning for osmotic distillation of tomato puree, *J. Membr. Sci.*, 87 (1994) 181.

Fane, A.G., C.J.D. Fell, A review of fouling and fouling control in ultrafiltration, *Desalination*, 62 (1987) 117.

Fane, A.G., K.J. Kim, Prospects for improved ultrafiltration membranes, in: *Proceedings IMTEC'88, International Membrane Technology Conference*, November 15–17, 1988, Sydney, New South Wales, Australia, pp. K10–K14.

Field, R.W., D. Wu, J.A. Howell, B.B. Gupta, Critical flux concept for microfiltration fouling, *J. Membr. Sci.*, 100 (1995) 259.

Field, R.W., D. Wu, J.A. Howell, B.B. Gupta, Critical flux concept for microfiltration fouling, *J. Membr. Sci.*, 100 (1995b) 250.

Finnigan, S.M., J.A. Howell, The effect of pulsed flow on ultrafiltration fluxes in a baffled tubular membrane system, *Desalination*, 79(2–3) (1990) 181.

Flath, R.A., D.R. Black, D.G. Guadagni, W.H. Mcfadden, T.H. Schultz, Identification and organoleptic evaluation of compounds in delicious apple essence, *J. Agric. Food Chem.*, 15 (1967) 29.

Galaverna, G., G. Di Silvestro, A. Cassano, S. Sforza, A. Dossena, E. Drioli, R. Marchelli, A new integrated membrane process for the production of concentrated blood orange juice: Effect on bioactive compounds and antioxidant activity, *Food Chem.*, 106 (2008) 1021.

Garaldes, V., V. Semiao, M.N. de Pinho, Flow management in nanofiltration spiral wound modules with ladder-type spacers, *J. Membr. Sci.*, 203(1–2) (2002) 87.

Ghosh, A.M., M. Balakrishnan, M. Dua, J.J. Bhagat, Ultrafiltration of sugarcane juice with spiral wound modules: On-site pilot trials, *J. Membr. Sci.*, 174 (2000) 205.

Girard, B., L. Fukumoto, Membrane processing of fruit juices and beverages: A review. *Crit. Rev. Biotechnol.*, 20(2) (2000a) 109.

Girard, B., L.R. Fukumoto, Membrane processing of fruit juices and beverages: A review, *Crit. Rev. Food Sci. Nutr.*, 40 (2000b) 91.

Gladdon, J.K., M. Dole, Diffusivity determination of sucrose and glucose solutions, *J. Am. Chem. Soc.*, 75 (1953) 3900.

Gore, D.W., Gore-Tex membrane distillation, *Proceedings of the Tenth Annual Convention of the Water Supply Improvement Association*, Honolulu, Hl, July 25–29, 1982.

Gostoli, C., Thermal effects in osmotic distillation, *J. Membr. Sci.*, 163(1) (1999) 75–91.

Gostoli, C., G.C. Sarti, S. Matulli, Low temperature distillation through hydrophobic membranes, *Sep. Sci. Technol.*, 22 (1987) 855.

Gryta, M., Effect of iron oxides scaling on the MD process performance, *Desalination*, 216(1–3) (2007) 88–102.

Guizard, C., F. Legault, N. Idrissi, A. Larbot, L. Cot, C. Gavach, Electronically conductive mineral membranes designed for electroultrafiltration, *J. Membr. Sci.*, 41 (1989) 127–142.

Gunko, S., S. Verbych, M. Bryk, N. Hilal, Concentration of apple juice using direct contact membrane distillation, *Desalination*, 190 (2006) 117.

Gupta, B.B., P. Blanpain, M.Y. Jaffrin, Permeate flux enhancement by pressure and flow pulsations in microfiltration with mineral membranes, *J. Membr. Sci.*, 70 (1992) 257–266.

He, Y., Z. Ji, S. Li, Effective clarification of apple juice using membrane filtration without enzyme and pasteurization pretreatment, *Sep. Purif. Technol.*, 57 (2007) 366.

Hermans, P.H., H.L. Bredee, Principles of the mathematical treatment of constant-pressure filtration, *J. Soc. Chem. Ind.*, 55T (1936) 1.

Hérmia, J., Constant pressure blocking filtration laws. Applications to power-law non-Newtonian fluids, *Trans. IChemE*, 60 (1982) 183.

Herron, J.R., E.G. Beaudry, C.E. Jochums, L.E. Medina, Osmotic concentration apparatus and method for direct osmotic concentration of fruit juice, U.S. Patent 5,281,430 (1994) January 25.

Ho, C.C., A.L. Zydney, A combined pore blockage and cake filtration model for protein fouling during microfiltration, *J. Colloid Interf. Sci.*, 232 (2000) 389.

Hofmann, R., C. Posten, Improvement of dead-end filtration of biopolymer with pressure electrofiltration, *Chem. Eng. Sci.*, 58 (2003) 3847–3858.

Howell, A., V. Sanchez, R.W. Fieldeds, *Membrane in Bioprocessing: Theory and Applications*, Chapman & Hall, London, U.K., 1993.

Hwang, S.T., K. Kammermeyer, *Membranes in Separations*, Wiley-Interscience, New York, 1975.

Hwang, K.J., C.Y. Liao, K.L. Tung, Analysis of particle fouling during microfiltration by use of blocking models, *J. Membr. Sci.*, 287 (2007) 287.

Jiao, B., A. Cassano, E. Drioli, Recent advances on membrane processes for the concentration of fruit juices: A review, *J. Food Eng.*, 63 (2004) 303.

Jiraratananon, R., A. Chanachai, A study of fouling in the ultrafiltration of passion fruit juice, *J. Membr. Sci.*, 111 (1996) 39.

Jiraratananon, R., D. Uttapap, C. Tangamornsuksun, Selfforming dynamic membrane for ultrafiltration of pineapple juice, *J. Membr. Sci.*, 129 (1997) 135.

Johnson, J.R., Technical and economical feasibility of a nonconventional method for concentrating orange juice, PhD thesis, University of Florida, Gainesville, FL, 1993.

Kilara, A., J.P. Van Buren, Clarification of apple juice, in: D.L. Downing (Ed.), *Processed Apple Products*, Van Nostrand Reinhold, New York, 1989, pp. 83–96, Chapter 4.

Kim, B.S., H.N. Chang, Effects of periodic backflushing on ultrafiltration performance, *Bioseparations*, 2 (1991) 23.

Kimura, S., S.I. Nakao, S. Shimatani, Transport phenomena in membrane distillation. *J. Membr. Sci.*, 33(3) (1987) 285.

Kimura, S., S. Sourirajan, Analysis of data in reverse osmosis with porous cellulose acetate membranes, *AIChE J.*, 13 (1967) 497.

Kobayashi, T., X. Chai, N. Fujii, Ultrasound enhanced cross-flow membrane filtration, *Sep. Purif. Technol.*, 17 (1999) 31.

Kokugan, T., S. Kaseno, S. Fufiwara, M. Shimizu, Ultrasonic effect on ultrafiltration properties of ceramic membrane, *Membrane*, 20 (1995) 213.

Kozak, A., E.B. Molnar, G. Vatai, Production of black-currant juice concentrate by using membrane distillation, *Desalination*, 241 (2009) 309.

Lagan, F., G. Barbieri, E. Drioli, Direct contact membrane distillation: Modelling and concentration experiments, *J. Membr. Sci.*, 166(1) (2000) 1.

Lakshminarayanaiah, N., *Equations of Membrane Biophysics*, Academic Press, New York, 1984.

Lamminen, M.O., H.W. Walker, L.K. Weavers, Mechanisms and factors influencing the ultrasonic cleaning of particle-fouled ceramic membranes, *J. Membr. Sci.*, 237 (2004) 213.

Lawhon, J.T., E.W. Lusas, Method of producing sterile and concentrated juices with improved flavor and reduced acid, U.S. Patent 4,643,902, 7 September, 1987.

Lawson, K.W., D.R. Lloyd, Review membrane distillation, *J. Membr. Sci.*, 124 (1997) 1–25.

Lee, S., M. Elimelech, Salt cleaning of organic-fouled reverse osmosis membranes, *Water Res.*, 41 (2007) 1134.

Li, Q.Y., R. Ghosh, S.R. Bellara, Z.F. Cui, D.S. Pepper, Enhancement of ultrafiltration by gas sparging with flat sheet membrane modules, *Sep. Purif. Technol.*, 14 (1998) 79.

Loeb, S., L. Titelman, E. Korngold, J. Freiman, Effect of porous support fabric on osmosis through a Loeb-Sourirajan type asymmetric membrane, *J. Membr. Sci.*, 129 (1997) 243–249.

Lonslade, H.K., Theory and practice of reverse osmosis and ultrafiltration, in R.E. Lacey and S. Loeb (Eds.), *Industrial Processing With Membranes*, Wiley-Interscience, New York, 1972.

Matsumoto, Y., T. Miwa, S.-I. Nakao, S. Kimura, Improvement of membrane permeation performance by ultrasonic microfiltration, *J. Chem. Eng. Jpn.*, 29(4) (1996) 561.

Matsuura, T., A.G. Baxter, S. Sourirajan, Studies on reverse osmosis for concentration of fruit juice, *J. Food Sci.*, 39 (1974) 704.

Matsuura, T., S. Sourirajan, Physicochemical criteria for reverse osmosis separation of monohydric and polyhydric alcohols and some related hydroxy compounds in aqueous solutions using porous cellulose acetate membranes, *J. Appl. Polym. Sci.*, 17 (1973) 1043.

Merson, R.L., A.I. Morgan, Juice concentration by reverse osmosis, *Food Technol.*, 22 (1968) 631.

Mondal, S., S. De, Generalized criteria for identification of fouling mechanism under steady state membrane filtration, *J. Membr. Sci.*, 344 (2009) 6.

Moulik, S.P., Normal and reverse field induced flow behaviours in electrofiltration, *Colloid Polym. Sci.*, 254 (1976) 39–44.

Muthukumaran, S., K. Yang, A. Seuren, S. Kentish, M. Ashok-kumar, G.W. Stevens, F. Grieser, The use of ultrasonic cleaning for ultrafiltration membranes in the dairy industry, *Sep. Purif. Technol.*, 39 (2004) 99.

Nabetani, H., Development of a membrane system for highly concentrated fruit juice, *J. Membr. (Japanese)*, 21(2) (1996) 102.

Nene, S., S. Kaurb, K. Sumod, B. Joshi, K.S.M.S. Raghavarao, Membrane distillation for the concentration of raw cane-sugar syrup and membrane clarified sugarcane juice, *Desalination*, 147 (2002) 157.

Onsekizoglu, P., K.S. Bahceci, M.J. Acar, Clarification and the concentration of apple juice using membrane processes: A comparative quality assessment, *J. Membr. Sci.*, 352(1–2) (2010) 160.

Oussedik, S., D. Belhocine, H. Grib, H. Lounici, D.L. Piron, N. Mameri, Enhanced ultrafiltration of bovine serum albumin with pulsed electric field and fluidized activated alumina, *Desalination*, 127 (2000) 59–68.

Pal, S., R. Bharihoke, S. Chakraborty, S.K. Ghatak, S. De, S. DasGupta, An experimental and theoretical analysis of turbulence promoter assisted ultrafiltration of synthetic fruit juice, *Sep. Purif. Technol.*, 62(3) (2008) 259.

Park, Y.G., Effect of electric field during purification of protein using microfiltration, *Desalination* 191 (2006) 404.

Peng, P., A.G. Fane, X.D. Li, Desalination by membrane distillation adopting a hydrophilic membrane, *Desalination*, 173 (2005) 45.

Petrotos, K.B., H.N. Lazarides, Osmotic concentration of liquid foods, *J. Food Eng.*, 49(2–3) (2001) 201.

Petrotos, K.B., P.C. Quantick, H. Petropakis, A study of the direct osmotic concentration of tomato juice in tubular membrane module configuration. I. The effect of certain basic process parameters on the process performance, *J. Membr. Sci.*, 150(1) (1998) 99.

Popper, K., W.M. Camirand, F. Nury, W.L. Stanley, Dialyzer concentrates beverages, *Food Eng.*, 38(4) (1966) 102.

Radovich, J.M., B. Behnam, C. Mullon, Steady state modeling of electro-ultrafiltration at constant concentration, *Sep. Sci. Technol.*, 20(4) (1985) 315–329.

Radovich, J.M., R.E. Sparks, Electrophoretic techniques for controlling concentration polarization in ultrafiltration, *Polym. Sci. Technol.*, 13 (1980) 249–268.

Rai, P., G.C. Majumdar, S. DasGupta, S. De, Modelling of Sucrose permeation through pectin gel during ultrafiltration of depectinized mosambi (*Citrus Sinensis* (L.) Osbeck) juice, *J. Food. Sci.*, 71(2) (2006a) E87.

Rai, P., G.C. Majumdar, G. Sharma, S. DasGupta, S. De, Effect of various cutoff membranes on permeate flux and quality during filtration of mosambi (*Citrus sinensis* (L.) Osbeck) juice, *Food Bioprod. Processing*, 84 (2006b) 213.

Ravindra Babu, B., N.K. Rastogi, K.S.M.S. Raghavarao, Effect of process parameters on transmembrane flux during direct osmosis, *J. Membr. Sci.*, 280 (2006) 185.

Riedl, K., B. Girard, R.W. Lencki, Influence of membrane structure on fouling layer morphology during apple juice clarification, *J. Memb. Sci.*, 139 (1998) 155.

Robinson, C.W., M.H. Siegel, A. Condemine, C. Fee, T.Z. Fahidy, B.R. Glick, Pulsed-electric-field cross flow ultrafiltration of bovine serum albumin, *J. Membr. Sci.*, 80 (1993) 209.

Sabri, N., P. Pirkonen, H. Sekki, *Proceedings of the 2nd Applications of Power Ultrasound in Physical and Chemical Processing*, Toulouse, France, 1997, p. 99.

Sarkar, B., S. DasGupta, S. De, Effect of electric field during gel-layer controlled ultrafiltration of synthetic and fruit juice, *J. Membr. Sci.*, 307 (2008a) 268.

Sarkar, B., S. DasGupta, S. De, Flux decline during electric field assisted cross flow ultrafiltration of mosambi (*Citrus sinensis* (L.) Osbeck) juice, *J. Membr. Sci.*, 331(1–2) (2009) 75.

Sarkar, B., S. De, Electric field enhanced gel controlled cross-flow ultrafiltration under turbulent conditions, *Sep. Purif. Technol.*, 74 (2010) 73–82.

Sarkar, B., S. De, S. DasGupta, Pulsed electric field enhanced ultrafiltration of synthetic and fruit juice, *Sep. Purif. Technol.*, 63 (2008b) 582.

Sarkar, B., S. Pal, T.B. Ghosh, S. De, S. DasGupta, A study of electric field enhanced ultrafiltration of synthetic fruit juice and optical quantification of gel deposition, *J. Membr. Sci.*, 311 (2008c) 112.

Shaw, P.E., M. Lebrun, M. Dornier, M.N. Ducamp, M. Courel, M. Reynes, Evaluation of concentrated orange and passionfruit juices prepared by osmotic evaporation, *Lebensmittel-Wissen und-Technologie*, 34(2) (2001) 60–65.

Sheng, J., R.A. Johnson, M.S. Lefebvre, Mass and heat transfer mechanisms in the osmotic distillation process, *Desalination*, 80 (1991) 113.

Sheu, M.J., R.C. Wiley, Preconcentration of apple juice by reverse osmosis, *J. Food Sci.*, 48 (1983) 422.

Sheu, M.J., R.C. Wiley, D.V. Schlimme, Solute and enzyme recoveries in apple juice clarification using ultrafiltration, *J. Food Sci.*, 52 (1987) 732.

Su, S.K., J.C. Liu, R.C. Wiley, Cross-flow microfiltration with gas backwash of apple juice, *J. Food Sci.*, 58 (1993) 638.

Sulaiman, M.Z., N.M. Sulaiman, L.S. Yih, Limiting permeate flux in the clarification untreated starfruit juice by membrane ultrafiltration, *Chem. Eng. J.*, 69 (1998) 145.

Tarleton, E.S., R.J. Wakeman, Microfiltration enhancement by electrical and ultrasonic force fields, *Filtr. Sep.*, 27 (1990) 192.

Thakur, B.R., R.K. Singh, A.K. Handa, Chemistry and uses of pectin—A review, *Crit. Rev. Food Sci. Nutr.*, 37(1) (1997) 47.

Thussen, H.A.C., Concentration processes for liquid foods containing volatile flavors and aromas, *J. Food Technol.*, 5 (1970) 211.

Todisco, S., P. Tallarico, E. Drioli, Modelling and analysis of the effects of ultrafiltration on the quality of freshly squeezed orange juice, *Ital. Food Bev. Technol.*, 12 (1998) 3–8.

Vaillant, F., E. Jeanton, M. Dornier, G.M. O'Brien, M. Reynes, M. Decloux, Concentration of passion fruit juice on an industrial scale using osmotic evaporation, *J. Food Eng.*, 47 (2001) 195.

Vladisavljevi'c, G.T., P. Vukosavljevi'c, B. Bukvi'c, Permeate flux and fouling resistance in ultrafiltration of depectinized apple juice using ceramic membranes, *J. Food Eng.*, 60 (2003) 241–247.

Wakeman, R.J., Electrically enhanced microfiltration of albumin suspensions, *Trans. IChemE* 76(C) (1998) 53–59.

Wakeman, R.J., E.S. Tarleton, Membrane fouling prevention in crossflow microfiltration by the use of electric fields, *Chem. Eng. Sci.*, 42 (1987) 829.

Walker, J.B., Membrane process for the production of superior quality fruit juice concentrate, in: *Proceedings of the 1990 International Congress on Membrane and Membrane Processes*, Chicago, IL, 1990.

Weigert, T., J. Altmann, S. Ripperger, Crossflow electrofiltration in pilot scale, *J. Membr. Sci.*, 159 (1999) 253.

Wrolstad, R.E., M.R. McDaniel, R. Durst, N. Michaels, K.A. Lampi, E.G. Beaudry, Composition and sensory characterization of red raspberry juice concentrated by direct-osmosis or evaporation, *J. Food Sci.*, 58(3) (1993) 633.

Wu, M.L., R.R. Zall, W.C. Tzeng, Microfiltration and ultrafiltration comparison for apple juice clarification, *J. Food Sci.*, 55 (1990) 1162.

Xu, J.B., S. Lange, J.P. Bartley, R.A. Johnson, Alginate-coated microporous PTFE membranes for use in the osmotic distillation of oily feeds, *J. Membr. Sci.*, 240 (2004) 81.

Youravong, W., Z. Li, A. Laorko, Influence of gas sparging on clarification of pineapple wine by microfiltration, *J. Food Eng.*, 96 (2010) 427.

Yu, Z.R., B.H. Chiang, Passion fruit juice concentration by ultrafiltration and evaporation, *J. Food Sci.*, 51(6) (1986) 1501.

Yuan, W., A. Kocic, A.L. Zydney, Analysis of humic acid fouling during microfilake filtration model, *J. Membr. Sci.*, 198 (2002) 51.

Yukawa, H., K. Shimura, A. Maniwa, Characteristics of cross flow electro-ultrafiltration for colloidal solution of protein, *J. Chem. Eng. Jpn.*, 16(3) (1983) 246.

Zumbusch, P.V., W. Kulcke, G. Brunner, Use of alternating electrical fields as anti-fouling strategy in ultrafiltration of biological suspensions—Introduction of a new experimental procedure for cross flow filtration, *J. Membr. Sci.*, 142 (1998) 75–86.

5 High-Intensity Pulsed Electric Field Applications in Fruit Processing

Ingrid Aguiló-Aguayo, Pedro Elez-Martínez,
Robert Soliva-Fortuny, and Olga Martín-Belloso

CONTENTS

5.1 INTRODUCTION

Nonthermal processes have gained importance in recent years due to the increasing consumer demand for food with a high nutritional value and "fresh-like" characteristics, representing an alternative to conventional thermal treatments. High-intensity pulsed electric fields (HIPEF) is a novel nonthermal technology that has been extensively studied for food processing. The application of electrical current for microbial inactivation and food treatment dates back to the beginning of the past century, but most of the advances were made in its last decade (Toepfl et al., 2005). First HIPEF research studies were carried out in buffer or modeling solutions and later in food matrices such as milk and fruit juices. Food-related applications of HIPEF have been focused on attaining microbial and enzyme inactivation. The potential of HIPEF for food preservation is investigated based on the impact of processing parameters in obtaining safe high-quality HIPEF-treated fluid foods with an acceptable shelf life. Successful results have evidenced the suitability of HIPEF treatments at electric field strengths higher than 10 kV/cm in processing several fruit juices since a high level of microbial destruction with little losses of flavor, color, taste, or nutrients has been achieved (Zhang et al., 1994; Barbosa-Cánovas et al., 1999; Barsotti and Chefter, 1999; Yeom et al., 2000a; Mosqueda-Melgar et al., 2008b; Aguiló-Aguayo et al., 2009a). Moreover, HIPEF has also received attention in the area of reversible and irreversible plant membrane permeabilization (Dörnerburg and Knorr, 1993; Knorr et al., 1994; Angersbach et al., 2000, 2002). The use of mild treatments (<10 kV/cm) as a pretreatment in food processing offers the potential as a complement to other processes (Olajide et al., 2006). Thus, the recovery and production of high-value metabolites (Brodelius et al., 1988; Dörnerburg and Knorr, 1993; Eshtiaghi and Knorr, 2002; Fincan et al., 2004), the improvement of fruit and vegetable juices yield (McLellan et al., 1991; Knorr, 1994; Bouzrara and Vorobiev, 2000; Ade-Omowaye et al., 2001a; Rastogi, 2003), and the acceleration of mass transport in drying processes (Angersbach and Knorr, 1997; Rastogi et al., 2000; Ade-Omowaye et al., 2001b; Taiwo et al., 2001; Tedjo et al., 2002) are some of the food processing areas in which HIPEF could offer great advantages.

5.1.1 HIPEF Technology

HIPEF involves the application of high-voltage energy (typically 0.5–80 kV/cm) in the form of very short pulses (μs to ms). HIPEF treatments are conducted at ambient, subambient, or slightly above ambient temperature, leading to a minimum heating of the product (Olajide et al., 2006). The principles of HIPEF treatments in food processing are explained through electroporation. When a large flux of electrical current flows through a food, the polarization of dipole molecules and bulk movement of charge carriers (such as ions) induce a capacitive and resistive current (Riley and Watson, 1987). This causes structural damages of the cell membrane by electric potential differences between cell plasma and the extracellular medium, inducing the formation of pores (Zimmermann, 1986; Ho and Mittal, 1996). Depending on the electric field strength applied, the induced membrane breakdown and subsequent permeabilization can be reversible or irreversible (Benz and Zimmermann, 1980). If critical field strength is applied for sufficient time, irreversible electroporation takes place, resulting in cell membrane disintegration as well as loss of cell viability, leading to the inactivation of microorganisms (Figure 5.1). However, when the electric field strength is applied below those critical field strength values, reversible permeabilization occurs allowing the cell membrane to recover its structure and functionality (Zimmermann, 1986).

Permeabilization induces an increase in the mass and heat transfer rates between the cells and their surroundings, enhancing the efficiency of subsequent processes such as fermentation, expression, extraction, and diffusion of plant metabolites (Knorr et al., 2001; Olajide et al., 2006). On the other hand, critical electric field strength to induce membrane permeabilization in cell cultures or plant systems is

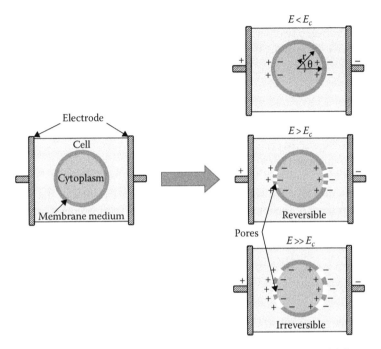

FIGURE 5.1   Biological cell in an electric field. E, electroporated area, which is represented by a dashed line; E_c, critical electric field strength. (From Donsì, F. et al., *Food Eng. Rev.,* 2, 109, 2010.)

also dependent on cell geometry and size (Neumann, 1996; Heinz et al., 2002). According to Heinz et al. (2002), critical electric field strength increases when the characteristic dimension of the cell is shifted to smaller values. Since the cell is rather large in plant tissues (~100 μm) when compared to microbial cells (~1–10 μm), the electric field strength required for electroplasmolysis in plant cells (0.5–5 kV/cm) is lower than that required for inactivation of microorganisms (10–50 kV/cm) (Barbosa-Cánovas et al., 1999; Knorr, 1999). Therefore, the required field strength must be high enough to trigger electroporation when applying HIPEF for food preservation, but moderate fields suffice to induce stress reactions and facilitate drying processes or the extraction of valuable components (Soliva-Fortuny et al., 2009).

5.1.2   HIPEF PARAMETERS

Technical factors including equipment design (batch or continuous systems) and operation characteristics are determinant extrinsic factors in the effectiveness of processing foods by HIPEF; however, both physical and chemical properties as well as the composition of the medium are also relevant intrinsic factors (Barbosa-Cánovas et al., 1999).

The main electrical parameters of HIPEF treatments reported as influential processing factors are electric field strength, treatment time, pulse width, pulse shape, pulse frequency, and pulse polarity (Martín-Belloso and Elez-Martínez, 2005;

Toepfl et al., 2005; Min et al., 2007; Aguiló-Aguayo et al., 2008a). When voltage is generated between two parallel-plate electrodes assuming that the permittivity is constant across the sample, the electric field strength (E) is defined as

$$E = \frac{V}{d},$$ (5.1)

where
  V is the voltage across the food sample
  d is the distance (m) between the electrodes

Capacitive and resistive currents are induced by polarization of the dipoles and the bulk movement of the ions when applying HIPEF. The load resistance (R) of the product in the treatment chamber is expressed by

$$R = \frac{\rho \cdot d}{S} = \frac{d}{\sigma \cdot S},$$ (5.2)

where
  ρ (Ωm) and σ (S/m) are the resistivity and the conductivity of the product, respectively
  S is the surface of the electrode (m^2)

The total duration of treatment time (t) is given by

$$t = n \cdot \tau,$$ (5.3)

where
  n is the number of pulses
  τ is the pulse width (μs)

Thus, the treatment time increases either with the number of pulses or with the pulse duration.

Energy may be applied in the form of exponentially decaying or square wave pulses.

Square-wave pulse appears to be more energy efficient and lethal than wave pulsed shape (De Haan and Willcock, 2002; Góngora-Nieto et al., 2002; Kotnik et al., 2003). Both pulse shapes can be applied in a mono- or bipolar fashion, which means only positive pulses (with respect to the ground level) or alternating positive and negative pulses, respectively (Ho et al., 1995). Bipolar pulses offer the advantage of reducing the deposition of solids on the electrodes of the treatment chamber and limiting the possibility of electrolytic reactions and resulting in a more uniform HIPEF treatment (Qin et al., 1994; Zhang et al., 1995). Moreover, the repeated changing in polarity causes movement of charges in the membrane, causing a mechanical oscillation that results in alternating stress and enhanced susceptibility to electric breakdown of cells (Ho and Mittal, 1996).

Another important parameter that has a role in determining the efficiency of the treatment is the pulse frequency defined through the number of pulses transmitted per second. Frequency defines the residence time of the product in the treatment chamber once the pulse width and treatment time are set (Hülsheger et al., 1981). Thus, a high pulse repetition frequency is preferred in order to apply more electric energy into the discharge.

5.1.3 HIPEF Equipment

A HIPEF food processing system consists of a treatment chamber, a pulse generator, and monitoring systems (Min et al., 2007). In general, a typical HIPEF generation schematic consists of a high-voltage power supply to charge the capacitors and a discharge switch that releases the stored electric energy from the capacitor(s) in the form of an electric field through the product (Vega-Mercado et al., 1997) (Figure 5.2).

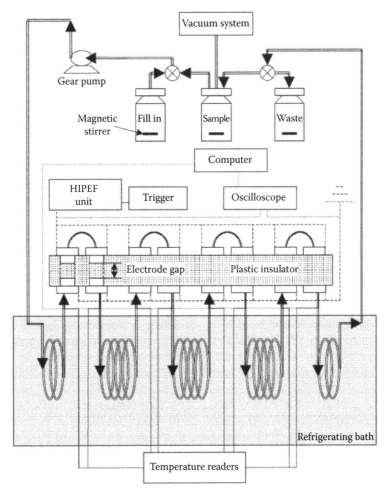

FIGURE 5.2 Diagram of the HIPEF bench-scale processing unit (OSU-4F). (From Elez-Martínez, P. et al., *J. Sci. Food Agric.,* 86, 71, 2006a.)

Treatment chambers contain the food material during HIPEF processing and house two discharging electrodes that deliver high-voltage energy to the food.

An insulating material holds the electrodes in fixed positions and forms the chamber containing the food to be processed. The insulator acts as a spacer between the electrodes in order to reduce electrochemical attack (Zhang et al., 1995; Mittal and Griffiths, 2005). A proper design of the treatment chamber is essential for the efficiency of the HIPEF processing (Alkhafaji and Farid, 2007). Materials selected to construct the treatment chamber need to be washable and autoclavable (Huang and Wang, 2009). Configuration of the treatment chambers should minimize the dielectric breakdown, which consists of the generation of sparks inside the chamber (Barbosa-Cánovas et al., 1999).

The treatment chamber can be either static (batch) or continuous. Common electrode configurations used in treatment chambers are parallel plates and coaxial or collinear cylinders. A large number of studies have been performed with parallel-plate systems in static and later on in continuous flow operation. In general, uniform electric field strength distribution can be achieved by parallel-plate geometry of the electrodes that should not oversize the electrode surface dimension (Dunn and Pearlmann, 1987; Zhang et al., 1995; Angersbach and Knorr, 1997; Rastogi et al., 1999; Angersbach et al., 2002; Fincan et al., 2004). On the other hand, electrical current flows perpendicular to the food flow in the coaxial configuration and in parallel to the food flow in the cofield electrodes (Zhang et al., 1995; Dunn, 2001). Moreover, disk-shaped and round-edged electrodes can minimize the electric field enhancement, also reducing the possibility of dielectric breakdown (Zhang et al., 1995).

Static chambers are mainly suitable for laboratory use to evaluate the influence of any relevant parameter critical to the process efficiency. Small volumes of treatment media are required, and treatment temperature is easy to maintain by cooling the electrodes and by slow repetition rates (Toepfl et al., 2005). Several static treatment chambers have been proposed for processing several foodstuffs (Sale and Hamilton, 1967; Dunn and Pearlman, 1987; Mizuno and Hori, 1988; Grahl et al., 1992; Zheng-Ying and Yan, 1993; Zhang et al., 1995; Angersbach and Knorr, 1997; Angersbach et al., 2000; Bazhal and Vorobiev, 2000; Fincan and Dejmek, 2002). The design of the static chamber must facilitate sample filling and removal. The expelling of air during filling must be completed since gas bubble is a potential trigger of dielectric breakdown. On the other hand, large-scale operations seem to be performed much better in continuous than in static chambers. Multiple treatment zones in line or baffled flow channels with fluid handling systems are used to allow the continuous treatment of liquid food without interruption (Zhang et al., 1995).

Pulse generators convert low voltage into high voltage and provide the high voltage to HIPEF chambers (Qin et al., 1994). Exponential decay or square pulses are generally used for HIPEF treatments. The generation of pulsed electric fields requires a fast discharge of electrical energy within a short period of time. The electric power is most commonly stored in a bank of capacitors connected in series or parallel and discharged into the treatment chamber across the high voltage switch and protective resistors within microseconds (Toepfl et al., 2005). On the other hand, temperature- and pulse-monitoring systems are the main equipment used to supervise HIPEF processing. Temperature is monitored with thermocouples, while pulses are monitored with high-voltage probes, current monitors, and oscilloscopes (Min et al., 2003a).

5.1.4 ENERGY CONSUMPTION AND COST

Economic costs of HIPEF are mainly associated with the energy consumption of the process. Depending on the type of product, experimental setup, treatment chamber geometry, and processing parameter (wave shape, pulse width, voltage), the specific energy input requirements vary in a broad range. This parameter depends on the number of pulses (n) and the energy delivered with each pulse (q). Provided that the shape of pulses is exponential (q_1) or squared (q_2), the input of total energy density per pulse (q, J/m³) can be expressed by Equations 5.4 and 5.5, respectively:

$$q_1 = \frac{C \cdot V_0^2}{2 \cdot v} \tag{5.4}$$

$$q_2 = \frac{C \cdot V_0^2}{v}, \tag{5.5}$$

where
  C is the capacitance of capacitor (F)
  V_0 is the initial charge voltage (V)
  v is the volume of the sample (m³)

Thus, the total input energy density supplied in the treatment chamber during HIPEF treatment Q (J/m³) is computed as

$$Q = n \cdot q. \tag{5.6}$$

The characteristic of the medium such as conductivity or the presence of solids may affect the manner in which the treatment is delivered, by altering the voltage or the pulse width and shape (Mittal and Griffiths, 2005). On the other hand, treatment intensity required for food preservation is much higher in terms of electric field strength and energy input than disintegration of plant or animal tissue. Energy input requirements of up to 8000 kJ/kg of HIPEF-treated product have been reported for an efficient pasteurization of liquid food with a peak electric field in the range of 30–40 kV/cm (Zhang et al., 1995; Heinz et al., 2002; Góngora-Nieto et al., 2003; Evrendilek and Zhang, 2005; Elez-Martínez et al., 2006b). In contrast, a total energy input of 10–20 kJ/kg is required for juice extraction considering electric field strength in the range of 1–2 kV/cm with an industrial-scale system (Toepfl et al., 2006).

The major concerns for the implementation of a new technology are the initial investment and the plant running cost. Although an industrial system is not yet available for food preservation, the investment cost can be scaled up from pilot-scale equipment. Commercial-scale HIPEF systems for food pasteurization processing between 600 and 2000 L of liquid food per hour with an average power of 30 kW and a peak voltage of 30 kV/cm have an operating cost in the range of 6.3 k€/ton (DIL, 2011). Moreover, the operation cost for food preservation by HIPEF seems to be lower than those observed in traditional thermal technologies. HIPEF is nearly a nonthermal process, and the cost for cleaning and downtime will be less. Faster cleaning at lower frequency than that spent by a thermal process and the reduction in downtime due to less complexity and number of used parts may represent significant savings per liter when processing a product by HIPEF (Hoogland and de Haan, 2007).

HIPEF is also configured for other applications involving cell disintegration. At present, HIPEF systems (DIL, 2011) can improve the extraction of cell contents such as fruit and vegetable juices, sugars, colors, or other active substances. The application of HIPEF is energy efficient, waste free, and commercially viable. The cost of investment for HIPEF equipment as tissue disintegration is estimated in the range of 155,000–62,000 €. Considering a system with a capacity of 36 tons/h with an average power of 5 kW and a peak voltage of 10 kV/cm, the operation cost is 1.011 €/ton (DIL, 2011). This process saves money in contrast to conventional cell disruption technologies, with an estimation cost of treatment of 8.48 €/ton. Moreover, for an application of HIPEF to improve drying or extraction with similar processing parameters (1–3 kV/cm, 10 kJ/kg), the total costs for the HIPEF treatment can also be expected in the same range. HIPEF could enhance the drying rates or obtain a higher quality product in terms of structure or solid content. In this way, HIPEF technology offers a reduction of air drying time, resulting in a significant reduction of operation cost or an increase in production capacity (Amami et al., 2005). On the other hand, a HIPEF-aided extraction is considered less detrimental than a heat extraction process since better maintenance of the product quality and minimum power consumption are achieved (Lebovka et al., 2004). Moreover, the application of HIPEF to facilitate extraction of oil from harder tissues has indicated that with an additional input of energy, the recovery rate of plant oils can be increased, resulting in a reduction of specific energy consumption (Guderjan et al., 2005).

5.2 HIPEF-ASSISTED PROCESSING OF FRUITS

The potential applications of HIPEF in solid food has been associated with drying operations, as a pretreatment to increase the yield in the production of fruit juices as well as with the extraction of valuable compounds such as antioxidants, colorants, or flavors (Donsì et al., 2010). These operations could include thermal treatment or mechanical grinding for disruption of cellular material that requires a high amount of thermal or mechanical energy. The application of HIPEF to cellular tissues is reported to enhance mass transfer from the inner part of the cells. Permeabilization of the membranes is easier to attain when lower electric field intensities are applied and lower energy is consumed (Knorr et al., 2001). Nevertheless, the degree of efficiency and the optimization of the HIPEF parameters are different in each application.

5.2.1 Drying Processes

One of the most energy-consuming unit operations during food processing is drying, as it entails the transport of large amounts of water from the inside of products to the surface and its removal. The permeabilization of vegetable tissues caused by HIPEF treatments induces an increase in the mass and heat transfer rates between the cells and their surroundings (Knorr et al., 2001). Therefore, tissue disintegration by HIPEF can be used to facilitate mass transport to enhance drying processes of several vegetable and fruit tissues (Toepfl and Knorr, 2006). In general, HIPEF has been configured as a pretreatment of the drying process, in order to either enhance the drying rates or to obtain a higher quality product in terms of structure or solid content (Donsì et al., 2010). Angersbach and Knorr (1997) showed that HIPEF pretreatment of potato tissue improved mass transfer during air drying in a fluidized bed drier (at 70°C, air velocity of 2 m/s). They obtained optimum conditions for potato permeabilization at field strengths between 1.5 and 3.0 kV/cm and number of pulses between 15 and 30. Therefore, the maximum liquid release (29%) was observed at a low specific energy of 6.4 kJ/kg with an increase in the temperature of the treated product up to 4.5°C.

Ade-Omowaye et al. (2001b) observed that HIPEF enhanced drying of red paprika in fluidized bed dryer at 60°C for 6 h and air velocity of 1 m/s. They compared different methods including water blanching, skin treatments, high hydrostatic pressure and HIPEF. The transfer coefficients obtained after pretreating red paprika slices with 10 exponential decay pulses at 2.4 kV/cm of 300 μs were comparable with those corresponding to the maximum value, measured after blanching. The permeabilization induced by HIPEF determined the enhancement of mass transport during drying. In fact, HIPEF pretreated samples showed a reduction of approximately 25% of drying time compared with control samples. The total specific energy consumption was 3.0 kJ/kg and the temperature increase due to HIPEF treatment was less than 1°C.

On the other hand, HIPEF has also been suggested as a method of pretreatment to accelerate mass transfer of moisture and solute diffusion during osmotic dehydration. In the case of carrots, applied energy of 2.25 kJ/kg increased the cell wall permeability, which resulted in increased mass transfer rates during osmotic dehydration. However, softening of the dehydrated product and a decrease in the firmness of the rehydrated carrots were observed (Rastogi et al., 1999). The influence of HIPEF and other pretreatment methods on the osmotic dehydration of apple slices was investigated by Taiwo et al. (2001, 2003a). Apple slices were pretreated with 20 exponential decay pulses of 800 μs at 1.4 kV/cm, and osmotic dehydration was performed by the immersion of samples in 50 °Bx sugar solution at 40°C for 6 h. HIPEF-treated samples showed a higher water loss than that in untreated tissue. Water loss after HIPEF was comparable to that obtained with other pretreatments studied such as high pressure and blanching. The enhanced water loss was attributed to the increased permeability of the cell membrane after HIPEF treatment (Taiwo et al., 2001). In other studies, Taiwo et al. (2003a) observed that cell membrane permeabilization increased with increasing field strength and pulse number. However, higher water loss and solids gain were obtained for the application of 20 pulses at the intermediate electric field of 1.0 kV/cm, whereas a lower rate of mass transfer was observed for the same number of pulses at 0.5 and 2 kV/cm. In the same way, the application of 0.9 kV/cm and pulse duration of 100 μs in combination with osmotic dehydration was shown to increase the rate of water loss and solid gain (Amami et al., 2005). The use of HIPEF (up to 2.5 kV/cm) was studied as a pretreatment to the osmotic dehydration of bell peppers at 35°C. The results showed that increasing field strength resulted in increased water loss from 36% to 50%, suggesting that HIPEF could be an attractive alternative to conventional thermal treatments (Ade-Omowaye et al., 2003). According to Bazhal et al. (2003), the optimal electric field strength depends on the type of tissue and is higher for cells with a developed secondary cell wall, such as aubergine (0.5–0.6 kV/cm), pear, and banana (0.9–1.1 kV/cm), while for apple, potato, carrot, and cucumber, the optimal electric field strength value was found to be in the range of 0.2–0.4 kV/cm. On the other hand, water loss is generally well correlated to the permeabilization achieved, regardless of the pretreatment applied. However, solid gain is reported to be affected by the pretreatment method. Studies on mango and strawberry showed that HIPEF treatment increased water loss while it minimized solids gain, thus causing a minimal alteration in the organoleptic and nutritional characteristics of the product (Tedjo et al., 2002; Taiwo et al., 2003b).

5.2.2 FRUIT JUICE EXPRESSION

The efficiency of HIPEF treatment in juice expression enhancement by cellular permeabilization from fruit and vegetables has been studied by many authors (Table 5.1). HIPEF induces permeabilization of food membranes within a very short time (microsecond to millisecond range), leaving the product matrix largely unchanged while positively affecting mass transfer in juice extraction (Ade-Omowaye et al., 2001b).

HIPEF treatment with energy between 0.06 and 0.7 kJ/kg was able to enhance juice expression from apple mash similarly to heating and increase yield up to 73% without affecting the color of the final juice (McLellan et al., 1991). However, Lebovka et al. (2004) showed that treatments combining mild temperatures (50°C for 10 min) and HIPEF (0.5 kV/cm for 10^{-2} s, pulse duration of 10 μs) resulted in an enhancement of juice extraction. Bazhal and Vorobiev (2000) investigated the effect of HIPEF treatment in combination with pressing on apple juice extraction yield. They proposed a new process of solid–liquid expression in which HIPEF was applied as an intermediate treatment after mechanical precompression of the samples. Thus, the use of HIPEF (0.1–0.5 kV/cm, pulse duration of 100 μs) in combination with pressing steps at 3 bar could be a suitable method of increasing apple juice yield, affecting the quality of extracted juice positively. Since the simultaneous pressure and HIPEF treatment application promote the damage of defective cells, it enhances diffusion migration of moisture and depresses the cell pressing (Praporscic et al., 2007).

Juice expression from carrots appears to be highly dependent on carrot mash particle size. HIPEF treatment at 2.6 kV/cm followed by expression (100 bar, room temperature, 5 min) resulted in higher juice yield for 1.5 mm (50%) than for 3 mm (30%) size. However, carrot juice expression resulted independent of size when applying 50 pulses of 2.6 kV/cm (Knorr et al., 1994). Evaluation of selected quality indexes showed that HIPEF-treated carrot mash resulted in less oxidation with higher β-carotene values than juices produced with traditional methods (Knorr et al., 1994).

TABLE 5.1

Summary of the Applications of PEF Treatment to Vegetable Tissue Expression

Tissue	E (kV/cm)	Number of Pulses	Pressing Conditions	Main Effects of PEF Treatment	Reference
Apple	0.1–0.5	1–1000	3 bar	Enhancement of juice yield and improvement of quality by simultaneous application of PEF and pressure	Bazhal et al. (2003)
	0.5	10	5 bar	Higher juice yield than nontreated and thermally pretreated samples	Lebovka et al. (2004)
	1–5	30	250 bar	Increase in juice yield from 2% to 8%, whereas enzymatic mash treatment resulted in a 4% growth	Schilling et al. (2007)
	0.25–0.4	100	5 bar	Permeation of cell membranes and increase in portion of juice from internal pores of particles	Praporscic et al. (2007)
Carrot	2.6	nd	100 bar	Increase in juice yield from 30% to 50%(depending on grating size) for untreated samples to 70% after PEF	Knorr et al. (1994)
	0.25–1	100	5 bar	Increase in juice yield from 51% to 67% after PEF, higher content of β-carotene	Grimi et al. (2007)
Sugar beet	1.2–2.5	1–200	20–50 bar	Increase in the solid concentration twofold in the obtained juice	Eshtiaghi and Knorr (2002)
	Up to 300	nd	32 bar	Same juice yield of thermal denaturation at 72°C (174 kJ/kg), but with significantly lower energy	Schultheiss et al. (2002, 2004)
	0.5	100	5 bar	Highest yield for intermediate PEF application	Praporscic et al. (2007)
	0.6	400	5 bar	Synergetic effect of ohmic heating and PEF: promotion of 85% of juice extraction from coarse cuts	Praporscic et al. (2005)

(continued)

TABLE 5.1 (continued)

Summary of the Applications of PEF Treatment to Vegetable Tissue Expression

Tissue	E (kV/cm)	Number of Pulses	Pressing Conditions	Main Effects of PEF Treatment	Reference
Black tea and mint leaves	Up to 0.125	nd	nd	Increase in extractability compared with heating alone and enhancement of the leaching of solute from fresh cellular material	Sensoy and Sastry (2004)
Alfalfa	1.25–2.5	nd	40 bar	Increase in extractable protein by 57% and mineral extraction by 73%	Gachovska et al. (2009)
	0.25–0.4	100	5 bar	Enhancement of juice yield and improvement in qualitative characteristics of the juice	Praporscic et al. (2007)
Potato	0.3	1–30,000	5 bar	Highest juice yield for PEF application to the precompressed samples	Lebovka et al. (2003)
	0.1–0.7	9	15–30 bar	Increase in yield with increased PEF intensity	Chalermchat and Dejmek (2005)
Pepper	1.7	20	100 bar	Juice yield 10% higher than control (comparable to enzyme treatments); 60% increase in β-carotene in the juice (compared with 44% for enzyme)	Ade-Omowaye et al. (2001c)

Source: Donsì, F. et al., *Food Eng. Rev.,* 2, 109, 2010.

nd: No data.

Treating sugar beet with HIPEF revealed that this technique is an alternative to high-temperature thermal degradation (70°C–120°C, 10–20 min), maintaining sugar quality and yield (Eshtiaghi and Knorr, 1999; Schultheiss et al., 2002). Further developments have led to a semiindustrial device for sugar beet pretreatment, with maximum electric field strength of 30 kV/cm and a repetition rate of 20 Hz (Schultheiss et al., 2004). Validation in a pilot-scale multiplate and frame pressing equipment (pressure of 5–15 bar and particles filling up to 15 kg) for the HIPEF-assisted cold pressing of sugar beet cossettes has been conducted, with an enhancement on the juice yield (Jeami and Vorobiev, 2003). On lab-scale, the combination of ohmic heating (60 kV/cm, 50 Hz) and HIPEF treatment (0.6 kV/cm, 0.04 s) lead to a synergetic effect, with an 85% enhancement of juice extraction, due to the combination of the electropermeabilization of cell membranes and the thermal softening of tissues (Praporscic et al., 2007).

Treatments of 1.25 and 2.5 kV/cm on alfalfa, followed by pressing at 40 bar for 2 min resulted in an increase in the extracted protein (57%) and minerals (73%) but the same juice yield was achieved for both treatments (Gachovska et al., 2009). HIPEF treatments (up to 0.68 kV/cm) and low deformation rates (below 0.1 mm/min) increased the juice yield obtained from the compression of potato (Chalermchat and Dejmek, 2005). Juice expression from paprika resulted in approximately 10% juice yield increase with a HIPEF treatment application (1.7 kV/cm) followed by a pressure of 10 MPa for 4 min.

5.2.3 EXTRACTION AND RECOVERY OF VALUABLE COMPOUNDS

The electroporation of plant cells can be used to enhance the extraction of intracellular metabolites of commercial interest due to the permeabilization of the cell membrane as well as of vacuoles, where some metabolites are contained (Toepfl and Knorr, 2006; Donsì et al., 2010). Several studies suggest that the application of HIPEF in combination with a mild thermal treatment could be used to enhance the recovery of soluble substances from different fruit and vegetable tissues such as apple slices (Jemai and Vorobiev, 2002), carrots (El-Belghiti et al., 2005), and chicory (Loginova et al., 2010). In the case of chicory, treatments between 0.4 and 0.6 kV/cm with a moderate electric energy consumption (<10 kJ/kg) induced a high level of tissue disintegration and a complete soluble matter extraction was obtained even at room temperatures within 3 h. An increase in extractability of black tea and mint leafs by moderate electric fields (up to 0.125 kV/cm) was reported by Sensoy and Sastry (2004). On the other hand, an increased extractability of valuable components provides an enormous potential for product development. The application of HIPEF to recover bioactive compounds with beneficial effects on human health represents other potentials of HIPEF technology. HIPEF induced better redness and higher amount of β-carotene in the juice than conventional enzyme pretreatments (Ade-Omowaye et al., 2001b). Irreversible disintegration on the cellular membrane of raspberry cells was caused by HIPEF treatments, and the extraction yield of anthocyanins from raspberry was linearly correlated to the number of pulses applied. Nevertheless, cyaniding-3-glucoside, the major anthocyanin in raspberry, was found to be significantly degraded when increasing the intensity of the treatment (Luo et al., 2008). The impact of HIPEF treatment on white grapes was investigated by Praporscic et al. (2007), showing an increase in juice yield as well as in the extraction of antioxidants represented by the polyphenol release from grape skins. Optimal treatment conditions with an energy input of 20 kJ/kg at electric field strength of 0.75 kV/cm increased the juice yield from 50% to 80% after 45 min pressing. During wine production, a juice yield of 87%, similar to that produced by enzymatic maceration, and an increased content of soluble solids and pigments were reported after cell disintegration by HIPEF (Eshtiaghi and Knorr, 2000; Balasa et al., 2006; Grimi et al., 2009). In other products such as artichoke, HIPEF treatments (1.5 kV/cm for 500 pulses of 10 μs) also enhanced the extraction yield of polyphenols in water.

Lopez et al. (2009) investigated the HIPEF-induced extractability of betanine, natural colorant, from red beet tissue in solid–liquid extraction process. Betanines were recovered through the aqueous extraction of shredded beetroots with the release of 90% of total betanine in McIlvaine buffer (pH 3.5) after applying treatments of 7 kV/cm and shorter pulse duration (2 μs) (Figure 5.3). HIPEF treatments can also be used to extract nonpolar compounds from plant origin raw materials. Hence, betulin yield increased by 20% when extracted with HIPEF (40 kV/cm, 2 pulses) from *Inonotus obliquus* mushrooms using a suitable solvent (Yin et al., 2008).

HIPEF treatments have also been related with the enhancement of the extraction of oil from harder tissues, such as maize, olives, and soybeans (Guderjan et al., 2005) as well as rapeseed (Guderjan et al., 2007). An increase in oil yield (7%) from maize germ was observed after applying 0.6 kV/cm, thus leading to an oil enriched in phytosterols and isoflavonoids (Guderjan et al., 2005). For olive oil, an increase of 7.4% in oil yield was found after applying 100 pulses of 1.3 kV/cm, with an energy input as low as 2 kJ/kg of olives or 18 kJ/kg oil output. The energy requirements for soy oil extraction have been estimated as 70 kJ/kg of oil output (Guderjan et al., 2005).

The application of HIPEF for sugar extraction from sugar beet has been studied by several authors (Schultheiss et al., 2002; Jemai and Vorobiev, 2003; El-Belghiti and Vorobiev, 2004; El-Belghiti et al., 2005; Lebovka et al., 2007; Lopez et al., 2009). The application of moderate HIPEF treatments (from 0.15 to 0.94 kV/cm) with specific energy requirements below 10 kJ/kg is required to obtain measurable enhancement of soluble transfer when extracting sugar from sugar beets (Jemai and Vorobiev, 2003; El-Belghiti and Vorobiev, 2004). A kinetic study of HIPEF-assisted sucrose extraction showed that the efficiency of the solid–liquid extraction is dependent on the electric field strength applied as well as the temperature of the extracting medium. The application of 20 pulses at 7 kV/cm (3.9 kJ/kg) increased the maximum yield by 7- and 1.6-fold, compared with non-HIPEF-treated samples (Lopez et al., 2009).

The effect of HIPEF on the extraction of coconut water was studied by Ade-Omowaye et al. (2001a). Minimum energy consumption (25.4 kJ/kg) with a satisfactory degree of permeabilization corresponded to a treatment of 20 pulses at 2.5 kV/cm. HIPEF treatment also induced an increase of 20% of coconut water yield with respect to control samples, and protein and fat contents analysis showed that HIPEF did not affect the quality of the extracted beverage. Avoiding an enzymatic treatment will reduce undesired side effects of enzymes, and in addition, retain the native structure of pectins within the tissue. Yin et al. (2009) observed the highest yield rate of 14.12% of pectin from apple pomace when applying 15 kV/cm and 10 pulses. Providing a potential to extract pectin from the pomace after liquid–solid separation

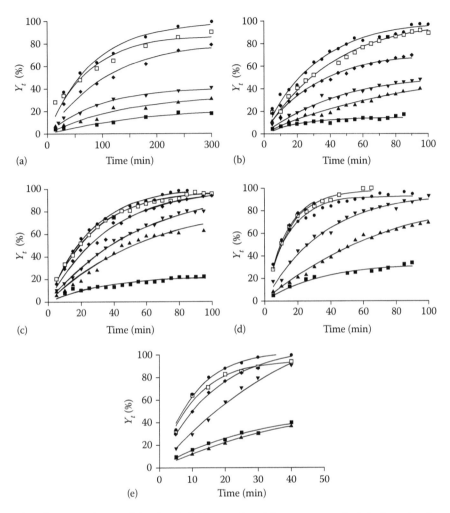

FIGURE 5.3 Influence of the electric field strength and the pressure on the betanine extraction by PEF. Treatment conditions: pressure applied during extraction: (a) 0 kg/cm², (b) 2 kg/cm², (c) 6 kg/cm², (d) 10 kg/cm², and (e) 14 kg/cm²; electric field strength applied: (■) 0 kV/cm, (▲) 1 kV/cm, (▼) 3 kV/cm, (♦) 5 kV/cm, (•) 7 kV/cm, and (Υ) 9 kV/cm. Extracting medium: McIlvaine buffer pH 3.5. Five pulses. (From Lopez, N. et al., *J. Food Eng.*, 90, 60, 2009.)

can help to additionally improve the utilization of resources and further cost-savings. Moreover, HIPEF treatments may also represent a viable option for enhancing the extraction of phenolic compounds from skin cells of grapes during the maceration steps in the winemaking process, without altering wine quality and with moderate energy consumption (Lopez et al., 2008; Boussetta et al., 2009). When applied to white grapes (Muscadelle, Sauvignon, and Semillon) prior or simultaneously to pressing (5 bar), HIPEF resulted beneficial not only in the increase in juice yield from 50% to 80% after 45 min of pressing but also in the significant improvement of juice quality (Praporscic et al., 2007).

5.3 PRESERVATION OF FRUIT JUICES BY HIPEF

HIPEF technology provides the potential of ensuring safety and maintaining the physicochemical quality of liquid food products without negatively affecting the content and composition of thermolabile compounds. This is especially relevant in the case of fruit juices, because some of the features that are currently most appreciated by consumers, such as aroma or bioactive compounds, are related to this heat-sensible fraction. Therefore, HIPEF has been suggested as an alternative technology to obtain safe and high-quality and nutritional foods with a shelf-life similar to that attained with conventional heat pasteurization.

5.3.1 FRUIT JUICE PASTEURIZATION

The effective inactivation of HIPEF on both pathogenic and spoilage microorganisms at pasteurization levels equivalent to those attained with conventional thermal processing has been proven in a broad variety of fruit juices (Evrendilek et al., 2000; Yeom et al., 2000a; Elez-Martínez et al., 2004; Molinari et al., 2004; Mosqueda-Melgar et al., 2007, 2008a,c; El-Hag et al., 2008). The mechanism of microorganism inactivation by HIPEF involves the formation of local instabilities in their membranes by electromechanical compression and electric-field-induced tension, which causes pores to form in the membrane (electroporation) and subsequent cellular death (Coster and Zimmermann, 1975; Tsong, 1991; Ho and Mittal, 1996; Weaver and Chizmadzhev, 1996; Barbosa-Cánovas et al., 1999).

Microbial inactivation is greatly conditioned by HIPEF processing variables such as electric field strength, treatment time or number of pulses, pulse width, frequency, and pulse polarity (Martín-Belloso and Elez-Martínez, 2005). Iu et al. (2001) and Liang et al. (2002) obtained a higher inactivation of *Salmonella typhimurium* and *Escherichia coli* O157:H7 populations in orange juice and apple cider, respectively, by increasing the number of pulses and electric field strength. Evrendilek et al. (2000), Zhong et al. (2005), and Mosqueda-Melgar (2007) reached higher microbial inactivation of *E. coli* O157:H7 and *Salmonella enteritidis* in several fruit juices when treatment time was increased. Altuntas et al. (2010) observed that inactivation of *E. coli* O157:H7, *Staphylococcus aureus*, *Listeria monocytogenes*, *Erwinia carotovora*, *Pseudomonas syringae*, *Botrytis cinerea*, and *Penicillium expansum* inoculated on sour cherry juice significantly increased with increasing electric field strength (up to 30 kV/cm) and treatment time (up to 210 μs) without adversely affecting important physical and quality parameters. Chen et al. (2010) studied the influence of pulse rise time on the inactivation of *S. aureus* inoculated on apple juice. The results showed that the pulse with a shorter rise time (200 ns, 35 kV/cm) had a higher effect on the inactivation of the microorganisms and resulted in a higher transmembrane potential of *S. aureus* cells and higher effect on intracellular material. On the other hand, Elez-Martínez et al. (2004, 2005) observed an increase in *Saccharomyces cerevisiae* and *Lactobacillus brevis* inactivation in orange juice when applying treatments of low frequency. With regard to polarity, bipolar pulses are generally reported to be slightly more efficient than monopolar pulses on the inactivation of microorganisms (Qin et al., 1994;

Ho et al., 1995; Elez-Martínez et al., 2004). Liang et al. (2002) reported more than 5.0 log reductions of *S. typhimurium* in fresh squeezed orange juice after applying 20 pulses of 90 kV/cm and an outlet temperature of 55°C. Likewise, McDonald et al. (2000) reached up to 5.0 log reductions in the counts of *E. coli* in orange juice by applying a treatment of 30 kV/cm for 0.43 μs with an outlet temperature of 54°C.

Several authors have developed the optimization of treatment conditions in order to standardize the process for obtaining the maximal destruction of microorganisms in fruit and vegetables juices. In this way, populations of *S. enteritidis, E. coli,* and *L. monocytogenes* in melon juice were reduced by up to 3.7 $\log_{10}$ cycles when applying 4 μs bipolar pulses of 35 kV/cm for 1709 μs at 193 Hz. On the other hand, reductions of up to 3.6 $\log_{10}$ units of the same microorganisms were reached in watermelon juice treated at 35 kV/cm for 1682 μs (Mosqueda-Melgar et al., 2007). A treatment time of 1000 μs was needed for 35-kV/cm pulses applied at 100 Hz to reduce *S. enteritidis* in tomato juice by 4.2 log cycles (Mosqueda-Melgar et al., 2008a).

Inactivation of microorganisms by HIPEF can also be influenced by the fluid medium, the target microorganism, or the microbial characteristics (Martín-Belloso and Elez-Martínez, 2005). The pH and electrical conductivity of juices have a great influence on the safety of the product. Mosqueda-Melgar et al. (2008c) observed that *E. coli* O157:H7 in acidic juices exhibits high sensibility to HIPEF (Figure 5.4). In the same way, the inactivation rates of microorganisms such as *L. brevis, E. coli, S. cerevisiae, Salmonella Dublin,* and *Listeria innocua* increased with decreasing conductivity of the treatment medium (Jayaram et al., 1992; Grahl and Märkl, 1996; Sensoy et al., 1997; Wouters et al., 2001). Juices usually have lower conductivity and pH than other liquid products, thus allowing a greater inactivation of microorganisms. Ferrer et al. (2007) reported that the electric field gained importance in the inactivation of *E. coli* in orange-carrot juice when increasing treatment intensity.

In general, Gram-negative bacteria are reported to be more susceptible to HIPEF inactivation than Gram-positive bacteria (Castro et al., 1993; Mazurek et al., 1995; Pothakamury et al., 1996; Qin et al., 1998; Dutreux et al., 2000). However, bacterial spores are resistant to HIPEF treatments, but after germination they become HIPEF sensitive (Marquez et al., 1997; Barbosa-Cánovas et al., 1998). Moreover, microbial inactivation rates have been shown to be dependent on the growth stage and the initial microbial concentration.

According to Rodrigo, Barbosa-Cánovas et al. (1998), cells in the logarithmic phase were found to be more sensitive to HIPEF treatments than those in the stationary phase. On the contrary, yeast cells in the stationary phase are reported to be more sensitive to HIPEF treatments than those in the logarithmic phase (Molinari et al., 2004). Molinari et al. (2004) studied the inactivation of three *S. cerevisiae* strains by HIPEF treatment (3 pulses at 8 kV/cm or 40 pulses at 12.5 kV/cm) at different growth phases and initial inoculum sizes in orange juice. They observed that yeast inactivation by HIPEF depended on the inoculum size and not on the growth phase with no significant differences among treatments.

On the other hand, the combination of HIPEF treatments with bacteriocins or other antimicrobials opens up innovative possibilities for application on low acidic products in a hurdle-type approach. Martínez-Viedma et al. (2009) observed a complete bacterial inactivation in freshly made orange and apple juices when combining

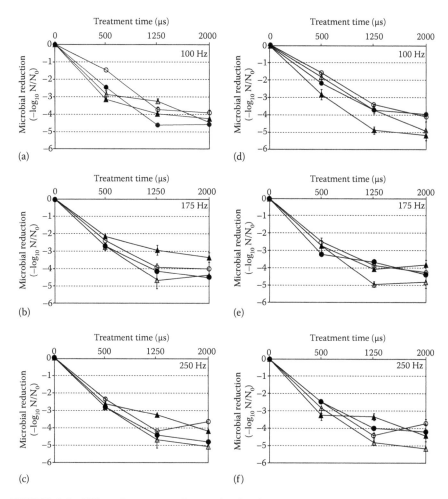

FIGURE 5.4 Effect of treatment time and pulse frequency on the microbial reductions of *S. enteritidis* (a through c) and *E.* coli O157:H7 (d through f) inoculated in fruit juices treated by HIPEF. Symbols of apple (o), pear (•), orange (△), and strawberry (▲) juices are the mean of four determinations ± SD. Treatment conditions: 35 kV/cm and 4 μs pulse length in bipolar mode without exceeding 40°C. (From Mosqueda-Melgar, J. et al., *Innov. Food Sci. Emerg. Technol.*, 9, 328, 2008b.)

the addition of subinhibitory AS-48 concentrations (0.175–60 μg/mL) with HIPEF treatments (35 kV/cm at 150 Hz and 4 μs pulse duration).

Application of essential oils and their bioactive phenolic compounds in food preservation has also been proposed to enhance the antimicrobial effect of HIPEF treatments in fruit juices, but it has been limited by the strong impact that they have on the organoleptic properties (Burt, 2004). Nevertheless, the strong bacterial effects reported for these antimicrobials opens up new possibilities due to the low concentration required for inhibition of target bacteria (Iu et al., 2001; Raybaudi-Massilia et al., 2006). Thus, combinations of HIPEF (35 kV/cm for 1000 μs at 100 Hz and 4 μs pulse length in bipolar mode) with 2.0% citric acid or 0.1% cinnamon bark oil were needed for inactivating *S. enteritidis* by more than 5.0 log units (Mosqueda-Melgar et al., 2008d). Moreover, populations of *E. coli* O157:H7, *S. enteritidis*, and *L. monocytogenes* were reduced by more than 5.0 log units in HIPEF-processed melon juice (35 kV/cm for 1709 µs at 193 Hz and 4 µs pulse duration) and watermelon juice (35 kV/cm for 1682 µs at 193 Hz and 4 µs pulse duration) containing 2.0% and 1.5% citric acid, respectively, or 0.2% cinnamon bark oil. These treatments were also able to inactivate mesophilic, psychrophilic, and mold and yeast populations, leading to a shelf life of more than 91 days in both juices stored at 5°C (Mosqueda-Melgar et al., 2008d).

An extension of the microbial shelf life in several fruit juices treated by HIPEF without antimicrobials has also been reported. HIPEF treatment (35 kV/cm for 1000 µs; bipolar 4 µs pulses at 200 Hz) proved effective in extending the shelf life of orange juice since microbial growth was not detected during 91 days of storage at 5°C (Elez-Martínez et al., 2006b). Nevertheless, Yeom et al. (2000a) and Min et al. (2003b) extended the microbial shelf life of the juice during 112 days, when applying 35 kV/cm for 59 µs or 40 kV/cm for 57 µs, respectively. Nguyen and Mittal (2007) kept the microbial counts below 1 $\log_{10}$ (CFU/mL) at least 28 days of storage at 4°C when processing tomato juice using HIPEF (87 kV/cm for 80 µs). Min et al. (2003a) reported a longer extension of the microbiological shelf life in tomato juice (>112 days at 4°C) processed by HIPEF (40 kV/cm for 57 µs). Moreover, Mosqueda-Melgar (2007) observed that the microbiological quality of strawberry, orange, apple, pear, and tomato juices processed by HIPEF (35 kV/cm for treatment times ranging 1000–1700 µs) was assured at least during 91 days of storage at 5°C, similar to thermal treatment.

5.3.2 EFFECTS ON ENZYME ACTIVITY

In general, enzymes are less sensible than microorganisms to HIPEF and their inhibition depends on the enzyme itself, the media where they are suspended, and the processing conditions (Martín-Belloso and Elez-Martínez, 2005). The mechanisms involved in the inactivation of enzymes by HIPEF are not fully understood. The effects of electric field on proteins include the association or dissociation of functional groups, movements or charged chains, and changes in alignment of α-helix (Tsong and Astumian, 1986). Some studies have evidenced that HIPEF treatments caused a loss of α-helix and an increase in β-sheet content, indicating that HIPEF affects the conformation in the secondary structure of enzymes (Zhang et al., 2007; Zhong et al., 2007). The impact of HIPEF on quality-related enzymes of fruit juices such as peroxidase (POD), polyphenoloxidase (PPO), lipoxygenase (LOX), hydroperoxide lyase (HPL), pectin methylesterase (PME), polygalacturonase (PG), and β-glucosidase (β-GLUC) has been studied (Table 5.2). In general, the effectiveness of enzyme inactivation in fruit juices by HIPEF is higher when electric field strength and treatment time increase (Ho et al., 1997; Yeom et al., 2000a, 2002; Van Loey et al., 2002; Elez-Martínez et al., 2007; Aguiló-Aguayo et al., 2008a). POD activities in grape juice treated by HIPEF were depleted as electric field strength and treatment time increased (Marsellés-Fontanet and Martín-Belloso, 2007). Elez-Martínez et al. (2006a) and Aguiló-Aguayo et al. (2008a) reported that POD inactivation in

TABLE 5.2

Effects of Pulsed Electric Fields on Food Quality–Related Enzymes in Fruits

Enzyme	Fruit	Treatment Conditions	Inactivation (%)	Reference
Peroxidase	Orange	35 kV/cm, 1,500 µs, 35°C	100	Elez-Martínez et al. (2006a)
	Grape	30 kV/cm, 5,000 µs, 40°C	50	Marselles-Fontanet and Martín-Belloso (2007)
	Apple	35 kV/cm, 74°C	100	Schilling et al. (2008)
		40 kV/cm, 100 µs, 50°C	68	Riener et al. (2009)
	Watermelon	35 kV/cm, 1,000 µs, 35°C	99	Aguiló-Aguayo et al. (2010b)
	Tomato	35 kV/cm, 2,000 µs, 35°C	100	Aguiló-Aguayo et al. (2008a)
Polyphenoloxidase	Apple	35 kV/cm, 74°C	93	Schilling et al. (2008)
		31 kV/cm, 40,000 µs, 60°C	32	Van Loey et al. (2002)
		24.6 kV/cm, 6,000 µs, 15°C	97	Giner et al. (2001)
		40 kV/cm, 100 µs, 50°C	71	Riener et al. (2009)
	Pear	24.6 kV/cm, 6,000 µs, 15°C	62	Giner et al. (2001)
	Peach	24.3 kV/cm, 5,000 µs, 25°C	70	Giner et al. (2002)
	Grape	30 kV/cm, 5,000 µs, 40°C	100	Marselles-Fontanet and Martín-Belloso (2007)
	Strawberry	35 kV/cm, 2,000 µs, 35°C	97.5	Aguiló-Aguayo et al. (2010a)

Enzyme	Product	Treatment conditions	%	Reference
Pectin methylesterase	Orange	35 kV/cm, 59 µs, 60.1°C	88	Yeom et al. (2000b)
		40 kV/cm, 97,000 µs, 45°C	88	Min et al. (2003a)
	Strawberry	35 kV/cm, 1,500 µs, 35°C	80	Elez-Martínez et al. (2007)
		35 kV/cm, 1,000 µs, 35°C	92	Aguiló-Aguayo et al. (2009c)
	Tomato	24 kV/cm, 8,000 µs, 15°C	94	Giner et al. (2000)
	Orange-carrot	35 kV/cm, 340 µs, 35°C	81.4	Rodrigo et al. (2003)
		25 kV/cm, 280 µs, 68°C	76	Rivas et al. (2006)
	Watermelon	35 kV/cm, 2,000 µs, 35°C	85	Aguiló-Aguayo et al. (2010d)
Polygalacturonase	Strawberry	35 kV/cm, 1,000 µs, 35°C	28	Aguiló-Aguayo et al. (2009c)
	Tomato	35 kV/cm, 1,000 µs, 35°C	62	Aguiló-Aguayo et al. (2009c)
	Watermelon	35 kV/cm, 1,000 µs, 35°C	40	Aguiló-Aguayo et al. (2010d)
Lipoxygenase	Tomato	35 kV/cm, 50 µs, 30°C	80	Min et al. (2003c)
		35 kV/cm, 1,000 µs, 35°C	20	Aguiló-Aguayo et al. (2009e)
	Strawberry	35 kV/cm, 1,000 µs, 40°C	35	Aguiló-Aguayo et al. (2008b)
	Watermelon	35 kV/cm, 1,000 µs, 35°C	52	Aguiló-Aguayo et al. (2010b)
	Green pea	20 kV/cm, 400 µs	0	Van Loey et al. (2002)
β-glucosidase	Strawberry juice	35 kV/cm, 1,000 µs, 40°C	27	Aguiló-Aguayo et al. (2008b)
Hydroperoxide lyase	Tomato juice	35 kV/cm, 1,000 µs, 35°C	93	Aguiló-Aguayo et al. (2009e)

orange and tomato juices increased when increasing treatment time. Treatments of longer duration also resulted in higher reductions of PPO activity in strawberry juice. Maximum PPO inactivation of 97.5% was achieved by selecting bipolar treatments at 35 kV/cm, frequencies higher than 229 Hz, and pulse widths between 3.23 and 4.23 μs for a constant treatment time of 2000 μs (Aguiló-Aguayo et al., 2010a). Moreover, Marsellés-Fontanet and Martín-Belloso (2007) observed that PPO activity in grape juice treated by HIPEF was lessened as electric field strength and treatment time increased. Consistently, Schilling et al. (2008) observed a similar trend in the inactivation of PPO of HIPEF-treated apple juice. In addition, other HIPEF critical variables such as pulse frequency, pulse width, and pulse polarity are also important in defining HIPEF treatment conditions necessary to adequately reduce enzyme activity (Elez-Martínez et al., 2006a; Aguiló-Aguayo et al., 2008a). Aguiló-Aguayo et al. (2008a) observed that it is feasible to maximize POD inactivation selecting pulse width higher than 5.5 μs in bipolar mode at a frequency of 200 Hz, keeping the electric field strength constant at 35 kV/cm and treatment time at 1500 μs. Higher POD inactivation in orange juice was achieved when pulse frequency increased, and residual activities of 6.9% were obtained when applying frequencies of 450 Hz (bipolar pulses of 4 μs at 35 kV/cm for 600 μs) (Elez-Martínez et al., 2006a). Elez-Martínez et al. (2007) reported 80% of PME inactivation when orange juice was treated at 35 kV/cm for 1500 μs by applying bipolar pulses of 4 μs at 200 Hz. Moreover, Aguiló-Aguayo et al. (2009c) observed that PME and PG activities in tomato juice were depleted by increasing pulse frequency and/or pulse width, irrespective of the treatment polarity. However, PME activity in strawberry juice was not affected when a HIPEF treatment was applied in mono- or bipolar mode.

The medium in which the enzyme is suspended has a significant effect on the inactivation level reached. POD from grape juice required stronger HIPEF treatment conditions than POD from watermelon juice to reach at least 50% of the degree of inactivation (Marsellés-Fontanet and Martín-Belloso, 2007; Aguiló-Aguayo et al., 2010b). An 80% reduction of LOX activity was observed when tomato juice was exposed to HIPEF at 35 kV/cm for 50 or 60 μs (3-μs pulse width) (Min et al., 2003c). In contrast, LOX in pea juice was not affected after applying 400 pulses of 1 μs at a field strength of 20 kV/cm and frequency of 1 Hz (Van Loey et al. 2002). Differences in the secondary or tertiary structure among enzymes may result in diverse sensitivity to HIPEF (Ho et al., 1997). Aguiló-Aguayo et al. (2008b) observed high resistance of LOX and β-GLUC to HIPEF in strawberry juice. They reported maximum LOX inactivation of 35% when processing the juice at 35 kV/cm for 1000 μs by applying 4 μs monopolar pulses at 150 Hz, whereas β-GLUC was activated up to 110% at the same treatment conditions. Tomato juice HPL was inactivated around 93% when juice was processed at 35 kV/cm for 1000 μs with 7 μs pulse width and 250 Hz, whereas the highest tomato juice LOX inactivation (20%) was obtained in monopolar mode at 150 Hz for 1 μs (Aguiló-Aguayo et al., 2009c).

Because of the resistance of several enzymes to HIPEF, treatments combining HIPEF with other hurdles such as mild heat or the use of some additives were studied on different fruit juices with promising results (Hodgins et al., 2002; Riener et al., 2008, 2009). The synergistic effects between HIPEF and thermal treatments in the inactivation of PME in orange juice have shown that an increase in electric field strength could cause greater levels of PME inactivation with an increase in temperature during the treatment (Yeom et al., 2000b).

Studies concerning enzyme activity stability during the storage of HIPEF-treated fruit juices revealed variations of the initial inactivation values reached after processing along storage.

A progressive decay of the initial residual values of LOX and PG has been observed in tomato, strawberry, and watermelon juices processed by HIPEF (35 kV/cm and treatment times up to 1727 μs) for 56 days of storage at 4°C (Aguiló-Aguayo et al., 2008c, 2009b, 2010c). However, other enzymes such as POD and PME were less affected than others by storage time. Thus, Elez-Martínez et al. (2006b) reported that POD activity of HIPEF-treated (35 kV/cm for 1000 μs; bipolar 4 μs pulses at 200 Hz) orange juice remained inactivated for 56 days, indicating that the changes induced in the enzyme structure were irreversible. In the same way, HIPEF-treated tomato and strawberry juices kept values of residual POD activities below 20% throughout storage (Aguiló-Aguayo et al., 2008c, 2010c). Residual PME activities in strawberry and orange juices remained below 10% and 20% during 56 days of storage (Elez-Martínez et al., 2006b; Aguiló-Aguayo et al., 2009b). Hence, the efficiency of HIPEF in inducing irreversible enzymatic changes has been demonstrated since no activation of enzyme has been reported in storage of HIPEF-treated fruit juices.

5.3.3 Influence on Product Quality

Studies on the effects of HIPEF processing on quality and nutritional-related compounds are more limited compared with those on microbial and enzyme inactivation. The ability of HIPEF processing to preserve fruit juices while maintaining their fresh-like characteristics has been observed by several authors (Min et al., 2003b; Elez-Martínez et al., 2006a; Rivas et al., 2006; Aguiló-Aguayo et al., 2009a,d; Akin and Evrendilek, 2009). Quality characteristics such as color, viscosity, acidity, soluble solids, and flavor, among others, have been studied in different HIPEF-processed fruit juices and compared with those heat-treated. Yeom et al. (2000a) reported that orange juices processed by HIPEF (35 kV/cm, 59 μs) exhibited less browning and minimal modification of pH and soluble solids content than those thermally treated (94.6°C for 30 s). Elez-Martínez et al. (2006a) also observed that HIPEF-treated (35 kV/cm, 1000 μs, bipolar 4 μs at 200 Hz) orange juices retained better color during storage than heat-pasteurized juices with no differences in pH, acidity, and °Bx between treated juices. In the same way, no differences in electrical conductivity, pH, soluble solids, viscosity, color, nonenzymatic browning index, hydroxymethyl-furfural content, and volatile flavor compounds were found between HIPEF-treated (28 kV/cm, 100 μs) and untreated citrus juices made from grapefruit, lemon, orange, and tangerine (Cserhalmi et al., 2006). Zárate-Rodríguez et al. (2000) also reported no differences between the soluble solids, pH, and acidity of HIPEF-treated and untreated apple juices. On the other hand, Aguiló-Aguayo et al. (2009c) and Aguiló-Aguayo et al. (2010d) observed that HIPEF processing parameters such as pulse frequency and pulse width affect the viscosity of tomato, watermelon, and strawberry juices. They reported that the combination of high frequencies and long pulse width increased the viscosity of tomato and watermelon juices, whereas high frequencies and short pulse widths promoted a rise in the viscosity of strawberry juices. They also observed that the application of HIPEF to those juices led to better preservation of color and less hydroxymethylfurfural formation than those thermally processed (Aguiló-Aguayo et al., 2009d).

The impact of HIPEF treatments on individual volatile compounds of fruit juices could affect their flavor profile and perception. In orange juice, HIPEF treatments (35 kV/cm, 59 μs) better retained volatile compounds (α-pinene, myrcene, octanal, D-limonene, and decanal) than heat-treated (94.6°C for 30 s) juices during storage at 4°C (Yeom et al., 2000a). Sensory evaluation of texture, flavor, and overall acceptability showed the highest score for the control and HIPEF-processed orange juices, followed by those thermally processed (Min et al., 2003a). Volatile compounds of apple juice were less affected by HIPEF treatments than heat treatments (Aguilar-Rosas et al., 2007). A sensory study showed that 37% of the participants preferred the apple cider processed by HIPEF, whereas the rest of the participants preferred the control sample or did not show any preference (Evrendilek et al., 2000). Several authors have reported that aroma and sensory characteristics of tomato juice are much better maintained when processing with HIPEF than with heat (Min and Zhang, 2003; Min et al., 2003b; Aguiló-Aguayo et al., 2008c). Moreover, HIPEF-treated tomato juice was preferred in terms of flavor and overall acceptability when a sensory test was conducted (Min et al., 2003b). Major volatile compounds determining strawberry aroma such as 2,5-dimethyl-4-hydroxy-3(2h)-furanone, ethyl butanoate, and 1-butanol were better retained in HIPEF-treated strawberry juices than in heat-treated juices (Aguiló-Aguayo et al., 2009a). Noticeable changes in odor, color, taste, sourness, and overall attributes were detected by panelists after treating watermelon juice with heat, whereas no significant changes were detected in HIPEF-processed juices (Mosqueda-Melgar et al., 2008d).

HIPEF processing could be an alternative to thermal treatment to obtain juices with a high nutritional quality. Reports concerning phenolic composition suggested that HIPEF have a slight effect on this fraction. Sánchez-Moreno et al. (2005b) reported no changes in the total flavanone content of orange juice after applying HIPEF treatment at 35 kV/cm for 450 μs with 4-μs bipolar pulses at 800 Hz. Odriozola-Serrano et al. (2008c) observed that anthocyanin retention in HIPEF-treated strawberry juice varied from 96.1% to 100.5% when increasing electric field strength from 20 to 35 kV/cm for up to 2000 μs. Tomato juices processed by HIPEF (35 kV/cm for 1500 μs with bipolar pulses of 4 μs at 100 Hz) exhibited better retention of the total phenolic fraction during 56 days than heat-treated juices (90°C for 60 s) (Odriozola-Serrano et al., 2008a). In the same way, HIPEF-treated strawberry juices (35 kV/cm for 1700 μs with bipolar pulses of 4 μs at 100 Hz) maintained higher amounts of phenolic acids (ellagic and p-coumaric acid) during the storage period than those thermally treated (Odriozola-Serrano et al., 2008b).

Among vitamins, vitamin C has been extensively studied due to its heat lability. In general, reports indicate that vitamin C is less affected by HIPEF treatments than by conventional treatments in very different juices such as orange juice (Min et al., 2003a; Sanchez-Moreno et al., 2005a; Elez-Martínez et al., 2006a), apple juice

(Evrendilek et al., 2000), strawberry juice (Odriozola-Serrano et al., 2008b), tomato juice (Min et al., 2003c; Odriozola-Serrano et al., 2008a), and orange-carrot juice (Torregrosa et al., 2006). Other vitamins, such as vitamin A in orange juice, have also shown a better stability when applying HIPEF treatments (30 kV/cm for 100 μs) compared with that achieved after a thermal treatment.

The effects of HIPEF processing on the antioxidant potential of various fruit juices have been evaluated through changes in the concentration of carotenoid compounds. Individual carotenoids such as β-carotene, β-cryptoxanthin, zeaxanthin, and lutein from orange juice did not change after applying bipolar treatments of 25–40 kV/cm for 30–340 μs (Cortés et al., 2006). Odriozola-Serrano et al. (2008d) observed an increase in the lycopene content of tomato juice after applying 35 kV/cm during 1000 μs with bipolar pulses of 7 μs at 250 Hz. They also observed that maximum values of antioxidant capacity in tomato juice were achieved when applying bipolar pulses at frequencies higher than 150 Hz and pulse widths lower than 5 μs. Consistently, Oms-Oliu et al. (2009) observed that bipolar pulses applied at higher frequencies led to a higher lycopene content in treated watermelon juice than in the untreated juice. Torregrosa et al. (2005) also reported that HIPEF treatments set up at 25 or 30 kV/cm caused a significant enhancement in the content of carotenoids in orange–carrot juice. However, Sánchez-Moreno et al. (2005a) did not observe differences between the antioxidant capacities of untreated and HIPEF-treated orange juices.

5.4 ADVANTAGES AND DISADVANTAGES OF HIPEF

The results obtained at an industrial scale indicate the potential of HIPEF to replace or enhance existing processes. However, before an industrial exploitation as a pretreatment for food plant or as a process for food preservation, the HIPEF technology should be economically interesting in terms of cost of operation and investment as well as product quality and consumer acceptance (Toepfl et al., 2005). The investment and the running cost of a HIPEF facility represent the main uncertainty. However, the application of HIPEF offers excellent perspectives regarding energy savings in different fields of food processing such as drying, juice expression or extraction, as well as for food preservation (Hoogland and de Haan, 2007; Donsi et al., 2010). HIPEF could be an alternative to conventional thermal treatments for fruit juice processing since the process results in an effective microbial inactivation with minimal changes in quality, including flavor, taste, aroma, color, nutrition, and appearance. However, the advantages that a new processing technology offers do not necessarily guarantee the success of a product in the market place. The attitudes toward HIPEF-treated products are based on general sociopolitical attitudes, on risk/benefit trade-offs of the product attributes, and on consumers' ability to evaluate the product (Sonne et al., 2010). Apart from consumer acceptance, the legislative situation is still unclear. For applications within the European Union, HIPEF treatment is subject to the Novel Food Regulation (NFR, 1997). The development of novel processing techniques at least requires proof of conformity, which is the principle of substantial equivalence that has been adapted to ensure that food safety is guaranteed after introduction of novel foods on the market. Moreover, the United States Food and Drug Administration (FDA) distinguish processes where physical treatments take place from processes where additives are used. Consumption of unpasteurized juices is of such high risk that the FDA has enforced a proven log 5 reduction of target pathogens during the production of juices, which is described in the recently installed Juice-HACPP (FDA, 2004). FDA accepts the efficiency of PEF in the food industry to remove pathogens from foods.

5.4.1 SCALE-UP CHALLENGES

Some technical challenges prelude the industrial implementation of HIPEF. First of all, differences in the HIPEF treatments among research groups in terms of building, method of operation, and construction materials make the comparison of the result difficult. Moreover, the monitoring of temperature is not always suitable and therefore, secondary effects on products are difficult to ascertain (Soliva-Fortuny et al., 2009). Other problems due to electrochemical reactions at the electrode/medium interfaces indicate that there is a challenge to replace commonly used stainless steel electrodes by other materials or to modify pulse generator systems to reduce the extent of electrochemical reactions (Toepfl et al., 2005). Furthermore, industrial equipment needs to be designed considering energy recovery systems. Optimization of treatment chambers and processing parameters will help to reduce the costs of operation for pasteurization of liquid food by HIPEF to meet the low energy requirements for thermal pasteurization, where heat recovery rates up to 95% can be reached (Toepfl et al., 2007). The improvement of HIPEF systems capable of adapting working conditions to different fluid characteristics and flow rates represents a significant design challenge that needs to be addressed before an industrial application.

5.5 CONCLUSIONS AND FUTURE TRENDS

HIPEF technology offers the potential to enhance mass transfer from cells, promoting significant energy savings in drying, to increased yield in juice expression or to the recovery of valuable cell metabolites with functional properties. However, not all the possible applications have been exploited yet, due to the lack of data about the mechanisms of electroporation and pore resealing. Moreover, the application of HIPEF to improve mass transfer rates requires low energy input in comparison to conventional techniques, but the industrial exploitation has not yet taken place. Therefore, more research and development activities are required to understand, optimize, and apply this complex process to its full potential.

On the other hand, HIPEF have been shown to be an interesting technology to process fruit juices since high-quality, safe, and shelf-stable products can be obtained without significant depletion of their nutritional and sensory properties. Future work will have to focus on understanding the mechanistic insight of the changes in HIPEF-treated fluid products and to validate their fresh-like quality in order to gain popularity and be commercialized. With the research carried out so far, it is expected that HIPEF technology will be industrially implemented as a potential support of many different applications in the food industry.

ACKNOWLEDGMENTS

Authors would like acknowledge to the financial support of the Commission of the European Communities, Framework 6, Priority 5 "Food Quality and Safety", Integrated Project NovelQ FP6-CT-2006-015710; the Spanish Institute of Agricultural and Food Research and Technology (INIA) through the project RTA2010-00079-C02-02; the Ministerio de Ciencia e Innovación (Spain) through the Project ALI 2005-05768; Generalitat de Catalunya (regional government); and the University of Lleida. I. Aguiló-Aguayo thanks the Ministerio de Ciencia e Innovación (Spain) for the pre-doctoral grant. ICREA Academia Award is also acknowledged by O. Martín-Belloso.

REFERENCES

Ade-Omowaye, B.I.O., Angersbach, A., Eshtiaghi, N.M., and Knorr, D. 2001a. Impact of high intensity electric field pulses on cell permeabilization and as pre-processing step in coconut processing. *Innovative Food Science and Emerging Technologies* 1, 203–209.

Ade-Omowaye, B.I.O., Angersbach, A., Taiwo, K.A., and Knorr, D. 2001b. The use of pulsed electric fields in producing juice from paprika (*Capsicum Annuum* L.). *Journal of Food Processing Preservation* 25, 353–365.

Ade-Omowaye, B.I.O., Angersbach, A., Taiwo, K.A., and Knorr, D. 2001c. Use of pulsed electric field pre-treatment to improve dehydration characteristics of plant based foods. *Trends in Food Science and Technology* 12, 285–295.

Ade-Omowaye, B.I.O., Talens, P., Angersbach, A., and Knorr, D. 2003. Kinetics of osmotic dehydration of red bell peppers as influenced by pulsed electric field pretreatment. *Food Research International* 36, 475–483.

Aguilar-Rosas, S.F., Ballinas-Casarrubias, M.L., Nevarez-Moorillon, G.V., Martín-Belloso, O., and Ortega-Rivas, E. 2007. Thermal and pulsed electric fields pasteurization of apple juice: Effects on physicochemical properties and flavour compounds. *Journal of Food Engineering* 83, 41–46.

Aguiló-Aguayo, I., Odriozola-Serrano, I., Quintão-Teixeira, L.J., and Martín-Belloso, O. 2008a. Inactivation of tomato juice peroxidase by high-intensity pulsed electric fields as affected by process conditions. *Food Chemistry* 107(2), 949–955.

Aguiló-Aguayo, I., Oms-Oliu, G., Soliva-Fortuny, R., and Martín-Belloso, O. 2009a. Flavor retention and related enzyme activities during storage of strawberry juices processed by high-intensity pulsed electric fields or heat. *Food Chemistry* 116, 59–65.

Aguiló-Aguayo, I., Oms-Oliu, G., Soliva-Fortuny, R., and Martín-Belloso, O. 2009b. Changes in quality attributes throughout storage of strawberry juice processed by high-intensity pulsed electric fields. *LWT—Food Science and Technology* 42(4), 813–818.

Aguiló-Aguayo, I., Sobrino-López, A., Soliva-Fortuny, R., and Martín-Belloso, O. 2008b. Influence of high-intensity pulsed electric field processing on lipoxygenase and β-glucosidase activities in strawberry juice. *Innovative Food Science and Emerging Technologies* 9, 455–462.

Aguiló-Aguayo, I., Soliva-Fortuny, R., and Martín-Belloso, O. 2008c. Comparative study con color, viscosity and related enzymes of tomato juice treated by high-intensity pulsed electric fields or heat. *European Food Research and Technology* 227, 599–606.

Aguiló-Aguayo, I., Soliva-Fortuny, R., and Martín-Belloso, O. 2009c. Changes on viscosity and pectolytic enzymes of tomato and strawberry juices processed by high-intensity pulsed electric fields. *International Journal of Food Science and Technology* 44, 2268–2277.

Aguiló-Aguayo, I., Soliva-Fortuny, R., and Martín-Belloso, O. 2009d. Avoiding non-enzymatic browning by high-intensity pulsed electric fields in strawberry, tomato and watermelon juices. *Journal of Food Engineering* 92, 37–43.

Aguiló-Aguayo, I., Soliva-Fortuny, R., and Martín-Belloso, O. 2009e. Effects of high-intensity pulsed electric fields on lipoxygenase and hydroperoxide lyase activities in tomato juice. *Journal of Food Science* 74(8), C595–C601.

Aguiló-Aguayo, I., Soliva-Fortuny, R., and Martín-Belloso, O. 2010a. High-intensity pulsed electric fields processing parameters affecting polyphenoloxidase activity of strawberry juice. *Journal of Food Science* 75(79), C641–C646.

Aguiló-Aguayo, I., Soliva-Fortuny, R., and Martín-Belloso, O. 2010b. Impact of high-intensity pulsed electric field variables affecting peroxidase and lipoxygenase activities of watermelon juice. *LWT—Food Science and Technology* 43(6), 897–902.

Aguiló-Aguayo, I., Soliva-Fortuny, R., and Martín-Belloso, O. 2010c. Color and viscosity of watermelon juice treated by high-intensity pulsed electric fields or heat. *Innovative Food Science and Emerging Technologies* 11(2), 299–305.

Aguiló-Aguayo, I., Soliva-Fortuny, R., and Martín-Belloso, O. 2010d. Optimizing critical high-intensity pulsed electric fields treatments for reducing pectolytic activity and viscosity changes in watermelon juice. *European Food Research and Technology* 231(4), 509–517.

Akin, E. and Evrendilek, G.A. 2009. Effect of pulsed electric fields on physical, chemical, and microbiological properties of formulated carrot juice. *Food Science and Technology International* 15, 275–282.

Alkhafaji, S.R. and Farid, M. 2007. An investigation on pulsed electric fields technology using new treatment chamber design. *Innovative Food Science and Emerging Technologies* 8, 205–212.

Altuntas, J., Evrendilek, G.A., Sangun, H.K., and Zhang, H.Q. 2010. Effects of pulsed electric field processing on the quality and microbial inactivation of sour cherry juice. *International Journal of Food Science and Technology* 45(5), 899–905.

Amami, E., Vorobiev, E., and Kechaou, N. 2005. Effect of pulsed electric field on the osmotic dehydration and mass transfer kinetics of apple tissue. *Dry Technology* 23, 581–595.

Angersbach, A., Heinz, V., and Knorr, D. 2000. Effect of pulsed electric fields on cell membranes in real food systems. *Innovative Food Science and Emerging Technologies* 1, 135–149.

Angersbach, A., Heinz, V., and Knorr, D. 2002. Evaluation of process induced dimensional changes in the membrane structure of biological cells using impedance measurement. *Biotechnology Progress* 18, 597–603.

Angersbach, A. and Knorr, D. 1997. Anwendung elektrischer Hochspannungsimpulse als Vorbehandlungsverfahren zur Beeinflussung der Trocknungscharakteristika und Rehydratationseigenschaften von Kartoffelwürfeln. *Nahrung* 41(4), 194–200.

Balasa, A., Toepfl, S., and Knorr, D. 2006. *Pulsed Electric Field Treatment of Grapes.* Food Factory of the Future 3, Gothenburg, Sweden.

Barbosa-Cánovas, G.V., Góngora-Nieto, M.M., Pothakamury, U.R., and Swanson, B.G. 1999. *Preservation of Foods with Pulsed Electric Fields.* San Diego, CA: Academic Press.

Barbosa-Cánovas, G.V., Pothakamury, U.R., Palou, E., and Swanson, B.G. 1998. Biological effects and applications of pulsed electric fields for the preservation of foods. In *Nonthermal Preservation of Foods*, pp. 73–112. New York: Marcel Dekker Inc.

Barsotti, L. and Cheftel, J.C. 1999. Food processing by pulsed electric fields. II. Biological aspects. *Food Reviews International* 15, 181–213.

Bazhal, M., Lebovka, N., and Vorobiev, E. 2003. Optimisation of pulsed electric field strength for electroplasmolysis of vegetable tissues. *Biosystems Engineering* 86, 339–345.

Bazhal, M. and Vorobiev, E. 2000. Electrical treatment of apple cossettes for intensifying juice pressing. *Journal of the Science of Food and Agriculture* 80, 1668–1674.

Benz, R. and Zimmermann, U. 1980. Relaxation studies on cell membranes and lipid bilayers in the high electric field range. *Bioelectrochemistry and Bioenergetics* 7, 723–739.

Boussetta, N., Lebovka, N., Vorobiev, E., Adenier, H., Bedel-Cloutour, C., and Lanoiselle, J.L. 2009. Electrically assisted extraction of soluble matter from chardonnay grape skins for polyphenol recovery. *Journal of Agricultural and Food Chemistry* 57, 1491–1497.

Bouzrara, H. and Vorobiev, E. 2000. Beet juice extraction by pressing and pulsed electric field. *International Sugar Journal* 102, 194–200.

Brodelius, P.E., Funk C., and Shillito, R.D. 1988. Permeabilization of cultivated plant cells by electroporation for release of intracellularly stored secondary products. *Plant Cell Report* 7, 186–188.

Burt, S. 2004. Essential oils: Their antibacterial properties and potential applications in foods—A review. *International Journal of Food Microbiology* 94, 223–253.

Castro, A.J., Barbosa-Cánovas, G.V., and Swanson, B.G. 1993. Microbial inactivation of foods by pulsed electric fields. *Journal of Process preservation* 17, 47–73.

Chalermchat, Y. and Dejmek, P. 2005. Effect of pulsed electric field pretreatment on solid-liquid expression from potato tissue. *Journal of Food Engineering* 71, 164–169.

Chen, J., Zhang, R., Xiao, J., Li, J., Wang, L., Guan, Z., and MacAlpine, M. 2010. Influence of pulse rise time on the inactivation of *Staphylococcus aureus* by pulsed electric fields. *IEEE Transactions on Plasma Science* 38(8 part 2), 1935–1941.

Cortés, C., Torregrosa, F., Esteve, M.J., and Frígola, A. 2006. Carotenoid profile modification during refrigerated storage in untreated and pasteurized orange juice and orange juice treated with high-intensity pulsed electric fields. *Journal of Agricultural and Food Chemistry* 54, 6247–6254.

Coster, H.G. and Zimmermann, U. 1975. The mechanism of electric breakdown in the membranes of *Valonia utricularis*. *Journal of Membrane Biology* 22, 73–90.

Cserhalmi, Zs., Sass-Kiss, A., Tóth-Markus, M., and Lechner, N. 2006. Study of pulsed electric field treated citrus juices. *Innovative Food Science and Emerging Technologies* 7, 49–54.

De Haan, S.W.H. and Willcock, P.R. 2002. Comparison of the energy performance of pulse generator circuits for HIPEF. *Innovative Food Science and Emerging Technologies* 3, 349–356.

DIL (German Institute for Food Technology). Generations and Technology (on line). Available at: <http://www.dil-ev.de/fileadmin/user_upload/pdf/elcrack_brochure.pdf> (January 19, 2011).

Donsì, F., Giovanna, F., and Gianpiero, P. 2010. Applications of pulsed electric field treatments for the enhancement of mass transfer from vegetable tissue. *Food Engineering Reviews* 2, 109–130.

Dörnerburg, H. and Knorr, D. 1993. Cellular permeabilization of cultured plant tissues by high electric field pulses or ultra high pressure for the recovery of secondary metabolites. *Food Biotechnology* 7(1), 35–48.

Dunn, J.E. 2001. Pulsed electric field processing: An overview. In *Pulsed Electric Fields in Food Processing: Fundamental Aspects and Applications*, eds. G.V. Barbosa-Cánovas and Q.H. Zhang, pp. 1–30. Lancaster, U.K.: Technomic.

Dunn, J.E. and Pearlman, J.S. 1987. Methods and apparatus for extending the shelf life of fluid food products. U.S. Patent 4695472, September 22.

Dutreux, N., Notermans, S., Witjez, T., Góngora-Nieto, M.M., Barbosa-Cánovas, G.V., and Swanson, B.G. 2000. Pulsed electric fields inactivation of attached and free-living *Escherichia coli* and *Listeria innocua* under several conditions. *International Journal of Food Microbiology* 54, 91–98.

El-Belghiti, K., Rabhi, Z., and Vorobiev, E. 2005. Effect of centrifugal force on the aqueous extraction of solute from sugar beet tissue pretreated by a pulsed electric field. *Journal of Food Processing Engineering* 28, 346–358.

El-Belghiti, K. and Vorobiev, E. 2004. Mass transfer of sugar from beets enhanced by pulsed electric field. *Food Bioproducts Processing* 82, 226–230.

El-Hag, A., Otunola, A., Jayaram, S.H., and Anderson, W.A. 2008. Reduction of microbial growth in milk by pulsed electric fields. In *2008 IEEE International Conference on Dielectric Liquids*, *(ICDL)*, Futurescope, Chasseneuil Cedex, France, June 30–July 3, 2008.

Elez-Martínez, P., Aguiló-Aguayo, I., and Martín-Belloso, O. 2006a. Inactivation of orange juice peroxidase by high-intensity pulsed electric fields as influenced by processing parameters. *Journal of the Science of Food and Agriculture* 86, 71–81.

Elez-Martínez, P., Escolà-Hernández, J., Espachs-Barroso, A., Barbosa-Cánovas, G.V., and Martín-Belloso, O. 2005. Inactivation of *Lactobacillus brevis* in orange juice by high intensity pulsed electric fields. *Food Microbiology* 22(4), 311–319.

Elez-Martinez, P., Escolá-Hernández, J., Soliva-Fortuny, R.C., and Martín-Belloso, O. 2004. Inactivation of *Saccharomyces cerevisiae* suspended in orange juice using high intensity pulsed electric fields. *Journal of Food Protection* 67, 2596–2602.

Elez-Martínez, P., Soliva-Fortuny, R., and Martín-Belloso, O. 2006b. Comparative study on shelf-life of orange juice processed by high intensity pulsed electric fields or heat treatment. *European Food Research and Technology* 222, 321–329.

Elez-Martínez, P., Suárez-Recio, M., and Martín-Belloso, O. 2007. Modeling the reduction of pectin methyl esterase activity in orange juice by high intensity pulsed electric fields. *Journal of Food Engineering* 78, 184–193.

Eshtiaghi, M.N. and Knorr, D. 1999. Process for treatment of sugar beet. European Patent, EP 99923708.

Eshtiaghi, M.N. and Knorr, D. 2000. Anwendung elektrischer Hochspannungsimpulse zum Zellaufschluss bei der Saftgewinnung am Beispiel von Weintrauben. *Lebensmitter-Wissenschaft and Technologie* 45, 23–27.

Eshtiaghi, M.N. and Knorr, D. 2002. High electric field pulse pretreatment: Potential for sugar beet processing. *Journal of Food Engineering* 53(2), 265–272.

Evrendilek, G.A., Jin, Z.T., Ruhlman, K.T., Qiu, X., Zhang, Q.H., and Ritcher, E.R. 2000. Microbial safety and shelf-life of apple juice and cider processed by bench and pilot scale HIPEF systems. *Innovative Food Science and Emerging Technologies* 1, 77–86.

Evrendilek, G.A. and Zhang, Q.H. 2005. Effects of pulse polarity and pulse delaying time on pulsed electric fields-induced pasteurization of *E. coli* 0157:H7. *Journal of Food Engineering* 68, 271–276.

FDA. 2004. *The Food and Drug Administration, Guidance for Industry: Juice HACCP; Small Entity Compliance Guide*, March 2004. Docket Number 02D-0333. Food and Drug Administration, Rockville, MD. http://www.cfsan.fda.gov/~dms/guidance.html

Ferrer, C., Rodrigo, D., Pina, M.C., Klein, G., Rodrigo, M., and Martínez, A. 2007. The Monte Carlo simulation is used to establish the most influential parameters on the final load of pulsed electric fields *E. coli* cells. *Food Control* 18(8), 934–938.

Fincan, M., De Vito, F., and Dejmek, P. 2004. Pulsed electric field treatment for solid-liquid extraction of red beetroot pigment. *Journal of Food Engineering* 64, 381–388.

Fincan, M. and Dejmek, P. 2002. In situ visualization of the effect of a pulsed electric field on plant tissue. *Journal of Food Engineering* 55(3), 223–230.

Gachovska, T.K., Adedeji, A.A., and Ngadi, M.O. 2009. Influence of pulsed electric field energy on the damage degree in alfalfa tissue. *Journal of Food Engineering* 95, 558–563.

Giner, J., Gimeno, V., Barbosa-Cánovas, G.V., and Martín, O. 2001. Effects of pulsed electric field processing on apple and pear polyphenoloxidases. *Food Science and Technology International* 784, 339–345.

Giner, J., Gimeno, V., Espachs, A., Elez, P., Barbosa-Cánovas, G.V., and Martín, O. 2000. Inhibition of tomato (*Licopersicon esculentum* mill.) pectin methylesterase by pulsed electric fields. *Innovative Food Science and Emerging Technologies* 1(1), 57–67.

Giner, J., Ortega, M., Mesegué, M., Gimeno, V., Barbosa-Cánovas, G.V., and Martín, O. 2002. Inactivation of peach polyphenoloxidase by exposure to pulsed electric fields. *Journal of Food Science* 67(4), 1467–1472.

Góngora-Nieto, M.M., Pedrow, P.D., Swanson, B.G., and Barbosa-Cánovas, G.V. 2003. Energy analysis of liquid whole egg pasteurized by pulsed electric fields. *Journal of Food Engineering* 57, 209–216.

Góngora-Nieto, M.M., Sepúlveda, D.R., Pedrow, P., Barbosa-Cánovas, G.V., and Swanson, B.G. 2002. Food processing by pulsed electric fields: Treatment delivery, inactivation level and regulatory aspects. *LWT—Food Science and Technology* 35, 375–388.

Grahl, T. and Märkl, H. 1996. Killing of microorganisms by pulsed electric fields. *Applied Microbiology and Biotechnology* 45, 148–157.

Grahl, T., Sitzmann, W., and Märkl, H. 1992. Killing of microorganisms in fluid media by high-voltage pulses. *Dechema Biotechnology Conference* Series 5B, 675–678.

Grimi, N., Lebovka, N.I., Vorobiev, E., and Vaxelaire, J. 2009. Effect of a pulsed electric field treatment on expression behavior and juice quality of chardonnay grape. *Food Biophysics* 4, 191–198.

Grimi, N., Praporscic, I., Lebovka, N., and Vorobiev, E. 2007. Selective extraction from carrot slices by pressing and washing enhanced by pulsed electric fields. *Separation and Purification Technology* 58, 267–273.

Guderjan, M., Elez-Martinez, P., and Knorr, D. 2007. Application of pulsed electric fields at oil yield and content of functional food ingredients at the production of rapeseed oil. *Innovative Food Science and Emerging Technologies* 8, 55–62.

Guderjan, M., Toepfl, S., Angersbach, A., and Knorr, D. 2005. Impact of pulsed electric field treatment on the recovery and quality of plant oils. *Journal of Food Engineering* 67, 281–287.

Heinz, V., Alvarez, I., Angersbach, A., and Knorr, D. 2002. Preservation of liquid foods by high intensity pulsed electric fields-basic concepts for process design. *Trends in Food Science and Technology* 12, 103–111.

Ho, S.Y. and Mittal, G.S. 1996. Electroporation of cell membranes: A review. *Critical Reviews in Biotechnology* 16(4), 349–362.

Ho, S.Y., Mittal, G.S., and Cross, J.D. 1997. Effects of high field electric pulses on the activity of selected enzymes. *Journal of Food Engineering* 31, 69–84.

Ho, S.Y., Mittal, G.S., Cross, J.D., and Griffiths, M.W. 1995. Inactivation of *Pseudomonas fluorescens* by high voltage electric pulses. *Journal of Food Science* 60, 1337–1340.

Hodgins, A.M., Mittal, G.S., and Griffiths, M.W. 2002. Pasteurization of fresh orange juice using low-energy pulsed electrical field. *Journal of Food Science* 67, 2294–2299.

Hoogland, H. and de Haan, W. 2007. Economic aspects of pulsed electric field treatment of food. In *Food Preservation by Pulsed Electric Fields: From Research to Application*, eds. H.L.M. Lelieveld, S. Notermans, and S.W.H. de Haan, pp. 257–265. Cambridge, U.K.: Woodhead publishing in Food Science, Technology and Nutrition.

Huang, K. and Wang, J. 2009. Designs of pulsed electric fields treatment chambers for liquid foods pasteurization process: A review. *Journal of Food Engineering* 95, 227–239.

Hülsheger, H., Potel, J., and Niemann, E.G. 1981. Killing of bacteria with electric pulses of high field strength. *Radiation and Environmental Biophysics* 20, 53–65.

Iu, J., Mittal, G.S., and Griffiths, M.W. 2001. Reductions in levels of *Escherichia coli* O157:H7 in apple cider by pulsed electric fields. *Journal of Food Protection* 64, 964–969.

Jayaram, S., Castle, G.S.P., and Margatiris, A. 1992. Kinetics of sterilization of *Lactobacillus brevis* cells by the application of high voltage pulses. *Biotechnology and Bioengineering* 40, 1412–1420.

Jemai, A.B. and Vorobiev, E. 2002. Effect of moderate electric field pulses on the diffusion coefficient of soluble substances from apple slices. *International Journal of Food Science and Technology* 37, 73–86.

Jemai, A.B. and Vorobiev, E. 2003. Enhanced leaching from sugar beet cossettes by pulsed electric field. *Journal of Food Engineering* 59, 405–412.

Knorr, D. 1994. Novel processes for the production of fruit and vegetables juices. *Flussiges Öbst* 61(10), 294–296.

Knorr, D. 1999. Novel approaches in food-processing technology: New technologies for preserving foods and modifying function. *Current Opinion in Biotechnology* 10, 485–491.

Knorr, D., Angersbach, A., Eshtiaghi, M.N., Heinz, V., and Lee, D. 2001. Processing concepts based on high intensity electric field pulses. *Trends in Food Science and Technology* 12, 129–135.

Knorr, D., Guelen, M., Grahl, T., and Sitzmann, W. 1994. Food application of high electric field pulses. *Trends in Food Science and Technology* 5, 71–75.

Kotnik, T., Pucihar, G., Rebersek, M., Miklavic, D., and Mir, M.L. 2003. Role of pulse shape in cell membrane electropermeabilization. *BBA-Biomembranes* 1614, 193–200.

Lebovka, N.I., Praporscic, I., and Vorobiev, E. 2003. Enhanced expression of juice from soft vegetable tissues by pulsed electric fields: Consolidation stages analysis. *Journal of Food Engineering* 59, 309–317.

Lebovka, N.I., Praporscic, I., and Vorobiev, E. 2004. Combined treatment of apples by pulsed electric fields and by heating at moderate temperature. *Journal of Food Engineering* 65, 211–217.

Lebovka, N.I., Shynkaryk, N.V., and Vorobiev, E. 2007. Pulsed electric field enhanced drying of potato tissue. *Journal of Food Engineering* 78, 606–613.

Liang, Z., Mittal, G.S., and Griffiths, M.W. 2002. Inactivation of *Salmonella* Typhimurium in orange juice containing antimicrobial agents by pulsed electric field. *Journal of Food Protection* 65, 1081–1087.

Loginova, K.V., Shynkaryk, M.V., Lebovka, N.I., and Vorobiev, E. 2010. Acceleration of soluble matter extraction from chicory with pulsed electric fields. *Journal of Food Engineering* 96, 374–379.

Lopez, N., Puertolas, E., Condon, S., Alvarez, I., and Raso, J. 2008. Effects of pulsed electric fields on the extraction of phenolic compounds during the fermentation of must of Tempranillo grapes. *Innovative Food Science and Emerging Technologies* 9, 477–482.

Lopez, N., Puertolas, E., Condon, S., Raso, J., and Alvarez, I. 2009. Enhancement of the extraction of betanine from red beetroot by pulsed electric fields. *Journal of Food Engineering* 90, 60–66.

Luo, W., Zhang, R., Wang, L., Chen, J., Guan, Z., Liao, X., and Mo, M. 2008. Effects of HIPEF-assisted extraction of anthocyanin in red raspberry. *Annual Report—Conference on Electrical Insulation and Dielectric Phenomena (CEIDP)*, Québec, Canada, IEEE, October 26–29, 2008, pp. 630–632.

Marquez, V.O., Mittal, G.S., and Griffiths, M.W. 1997. Destruction and inhibition of bacterial spores by high voltage pulsed electric field. *Journal of Food Science* 62, 399–401.

Marselles-Fontanet, A.R. and Martín-Belloso, O. 2007. Optimization and validation of HIPEF processing conditions to inactivate oxidative enzymes of grape juice. *Journal of Food Engineering* 83, 452–462.

Martín-Belloso, O. and Elez-Martínez, P. 2005. Food Safety aspects of pulsed electric fields. In *Emerging Technologies for Food Processing*, ed. D.W. Sun, pp. 183–208. Oxford, U.K.: Elsevier.

Martínez-Viedma, P., Abriouel, H., Sobrino-López, A., Ben-Omar, N., Lucas-López, R., Valdivia, E., Martín-Belloso, O., and Gálvez, A. 2009. Effect of enterocin AS-48 in combination with high-intensity pulsed-electric field treatment against the spoilage bacterium *Lactobacillus diolivorans* in apple juice. *Food Microbiology* 26859, 491–496.

Mazurek, B., Lubicki, P., and Staroniewicz, Z. 1995. Effect of short HV pulses on bacteria and fungi. *IEEE Transactions on Dielectrics and Electrical Insulation* 2, 418–425.

McDonald, C.J., Lloyd, S.W., Vitale, M.A., Perterson, K., and Inning, F. 2000. Effects of pulsed electric fields on microorganisms in orange juice using electric fields strengths of 30 and 50 kV/cm. *Journal of Food Science* 65, 984–989.

McLellan, M.R., Kime, R.L., and Lind, L.R. 1991. Electroplasmolysis and other treatments to improve apple juice yield. *Journal of the Science of Food and Agriculture* 57, 303–306.

Min, S., Evrendilek, G.A., and Zhang, Q.H. 2007. Pulsed electric fields: Processing system, microbial and enzyme inhibition, and shelf-life extension of foods. *IEEE Transactions of Plasma Science* 35, 59–73.

Min, S., Jin, Z.T., Min, S.K., Yeom, H., and Zhang, Q.H. 2003a. Commercial-scale pulsed electric field processing of orange juice. *Journal of Food Science* 68, 1265–1271.

Min, S., Jin, Z.T., and Zhang, Q.H. 2003b. Commercial scale pulsed electric field processing of tomato juice. *Journal of Agricultural and Food Chemistry* 51, 3338–3344.

Min, S., Min, S.K., and Zhang, Q.H. 2003c. Inactivation kinetics of tomato juice lipoxygenase by pulsed electric fields. *Journal of Food Science* 68, 1995–2001.

Min, S. and Zhang, Q.H. 2003. Effects of commercial-scale pulsed electric field processing on flavor and color of tomato juice. *Journal of Food Science* 68, 1600–1606.

Mittal, G.S. and Griffiths, M.W. 2005. Pulsed electric field processing of liquid foods and beverages. In *Emerging Technologies for Food Processing*, ed. D.W. Sun, pp. 99–139. Amsterdam, the Netherlands: Elsevier.

Mizuno, A. and Hori, Y. 1988. Destruction of living cells by pulsed high voltage application. *IEEE Transactions on Industry Applications* 24(3), 387–394.

Molinari, P., Pilosof, A.M.R., and Jagus, M.J. 2004. Effect of growth phase and inoculum size on the inactivation of *Saccharomyces cerevisiae* in fruit juices, by pulsed electric fields. *Food Research International* 37, 793–798.

Mosqueda-Melgar, J. 2007. Aplicación de pulsos eléctricos de alta intensidad de campo en combinación con sustancias antimicrobianas para garantizar la calidad e inocuidad microbiológica de zumos de fruta. Thesis document. University of Lleida, Lleida, Spain.

Mosqueda-Melgar, J., Elez-Martínez, P., Raybaudi-Massilia, R.M., and Martín-Belloso, O. 2008a. Effects of pulsed electric fields on pathogenic microorganisms of major concern in fluid foods. *Critical Reviews in Food Science and Nutrition* 48, 747–759.

Mosqueda-Melgar, J., Raybaudi-Massilia, R.M., and Martín-Belloso, O. 2007. Influence of treatment time and pulse frequency on *Salmonella* Enteritidis, *Escherichia coli* and *Listeria monocytogenes* populations inoculated in melon and watermelon juices treated by pulsed electric fields. *International Journal of Food Microbiology* 117, 192–200.

Mosqueda-Melgar, J., Raybaudi-Massilia, R.M., and Martín-Belloso, O. 2008b. Non-thermal pasteurization of fruit juices by combining high-intensity pulsed electric fields with natural antimicrobials. *Innovative Food Science and Emerging Technologies* 9, 328–340.

Mosqueda-Melgar, J., Raybaudi-Massilia, R.M., and Martín-Belloso, O. 2008c. Inactivation of *Salmonella enterica* Ser. Enteritidis in tomato juice by combining of high-intensity pulsed electric fields with natural antimicrobials. *Journal of Food Science* 73, M47–M53.

Mosqueda-Melgar, J., Raybaudi-Massilia, R.M., and Martín-Belloso, O. 2008d. Combination of high-intensity pulsed electric fields with natural antimicrobials to inactivate pathogenic microorganisms and extend the shelf-life of melon and watermelon juices. *Food Microbiology* 25, 479–491.

Neumann, E. 1996. Gene delivery by membrane electroporation. In *Electrical Manipulation of Cells*, eds. P.T. Lynchm and M.R. Davey, pp. 157–184. New York: Chapman and Hall.

NFR. 1997. EU Council Directive EC (258/97), Novel Food Regulation (1997).

Nguyen, P. and Mittal, G.S. 2007. Inactivation of naturally occurring microorganisms in tomato juice using pulsed electric field (PEF) with and without antimicrobials. *Chemical Engineering and Processing* 46, 360–365.

Odriozola-Serrano, I., Soliva-Fortuny, R., Gimeno-Añó, V., and Martín-Belloso, O. 2008a. Modeling changes in health-related compounds of tomato juice treated by high-intensity pulsed electric fields. *Journal of Food Engineering* 89, 210–216.

Odriozola-Serrano, I., Soliva-Fortuny, R., Hernández-Jover, T., and Martín-Belloso, O. 2008b. Carotenoid and phenolic profile of tomato juices processed by high intensity pulsed electric fields compared with conventional thermal treatments. *Food Chemistry* 112, 258–266.

Odriozola-Serrano, I., Soliva-Fortuny, R., and Martín-Belloso, O. 2008c. Changes of health-related compounds throughout cold storage of tomato juice stabilized by thermal or high intensity pulsed electric field treatments. *Innovative Food Science and Emerging Technologies*, 9, 272–279.

Odriozola-Serrano, I., Soliva-Fortuny, R., and Martín-Belloso, O. 2008d. Phenolic acids, flavonoids, vitamin C and antioxidant capacity of strawberry juices processed by high intensity pulsed electric fields or heat treatments. *European Food Research and Technology* 228, 239–248.

Olajide, J.O., Adedeji, A.A., Ade-Omowaye, B.I.O., Otunola, E.T., and Adejuyitan, J.A. 2006. Potentials of high intensity electric field pulses (HELP) to food processors in developing countries. *Nutrition and Food Science* 3684, 248–258.

Oms-Oliu, G., Odriozola-Serano, I., Soliva-Fortuny, R., and Martín-Belloso, O. 2009. Effects of high-intensity pulsed electric field processing conditions on lycopene, vitamin C and antioxidant capacity of watermelon juice. *Food Chemistry* 115, 1312–1319.

Pothakamury, U.R., Vega, H., Zhang, Q., Barbosa-Cánovas, G.V., and Swanson, B.G. 1996. Effect of growth stage and processing temperature on the inactivation of *E. coli* by pulsed electric fields. *Journal of Food Protection* 59, 1167–1171.

Praporscic, I., Ghnimi, S., and Vorobiev, E. 2005. Enhancement of pressing of sugar beet cuts by combined ohmic heating and pulsed electric field treatment. *Journal of Food Processing* 29, 378–389.

Praporscic, I., Lebovka, N., Vorobiev, E., and Mietton-Peuchot, M. 2007. Pulsed electric field enhanced expression and juice quality of white grapes. *Separation and Purification Technology* 52, 520–526.

Qin, B.L., Barbosa-Cánovas, G.V., Swanson, B.G., Pedrow, P.D., and Olsen, R.G. 1998. Inactivating microorganisms using a pulsed electric field continuous treatment system. *IEEE Transactions on Industry Applications* 34, 43–50.

Qin, B., Zhang, Q., Barbosa-Cánovas, G.V., Swanson, B.G., and Pedrow, P.D. 1994. Inactivation of microorganisms by pulsed electric fields of different voltage waveforms. *IEEE Transactions on Dielectrics and Electrical Insulation* 1, 1047–1057.

Rastogi, N.K. 2003. Application of high intensity pulsed electric fields in food processing. *Food Reviews International* 19(3), 229–251.

Rastogi, N.K., Anghersbach, A., and Knorr, D. 2000. Evaluation of mass transfer mechanism during osmotic treatment of plant materials. *Journal of Food Science* 65(6), 1016–1019.

Rastogi, N.K., Eshtiaghi, M.N., and Knorr, D. 1999. Accelerated mass transfer during osmotic dehydration of high intensity electrical field pulse pretreated carrots. *Journal of Food Science* 64, 1020–1023.

Raybaudi-Massilia, R.M., Mosqueda-Melgar, J., and Martín-Belloso, O. 2006. Antimicrobial activity of essential oils on *Salmonella* enteritidis, *Escherichia coli*, and *Listeria innocua* in fruit juices. *Journal of Food Protection* 69(7), 1579–1586.

Riener, J., Noci, F., Cronin, D.A., Morgan, D.J., and Glyng, J.G. 2008. Combined effect of temperature and pulsed electric fields on apple juice peroxidase and polyphenoloxidase inactivation. *Food Chemistry* 109, 402–407.

Riener, J., Noci, F., Cronin, D.A., Morgan, D.J., and Glyng, J.G. 2009. Combined effect of temperature and pulsed electric fields on pectin methyl esterase inactivation in red grape fruit juice (*Citrus paradisi*). *European Food Research Technology* 228, 373–379.

Riley, T. and Watson, A. 1987. *Polarography and Other Voltammetric Methods*. New York: John Wiley & Sons.

Rivas, A., Rodrigo, D., Martínez, A., Barbosa-Cánovas, G.V., and Rodrigo, M. 2006. Effect of HIPEF and heat pasteurization on the physical-chemical characteristics of blended orange and carrot juice. *LWT—Food Science and Technology* 39, 1163–1170.

Rodrigo, D., Barbosa-Cánovas, G.V., Martínez, A., and Rodrigo, M. 2003. Pectin methyl esterase and natural microflora of fresh mixed orange juice and carrot juice treated with high intensity pulsed electric fields. *Journal of Food Protection* 66, 2336–2342.

Sale, A.J.H. and Hamilton, V.A. 1967. Effects of high electric fields on microorganisms. I. Killing of bacteria and yeasts. *Biochimica et Biophysica Acta* 148, 781–788.

Sánchez-Moreno, C., Cano, M.P., De Ancos, B., Plaza, L., Olmedilla, B., Granado, F., Elez-Martínez, P., Martín-Belloso, O., and Martín, A. 2005a. Intake of Mediterranean vegetable soup treated by pulsed electric fields affects plasma vitamin C and antioxidant biomarkers in humans. *International Journal of Food Science and Nutrition* 56, 115–124.

Sánchez-Moreno, C., Plaza, L., Elez-Martínez, P., De Ancos, B., Martín-Belloso, O., and Cano, M.P. 2005b. Impact of high pressure and pulsed electric fields on bioactive compounds and antioxidant activity of orange juice in comparison with traditional thermal processing. *Journal of Agricultural and Food Chemistry* 53, 4403–4409.

Schilling, S., Alber, T., Toepfl, S., Neidhart, S., Knorr, D., Schieber, A., and Carle, R. 2007. Effects of pulsed electric field treatment of apple mash on juice yield and quality attributes of apple juices. *Innovative Food Science and Emerging Technologies* 8, 127–134.

Schilling, S., Schmid, S., Jäger, H., Ludwig, M., Dietrich, H., Toepfl, S., Knorr, D., Neidhart, S., Schieber, A., and Carle, R. 2008. Comparative study of pulsed electric field and thermal processing of apple juice with particular consideration of juice quality and enzyme deactivation. *Journal of Agricultural and Food Chemistry* 56, 4545–4554.

Schultheiss, C., Bluhm, H.J., Mayer, H.G., Kern, M., Michelberger, T., and Witte, G. 2002. Processing of sugar beets with pulsed electric fields. *IEEE Transactions on Plasma Science* 30(4), 1547–1551.

Schultheiss, C., Bluhm, H.J., Sack, M., and Kern, M. 2004. Principle of electroporation and development of industrial devices. *Zuckerindustrie* 129(1), 40–44.

Sensoy, I. and Sastry, S.K. 2004. Extraction using moderate electric fields. *Journal of Food Science* 69(1), 7–13.

Sensoy, I., Zhang, Q.H., and Sastry, S.K. 1997. Inactivation kinetics of *Salmonella dublin* by pulsed electric field. *Journal of Food Processing Engineering* 20(5), 367–381.

Soliva-Fortuny, R., Balasa, A., Knorr, D., and Martín-Belloso, O. 2009. Effects of pulsed electric fields on bioactive compounds in foods: A review. *Trends in Food Science and Technology* 20, 544–556.

Sonne, A.M., Grünert, K.G., Olsen, N.V., and Bánáti, D. 2010. Consumers' perceptions of HHP and HIPEF food products. *The British Food Journal*, 52(1), 115–126.

Taiwo, K.A., Angersbach, A., Ade-Omowaye, B.I.O., and Knorr, D. 2001. Effects of pretreatments on the diffusion kinetics and some quality parameters of osmotically dehydrated apple slices. *Journal of Agricultural and Food Chemistry* 49, 2804–2811.

Taiwo, K.A., Angersbach, A., and Knorr, D. 2003a. Effects of pulsed electric field on quality factors and mass transfer during osmotic dehydration of apples. *Journal of Food Processing and Engineering* 26, 31–48.

Taiwo, K.A., Eshtiaghi, M.N., Ade-Omowaye, B.I.O., and Knorr, D. 2003b. Osmotic dehydration of strawberry halves: Influence of osmotic agents and pretreatment methods on mass transfer and product characteristics. *International Journal of Food Science and Technology* 38, 693–707.

Tedjo, W., Taiwo, K.A., Eshtiaghi, N., and Knorr, D. 2002. Comparison of pretreatment methods on water and solid diffusion kinetics of osmotically dehydrated mangos. *Journal of Food Engineering* 53, 35–43.

Toepfl, S., Heinz, V., and Knorr, D. 2005. Overview of pulsed electric field processing of foods. In *Emerging Technologies for Food Processing*, ed. D.W. Sun, pp. 67–97. Oxford, U.K.: Elsevier.

Toepfl, S., Heinz, V., and Knorr, D. 2006. Applications of pulsed electric field technology for the food industry. In *Pulsed Electric Field Treatment of Foods*, eds. J. Raso and V. Heinz, pp. 197–221. Oxford, U.K.: Elsevier.

Toepfl, S., Heinz, V., and Knorr, D. 2007. High intensity pulsed electric fields applied for food preservation. *Chemical Engineering and Process* 46, 537–546.

Toepfl, S. and Knorr, D. 2006. Pulsed electric fields as a pretreatment in drying processes. *Stewart Postharvesting Reviews* 3(4), 1–6.

Torregrosa, F., Cortés, C., Esteve, M.J., and Frígola, A. 2005. Effects of high-intensity pulsed electric fields processing and conventional heat treatment on orange-carrot juice carotenoids. *Journal Agriculture and Food Chemistry* 53, 9519–9525.

Torregrosa, F., Esteve, M.J., Frigola, A., and Cortés, C. 2006. Ascorbic acid stability during refrigerated storage of orange-carrot juice treated by high pulsed electric field and comparison with pasteurized juice. *Journal of Food Engineering* 73, 339–345.

Tsong, T.Y. 1991. Electroporation of cell membranes. *Biophysical Journal* 60, 297–306.

Tsong, T.Y. and Astumian, R.D. 1986. Absorption and conversion of electric field energy by membrane bound ATPases. *Bioelectric and Bioenergetics* 15, 457–476.

Van Loey, A., Verachtert, B., and Hendrickx, M. 2002. Effects of high electric field pulses on enzymes. *Trends in Food Science and Technology* 12, 94–102.

Vega-Mercado, H., Martín-Belloso, O., Qin, B.L., Chang, F.J., Góngora-Nieto, M.M., Barbosa-Cánovas, G.V., and Swanson, B.G. 1997. Non-thermal food preservation: Pulsed electric fields. *Trends in Food Science and Technology* 8(5), 151–157.

Weaver, J.C. and Chizmadzhev, Y.A. 1996. Theory of electroporation: A review. *Bioelectrochemistry and Bioenergetics* 41, 135–160.

Wouters, P.C., Alvarez, I., and Raso, J. 2001. Critical factors determining inactivation kinetics by pulsed electric field food processing. *Trends in Food Science and Technology* 12, 112–121.

Yeom, H.W., Streaker, C.B., Zhang, Q.H., and Min, D.B. 2000a. Effects of pulsed electric fields on the quality of orange juice and comparison with heat pasteurization. *Journal of Agricultural and Food Chemistry* 48, 4597–4605.

Yeom, H.W., Streaker, C.B., Zhang, Q.H., and Min, D.B. 2000b. Effects of pulsed electric fields on the activities of microorganisms and pectin methyl esterase in orange juice. *Journal of Food Science* 65, 1359–1363.

Yeom, H.W., Zhang, Q.H., and Chism, G.W. 2002. Inactivation of pectin methyl esterase in orange juice by pulsed electric fields. *Journal of Food Science* 67, 2154–2159.

Yin, Y.G., Cui, Y.R., and Ding, H.W. 2008. Optimization of betulin extraction process from *Inonotus obliquus* with pulsed electric fields. *Innovation Food Science and Emerging Technologies* 9, 306–310.

Yin, Y.G., Fan, X.D., Liu, F.X., Yu, Q.Y., and He, G.D. 2009. Fast extraction of pectin from apple pomace by high intensity pulsed electric field. *Journal of Jilin University (Engineering and Technology Edition)* 39(5), 1224–1228.

Zárate-Rodríguez, E., Ortegas-Rivas, E., and Barbosa-Cánovas, G.V. 2000. Quality changes in apple juice as related to nonthermal processing. *Journal of Food Quality* 23, 337–349.

Zhang, Q., Barbosa-Cánovas, G.V., and Swanson, B.G. 1995. Engineering aspects of pulsed electric fields pasteurization. *Journal of Food Engineering* 25, 261–281.

Zhang, Q., Chang, F.J., Barbosa-Cánovas, G.V., and Swanson, B.G. 1994. Inactivation of microorganisms in a semisolid model food using high voltage pulsed electric fields. *LWT—Food Science and Technology* 27, 538–543.

Zhang, Y., Liao, X., Ni, Y., Wu, J., Hu, X., Wang, Z., and Chen, F. 2007. Kinetic analysis of the degradation and its color change of cyanidin-3-glucoside exposed to pulsed electric field. *European Food Research and Technology* 224, 597–603.

Zheng-Ying, L. and Yan, W. 1993. Effects of high voltage pulse discharges on micro-organisms dispersed in liquid. In *Eighth International Symposium in High Voltage Engineering*, August 23–27, Yokohama, Japan.

Zhong, K., Hu, X., Zhao, G., Chen, F., and Liao, X. 2005. Inactivation and conformational change of horseradish peroxidase induced by pulsed electric field. *Food Chemistry* 92, 473–479.

Zhong, K., Wu, J., Wang, Z., Chen, F., Liao, X., Hu, X., and Zhang, Z. 2007. Inactivation kinetics and secondary structural change of HIPEF-treated POD and PPO. *Food Chemistry* 100, 115–123.

Zimmermann, U. 1986. Electric breakdown, electropermeabilization and electrofusion. *Reviews on Physiological Biochemical Pharmacology* 105, 175–256.

6 Applications of Ozone in Fruit Processing

Patrick J. Cullen and Brijesh K. Tiwari

CONTENTS

6.1 INTRODUCTION

Ozone (O_3) finds a wide range of applications in the food industry from food surface decontamination to sanitation of plant equipment to wastewater treatment. In recent years, there has been renewed interest in the use of O_3 within the food industry as a potential substitute to chlorine. The fresh produce industry is showing interest in O_3 applications due to growing consumer preference for minimally processed foods, frequent outbreaks of food-related illnesses, identification of new food pathogens, and the passage of legislation governing food quality and safety. Global legislation governing ozonation typically has developed in response to the evolving use of O_3 from initial applications for water treatment, to surface and equipment cleaning, to food produce washes, controlled atmospheric storage, and more recently as a direct food additive. O_3 is approved by the U.S. Food and Drug Administration (FDA) for use in the United States and has been employed successfully for applications including surface decontamination to extend the shelf life of fresh produce, decontamination of packaging materials, disinfecting of process water, and sanitizing of processing equipment and food storage areas, among others. Microbial safety is of paramount importance for the minimally processed fresh food industry. O_3 has been examined as an alternative sanitizing technology in the fresh produce industry. The FDA affirmation of O_3 with generally recognized as safe (GRAS) status triggered broad usage of O_3 gas in the food industry. The introduction of legislation governing ozonation applications in the United States, Canada, Australia, and New Zealand in food processing will also encourage the adoption of ozonation processes in industry. This chapter discusses the application of O_3 in fruit processing, state-of-art case studies, and challenges faced for industrial adoption of this technology.

6.2 PROPERTIES OF O_3

O_3 is a triatomic form of oxygen where the three atoms of oxygen in the O_3 molecule are arranged at an obtuse angle whereby a central oxygen atom is attached to two equidistant oxygen atoms; the included angle is approximately $116°49'$ with a bond length of $1.278\,\text{Å}$. The molecular weight of O_3 is $47.98\,\text{g/mol}$. At normal pressure of 1.1013 bar, liquid O_3 density is $1352\,\text{kg/m}^3$ with boiling point of $-111.3°C$ and latent heat of vaporization of $316.3\,\text{kJ/kg}$, whereas in the gaseous phase, O_3 has a density of $2.141\,\text{kg/m}^3$ at 1.013 bar pressure and $0°C$ temperature with a specific gravity and volume of 1.612 (air $= 1$) and $0.519\,\text{m}^3\text{/kg}$ at 1.013 bar and $21°C$, respectively. O_3 is characterized by a high oxidation potential that conveys bactericidal and viricidal properties (Kim et al., 1999b). Interest in O_3 as an alternative to chlorine and other chemical disinfectants in cleaning and disinfection operations is based on high oxidation reduction potential of O_3 ($2.07\,\text{V}$) compared to chlorine ($1.36\,\text{V}$) and chlorine dioxide ($1.27\,\text{V}$). O_3 solubility in water is affected by temperature with solubility decreasing with increasing temperature. O_3 solubility in water is 13 times that of oxygen at $0°C-30°C$. At $0°C$, O_3 solubility is $0.640\,\text{L}$ O_3/L water, whereas at $60°C$, it is insoluble in water (Hill and Rice, 1982). At room temperature, O_3 is an unstable gas and readily degrades but has a longer half-life in the gaseous state than in aqueous solution (Rice, 1986). Although O_3 in pure water degrades rather quickly to oxygen, it degrades even more rapidly in impure solutions. Hill and Rice (1982) reported that approximately 50% of O_3 is destroyed in 20 min at $20°C$ in distilled or tap water, whereas only 10% of O_3 breaks down in 85 min in $20°C$ double distilled water. O_3 decomposition is faster at higher water temperatures.

O_3 inactivates microorganisms through oxidization, and residual O_3 spontaneously decomposes to nontoxic products (i.e., oxygen), making it an environmentally friendly antimicrobial agent for use in the food industry (Kim et al., 1999). O_3 is widely used as a disinfectant in drinking water and wastewater treatment. The strong biocidal characteristics of O_3 are due to a combination of its high oxidizing potential and its ability to diffuse through biological membranes (Hunt and Marinas, 1997). The effects of O_3 on various microorganisms have received much attention because of its increasing use in water and sewage disinfection (Katzenelson and Biedermann, 1976; Boyce et al., 1981). O_3 has distinct advantage over other disinfectant agents because of (1) high antimicrobial activity, (2) short contact time, (3) GRAS substance, (4) no residue problems, (5) no formation of hazardous DBP as in case of chlorine, and (6) on-site production reducing the risk of storage. These advantages coupled with legislative clearance will promote the use of O_3 in fruit processing and preservation as a greener alternative to conventional disinfectants.

6.3 O_3 GENERATION

O_3 is generated by reaction of free oxygen radicals with diatomic oxygen to form triatomic oxygen molecules. Generation of the free oxygen radical occurs by breakage of strong O–O bonds, requiring a significant energy input. O_3 results from the rearrangement of atoms when oxygen molecules are subjected to high-voltage electric discharge. The product is a bluish gas with pungent odor and strong oxidizing properties (Cullen et al., 2009). In nature, O_3 generation occurs when oxygen molecules react in the presence of electric discharges, for example, lightning, and by action of high-energy electromagnetic radiation. There have been significant developments in the methodologies of O_3 production including corona discharge/plasma and UV radiation which make ozonation a more attractive approach for food processing. In addition to photochemical (UV radiation) and electric discharge methods, O_3 can be produced by chemical, thermal, chemonuclear, and electrolytic methods (Kim et al., 1999). Briefly, in electric (corona) discharge method, relatively ambient feed gas (dry air or O_2) is passed between two high-voltage electrodes separated by a dielectric material, which is usually glass. The O_3/gas mixture discharged from the ozonator normally contains 1%–3% O_3 when using dry air, and 3%–6% O_3 when using high-purity oxygen as the feed gas (Tapp and Rice, 2012). In electrochemical (cold plasma) method, an electric current is applied between an anode and cathode in an electrolytic solution containing water and a solution of highly electronegative anions. A mixture of oxygen and O_3 is produced at the anode. The advantages associated with this method are the use of low-voltage DC current, no feed gas preparation, reduced equipment size, possible generation of O_3 at high concentration, and generation in water (Tapp and Rice, 2012). Although this technique is not yet used commercially in potable or wastewater treatment, it is anticipated that in the future, this will find much more widespread application (Jun et al., 2009). In the UV method, O_3 is formed when O_2 is exposed to UV light of 140–190 nm wavelength. This splits the oxygen molecules into oxygen atoms, which then combine with other oxygen molecules to form O_3. Although a wide range of UV bulbs are available, an appropriate wavelength range varies from 180 to 254 nm. However, with currently available technology, these bulbs are still not an economical or efficient way to generate O_3, producing only up to 0.3%–0.4% by weight. Novel systems for generation of O_3 within sealed packages under air or modified atmospheres offer potential for fruit decontamination, and shelf-life extension postpackaging has recently been developed (Klockow and Keener, 2009).

6.4 APPLICATIONS IN FRUIT PROCESSING

O_3 is one of the most potent sanitizers and is typically viewed as a greener alternative to chlorine (Cullen et al., 2010). The potency of O_3 as an effective agent against a wide range of both pathogenic and spoilage microorganisms at relatively low

TABLE 6.1

Industrial Advantages and Disadvantages of O$_3$ in Fruit Processing

Advantages
Broad spectrum antimicrobial agent
In situ production
GRAS status
Reduces and degrades pesticide residues
Wastewater with reduced biological/chemical oxygen demand
Environmental friendly

Disadvantages
Limited solubility in water
High initial capital cost but no chemical cost
Health and safety (OSHA limits)
Short half-life
Oxidative damage to equipment materials such as rubber, some plastics, metal corrosions

concentrations makes it attractive for food applications. Table 6.1 lists some advantages and disadvantages of O$_3$ application in fruit processing. Application of O$_3$ in fruits is accomplished either by the washing using O$_3$-containing water or the use of gaseous O$_3$ during storage. O$_3$ as a disinfectant agent has been applied to fresh fruits for disinfection purposes by reducing microbial populations and extending the shelf life (Cullen et al., 2009, 2010). O$_3$-containing water can be obtained by adding gaseous O$_3$ in water, which can be accomplished either by venturi injection or fine bubble diffusion. Application of gaseous O$_3$ in storage systems can be achieved by generating and applying O$_3$ produced from ambient air or oxygen and by passing into controlled atmospheric storage room as shown in Figure 6.1.

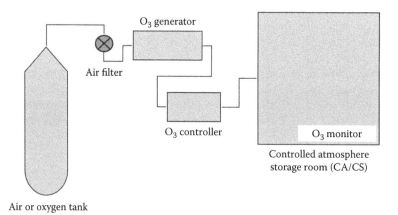

FIGURE 6.1   Schematic diagram showing the application of O$_3$ CA/CS.

6.4.1 AQUEOUS O₃ APPLICATIONS

Currently, researchers and industry alike are examining O_3's potential for directly cleaning fruit surfaces. O_3-containing water has been applied to fruits for sanitation purposes, reducing microbial populations and extending the shelf life of some of these products. Application of O_3 by washing of whole fruits for microbial decontamination has been reported for apples (Achen and Yousef, 2001), watermelon (Fonseca and Rushing, 2006), blueberries (Bailka and Demirci, 2007), tangerine (Whangchai et al., 2010), kiwifruit (Hur et al., 2005; Barboni et al., 2010), figs (Oztekin et al., 2006; Akbas and Ozdemir, 2008) and longan fruit (Whangchai et al., 2006). Table 6.2 lists some examples showing the application of O_3 in fruit processing.

Achen and Yousef (2001) compared the efficacy of O_3 against *Escherichia coli* O157:H7 on apples by bubbling of O_3 during apple washing and dipping apples in preozonated water. Bubbling was found to be more effective than dipping. Wang et al. (2004) employed tap water, acidic electrolyzed water (AEW), aqueous O_3, chlorinated water, and aqueous O_3 followed by AEW (sequential wash) for treatment of fresh-cut cilantro samples. They observed that the sequential wash, that is, aqueous O_3 followed by AEW, is more effective in initial microbial count reduction and maintains low microbial growth during storage at 0°C for 14 days. However, the combination of O_3 and AEW led to more tissue injury, which influenced the overall quality, whereas O_3 treatment alone achieved the highest overall quality and maintained the typical cilantro aroma. For effective washing of fruits, a system should include a prewashing step to remove dirt and cell exudates from the cut surfaces of fruits followed by the immersion of the fruits in a tank containing O_3 water (Gil et al., 2009).

Washing of fruits with O_3-containing water is also reported to degrade pesticide residues. Wu et al. (2007) reported that rinsing at a dissolved O_3 concentration of 1.4 mg/L for 15 min effectively removes 27%–34% of residual pesticide from vegetables. However, higher degradation of pesticides residues can be obtained with an increase in O_3 concentration (Ong et al., 1996; Ou-Yang et al., 2004).

6.4.2 GASEOUS O₃ APPLICATIONS

The majority of gaseous O_3 treatments are reported to be during storage. Storage of fruits in O_3-rich atmospheres has been reported to reduce or eliminate odor and to control spoilage caused by microbial and fungal pathogens. Continuous exposure of fresh commodities to O_3 during storage is reported to reduce postharvest decay and to reduce microbial spoilage of fruits and vegetables (Liew and Prange, 1994; Barth et al., 1995; Sarig et al., 1996; Perez et al., 1999; Palou et al., 2002; Aguayo et al., 2006; Tzortzakis et al., 2008). Applications of O_3-rich atmospheres have been studied for apples, cherries, kiwi, peach, plum, table grapes, tomatoes, blackberries, and strawberries. O_3 can be used as a relatively brief prestorage treatment in air, or it can be added continuously or intermittently to the storage room atmosphere throughout the storage period to prevent or delay fruit decay (Skog and Chu, 2001; Palou et al., 2003; Cayuela et al., 2009). Selma et al. (2008b) reported that gaseous O_3 treatment

TABLE 6.2
Selected Examples of O$_3$ Application in Fruit Processing and Preservation

Fruit Product	Processing Conditions	Target Microbial Population	Salient Findings	Reference
Kiwifruit (*Actinidia deliciosa*)	Conventional cold storage (0°C, RH 95%), storage with continuous supply of O$_3$ (0.3 μL/L)	*Botrytis cinerea*	Disease delay incidence by 56%	Minas et al. (2010)
Fresh-cut pineapple, banana, guava	O$_3$ at a flow rate of 8 (± 0.2) mL/s for 0, 10, 20, and 30 min		*Pineapple*: total phenol (↑) and flavonoid (↑) and antioxidant activity (↑), vitamin C (↓) *Banana*: total phenol (↑) and flavonoid (↑) and antioxidant activity (↑), vitamin C (↓) *Guava*: total phenol (↓) and flavonoid (↓) and antioxidant activity (↓), vitamin C (↓)	Alothman et al. (2010)
Tangerine	Washing with electrolyzed oxidizing (EO) followed by continuous O$_3$ exposure (200 mg/L for 2 h/day) for 5 days	Postharvest decay (*Conidia* and *Penicillium digitatum*)	Total soluble solids (∼), titratable acidity (∼), weight loss (∼), and peel color (∼)	Whangchai et al. (2010)
Kiwi fruits	O$_3$ gas treatment chamber of 2 m^3 (temperature 0°C). Gaseous O$_3$ concentration: 4 mg/h and a humidity of 90%–95%	*Botrytis cinerea*	Fungicidal effect on *Botrytis cinerea* kiwi mass (↓), firmness (↓), and acidity (↓), reducing sugar (↑), soluble solids (↑), and pH (↑), nonvolatile organic acids (↓)	Barboni et al. (2010)
Grapes	O$_3$ as fumigant, concentration: 10,000 μL/L. The temperature during fumigation was 5 ± 2°C	*Botrytis cinerea*	Gray mold incidence (↓)	Gabler et al. (2010)

TABLE 6.2 (continued)
Selected Examples of O₃ Application in Fruit Processing
and Preservation

Fruit Product	Processing Conditions	Target Microbial Population	Salient Findings	Reference
Grapes	Continued and intermittent (12 h/day) O₃ treatments (2 ppm) at 5°C for 72 days		Postharvest decay (↓), postharvest losses (↓), resveratrol content (↓), sensory attributes (↓), weight loss (↑)	Cayuela et al. (2009)
Apple	Bubbled O₃ in water and apple dipping in O₃-containing water	*E. coli*	Bubbling: *E. coli* O157:H7 (3.7 LR) Dipping: *E. coli* O157:H7 (2.6 LR)	Achen and Yousef (2001)
Fresh-cut watermelon	Dipping in O₃-containing water (0.4 μL/L)	APC (1–1.5 LR)	Color (↓), overall quality (↓)	Fonseca and Rushing (2006)
Strawberry	Stored at 2°C in an atmosphere containing O₃ (0.35 ppm). After 3 days at 2°C, fruits were moved to 20°C to mimic retail conditions	Fungal mycelial development	Fungal decay (delayed), sucrose (↓), glucose (↑), fructose (↑), vitamin C (↑), aroma quality (↑)	Perez et al. (1999)

of 5,000 and 20,000 ppm for 30 min reduced total coliforms, *Pseudomonas fluorescens*, yeast, and lactic acid bacteria recovery from fresh-cut cantaloupe while gaseous O₃ (10,000 ppm for 30 min) under partial vacuum resulted in a reduction of viable, recoverable *Salmonella* from inoculated physiologically mature nonripe and ripe melons. Akbas and Ozdemir (2008) concluded that ozonation was a promising method for the decontamination of dried figs and found to be effective especially in reduction of vegetative cells (*E. coli*, *Bacillus cereus*). Oztekin et al. (2006) reported on the effects of O₃ treatment on the microflora of dried figs, where the application of gaseous O₃ at 5 or 10 ppm for 3–5 h resulted in significant reductions in total bacteria, coliform, and yeast/mold counts. Najafi and Khodaparast (2009) concluded that a minimum of 1 h O₃ treatment at 5 ppm could be successfully used for reducing both coliform and *Staphylococcus aureus* populations on date fruits, but that longer exposure times are required for elimination of the total mesophilic bacteria as well as yeast/mold.

An increase in the shelf life of apples and oranges by O₃ treatment has been attributed to the oxidation of ethylene. Fungal deterioration of blackberries and grapes was decreased by ozonation of the fruits (Beuchat, 1992). The application of gaseous O_3 for control of green and blue mold development on cold stored citrus fruit has also been investigated, which showed that sporulation of *P. italicum* was reduced (Palou et al., 2001). Selma et al. (2008b) reported the combination of hot water and gaseous O_3 to be an efficient treatment for controlling microbial growth and maintaining sensory quality of melons. Artes-Hernandez et al. (2007) observed that the sensory quality was preserved with both continuous and intermittent applications of O_3 gas treatments applied during cold storage of "autumn seedless" table grapes. Barboni et al. (2010) observed a significant decrease in *Botrytis* contamination during O_3-rich atmospheric storage of kiwi fruits which are often stored for 4–6 months. They observed a significant increase in fructose, glucose, and sucrose during storage in both air at 0°C and O_3-enriched air. However, a decrease in nonvolatile organic acids after 25 weeks of storage in O_3-rich atmosphere was observed. Barboni et al. (2010) also compared the effect of O_3-rich storage and air storage over a period of 7 months on the vitamin C content of kiwi fruit. A gaseous O_3 concentration was 4 mg/h in a 2 m³ chamber at a temperature of 0°C, and a humidity of 90%–95% was employed. The authors did not observe any significant change in ascorbic acid content over the storage period. However, O_3 would be expected to cause the loss of antioxidant bioactive compounds because of its strong oxidizing activity. Contradictory reports are found in the literature regarding ascorbic acid. Perez et al. (1999) studied the effect of O_3 treatment on the postharvest quality of strawberry stored at 2°C in an atmosphere containing O_3 (0.35 ppm) and reported an increase in ascorbic acid levels in response to O_3 exposure. O_3 treatments were also reported to have minor effects on anthocyanin contents of strawberries (Perez et al., 1999) and blackberries (Barth et al., 1995). Alothman et al. (2010) investigated the effect of O_3 treatment on total phenol, flavonoid, and vitamin C content of fresh-cut honey pineapple, banana "pisang mas," and they observed a significant increase in total phenol and flavonoid contents of pineapple and banana when exposed to O_3 for up to 20 min, with a concomitant increase in antioxidant activity. However, they observed a decrease for guava with a significant decreased in the vitamin C content of all three fruits. The main factor that could have contributed to the degradation of ascorbic acid in the three O_3-treated fruit types is the activation of ascorbate oxidase. This enzyme is activated under stress conditions, such as chemical exposure. Ascorbate oxidase has been reported to promote the degradation of ascorbic acid to dehydroascorbic acid (Lee and Kader, 2000). In contrast, this study showed promising results for enhancing antioxidant capacity of some fresh fruits by O_3 treatment although the positive effect is compromised by a reduction in vitamin C content.

Minas et al. (2010) studied the effect of gaseous O_3 (0.3 µL/L) on the development of stem-end rot disease which is caused by *Botrytis cinerea* on kiwifruit during storage. Figure 6.2 shows the appearance of infected kiwifruit by *Botrytis cinerea* after 4 month storage. They observed that O_3 treatment delayed and simultaneously decreased disease incidence by 56% as shown in Figure 6.3. The authors also observed that the exposure of kiwifruit in an O_3-enriched atmosphere (0.3 µL/L, 0°C, RH 95%) for 2 h significantly reduced the total carotenoid content by 9.6%; however, total carotenoid content was found to increase by 5.7%

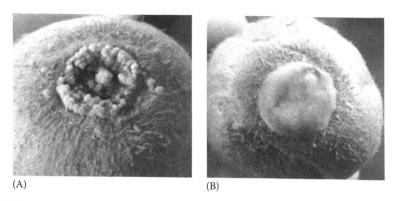

(A)   (B)

FIGURE 6.2   Appearance of infected kiwifruit by *Botrytis cinerea* after 4 months of cold storage (0°C, RH 95%) with catalytic oxidation of ethylene plus continuous supply of O_3 (0.3 µL/L) (A, sclerotia formation) or with catalytic oxidation of ethylene (conventional cold storage) (B, hyphae development and sporulation). (From Minas, I.S. et al., *Postharvest Biol. Technol.*, 58(3), 203, 2010.)

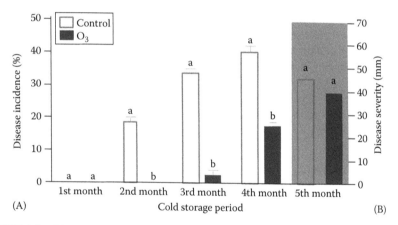

FIGURE 6.3   Disease incidence (%) (A) and severity (decay depth, mm) (B) of stem-end rot caused by *Botrytis cinerea* on artificially infected "Hayward" kiwifruit during a 4 month cold storage period (0°C, RH 95%) in a storage room with catalytic oxidation of ethylene (control) or catalytic oxidation of ethylene plus continuous supply of O_3 (0.3 µL/L). Values followed by the same letters for each assessment time did not differ significantly according to pair-wise student's *t*-test ($P < 0.05$). Vertical lines represent the standard error of the means. (From Minas, I.S. et al., *Postharvest Biol. Technol.*, 58(3), 203, 2010.)

and 8.7% when the exposure time was increased to 72 and 144 h, respectively, compared to the control.

O_3 in combination with citric or oxalic acid as a fumigant is reported to reduce postharvest decay and pericarp browning of longan fruit. Whangchai et al. (2006) observed that longan fruit treated with O_3 in combination with oxalic or citric acid reduces enzymatic browning and could be a partial alternative to sulfur dioxide fumigation for control of postharvest decay and browning. Pericarp browning has been attributed to oxidation of phenolics by polyphenol oxidase (PPO), producing brown-colored by-products (Ferrar and Walker, 1996).

Reported studies show that the effect of O_3 during storage is variable and strongly depends on the type of microorganisms, commodities, and storage conditions. Apart from reduction in microbial decay, O_3 is reported to be an effective agent in removing ethylene from the atmosphere during apple and pear storage room without a significant change in quality attributes (Skog and Chu, 2001). Exposure of horticulture crops to O_3 can reduce postharvest decay and may be effective in reducing application of field-applied fungicides used to control these pathogens. Studies have shown reductions in decay of blackberries during continuous O_3 treatment (Barth et al., 1995).

Gabler et al. (2010) investigated the efficacy of O_3 to control postharvest decay of table grapes and for the potential replacement of sulfur dioxide which is used as a commercial fumigant. They observed that grapes stored in O_3-rich atmospheres may develop thin longitudinal darkened lesions. This injury is reported to be irregular and was not always associated with an O_3 dose or cultivar (Gabler et al., 2010). Similarly, Tzortzakis et al. (2007) observed a substantial decline in spore production and development of visible lesions mainly due to *Botrytis cinerea* (gray mold) during the storage of tomatoes, strawberries, table grapes, and plums during chilled storage (13°C) with low level of O_3 enrichment (0.1 µmol/mol). Martínez-Sánc et al. (2006) investigated the effect of several sanitizers on the visual quality and color of rocket leaves during storage in air and low O_2 (1–3 kPa) + high CO_2 (11–13 kPa) for 15 days at 4°C. They observed that O_3 effects were comparable with other sanitizers except for purac- (lactic acid–) treated samples. O_3 was also reported to change the surface color of products such as peaches (Badiani et al., 1995) and carrots (Liew and Prange, 1994).

6.4.3 FRUIT JUICES

The practical application of ozonation to fruit juices is still in its infancy as of this writing. The use of O_3 has been reported for processing various fruit juices (Steenstrup and Floros, 2004; Tiwari et al., 2009a,b). Ozonation of liquid phases is most frequently accomplished by injecting O_3 gas (mixtures of air/O_3 or oxygen/O_3) through a sparger into a liquid. Usually the studies on O_3 absorption in the aqueous systems are carried out in stirred-tank reactors or bubble columns (Cullen et al., 2009). The approval of O_3 as a direct food additive (U.S. FDA, 2004) led to the application of O_3 for processing of apple cider (Steenstrup and Floros, 2004; Choi and Nielsen, 2005). Other studies have reported on the effects of ozonation of orange juice (Angelino et al., 2003; Tiwari et al., 2008; Patil et al., 2009a), blackberry juice (Tiwari et al., 2009a), and strawberry juice (Tiwari et al., 2009b). Ozonation of fruit juices is reported to meet the U.S. FDA's requirement of a mandatory 5 log reduction of the most resistant pathogens (*E. coli, Salmonella, Listeria monocytogenes*). Patil et al. (2010) found fast inactivation times for *E. coli* strains in apple juice with 5 log reductions achieved within 5 min. Significant changes in color values (L^*, a^*, b^*), total phenol, and polyphenolic content of apple juice were observed. Thus, the effects of ozonation on the nutritional properties of liquid foods such as fruit juice or juice products should also be considered prior to its adoption as a preservation technique. Patil et al. (2011) showed that the shelf life of apple juice could be extended with ozonation with respect to control of *S. cerevisiae*.

Apart from microbial inactivation, O_3 treatment of apple juice has also been reported for destruction of mycotoxins (Cataldo, 2008). Mycotoxins have been found to occur in a number of foods including apple juice. Patulin is a predominant mycotoxin in apple juice with a U.S. FDA action level of 50 µg/L. Patulin has carcinogenic properties, and it survives conventional pasteurization processes. Cataldo (2008) reported that a moderate O_3 treatment of apple juice may become a standard industrial practice to reduce or eliminate the patulin toxin from juices. Similarly, Ashirifie-Gogofio et al. (2009) reported that an apple juice model system (0.5% malic acid, buffered at pH 3.5–4.0) with an initial level of 1000 ppb patulin was degraded to <50 ppb in 20, 15, and 10 min for 350, 1500, and 2500 ppm O_3, respectively. The efficiency of O_3 in mycotoxin degradation is due to the presence of a polyketide lactone, which makes it highly vulnerable to oxidation; consequently, O_3 treatment is found to effectively degrade mycotoxins.

6.4.4 SENSORY PROPERTIES OF O_3-TREATED FRUITS

Applying O_3 at doses that are large enough for effective decontamination may change the sensory qualities of food. O_3 is not universally beneficial and, in some cases, may promote oxidative spoilage in foods. Surface oxidation, discoloration, or development of undesirable odors may occur in substrates from excessive use of O_3 (Khadre et al., 2001). O_3 treatment of fruit either by washing or in storage consisting of O_3 gas is reported to have significant effects on texture. Firmness of fresh cilantro leaves was reported to decrease through washing with O_3-containing water compared to control. A decrease in firmness also was reported by washing with chlorinated water (Wang et al., 2004). Another study conducted by Selma et al. (2008b) reported nonsignificant changes in firmness of fresh-cut cantaloupe irrespective of gaseous O_3 concentration (5000 or 2000 ppm) for 30 min during storage compared to control. Change in texture during ozonation and subsequent storage possibly is due to postharvest changes in cellulose and hemicellulose contents due to O_3 application during modified atmosphere packaging (MAP). This could be due to polymerization and epimerization of cellulose and hemicelluloses contents of cell walls inducing thickening of the cell walls, causing textural changes in fresh-cut green asparagus during storage after O_3 treatment (An et al., 2007). They reported an increase in cellulose, hemicelluloses, and lignin content during MAP storage after pretreatment with aqueous O_3. Ozonation of fruits has also been reported to enhance firmness of citrus fruits and cucumbers compared to controls (Skog and Chu, 2001). O_3 is reported to delay softening in strawberries during cold room storage and storage at room temperature (Nadas et al., 2003). Wang et al. (2004) compared five washing treatment systems including tap water, AEW, O_3-containing water, chlorinated water, and aqueous O_3 followed by AEW (sequential wash) on the firmness of cilantro packaged in polyethylene bags and stored at 0°C for 14 days. They observed a slight decrease in firmness of treated samples compared to the control on day 0 with no significant differences in firmness among chlorine, O_3, AEW, and the sequential treatments at day 0. They observed a gradual decrease in firmness during storage which may be attributed to the tissue injury caused by the treatments.

The most notable effect of O_3 on the sensory quality of fruits reported in the literature is the loss of aroma. O_3-enriched cold storage of strawberries resulted in reversible losses of fruit aroma (Perez et al., 1999; Nadas et al., 2003). This behavior probably is due to the oxidation of volatile compounds. However, Tzortzakis et al. (2007) did not observe any significant changes in tomato fruit weight, antioxidant status, CO_2/H_2O exchange, ethylene production or organic acid, vitamin C (pulp and seed), and total phenolic content when exposed to O_3 concentrations ranging between 0.005 and 1.0 µmol/mol at 13°C and 95% RH. Similar results were reported by Kute et al. (1995) for strawberry exposed to O_3 concentrations between 0.3 and 0.7 µmol/mol for up to 1 week. Applying O_3 at doses that are large enough for effective decontamination may change the sensory qualities of these products.

6.5 FACTORS AFFECTING O_3 EFFICACY

Several factors such as O_3 concentration, temperature and pH of medium, and form of O_3 application (washing, dipping, or gaseous) are some of the important factors affecting/influencing the diffusion and solubility of O_3. Solubility of O_3 is a critical factor which is important for O_3 washes of fruits. Residual O_3 is the concentration of O_3 that can be detected in the medium after application to the target surface. Both the instability of O_3 under certain conditions and the presence of O_3-consuming materials affect the level of residual O_3 available in the medium. Therefore, it is important to distinguish between the concentration of applied O_3 and residual O_3 necessary for effective disinfection. Presence of other organic matter on fruit surfaces or in the medium may compete with microorganisms for O_3 (Khadre et al., 2001). Consequently, the presence of organic matter or dissolved solids in water intended for washing of fruits may increase O_3 demand and may form undesirable by-products due to reaction with O_3.

Ozonation of liquid foods is also influenced by the factors discussed earlier. Apart from O_3 concentration, temperature, and pH of the medium, factors such as flow rate and bubble size also influence the efficacy of O_3 for microbial inactivation. Bubble size has been shown to have an effect on O_3's solubilization rate and disinfection efficacy. The dissolution efficiency varies rapidly when the initial bubble size reaches a critical value while the change of efficiency is much slower at other bubble sizes. Ahmad and Farooq (1985) reported that O_3 mass transfer and disinfection efficacy increased as bubble size decreased (O_3 bubble size was varied while all other factors were kept constant). Gong et al. (2008) showed a nonlinear dependence of the O_3 dissolution efficiency on the initial bubble size in a study of the mass transfer process of O_3 dissolution in a bubble plume (bubble plumes are an interacting collections of bubbles formed by some event) inside a rectangular water tank.

O_3 concentration present or available in the medium is another parameter that determines O_3 efficacy. Faster inactivation with a shorter lag time and smaller D values at the highest O_3 concentrations was reported by Steenstrup and Floros (2004) during inactivation of *E. coli* O157:H7 in apple cider by O_3. Temperature influences O_3 solubility increase with decrease in temperature. As temperature increases, O_3 becomes less soluble and less stable with an increase in the decomposition rate (Rice et al., 1981). However, there is no consensus on the effect of temperature on the biocidal efficacy of O_3, for example, a drop in the temperature of the aqueous medium increases O_3 solubility and stability, augmenting its availability in the medium, and consequently efficacy rises. Studies to date show that extrinsic factors listed below governs O_3 efficacy.

1. Water quality used for washing (pH, temperature, turbidity, organic matter, oxidizable inorganic)
2. O_3 parameters (concentration, contact time decontamination treatment)
3. Application method (dipping, spraying and agitated, rubbed, or static condition during exposure)
4. Produce/water ratio, single or multiple batches, rinse after sanitation, multiple washings
5. Microbial load (characteristics of microbial strain, physiological states of the bacterial cells, natural or inoculated microorganisms, population size)
6. Fruit product (type of fruits, characteristics of the product surfaces [cracks, crevices, hydrophobic tendency, and texture], relation weight, and surface area)

6.6 CASE STUDY: FRUIT AND VEGETABLE DISINFECTION AT SAMRO LTD., SWITZERLAND

The use of O_3 for disinfection has successfully been employed at commercial scale for fruits and vegetable using Ventafresh technology at the SAMRO AG plant in Burgdorf, Switzerland. This technology includes washing fruits and vegetables with O_3-containing water, then treatment with UV radiation in a specially designed "disinfection tunnel" (Steffen et al., 2010). Both O_3 and UV radiation help in disinfecting fruits and vegetable in a sequential manner. O_3-containing water is used to disinfect, and subsequent UV radiation continues the disinfection and simultaneously destroys remaining O_3 with the treated produce stored at high humidity in an atmosphere containing a lower concentration of O_3 (Steffen et al., 2010). The Ventafresh process is applicable to all root vegetables, plus root celery and asparagus, and also to certain fruits (apples, pears, kiwis, tomatoes, and Sharon fruits). The Ventafresh technology has attained commercial application for potatoes at SAMRO. Steffen et al. (2010) also compared a cost for the use of O_3 for processing and storing potatoes by the Ventafresh technology with costs for cold and natural storage at SAMRO plant. They observed a significant savings of about 308,000 and 158,000 Sfr compared to cold storage and natural storage, respectively (Table 6.1). The cost–benefit analysis shown in Table 6.3 was based on the rental of two Ventafresh disinfection tunnels with the process efficiency of 15 ton of potatoes per hour.

TABLE 6.3
Cost Benefits of Processing/Storage of Potatoes
by Ventafresh Technology

Cost Item	Ventafresh	Cold Storage	Natural Storage
Rental of Ventafresh equipment	Sfr. 172,000	—	—
Storage cost per tonne	Sfr. 50	Sfr. 100	Sfr. 50
Disinfection cost per tonne	Sfr. 5	Sfr. 10	Sfr. 10
Shrinkage losses	4%	5%	8%
Shrinkage costs per tonne	Sfr. 20	Sfr. 25	Sfr. 40
Storage damage losses	4%	8%	10%
Storage damage loss costs/tonne	Sfr. 20	Sfr. 40	Sfr. 50
Total	Sfr. 742,000	Sfr. 1,050,000	Sfr. 900,000
Savings	—	Sfr. 308,000	Sfr. 158,000

Source: Steffen, H. et al., *Ozone Sci. Eng.*, 32(2), 144, 2010.

6.7 INDUSTRIAL HEALTH AND SAFETY

Though O_3 is a GRAS chemical, it is also a toxic gas; toxicity is dependent on concentration and length of exposure (Pascual et al., 2007). A large number of factors come into play when considering health and safety aspects of O_3 for fruit processing. Aqueous O_3 applications such as washing of fruit produce the safety item of most concern which is exposure of personnel to quantities of O_3 gas that might escape from the washer due to degassing from O_3-containing water during bubbling or agitation. Gaseous application during storage of fruits may cause direct O_3 exposure of personnel inside the storage chamber. Like other oxidizing agent, O_3 is also potentially harmful to humans if exposed to sufficient O_3 dose for extended duration. In the United States, the current permissible exposure limit (PEL) to O_3 allowed by OSHA regulations is 0.1 ppm_v, time-weighted average over an 8 h work day, 5 days/week. This level is rather high considering the fact that the olfactory senses of an average human being can detect O_3 gas at levels as low as 0.01 ppm_v (0.02 mg/m^3). The OSHA also has a short-term exposure limit (STEL) to O_3 of 0.3 ppm_v (0.6 mg/m^3), defined as a 15 min exposure to O_3, not to be exceeded more than four times per day (Pryor and Rice, 2000). In nutshell, at short-term exposure rates of 0.1–1.0 ppm, symptoms include headaches, nosebleeds, eye irritation, dry throat, and respiratory irritation. At higher exposure levels (1.0–100 ppm), symptoms become more severe and include asthma-like symptoms, tiredness, and loss of appetite. Hemorrhage and pulmonary congestion are possible at high O_3 exposure (Pascual et al., 2007).

REFERENCES

Achen, M. and Yousef, A. E. (2001). Efficacy of ozone against *Escherichia coli* O157: H7 on apples. *J. Food Sci.*, 66(9), 1380–1384.
Aguayo, E., Escalona, V. H., and Artes, F. (2006). Effect of cyclic exposure to ozone gas on physicochemical, sensorial and microbial quality of whole and sliced tomatoes. *Postharvest Biol. Technol.*, 39, 169–177.

Ahmad, M. and Farooq, S. (1985). Influence of bubble sizes on ozone solubility utilization and disinfection. *Water Sci. Technol.*, 17, 1081–1090.

Akbas, M. Y. and Ozdemir, M. (2008). Application of gaseous ozone to control populations of *Escherichia coli, Bacillus cereus* and *Bacillus cereus* spores in dried figs. *Food Microbiol.*, 25(2), 386–391.

Alothman, M., Kaur, B., Fazilah, A., Bhat, R., and Karim, A. A. (2010). Ozone-induced changes of antioxidant capacity of fresh-cut tropical fruits. *Innov. Food Sci. Emerg. Technol.*, 11(4), 666–671.

An, J., Zhang, M., and Lu, Q. (2007). Changes in some quality indexes in fresh-cut green asparagus pretreated with aqueous ozone and subsequent modified atmosphere packaging. *J. Food Eng.*, 78(1), 340–344.

Angelino, P. D., Golden, A., and Mount, J. R. (2003). Effect of ozone treatment on quality of orange juice. In: *IFT Annual Meeting Book of Abstract*, 2003, Abstract No. 76C–2. Institute of Food Technologists, Chicago, IL.

Artes-Hernandez, F., Aguayo, E., Artes, F., and Tomas-Barberan, F.A. (2007). Enriched ozone atmosphere enhances bioactive phenolics in seedless table grapes after prolonged shelf life. *J. Sci. Food Agric.*, 87, 824–831.

Ashirifie-Gogofio, J., Floros, J. D., and LaBorde, L. F. (2009). Ozone degradation of patulin in model apple juice system. *Annual Meeting and Food Expo, IFT 09*, June 2009, Anaheim, CA.

Badiani, M., Fuhrer, J., Paolacci, A. R., and Sermanni, G. G. (1995). Deriving critical levels for ozone effects on peach trees [*Prunus persica* (L) Batsch] grown in open-top chambers in central Italy. *8th International Symposium on Environmental Pollution and Its Impact on Life in the Mediterranean Region*, Rhodes, Greece, Inst Lebensmitteltechnologie Analytische Chemie.

Badiani, M., Fuhrer, J., Paolacci, A. R., and Sermanni, G. G. (1995). Deriving critical levels for ozone effects on peach trees [*Prunus persica* (L.) Batsch] grown in open-top chambers in central Italy. *Fresenius Environ. Bull.*, 5(1996), 592–597.

Barboni, T., Cannac, M., and Chiaramonti, N. (2010). Effect of cold storage and ozone treatment on physicochemical parameters, soluble sugars and organic acids in *Actinidia deliciosa*. *Food Chemistry*, 121(4), 946–951.

Barth, M. M., Zhou, C., Mercier, J., and Payne, F. A. (1995). Ozone storage effects on anthocyanin content and fungal growth in blackberries. *J. Food Sci.*, 60(6), 1286–1288.

Beuchat, L. R. (1992). Surface disinfection of raw produce. *Dairy Food Environ. Sanit.*, 12(1), 6–9.

Bialka, K. L. and Demirci, A. (2007). Decontamination of *Escherichia coli* O157:H7 and *Salmonella enterica* on blueberries using ozone and pulsed UV-light. *J. Food Sci.*, 72(9), 391–396.

Cataldo, F. (2008). Ozone decomposition of Patulin—A micotoxin and food contaminant. *Ozone Sci. Eng.*, 30(3), 197–201.

Cayuela, J. A., Vazquez, A., Perez, A. G., and Garcia, J. M. (2009). Control of table grapes postharvest decay by ozone treatment and resveratrol induction. *Food Sci. Technol. Int.*, 15(5), 495–502.

Choi, L. H. and Nielsen, S. S. (2005). The effects of thermal and nonthermal processing methods on apple cider quality and consumer acceptability. *J. Food Qual.*, 28, 13–29.

Cullen, P. J., Tiwari, B. K., O'Donnell, C. P., and Muthukumarappan, K. (2009). Modelling approaches to ozone processing of liquid foods. *Trends Food Sci. Technol.*, 20(3–4), 125–136.

Cullen, P. J., Valdramidis, V. P., Tiwari, B. K., Patil, S., Bourke, P., and O'Donnell, C. P. (2010). Ozone processing for food preservation: An overview on fruit juice treatments. *Ozone: Science & Engineering*, 32(3), 166–179.

FDA. (2004). *FDA Guidance to Industry, 2004: Recommendations to Processors of Apple Juice or Cider on the Use of Ozone for Pathogen Reduction Purposes*. http://www.fda.gov/Food/GuidanceComplianceRegulatoryInformation/GuidanceDocuments/Juice/ucm072524.htm

Ferrar, P. H. and Walker, J. R. L. (1996). Inhibition of diphenol oxidases: A comparative study. *J. Food Biochem.*, 20(1), 15–30

Fonseca, J. M. and Rushing, J. W. (2006). Effect of ultraviolet-C light on quality and microbial population of fresh-cut watermelon. *Postharvest Biol. Technol.*, 40, 256–261.

Gabler, F. M., Mercier, J., Jimenez, J. I., and Smilanick, J. L. (2010). Integration of continuous bio-fumigation with *Muscodor albus* with pre-cooling fumigation with ozone or sulfur dioxide to control postharvest gray mold of table grapes. *Postharvest Biol. Technol.*, 55(2), 78–84.

Gil, M. I., Selma, M. V., López-Gálvez, F., and Allende, A. (2009). Fresh-cut product sanitation and wash water disinfection: Problems and solutions. *Int. J. Food Microbiol.*, 134(1–2), 37–45.

Gong, J., Liu, Y., and Sun, X. (2008). O_3 and UV/O_3 oxidation of organic constituents of biotreated municipal wastewater. *Water Res.*, 42(4–5), 1238–1244.

Hill, A. G. and Rice, R. G. (1982). Historical background, properties and applications. In: Rice, R. G., Ed. *Ozone Treatment of Water for Cooling Application*, Ann Arbor Science Publishers, Ann Arbor, MI, pp. 1–37.

Hur, J. S., Oh, S. O., Lim, K. M., Jung, J. S., Kim, J. W., and Koh, Y. J. (2005). Novel effects of TiO_2 photocatalytic ozonation on control of postharvest fungal spoilage of kiwifruit. *Postharv. Biol. Tech.*, 35(1), 109–113.

Khadre, M. A., Yousef, A. E., and Kim, J. (2001). Microbiological aspects of ozone applications in food: A review. *J. Food Sci.*, 6, 1242–1252.

Kim, J. G., Yousef, A. E., and Chism, G. W. (1999a). Use of ozone to inactivate microorganisms on lettuce, *J. Food Saf.*, 19(1), 17–34.

Kim, J. G., Yousef, A. E., and Dave, S. (1999b). Application of ozone for enhancing the microbiological safety and quality of foods: A review. *J. Food Prot.*, 62(9), 1071–1087.

Klockow, P. A. and Keener, K. M. (2009). Safety and quality assessment of packaged spinach treated with a novel ozone-generation system. *LWT—Food Sci. Technol.*, 42, 1047–1053.

Kute, K. M., Zhou, C., and Barth, M. M. (1995). The effect of ozone exposure on total ascorbic acid activity and soluble solids contents in strawberry tissue. In: *Proceedings of the Annual Meeting of the Institute of Food Technologists (IFT)*, New Orleans, LA, January 28–February 2, p. 82.

Lee, S. K. and Kader, A. A. (2000). Preharvest and postharvest factors influencing vitamin C content of horticultural crops. *Postharv. Biol. Technol.*, 20(3), 207–220.

Liew, C. L. and Prange, R. K. (1994). Effect of ozone and storage-temperature on postharvest diseases and physiology of carrots (*Daucus carota* L). *J. Am. Soc. Hortic. Sci.*, 119(3), 563–567.

Martínez-Sánchez, A., Allende, A., Bennett, R. N., Ferreres, F., and Gil, M. I. (2006). Microbial, nutritional and sensory quality of rocket leaves as affected by different sanitizers. *Postharvest Biol. Technol.*, 42(1), 86–97.

Minas, I. S., Karaoglanidis, G. S., Manganaris, G. A., and Vasilakakis, M. (2010). Effect of ozone application during cold storage of kiwifruit on the development of stern-end rot caused by *Botrytis cinerea*. *Postharvest Biol. Technol.*, 58(3), 203–210.

Nadas, A., Olmo, M., and Garcia, J. M. (2003). Growth of *Botrytis cinerea* and strawberry quality in ozone-enriched atmospheres. *J. Food Sci.*, 68(5), 1798–1802.

Najafi, M. B. H. and Khodaparast, M. H. H. (2009). Efficacy of ozone to reduce microbial populations in date fruits. *Food Control*, 20(1), 27–30.

Ong, K. C., Cash, J. N., Zabik, M. J., Siddiq, M., and Jones, A. L. (1996). Chorine and ozone washes for pesticide removal from apples and processed apple sauce. *Food Chem.*, 55(2), 153–160.

Ou-Yang, X. K., Liu, S. M., and Ying, M. (2004). Study on the mechanism of ozone reaction with parathion-methyl. *Saf. Environ. Eng.*, 11(2), 38–41.

Oztekin, S., Zorlugenc, B., and Zorlugenc, F. K. (2006). Effects of ozone treatment on microflora of dried figs. *J. Food Eng.*, 75(3), 396–399.

Palou, L., Crisosto, C. H., Smilanick, J. L., Adaskaveg, J. E., and Zoffoli, J. P. (2002). Effects of continuous 0.3 ppm ozone exposure on decay development and physiological responses of peaches and table grapes in cold storage. *Postharvest Biol. Technol.*, 24(1), 39–48.

Palou, L., Smilanick, J. L., Crisosto, C. H., and Mansour, M. (2001). Effect of gaseous ozone exposure on the development of green and blue molds on cold stored citrus fruit. *Plant Dis.*, 85(6), 632–638.

Palou, L., Smilanick, J. L., Crisosto, C. H., Mansour, M., and Plaza, P. (2003). Ozone gas penetration and control of the sporulation of *Penicillium digitatum* and *Penicillium italicum* within commercial packages of oranges during cold storage. *Crop Prot.*, 22(9), 1131–1134.

Pascual, A., Llorca, I., and Canut, A. (2007). Use of ozone in food industries for reducing the environmental impact of cleaning and disinfection activities. *Trends Food Sci. Technol.*, 18, S29–S35.

Patil, S., Bourke, P., Frias, J. M., Tiwari, B. K., and Cullen, P. J. (2009a). Inactivation of *Escherichia coli* in orange juice using ozone. *Innov. Food Sci. Emerg. Technol.*, 10(4), 551–557.

Patil, S., Cullen, P. J., Kelly, B., Frías, J. M., and Bourke, P. (2009b). Extrinsic control parameters for ozone inactivation of *Escherichia coli* using a bubble column. *J. Appl. Microbiol.*, 107(3), 830–837.

Patil, S., Torres, B., Tiwari, B. K., Hilde, H. W., Bourke, P., Cullen, P. J., O'Donnell, C. P., and Valdramidis, V. P. (2010). Safety and quality assessment during the ozonation of cloudy apple juice. *J. Food Sci.*, 75, M437–M443.

Patil, S., Valdramidis, V., Cullen, P. J., and Bourke, P. (2011). Quantitative assessment of the shelf-life of ozonated apple juice. *Eur. Food Res. Technol.*, 232, 469–477.

Perez, A. G., Sanz, C., Rios, J. J., Olias, R., and Olias, J. M. (1999). Effects of ozone treatment on postharvest strawberry quality. *J. Agric. Food Chem.*, 47, 1652–1656.

Pryor, A. and Rice, R. G. (2000). Ozone toxicology and guidelines for safe use in food processing systems, *Ozone News*, 28(4), 19–28.

Rice, R.G. (1986). Application of ozone in water and waste water treatment. In: Rice, R.G. and Browning, M. J., Eds. *Analytical Aspects of Ozone Treatment of Water and Waste Water*, The Institute, Syracuse, NY, pp. 7–26.

Rice, R. G., Robson, C. M., Miller, G. W., and Hill, A. G. (1981). Uses of ozone in drinking water treatment. *J. Am. Water Works Assoc.*, 73(1), 44–57.

Sarig, P., Zahavi, T., Zutkhi, Y., Yannai, S., Lisker, N., and BenArie, R. (1996). Ozone for control of post-harvest decay of table grapes caused by *Rhizopus stolonifer. Physiol. Mol. Plant Pathol.*, 48(6), 403–415.

Selma, M. V., Allende, A., Lopez-Galvez, F., Conesa, M. A., and Gil, M. I. (2008a). Disinfection potential of ozone, ultraviolet-C and their combination in wash water for the fresh-cut vegetable industry. *Food Microbiol.*, 25(6), 809–814.

Selma, M. V., Ibanez, A. M., Cantwell, M., and Suslow, T. (2008b). Reduction by gaseous ozone of Salmonella and microbial flora associated with fresh-cut cantaloupe. *Food Microbiol.*, 25, 558–565.

Skog, L. J. and Chu, C. L. (2001). Effect of ozone on qualities of fruits and vegetables in cold storage. *Can. J. Plant Sci.*, 81(4), 773–778.

Steenstrup, L. D. and Floros, J. D. (2004). Inactivation of *E. coli* 0157:H7 in apple cider by ozone at various temperatures and concentrations. *J. Food Process. Preserv.*, 28, 103–116.

Steffen, H., Zumstein, P., and Rice, R. G. (2010). Fruit and vegetables disinfection at SAMRO, Ltd. Using hygienic packaging by means of ozone and UV radiation. *Ozone: Sci. Eng.*, 32(2), 144–149.

Tapp, C. and Rice, R. G. (2012). Generation, application and control of ozone in food processing plants. In O'Donnel, C., Tiwari, B. K., Cullen, P. J., Rice, R. G., Eds. *Ozone in Food Processing*, Wiley Blackwell Publishing, New York, ISBN: 978-1-4443-3442-5.

Tiwari, B. K., Muthukumarappan, K., O'Donnell, C. P., and Cullen, P. J. (2008). Kinetics of freshly squeezed orange juice quality changes during ozone processing. *J. Agric. Food Chem.*, 56(15), 6416–6422.

Tiwari, B. K., O'Donnell, C. P., Muthukumarappan, K., and Cullen, P. J. (2009a). Anthocyanin and colour degradation in ozone-treated blackberry juice. *Innov. Food Sci. Emerg. Technol.*, 10(1), 70–75.

Tiwari, B. K., O'Donnell, C. P., Patras, A., Brunton, N., and Cullen, P. J. (2009b). Effect of ozone processing on anthocyanins and ascorbic acid degradation of strawberry juice. *Food Chem.*, 113(4), 1119–1126.

Tzortzakis, N., Borland, A., Singleton, I., and Barnes, J. (2007). Impact of atmospheric ozone-enrichment on quality-related attributes of tomato fruit. *Postharvest Biol. Technol.*, 45(3), 317–325.

Tzortzakis, N., Singleton, I., and Barnes, J. (2008). Impact of low-level atmospheric ozone-enrichment on black spot and anthracnose rot of tomato fruit. *Postharvest Biol. Technol.*, 47(1), 1–9.

Wang, H., Feng, H., and Luo, Y. (2004). Microbial reduction and storage quality of fresh-cut cilantro washed with acidic electrolyzed water and aqueous ozone. *Food Res. Int.*, 37(10), 949–956.

Whangchai, K., Saengnil, K., Singkamanee, C., and Uthaibutra, J. (2010). Effect of electrolyzed oxidizing water and continuous ozone exposure on the control of Penicillium digitatum on tangerine cv. 'Sai Nam Pung' during storage. *Crop Prot.*, 29(4), 386–389.

Whangchai, K., Saengnil, K., and Uthaibutra, J. (2006). Effect of ozone in combination with some organic acids on the control of postharvest decay and pericarp browning of longan fruit. *Crop Prot.*, 25(8), 821–825.

Wu, J. G., Luan, T. G., Lan, C. Y., Lo, W. H., and Chan, G. Y. S. (2007). Efficacy evaluation of low-concentration of ozonated water in removal of residual diazinon, parathion, methyl-parathion and cypermethrin on vegetable. *J. Food Eng.*, 79(3), 803–809.

7 Irradiation Applications in Fruit and Other Fresh Produce Processing

Rosana G. Moreira and Elena M. Castell-Perez

CONTENTS

7.1 INTRODUCTION

Treatment of fresh produce (fruits and vegetables) using ionizing radiation (food irradiation) has a significant strategic importance for the future of food safety worldwide. This is because food irradiation is the most researched nonthermal food process technology and has been proven to be safe (when done properly).

Food irradiation is a process in which food products are exposed to radiant energy (ionizing radiation) to inactivate pathogens such as *Escherichia coli*, *Listeria*, and *Salmonella*. Many studies have shown that irradiation kills pathogens or markedly reduces pathogen counts. The process can also control insects and parasites, reduce spoilage, and inhibit ripening and sprouting.

Ionizing radiation, defined as a radiation with enough energy to remove tightly bound electrons from atoms, thus creating ions, has been utilized mainly to eliminate potentially harmful organisms from foods. The Food and Agriculture Organization (FAO) of the United Nations, the International Atomic Energy Agency (IAEA), and

TABLE 7.1

Absorbed Dose Requirements for Different Irradiation Treatment

Treatment	Dose (kGy)
Sprouting inhibition	0.1–0.2
Insect disinfestations	0.3–0.5
Parasite control	0.3–0.5
Delay of ripening	0.5–1.0
Fungi control	1.5–3.0
Bacteria control	1.5–3.0
Sterilization	15.0–30.0

the World Health Organization (WHO) maintain that irradiated food is both safe to consume and nutritionally adequate (WHO, 1999). These agencies also set the limits for maximum absorbed dose for the different types of foods. The quantity of energy (in Joules) absorbed by 1 kg of the food product is referred to as dose measured in units of Gray (Gy). Dose in food processing operations is measured in kilogray (1 kGy = 1000 Gy). Depending on the food irradiation application, the required dose varies. Table 7.1 shows the absorbed dose requirements for different types of irradiation treatments (Miller, 2005).

Food irradiation entails exposing the food, either in bulk or packaged, to controlled amounts of ionizing radiation for a specific time to attain certain goals. Many foods are already irradiated to inactivate bacteria and keep the foods longer without spoiling. These include red meat, spices, poultry, and molluscan shellfish (e.g., oysters, clams, mussels, and scallops). On August 22, 2008, the Food and Drug Administration (FDA) (FDA, 2008) approved the use of irradiation to make leafy vegetables such as fresh iceberg lettuce and fresh spinach safer. Irradiation of fruits is being done in several countries including South Africa, France, and the United States (Hawaii), mainly for quarantine purposes. However, the suitability of irradiation for pathogen decontamination in fruits has become more evident in recent years.

The practicability of applying ionizing radiation depends on the tolerance of the target food to the irradiation dose. As irradiation doses increase, more microorganisms are affected (bacteria have different levels of resistance to ionizing radiation). However, a higher dose, while not creating any harmful products, can induce changes in the organoleptic and chemical properties of the food. This is of particular importance in the case of fresh fruits and vegetables due to their more delicate structures. Thus, a balance must be achieved between the quality of the irradiation treatment and the dose received by the food. Furthermore, since not all pathogens are equally sensitive to radiation, a microorganism that is radio-resistant must be treated with a higher dose, and the consequent side effects are unavoidable.

This chapter addresses the needs for irradiation treatment of fresh produce, describes the irradiation technologies commonly used to treat fresh fruits and vegetables, and provides examples of the challenges and benefits of using this technology in fresh produce. Vegetables are included due to the fact that most of the studies on irradiation of fresh produce have focused on vegetables.

7.2 FOOD IRRADIATION PROCESSES

7.2.1 FOOD IRRADIATION SOURCES

There are three types of ionizing radiation used to decontaminate foods: gamma (γ) rays, x-rays, and electron beams. Gamma and x-rays are electromagnetic radiations (called photons) with the same physical properties at any given quantum energy. The difference between the two types is their modes of origin. Gamma rays originate in the nucleus of a radionuclide (cobalt-60), while x-rays originate in the electron fields surrounding the nucleus and are machine produced. Unlike gamma and x-rays, electrons are charged particles, which can be emitted from a nucleus or come from a charged particle collision. Gamma radiation emission occurs when the nucleus of a radioactive atom has too much energy (Turner, 2007). The maximum energies for x-rays and e-beams to be used in food processing are 7 and 10 MeV (megaelectron volts), respectively (Attix, 1986). These energy limits are set to prevent the possibility of inducing radioactivity in the irradiated food via photonuclear reactions (Alpen, 1998).

7.2.2 FOOD IRRADIATION TECHNOLOGIES

Table 7.2 compares the radiation technologies used to treat food products.

7.2.2.1 Gamma: Cobalt-60

Gamma facilities store the source in stainless steel capsules ("pencils") in underwater tanks. Cobalt-60 has several advantages: penetrates deeply, yields substantial

TABLE 7.2
Food Irradiation Technologies

	E-beam	X-Ray	Co-60 Gamma Rays
Energy (MeV)	10	5 or 7	1.17 or 1.33
Penetration (cm)	<10	100	70
Irradiation on demand	Power stop	Power stop	Not possible
Throughput	+++	++	+
Dose uniformity	+	+++	++
Administration process	Authorization required—standard registration	Authorization required—standard registration	Authorization required—complex and difficult process
Treatment time	Seconds	Minutes	Hours
Dose rate	Very high	Medium	Low

Source: Ion Beam Application Group (IBA), eXelis® Food X-rays for clean and safe food treatment, IBA S.A, www.iba-food.com, 2009.

uniformity of the dose in the food product, and up to 95% of its emitted energy is available for use. However, its 5.3 year half-life offers disadvantages. Cobalt-60 "pencils" require frequent replenishment, and treatment of the food is relatively slow (EPA, 2010). The main issues driving the industry to look for alternatives are cobalt price fluctuations, future availability of cobalt-60, increasing regulations on cobalt-60 transportation, and difficulty related to the disposal of radioactive sources (IBA, 2009).

7.2.2.2 E-Beam

E-beam facilities generate e-beams with an e-beam linear accelerator (LINAC). The electrons are concentrated and accelerated to 99% of the speed of light and energies of up to 10 MeV. Because e-beams are generated electrically, they offer certain advantages: they can be turned on only as needed; they do not require replenishment of the source as does cobalt-60; there is no radioactive waste (EPA, 2010).

E-beam is the most efficient (high-throughput) cold process utilizing energy from accelerated electrons. However, an e-beam can only penetrate a few centimeters depending on food density. Thicker products can be treated by beaming electrons from both sides. For deeper penetration applications, thicker products, or food packaged in pallets, x-ray is the answer (IBA, 2009).

E-beam technology has other disadvantages including high electric power consumption, complexity, and potentially high maintenance.

7.2.2.3 X-Ray Accelerators

X-ray facilities use an e-beam accelerator to target electrons on a dense metal plate, such as tungsten. Like gamma rays, x-rays can penetrate up to a 100 cm deep target. This allows food to be processed in a shipping container. X-rays share the e-beam technology disadvantages (EPA, 2010).

The development of high-energy electron accelerators with very high power has recently made x-ray processing a practical alternative to gamma ray for food preservation. The feasibility of radiation processing with high-energy x-rays has been demonstrated in various industrial facilities around the world. The forward peaked emission of x-rays facilitates the treatment of single pallet loads of product. X-ray dose rates are at least twice as high as gamma rays (IBA, 2009).

7.2.3 Dose and Dosimetry

The best procedure for food irradiation processing depends largely on accurate and reproducible measurement of radiation quantities. When using e-beams, a clear understanding of how the electrons transport within the target food allows for accurate prediction of where the radiation deposits energy in the food. As a result, advances in precise dose calculation methods are critical for production of higher quality and safer irradiated fresh produce.

The primary physical quantity used in dosimetry is the absorbed dose, defined as the energy absorbed per unit mass from any kind of ionizing radiation in any target. The unit dose commonly used is the Gray (Gy) or Joules per kilogram.

Food irradiation facilities must be designed to yield an absorbed dose in the product within the minimum and maximum limits in accordance with process specifications and government regulatory requirements (Codex, 1984). The actual minimum and maximum dose, D_{min} and D_{max}, as measured in the product, must be within these limits. The dose uniformity ratio (DUR) is then defined as D_{max}/D_{min}, a notion used by irradiator designers and food engineers. For research applications, this ratio should be as close to 1 as possible; this simply means that the dose should be very uniform in a small sample (the food) so that experimental results can clearly demonstrate the dose–effect relationship. For industrial applications where large process loads are irradiated, wider dose variation is inevitable, and a great portion of the food item will receive a dose significantly greater than the minimum absorbed dose required for safety. This can be sometimes as much as double the required dose for inactivation of pathogens. Although low DUR (less or equal to 1.5) is the main target in food irradiation facilities, many food product applications can tolerate a higher DUR of 2 or yet 3 (IAEA, 2002).

Dosimeters, in radiation research and commercial processing, are used for quality and process control. In food irradiation applications, doses are measured either with alanine or radiochromic film dosimeters placed at the food's surface (Sharpe, 1990). Yet these dosimeters do not suitably fit the irregular and complex shapes of some fresh produce (e.g., a head of broccoli, baby spinach leaves in a bag) that may change the absorption of the radiation energy, which is monoenergetic, and might introduce variability in the absorbed dose. Therefore, the energy distribution in a fresh product is strongly related to the electron's entrance region at the surface of the product and the density of the produce.

Thus, inaccurate interpretation of the measured dose can result in the incorrect irradiation treatment. To complicate things, dosimeters for measuring doses in the internal regions of the produce are currently nonexistent.

Fresh produces in general are very susceptible to overdosage (DUR > 2–3), which usually results in damage to their tissues thus affecting their quality attributes (color, flavor, aroma, and texture). Hence, the need for the dose distribution in fresh produces to be as uniform as possible and high enough to inactivate pathogens without deteriorating produce quality. This makes irradiation of fresh fruits and vegetables a challenging task. In that sense, e-beam sources (LINAC) are not the best alternative for fresh produces because of their short penetration depths in addition to the potential over- and underdosage when the dual mode (top and bottom beams) is used. X-rays and gamma rays, on the other hand, are useful to irradiate large quantities of fresh produce in boxes; however, these treatments must be well tuned to reduce the DUR.

Figure 7.1 illustrates the simulation of the dose distribution in different fresh produces irradiated with various irradiation sources (e-beam and gamma ray). A 10 MeV e-beam source has a penetration power of 4–5 cm, depending on the product density, as shown by the dose distribution in the cantaloupe (Figure 7.1a) and broccoli head (Figure 7.1d). When two 10 MeV sources are used (top and bottom), the penetration depth will double, as is seen in the mango (Figure 7.1b). However, the dose distribution inside the mango shows that some parts of the fruit are overirradiated (center) and other parts are underirradiated (surface). Gamma-ray irradiation

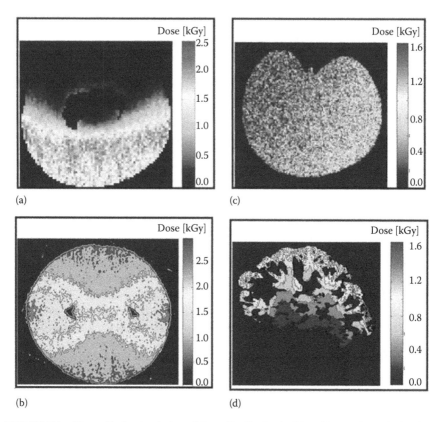

FIGURE 7.1 Monte Carlo simulation of dose distribution inside a (a) cantaloupe being irradiated with a 10 MeV e-beam source on the bottom, (b) mango with 10 MeV sources from top and bottom, (c) peach being irradiated with a gamma-ray source from the right, and (d) broccoli head with a 10 MeV e-beam source from the top. (Adapted from Kim, J. et al., *J. Food Eng.*, 97(3), 425, 2010; Kim, J. et al., *J. Food Eng.*, 86(4), 595, 2008; Moreno, M. et al., *J. Food Process Eng.*, 31(2), 120, 2008.)

of a peach (Figure 7.1c) shows the deeper penetration of the radiant energy, but the treatment results in an overdose to the right of the fruits compared to the left side. Therefore, it is very important to understand how energy is absorbed in fruits and vegetables so the best treatment can be used correctly and efficiently. The simulation was carried out using Monte Carlo simulation software (Kim et al., 2008, 2010).

7.3 FRESH FRUITS AND VEGETABLES IRRADIATION TREATMENTS

7.3.1 DISINFESTATION AND SHELF-LIFE EXTENSION

The U.S. FDA has approved radiation doses up to 1 kGy for preservation and disinfestation of fruits and vegetables (FDA, 1986). Ionizing radiation breaks chemical bonds within DNA and other molecules, thereby disrupting normal cellular function in the infesting insect (Ducoff, 1972; Koval, 1994). Insects and other living organisms are able to repair molecular damage done by ionizing energy (Alpen, 1998), but large irradiation doses are fatal or cause permanent sterility. Unlike other disinfestation techniques, irradiation does not need to kill the pest immediately to provide quarantine security, and therefore live (but sterile or not viable) insects may occur within the exported commodity. The goal of a quarantine treatment is to prevent reproduction (Follett, 2009).

Radiation treatments for extension of the shelf life of fresh fruits and vegetables have been widely studied (Lacroix and Ouattara, 2000). The application of irradiation processing with doses less than 1 kGy inhibits germination, slows ripening of fruit, and eliminates insect infestation in many fruits and vegetables.

Doses between 1 and 3 kGy can be used to eliminate molds in fresh produces. Most fruits and vegetables tolerate irradiation treatments with a dose of 0.25 kGy without changing their quality attributes (firmness, flavor, physiological characteristics, and respiration rates). A dose of 2.25 kGy is usually the optimal dose that fruits and vegetables may tolerate to keep their quality intact (Lacroix and Ouattara, 2000). Strawberries treated with doses as high as 3 kGy showed mold reduction without affecting their quality for a period of 14 days at 5°C (Lacroix and Ouattara, 2000).

However, only few commodities can tolerate the use of irradiation for postharvest disease control. The doses required for effective control of fungal spoilage frequently result in quality degradation of fruits (Mahrour et al., 1998). The tolerance to irradiation varies among species and varieties and is influenced by the degree of ripeness at the time of treatment (Thomas, 1986).

For example, for effective irradiation of mangoes, it is recommended to treat fully mature fruits in the hard-green preclimacteric state. Irradiation of immature mangoes may result in increased shriveling and uneven ripening. The optimum dose to delay ripening and the maximum dose that can be tolerated by the fruit depend on the fruit variety (Thomas, 1986).

Several investigations have demonstrated, however, the usefulness of a pretreatment, such as dipping the fruit in warm water, to reduce the dose required to extend the shelf life of certain fresh fruits without affecting their quality attributes (Gagnon et al., 1993). For fungal treatment of mangoes, a dose between 0.75 and 1.5 kGy is required; however, it results in unacceptable injuries to the fruit. The best treatment for mangoes, which controls insects' infestation and increases storage life (delayed ripening and reduced fungal spoilage), is a combination of hot water (50°C for 10 min) and irradiation (0.25–0.75 kGy) (Thomas, 1986).

The irradiation process produces a great deal of ozone, which could be harmful to the tissue cells of many fruits and vegetables. Modified atmospheric packages (MAPs) have been used to reduce radiation damages in mangoes (Dharkar et al., 1966). Carbon dioxide produced a minimum chlorophyll disappearance, but a nitrogen atmosphere produced the best results in terms of organoleptic qualities, spoilage data, and changes in ascorbic acid and carotenoids, with delay in ripening.

A combination of hot-water pasteurization of whole cantaloupe and low-dose irradiation of packaged fresh-cut melon showed reduced population of native microflora while maintaining their quality (Fan et al., 2006). Boyton et al. (2006) studied the combination of e-beam irradiation of fresh-cut cantaloupe in MAP and showed that the produce shelf life extended to about 2–3 weeks.

With fresh fruits and vegetables, it is important to establish the exact parameters such as proper ripening stage, proper pretreatment, posttreatment, and transport conditions yielding optimum results for shelf life extension.

7.3.2 Inactivation of Pathogens

Consumers are becoming more aware of the importance of eating fresh fruits and vegetables. However, these minimally processed foods have repeatedly become a source of foodborne illnesses in the United States. Thus, *E. coli* O157:H7, *Salmonella* spp., *Listeria* spp., and *Shigella* spp. have become a public health concern since these microorganisms have been associated with foodborne outbreaks from consumption of fresh produce including cantaloupes, bean sprouts, tomatoes, green onions, spinach, lettuce, bagged salad mixes, unspecified fruits, strawberries, raspberries, and even fruit salads. In the United States, approximately 20.8% of the total fresh produce outbreaks were due to contaminated fruit consumption in the 1990–1998 periods (CDC, 2000). In 2002, a *Salmonella* outbreak due to contaminated mangoes was reported (Sivapalasingham et al., 2003).

Another significant problem has been identified on recent studies indicating that pathogenic microorganisms may internalize into the core of leafy vegetables in addition to contaminating the exposed surface of the leaves. This bacterial mobility renders traditional surface treatments to reduce pathogens unsuccessful. For instance, most of the commercially used interventions employ chemical agents, such as washing with 2% chlorinated water, which cannot either wash these pathogens off the produce or inactivate them. In addition, this chemical treatment produces detrimental effects on the organoleptic properties of the food.

Since current production and processing practices cannot ensure pathogen-free fresh and fresh-cut produce and with almost 25% of food production after harvest in the United States lost due to damage caused by bacteria, mold, insects, and contamination with spoilage microorganisms, effective food safety interventions are needed for implementation throughout the production, processing, and distribution of these food items. The only viable option is the use of ionizing radiation. The challenge for irradiator operators is how to reduce the damage to produce quality while applying sufficient doses of radiation to inactivate the pathogen.

Correct food irradiation depends on accurate and reproducible measurement of the absorbed dose, also called dosimetry (Kim et al., 2005, 2006, 2007, 2008; Gomes et al., 2008a,b). Problems with dosimetry for fresh produce include accuracy and uniformity of dose measured all over the product, and lack of standards to validate the measurements. Inaccurate interpretation of dose measurements can result in misleading D_{10} values for a target pathogen in a particular produce. The D_{10} value is defined as the dose (in kGy) required to reduce the microbial population by 90%. Overdosing the produce is obviously costly, and underdosing may have severe safety implications.

Recent studies have evaluated the feasibility of irradiation as a pathogen decontamination treatment. Results from sensory evaluation of produce overall quality suggest a potential fungicidal effect of low-dose irradiation (1.0 kGy) of packaged romaine lettuce hearts without altering the overall quality of the produce as well as the low-density polyethylene (LDPE) packaging characteristics. Romaine lettuce hears were irradiated using a 10 MeV LINAC at room temperature. Irradiated romaine lettuce hearts showed slight changes in color compared with nonirradiated samples (Han et al., 2004). As the dose increased, lettuce firmness decreased by 49.58% (leaves) and 29.13% (ribs). Sensory attributes such as overall quality, color, sogginess, and off-flavor were less acceptable at the higher dose level. Irradiation also affected the respiration rates inside the lettuce LDPE bags. Irradiation at 1.5 and 3.2 kGy dose levels improved the oxygen barrier capability of the bags (7.67% and 4.48%, respectively). Water vapor permeability was unaffected at all the irradiation dose levels. The stiffness of LPDE films did not change due to irradiation treatment. Thus, irradiation carried out at the levels required for pathogen inactivation in fresh fruits and vegetables does not have a negative effect on the polymeric materials used in the packaging.

Irradiation of cantaloupes using e-beams, as whole fruits with dose up to 1.0 kGy, caused no significant changes on the fruit's physical and nutritional quality attributes. Irradiating at higher doses had an undesirable effect on product quality (Castell-Perez et al., 2004). In contrast, fresh-cut packaged cantaloupe may be irradiated up to 1.5 kGy without worsening the product quality attributes.

A suitable irradiation process can be designed for effective decontamination of bagged, ready-to-eat baby spinach leaves (Gomes et al., 2008b). Treatment with up to 1.0 kGy did not cause significant changes in produce quality. When using a 10 MeV LINAC, the uniformity of dose distribution in a bag of spinach leaves depends on the arrangement of the leaves inside the bag (relative to the direction of the e-beam). Thus, treatment of leafy vegetables with a 10 MeV LINAC requires a well-controlled operating process as well as package design and positioning to ensure reduction of a population of a particular pathogen to the desired level using the lowest dose possible without detriment to produce quality.

The viability of using ionizing radiation as a produce decontamination technique at doses greater than 1.0 kGy depends on the tolerance of the produce to dose. Irradiation did not affect color, firmness, pH, and weight loss of fresh head of broccoli irradiated at doses of 1–3 kGy (Gomes et al., 2008a). Irradiation affected the respiration rates on the first 5 days of storage; vitamin C content of all samples decreased with storage time; chlorophyll and total carotenoids content followed the same trend. In terms of overall acceptability, color, odor, and texture, all irradiated samples were highly accepted by the panelists. By the end of shelf life, only the non-irradiated samples showed significant quality decline (yellow color, off-odor) due to microbial spoilage. In summary, e-beam treatment up to 3.0 kGy maintains the overall quality of fresh broccoli.

"Tommy Atkins" mangoes exposed to 1.5 and 3.1 kGy were softer than nonirradiated fruits throughout storage at 15°C (Moreno et al., 2006). The radiation-induced softening of the fruits may be associated with changes in the structural cell such as cracks and depressions on the surface and the breakdown of the cells and its components. Irradiation at 3.1 kGy affected the color of mangoes by the end of storage. Doses up to 1.5 kGy kept respiration rates at a normal level. Irradiation did not affect the specific gravity of mangoes, a parameter associated with fruit maturity levels. No effect of irradiation on pH, water activity, moisture content, acidity, and juiciness of mangoes was detected at the dose levels used in this study. Only fruits irradiated at 3.1 kGy were unacceptable to the sensory panelists in terms of overall quality, texture, and aroma. E-beam irradiation of "Tommy Atkins" mangoes at 1.0 kGy is therefore the recommended treatment to maintain the overall fruit quality attributes.

Irradiation at doses higher than 1.1 kGy did affect the texture of blueberries as the fruits became considerably softer and less acceptable throughout storage at 15°C (Moreno et al., 2007). Only irradiation at 3.2 kGy affected the color of blueberries by the end of storage. Irradiation slightly reduced the respiration rates of the blueberries by the end of storage. In terms of overall quality, texture, and aroma, only fruits exposed to 3.2 kGy were found unacceptable by the sensory panelists. Irradiation did not affect the density, pH, water activity, moisture content, acidity, and juiciness of blueberries. E-beam irradiation of blueberries up to 1.6 kGy is a feasible decontamination treatment that maintains the overall fruit quality attributes.

7.4   RADIOSENSITIZATION STRATEGIES

To ensure the high quality of irradiated fresh fruits and vegetables, methods to reduce the applied dose must be developed such as MAP and the use of radiosensitizers.

MAP was successfully used to reduce the dose required to inactivate microorganisms in fresh or frozen (−5°C) baby spinach leaves (Gomes et al., 2011a). Increased concentrations of oxygen in the packaging bags significantly increased the radiation sensitivity of *Salmonella* spp. and *Listeria* ssp. microorganisms, ranging from 7% up to 25% reduction in D_{10} values. In particular, radiation sensitization could be effected by production of ozone, which increases with increasing dose rate and oxygen concentration and reducing temperatures. These results suggest that low-dose (below 1.0 kGy) e-beam irradiation under MAP (100% O_2 and N_2:O_2 [1:1]) may be a viable tool for reducing microbial populations or eliminating *Salmonella* spp. and *Listeria* spp. from baby spinach leaves. A suggested treatment to achieve a 5 log reduction of the test organisms would be irradiation at room temperature under 100% O_2 atmosphere at a dose level of 0.7 kGy (Gomes et al., 2010).

Natural compounds and extracts (*trans*-cinnamaldehyde, eugenol, garlic extract, propolis extract, and lysozyme with EDTA [ethylenediaminetetraacetic acid disodium salt dihydrate]) have shown bacteriostatic effect against *Salmonella* spp. and *Listeria* spp. To mask odor and off-flavor inherent of several compounds, and to increase their solubility, complexes of these natural compounds and extracts with β-cyclodextrin (β-CD) are generally micro-/nanoencapsulated. These microencapsulated compounds can be effective as radiosensitizers when combined with irradiation treatments to inactivating pathogens in fruits and vegetable. The increase in radiation sensitivity (up to 40%) can vary with the antimicrobial compound (Gomes et al., 2011b). The combination of spraying of microencapsulated antimicrobials with e-beam irradiation at the surface of baby spinach is effective in increasing the killing effect of irradiation (increased bacterial radiosensitization), demonstrating the potential of this technology to reduce the required radiation dose to control microbial contamination thus minimizing produce quality detriment. These technologies can be easily applied to irradiation treatment of a wide variety of fruits.

7.5   ECONOMIC ASPECTS

Food irradiation is not inexpensive. However, the actual cost of food irradiation operations for treatment of fresh produce depends on how much dose must be applied to the fresh produce, how sensitive is the fresh produce to ionizing radiation, any packaging and stacking requirements (bagged versus in bulk, pallets, etc.), in addition to the construction costs inherent of building, maintaining, and operating an irradiation facility. Initial investments range from $4 to $10 million (Forsythe and Evangelou, 1993; Kundstat and Steeves, 1993). Nevertheless, the many benefits of irradiation such as extended shelf life and reduced quality detriment may offset the cost of the technology. Low-dose applications such as disinfestation of fruits ($<1.0\,kGy$) range between $0.01/lbs and $0.08/lbs ($0.022/kg to $0.18/kg) while higher-dose applications can cost as much as $0.125/lbs ($0.28/kg) (Forsythe and Evangelou, 1993; ANI, 2010). Other studies show irradiation cost of strawberries, oranges, grapefruit, mushrooms, and tomatoes as low as $0.02/lbs ($0.04/kg) (Corrigan, 1993). These costs are very competitive when food irradiation is compared with fumigation with methyl bromide (Table 7.3) or mild thermal treatments. Since the use of the fumigant has been restricted in many countries, the economic feasibility of quarantine treatment of fresh fruits and vegetables using irradiation is obvious. Morrison (1989) conducted an in-depth economic analysis of irradiation of papayas and strawberries when using electron accelerators and cobalt-60 sources. A similar analysis should be done now to address the advances in the technology as well as consumer acceptance issues.

TABLE 7.3
Comparison of Estimated Postharvest Treatment Costs
for Selected Fruits

Crop	Methyl Bromide (Cents per Pound)	Irradiation (Cents per Pound)
Strawberries	0.88–0.94	2.5–8.1
Papaya	0.88–0.94	0.9–4.2
Mango	0.88–0.94	0.8–7.2

Source: Modified from Forsythe, K.W. and Evangelou, P., Costs and benefits of irradiation and other selected quarantine treatments for fruit and vegetable imports to the United States of America, Issue Paper, *Proceedings of an International Symposium on Cost-Benefit Aspects of Food Irradiation Processing* jointly organized by the International Atomic Energy Agency, The Food and Agricultural Organization of the United Nations, and the World Health Organization, March 1–5, 1993, Aix-En-Provence, Vienna, 1993; Forsythe, K.W. and Evangelou, P., *Costs and Benefits of Irradiation versus Methyl Bromide Fumigation for Disinfestation of U.S. Fruit and Vegetable Imports*, U.S. Department of Agriculture, Economic Research Service, Agriculture and Trade Analysis Division, Washington, DC, March 1994; Morrison, R.M., An economic analysis of electron accelerators and cobalt-60 for irradiating food, U.S. Department of Agriculture, Economic Research Service, Technical Bulletin #1762, June 1989.

7.6 ACCEPTANCE OF IRRADIATED FRUITS AND VEGETABLES

In general, consumer acceptance of irradiated fruits is higher than expected. Strawberries, papayas, and mangoes have been found acceptable by consumers in Europe, South Africa, and the United States (Bruhn, 1995). It is fair to say that if science-based information on the benefits of irradiation is made available to the public, consumers may be more than ready to buy irradiated fruits and vegetables.

REFERENCES

Alpen, E.N. 1998. *Radiation Biophysics*. Academic, San Diego, CA.

American Nuclear Institute (ANI). 2010. Nuclear science and technology and how it influences your life. Food. http://www.aboutnuclear.org/view.cgi?fC=Food,Benefits_^_ Effects (accessed January 2011).

Attix, F.H. 1986. *Introduction to Radiological Physics and Radiation Dosimetry*. Wiley-Interscience Publication, New York

Boyton, B.B., Welt, B.A., Sims, C.A., Balaban, M.A., Brecht, J.K., and Marshall, M.R. 2006. Effects of low-dose electron-beam irradiation on respiration, microbiology, texture, color, and sensory characteristics of fresh-cut cantaloupe stored in modified-atmosphere packages. *Journal of Food Science* 71, S149–S155.

Bruhn, C.M. 1995. Consumer attitudes and market response to irradiated food. *Journal of Food Protection* 58(2), 175–181.

Castell-Perez, M.E., Moreno, M., Rodriguez, O., and Moreira, R.G. 2004. Electron-beam irradiation treatment of cantaloupes: Effect on product quality. *Food Science and Technology International* 10(6), 383–390.

Centers for Disease Control and Prevention (CDC). 2000. CDC. Surveillance for foodborne-disease outbreaks—United States, 1993–1997. Surveillance Summaries, *MMWR*, 49(No. SS-1), 64.

Codex Alimentarius Commission. 1984. *Codex General Standard for Irradiated Foods and Recommended International Code of Practice for the Operation of Radiation Facilities used for the Treatment of Foods*. CAC Vol. XV, Edition 1, Rome.

Corrigan, J.P. 1993. Experiences in selling irradiated foods at the retail level. In *Cost-benefit Aspects of Food Irradiation Processing. Proceedings of an IAEA/FAO/WHO International Symposium*, Aix-en-Provence, France, pp. 447–453.

Dharkar, S.D., Savagaon, K.A., Srirangaraja, A.N., and Sreenivasan, A. 1966. Irradiation of Mangoes. I. Radiation-induced delay in ripening of Alphonso mangoes. *Journal of Food Science* 31(6), 863–869.

Ducoff, H.S. 1972. Causes of cell death in irradiated adult insects. *Biology Review* 47, 211–240.

Environmental Protection Agency (EPA). 2010. Food irradiation. Environmental protection agency. http://www.epa.gov/radiation/sources/food_irrad.html#cobalt60. Last updated on Friday, October 01, 2010.

Fan, X., Annous, B.A., Sokorai, K.J.B., Burke, A., and Mattheis, J.P., 2006. Combination of hot-water surface pasteurization of whole fruit and low-dose gamma irradiation of fresh-cut cantaloupe. *Journal of Food Protection* 69, 912–919.

Follet, P.A. 2009. Generic radiation quarantine treatments: The next steps. *Journal of Economic Entomology* 102(4), 1399–1406.

Food and Drug Administration (FDA). April 18, 1986. Irradiation in the production, processing, and handling of food. Rules and Regulations. *Federal Register* 51(75): 13376. http://www.fda.gov/consumer/updates/irradiation082208.html (accessed January 2009).

Food and Drug Administration. 2008. *Irradiation in the Production, Processing and Handling of Food*, 21 CFR Part 179. Department of Health and Human Services. FDA Docket no. FDA-1999-F-2405 (formerly 1999F–5522), pp. 49593–49603.

Forsythe, K.W. and Evangelou, P. 1993. Costs and benefits of irradiation and other selected quarantine treatments for fruit and vegetable imports to the United States of America. Issue Paper. *Proceedings of an International Symposium on Cost-Benefit Aspects of Food Irradiation Processing* jointly organized by the International Atomic Energy Agency, The Food and Agricultural Organization of the United Nations, and the World Health Organization, March 1–5, 1993, Aix-En-Provence, Vienna, Austria.

Forsythe, K.W. and Evangelou, P. 1994. *Costs and Benefits of Irradiation versus Methyl Bromide Fumigation for Disinfestation of U.S. Fruit and Vegetable Imports*. U.S. Department of Agriculture, Economic Research Service, Agriculture and Trade Analysis Division, Washington, DC.

Gagnon, M., Lacroix, M., Pringsulaka, V., Latreille, B., Jobin, M., Nouchpramool, K., Prachasitthisak, Y., Charoen, S., Adulyatham, P., Lettre, J., and Grad, B. 1993. Effect of gamma irradiation combined with hot water dip and transportation from Thailand to Canada on biochemical and physical characteristics of Thai mangoes (Nahng Glahng Wahn variety). *Radiation Physics and Chemistry* 42, 283–287.

Gomes, C., Da Silva, P., Chimbombi, E., Castell-Perez, E., and Moreira, R.G. 2008a. Effects of low-dose electron-beam irradiation and storage on quality of broccoli heads (*Brassica oleracea* L var Italica). *LWT—Food Science and Technology* 10, 1828–1833.

Gomes, C., Moreira, R., and Castell-Perez, E. 2011a. Radiosensitization of *Salmonella* spp. and *Listeria* spp. in ready-to-eat baby spinach leaves. *Journal of Food Science* 76, E141–E148. doi: 10.1111/j.1750-3841.2010.01904.x.

Gomes, C., Moreira, R., and Castell-Perez, E. 2011b. Microencapsulated antimirobial compounds as a means to enhance electron beam irradiation treatment of fresh produce. *Journal of Food Microbiology* 76(6), E479–E488. doi: 10.1111/j.1750-3841.2011.02264.x

Gomes, C., Moreira, R., Castell-Perez, M.E., Kim, J., Da Silva, P., and Castillo, A. 2008b. Effects of low-dose electron-beam irradiation on quality and safety of ready-to-eat spinach leaves (*Spinacea oleracea*). *Journal of Food Science* 73, 95–102.

Han, J., Gomes-Feitosa, C.L., Castell-Perez, E., Moreira, R.G., and Da Silva, P. 2004. Quality of packaged romaine lettuce hearts exposed to low-dose electron-beam irradiation. *Lebensmittel-Wissenschaft und-Technologie/Food Science and Technology* 37(7), 705–715.

International Atomic Energy Agency (IAEA). 2002. Dosimetry for food irradiation. Technical Reports Series No. 409. IAEA, Vienna, Austria.

Ion Beam Application Group (IBA). 2009. eXelis® Food X-rays for clean and safe food treatment. IBA S.A. www.iba-food.com (accessed January 15, 2011).

Kim, J., Moreira, R.G., and Castell-Perez, M.E. 2008. Validation of irradiation of Broccoli with a 10 MeV electron-beam accelerator. *Journal of Food Engineering* 4, 595–603.

Kim, J., Moreira, R., and Castell-Perez, E. 2010. Simulation of pathogen inactivation in whole and fresh-cut cantaloupe (*Cucumis melo*) using electron-beam treatment. *Journal of Food Engineering* 97(3), 425–433.

Kim, J., Moreira, R.G., Huang, Y., and Castell-Perez, M.E. 2007. 3-D dose distributions for optimum radiation treatment planning of complex foods. *Journal of Food Engineering* 79, 312–321.

Kim, J., Moreira, R.G., Rivadeneira, R., and Castell-Perez, M.E. 2005. Monte Carlo based food irradiation simulator. *Journal of Food Processing Engineering* 29(1), 72–88.

Kim, J., Rivadeneira, R., Castell-Perez, M.E., and Moreira, R.G. 2006. Development and validation of a methodology for dose calculation in electron-beam irradiation of complex-shaped foods. *Journal of Food Engineering* 74, 359–369.

Koval, T.M. 1994. Intrinsic stress resistance of cultured lepidopteran cells. In K. Marmarorosh and A. McIntosh (eds.) *Insect Cell Biotechnology*. CRC, Boca Raton, FL, pp. 157–185.

Kundstat, P. and Steeves, C. 1993. Economics of food irradiation. In *Cost-benefit Aspects of Food Irradiation Processing. Proceedings of an IAEA/FAO/WHO international symposium*, Aix-en-Provence, France, pp. 395–415.

Lacroix, M. and Ouattara, B. 2000. Combined industrial processes with irradiation to assure innocuity and preservation of food products: A review. *Food Research International* 33, 719–724.

Mahrour, A., Lacroix, M., Nketsa-Tabiri, J., Calderon, N., and Lacroix, M. 1998. Antimicrobial properties of natural substances in irradiated fresh poultry. *Radiation Physics and Chemistry* 52, 81–84.

Miller, R., 2005. *Electronic Irradiation of Foods: An Introduction to the Technology*. Springer, New York.

Moreno, M., Castell-Perez, M.E., Gomes, C., Da Silva, P., Kim, J., and Moreira, R.G. 2008. Optimizing electron-beam irradiation of Tommy Atkins Mangoes (*Mangifera indica* L.). *Journal of Food Process Engineering* 31(2), 120–134.

Moreno, M., Castell-Perez, M.E., Gomes, C., Da Silva, P., and Moreira, R.G. 2006. The effects of electron-beam irradiation on physical, textural and micro structural properties of Tommy Atkins Mangoes (*Mangifera indica* L.). *Journal of Food Science* 71(2), E80–E86.

Moreno, M., Castell-Perez, M.E., Gomes, C., Da Silva, P., and Moreira, R.G. 2007. Quality of electron-beam irradiation of blueberries (*Vaccinium corymbosum* L.) at medium dose levels (1.0–3.2 kGy). *LWT—Food Science and Technology* 40, 1123–1132.

Morrison, R.M. 1989. An economic analysis of electron accelerators and cobalt-60 for irradiating food. U.S. Department of Agriculture, Economic Research Service, Technical Bulletin #1762, June 1989.

Sharpe, P.H.G. 1990. Dosimetry for food irradiation. In D.E. Johnston and Stevenson, M.H. (eds.) *Food Irradiation and the Chemist* Spec. Pub. 86, Royal Society of Chemistry, Cambridge, U.K., pp. 109–123.

Sivapalasingham, S., Barret, E., Kimura, A. et al., 2003. A multistate outbreak of *Salmonella enterica* Serotype Newport infection linked to mango consumption: Impact of water-dip disinfestation technology. *Clinical Infectious Diseases* 37, 1585–1590.

Thomas, P. 1986. Radiation preservation of foods of plant origin. Part III. Bananas, mangoes, and papayas. *CRC Critical Review of Food Science and Nutrition* 23, 147–205.

Turner, J.E. 2007. *Atoms, Radiation, and Radiation Protection*, 3rd edn., Wiley-VCH, Weinheim, Germany.

World Health Organization (WHO). 1999. High-dose irradiation: Wholesomeness of food irradiated with doses above 10 kGy. Report of a Joint FAO/IAEA/WHO Study Group. World Health Organization Technical Report Series 890, pp. i–vi, 1–197.

8 Minimal Processing
Fruits and Vegetables

Ebenézer de Oliveira Silva,
Maria do Socorro Rocha Bastos,
Nédio Jair Wurlitzer, Zoraia de Jesus Barros,
and Frank Mangan

CONTENTS

8.1 INTRODUCTION

Eating habits are diverse in different parts of the world; however, some aspects in the production and consumption of food are universal, such as the consumption of minimally processed vegetables. According to market research conducted by ACNielsen Global Services in 66 countries, representing over 75% of consumers in the world, the global demand for food ready for consumption, among which are minimally processed vegetables, increased by 4% between 2005 and 2006. During that same period, the growth in sales of fresh, ready-to-eat salads increased by 10%. This increase is due in part to two key trends observed in this comprehensive study: a growing focus on health, which includes the consumption of fresh vegetables, and the increased preference for convenience. To the extent possible, consumers want healthy products for their meals, available at a good value, convenient, safe, and with good quality.

In the fresh produce industry, especially with fruits and vegetables, mini-mally processed products have been defined as "any fruit or vegetable, or any combination thereof, which has been physically altered from its original form, but has remained in its fresh condition" (IFPA, 1999). However, the activities implemented to minimally process fresh produce can cause damage to plant tissues, which accumulate during the succeeding stages of processing. Thus, the damage to the products during this processing will adversely affect the shelf life for commercial distribution, which will be shorter compared to the same products that are not minimally processed.

The plant tissues of these minimally processed products, according to Brecht (1995), respond physiologically similar to being subjected to stress. The physiological responses by the plants due to the minimal processing can accelerate the loss of quality and modify sensory attributes unique to each product. The main changes in the products are increased metabolic activity, enzymatic browning, and presence of microorganisms and pathogens in the plant tissues that have been processed.

8.2 EFFECTS OF MINIMAL PROCESSING OF FRUITS AND VEGETABLES

8.2.1 ACCELERATION OF PLANT METABOLISM AND BROWNING

The physical damage accrued during minimal processing, especially with wounding, causes disruption of the cellular membrane, putting enzymes and their substrates in direct contact which accelerates the loss of quality. Among these reactions, the most important ones related to loss of quality are the increase in the respiration rate (RR), the production of ethylene (EE), total soluble phenolics and phenylalanine ammonia-lyase activity (PAL), peroxidase (POD), catalases (CATs), and polyphenol oxidase (PPO), among others. These physiological changes, resulting from minimal processing, as well as the implications for shelf life and quality of the processed products have been described in detail for various fruits and vegetables (Priepke et al., 1976; Rolle and Chism, 1987; Avena-Bustillos et al., 1993; Kim et al., 1994; Nicoli et al., 1994; Brecht, 1995; Ahvenainen, 1996; Wiley, 1997; Watada and Qi, 1999; Soliva-Fortuny and Martín-Belloso, 2003; Oms-Oliu et al., 2010).

The increases in RR and EE are physiological and biochemical effects that are inversely related to the shelf life of minimally processed products (Watada et al., 1990). Increases in respiratory activity have been reported by several authors in melon (Durigan and Sargent, 1999), pineapple (Sarzi et al., 2001b), papaya (Sarzi et al., 2001a), guava (Mattiuz et al., 2001), onion (Cassaro, 1999), and potato (Rolle and Chism, 1987).

The increase in EE from mechanical injury accelerates senescence in plant tissues (Abeles et al., 1992). Ethylene resulting from the physical action of minimally processing was sufficient to accelerate the loss of chlorophyll from spinach, but not in broccoli (Abe and Watada, 1991). In spinach, this is because the increase in chlorophyllase activity is directly related to increased synthesis of ethylene (Sabater and Rodriguez, 1978; Rodriguez et al., 1987; Watada et al., 1990; Yamauchi and Watada, 1991). Increases in ethylene production have also been reported, mainly for climacteric fruits such as melon (Abeles et al., 1992; Durigan and Sargent, 1999), tomato (Brecht, 1995), banana, kiwi (Abe and Watada, 1991), and zucchini (Abeles et al., 1992).

During minimal processing, parts of the surface membrane systems in the plant are negatively affected upon being cut (Rolle and Chism, 1987), which occurs after a more extensive enzymatic degradation (Watada et al., 1990; Brecht, 1995). In plant tissues, the cellular compartmentalization provides better contact between the ethylene-generating systems (Watada et al., 1990) and also an increase in the synthesis and activity of 1-aminocyclopropane-1-carboxylic acid (ACC) synthase, which culminates in the accumulation of acid ACC, a precursor of ethylene (Hyodo et al., 1985), in the plant tissues. In the presence of O_2, ACC can be rapidly oxidized to ethylene, a reaction catalyzed by the enzyme ACC oxidase (Abeles et al., 1992). The ethylene produced in these tissues accelerates the degradation of other cellular membranes, disrupting and destroying the plant tissues (Brecht, 1995).

Enzymatic browning in cut tissues occurs as a result of decompartmentalization substrates and oxidative enzymes, because the tissue experiences a larger exposure to oxygen (Rolle and Chism, 1987). The process of injury and the increase of EE induce increases in PAL activity, which catalyzes the biosynthesis of phenylpropanoids. Browning occurs when products of phenylpropanoid metabolism, such as phenols and possibly other substrates, are oxidized in reactions catalyzed by fenolases such as PPO and POD (Brecht, 1995). In minimally processed lettuce, increased levels of EE led to higher levels of oxidative browning, through the induction of PAL and PPO. In this trial, the browning started 3 days after processing, but the visual quality of lettuce was completely degraded 6 days after processing, during storage at 2.5°C (Couture et al., 1993).

Ethylene and the occurrence of injury induced PAL activity, but apparently by different mechanisms (Abeles et al., 1992). The use of ethylene absorbers, however, did not prevent the onset of browning (Howard and Griffin, 1993), demonstrating, in accordance with Watada and Qi (1999), that the reductions of TR and EE at low temperatures (Kim et al., 1993; Howard et al., 1994) associated with modified atmosphere (Barth et al., 1993; Nicoli et al., 1994) reduce the enzymatic metabolism of minimally processed products, thus slowing the development of these undesirable symptoms in the processed products. Moreover, this enzymatic metabolism responsible for the undesirable symptoms of browning can be reduced by antioxidants, as discussed in the following text.

8.2.2 Pathogenic and Degrading Microorganisms

The presence and activity of pathogens that can adversely affect the quality of minimally processed and packaged products is another critically important consideration in the processing and marketing chain. The exudates from the cut tissue are an excellent medium for the growth of fungi and bacteria, and the subsequent handling of the processed products creates the potential for the development of microorganisms (Burns, 1995). The occurrence of food-borne diseases (FBDs), due to pathogens in minimally processed products, increases the risk of food poisoning due to the fact that in most cases these products are consumed without any subsequent heat treatment (Nguyen-The and Carlin, 1994). The development of microbial contaminants in minimally processed products can be controlled by pH (O'Connor-Shaw et al., 1994), by low temperatures (Bolin and Huxsoll, 1991), by modified atmosphere (Priepke et al., 1976; López-Malo et al., 1994), and by sanitation (Hurst, 1995).

The reports of human infections associated with minimally processed foods have heightened the concern of public health agencies and consumers (Vanetti, 2004). FBDs of a biological origin have increased significantly, even in developed countries (Feitosa et al., 2008). According to the World Trade Organization, the majority of FBDs occurring in Latin American countries are caused by eating food contaminated with pathogenic microorganisms; in Brazil, over 60% of these diseases are caused by *Salmonella* sp., *Staphylococcus aureus*, *Clostridium perfringens*, *Bacillus cereus*, and *Clostridium botulinum* (Feitosa et al., 2008). In most of these cases, the cause was the inadequate process of washing and sanitizing the surface of fruits and vegetables resulting in contamination of the edible part, either in consuming the fresh product without processing or after the processing (i.e., minimally processed). The reduction of microbiological risk from the consumption of contaminated fruit or vegetables, either raw or minimally processed, must be subject to procedures that reduce or, in the best case scenarios, eliminate the surface contamination (Yaun et al., 2004).

8.3 ALTERNATIVES FOR THE MANAGEMENT OF PLANT METABOLISM AND BROWNING

The increase in metabolic activity in minimally processed vegetables promotes rapid chemical and biochemical reactions responsible for changes in sensory quality (color, flavor, aroma, and texture) and the reduction of its life as a food (Cantwell, 1992; Luengo and Lana, 1997; Chitarra, 1999). The senescence process can be accelerated when the temperature and atmospheric compositions in the packages are outside of the desired levels.

8.3.1 COOLING

Maintaining the cold chain at appropriate temperatures for each fresh product, from the processing to markets, is undoubtedly the most important technique available to slow the adverse effects of minimal processing; lowering the temperature reduces enzymatic processes, such as RT and EE (Wills et al., 1998), and thus slows the processes of senescence and increases the shelf life of minimally processed products. The drop in temperature, however, must be maintained at levels sufficient to keep the cells alive, but also to preserve product quality during storage and marketing.

With vegetables such as lettuce and chicory, after being cut, packaged, and stored at 4°C, the RT increased compared to control and intact leaves stored under the same conditions (Priepke et al., 1976). This response was also observed in minimally processed melons (McGlasson and Pratt, 1964), tomatoes (Lee et al., 1970), and kiwi (Watada et al., 1990).

8.3.2 Modified Atmosphere

Modified atmosphere can be defined as a packaging system designed to change the composition of normal air (78% nitrogen, 21% oxygen, 0.03% carbon dioxide, in addition to trace amounts of noble gases) in order to provide an internal atmosphere that facilitates an extended shelf life and maintain the quality of the fresh products in these conditions (Moleyar and Narasimham, 1994; Phillips, 1996).

Thus, the packaging material must have selective permeability to gases, allowing desirable changes to occur within the gaseous composition of the internal atmosphere. The concentrations of oxygen (O_2), carbon dioxide (CO_2), ethylene (C_2H_4), and water vapor are used to form packaging systems of modified atmosphere that are passive or active.

In the first case, the atmosphere is altered passively with respiration of the product in the packaging, either increasing the partial pressure of CO_2 or reduction of O_2 in the interior of the packaging until an atmospheric balance is reached. In this atmospheric balance, the CO_2 produced by respiration diffuses through the selective packaging into the environment at the same rate that the O_2 outside the packaging enters the packaging to compensate for the oxygen consumed during respiration. In the case of modified atmosphere, a gas mixture with defined concentrations of O_2 and CO_2 is injected into the package so that the atmospheric equilibrium is reached quickly. However, the concentrations at equilibrium will be maintained by the RR of the product and the level of permeability of the packaging to these gases.

With passive or active modified atmosphere, the reduction in O_2 partial pressure reduces respiratory metabolism (Wills et al., 1998) and ethylene production (Yang and Hoffman, 1984; Abeles et al., 1992). Moreover, the CO_2 accumulating inside the packaging acts as an inhibitor of respiration (Wills et al., 1998) and of ethylene action (Yang and Hoffman, 1984; Abeles et al., 1992). Thus, there is simultaneously the effect of reducing both respiration and ethylene production, coupled with lower activity of this hormone, dramatically increasing the marketing period of these products without the accelerated loss of quality (Saltveit, 1999). The atmosphere created inside the package can easily be transported with the product, ensuring that the increase in CO_2 concentration does not reach undesirable concentrations, nor that the reduction of O_2 concentration will facilitate anaerobic respiration.

Atmospheres with 5%–15% CO_2 and 2%–8% O_2 have been shown to be effective in maintaining the quality of minimally processed products; however, for each fresh product, there is a specific atmosphere that maximizes its shelf life (Cantwell, 1992).

Minimally processed collard green packaged in polyolefin plastic film with both passive and active modified atmosphere (0% O_2 + 5% CO_2) maintained vitamin C levels and good consumer acceptance for 12 days when stored at 10°C and then for 20 days at 5°C (Teles, 2001).

Minimally processed cantaloupe melons in packaging injected with 4% O_2 plus 10% CO_2 retained better color and reduced respiratory activity more effectively than passive modification of the atmosphere (Bai et al., 2001). This same product stored at 3°C, injected with 5% O_2 + 20% CO_2, was effective in controlling microbial activity, and under these conditions, the melons maintained good quality for 12 days, while melons not packaged using modified atmosphere maintained good quality for only 6 days (Arruda et al., 2004).

Pineapples minimally processed and stored at 5°C with active modified atmosphere (2% O_2 + 10% CO_2 and 5% O_2 and 5% CO_2) using polyethylene plastic film laminated with polypropylene (PP) provided greater firmness compared to pineapples not stored in modified atmosphere (Prado et al., 2003). However, according to the authors, these gas compositions promoted the browning of pineapple tissues, stimulating the activity of PPO.

According to Kader (1995), the tolerances to low concentrations of O_2 and high CO_2 concentrations with minimally processed vegetables can vary considerably; very low partial pressures of O_2 and/or very high CO_2 can cause physiological damage in products such as anaerobic respiration. Lettuce quality, for example, is negatively affected from exposure to high levels of CO_2 (Ke and Saltveit, 1989). Cameron et al. (1995) argue that CO_2 concentrations above 2% can lead to browning of veins in lettuce; however, Ballantyne et al. (1988) found no change in color of lettuce for up to 20 days when placed in packaging with CO_2 concentrations above 10%. Pirovani et al. (1998) found that the visual quality of lettuce was better when packed in PP film and stored at 4°C at ambient concentrations of O_2 and 12% CO_2. The concentrations of CO_2 and O_2 reached on the eighth day were 1.5% O_2 and 12% CO_2, while in the other packages tested (polyethylene, polyolefin, and polyvinyl chloride [PVC]), the average internal concentrations of O_2 and CO_2 were 17% and 2%, respectively. Based on these studies, there is no consensus on the effect of different partial pressures of CO_2 and O_2 in terms of damage to the fresh-cut products. Thus, more detailed scientific studies should be conducted in this area.

In the selection of packaging used with modified atmosphere, the following characteristics should be considered: respiratory activity of the product, storage temperature, film permeability to gases, surface and film thickness, and the mass of the product being packaged. It is important to take into consideration that both respiratory activity and permeability of the film are sensitive to temperature changes and respond to these changes differently. It is expected then that the system (product + packaging) under modified atmosphere will maintain the desired atmospheric balance within a specified temperature range (Zagory, 2000).

8.3.3 Trends in Packaging for Minimally Processed Products

The use of packages made of plastic or polymer films is a common way to control the atmosphere around the product. In selecting the type of packaging to be used, it is important to take into consideration the characteristics of the products to be packaged, such as quantity, weight, use of clear packaging, and the biochemical composition as related to RR in a specified space according to the shelf life desired. This knowledge permits the practitioner to know whether they should use a packaging with low or high barriers to gas exchange, moisture, or light. In addition, the information related to importance of gas exchanges is important to know.

The plastic packages shaped in bags, pots, cups, and trays are the ones with easiest application and serve various functions, such as containing the product, being attractive to the consumer, and facilitating consumption. There are many types of plastic materials available. The most common ones with their desirable properties are highlighted as follows:

- Low-density polyethylene: lower cost, high gas and water vapor permeability, and sturdy which allows the packaging to be used as bags to create layers for products to be stored in multilayers. These are also used for microperforated films to facilitate water vapor exchange.
- Expanded polystyrene: used in the manufacture of trays, often with a PVC coating, thinner film, elasticity, and with good weldability. Polystyrene can also be used to make pots.
- PVC, PP, and polyethylene terephthalate (PET): used for production of trays with or without lids.

When you want to control the gaseous environment in the space between the product and packaging with the use of active modified atmosphere, there should be multilayered coextruded film, such as PP-nylon, PP-ethylene vinyl alcohol-PP, and polyvinyl vinylidene, among others. These materials have low permeability to gases, making it necessary to extract the air with a modest vacuum before injecting the desired gas mixture. Gorny (2001) lists the ideal concentrations of O_2 and CO_2 to be used in modified atmosphere systems for various minimally processed fruits and vegetables.

The use of low temperatures, associated with modified atmosphere during storage, reduces the elevation of respiration and ethylene synthesis in minimally processed vegetables such as lettuce (Singh et al., 1972a,b), broccoli (Barth et al., 1993), and melon (O'Connor-Shaw et al., 1994), and also in fruits such as apple (Kim et al., 1993; Nicoli et al., 1994), kiwi, pineapple, and papaya (O'Connor-Shaw et al., 1994).

Another system using active modified atmosphere is realized by creating a partial vacuum inside the package; this is especially recommended for roots and tubers. In this system, the film used should allow only low levels of permeability to gases, such as those found in nylon multilayer films (Durigan et al., 2007).

Moreover, there is interest to replace petroleum-based packaging materials with biodegradable materials, characterized by edible films, the purpose of which is to mimic the plant cuticle that can be removed during the stages of processing.

The most commonly used materials in these edible coatings are lipids (oil or paraffin wax, beeswax, carnauba wax, vegetable oil, mineral oil, etc.), polysaccharides (alginate, starch, cellulose, gums, pectin, chitosan, etc.), and proteins (casein, gelatin, collagen, albumin, etc.) (Cuq et al., 1995; Guilbert and Biquet, 1996; Cutter, 2006; Freire, 2008). The use of such coatings on minimally processed fruits and vegetables requires very specific considerations which depend on several factors, such as the ability to both protect and positively interact with processed product in terms of their sensory properties and functional stability of biochemical, physicochemical, and microbiological aspects (Vilas Boas et al., 2007).

Films made from polysaccharides, particularly starch, are those which currently have the greatest potential for use (Davis and Song, 2006). They have lower costs when compared to traditional films currently being used and are as such considered a low-cost technique. The films made with starches possess physical characteristics similar to synthetic polymers which allow transparent, odorless, flavorless, and a median permeability to CO_2, and low permeability to O_2. Studies on the production and use of edible films with starch incorporated with antimicrobial agents have been found to be effective in extending shelf life of minimally processed vegetables (Garcia et al., 2000; Durango et al., 2006). In minimally processed carrots, coating with starch (4%), associated with chitosan (0.5% and 1.5%), reduced the population of *S. aureus* and *Escherichia coli* at 2.56 and 4.18 log cycles, respectively (Durango et al., 2006). Minimally processed garlic coated with polymers of agar-agar, embedded with antimicrobial substances (acetic acid, chitosan, acetic acid + chitosan), had low counts of fungi and aerobic mesophiles (Geraldine et al., 2008).

The use of alginate as polymer matrix for edible coatings has also been studied in several minimally processed fruits; for example, the shelf life of apples was extended by 2 weeks when using alginate as a polymer matrix in the processing (Rojas-Graü et al., 2008). Alginates are linear polymers formed by residues of α-L-gulurônico (G) and β-D-mannuronic (M) present in varying proportions and sequences in the cell wall and intercellular spaces of brown algae, such as *Laminaria digitata* and *Macrocystis pyrifera* (King, 1983). To produce the gel which is used in the film formation, the alginate must react with polyvalent cations, with calcium ions being the most effective gelling agents (Allen et al., 1963). Calcium ions play a critical role in keeping the alginate chains together through ionic interactions after the formation of hydrogen bonds between the chains, which produce a gel with a three-dimensional network structure (King, 1983).

There have been a number of studies implemented on the incorporation of antimicrobial agents in edible films. Some of these materials, such as essential oils and organic acids, have been evaluated for their effectiveness in controlling the quality and safety in food products. These combinations have been used in food packaging as active packaging, which provide direct interaction with the product. It is important that the use of these products be implemented in accordance with the various laws related to the use of antimicrobials in food. Minimally processed fruits using these formulations are emerging as alternatives to improve their quality and shelf life (Rojas-Graü et al., 2009). In minimally processed melon "Pele de Sapo," Raybaudi-Massilia et al. (2008) studied the effect of malic acid and essential oils of cinnamon, palmarosa, and lemongrass incorporated into edible alginate films. The incorporation of essential oils in the film used in this study provided 21 days of shelf life due to the reduction in of microbiological activity. In terms of the sensory studies, the palmarosa had the highest acceptance by consumers.

The use of fruit purees can also be an alternative in the preparation of edible coatings for minimally processed fruit (McHugh et al., 1996; Senesi and McHugh, 2002). The use of fruit purees is related to the presence of biopolymers, especially polysaccharides such as starch, pectin, and cellulose derivatives in their structure (Kaya and Maskan, 2003). The edible coatings made from apple puree provided an excellent barrier to oxygen, but not to water vapor (McHugh et al., 1996). However, the addition of a hydrocolloid, such as alginate, improved the barrier properties of these coatings (Mancini and McHugh, 2000). Minimally processed apples ("golden") with edible coatings showed less browning and less moisture loss and thus maintained the characteristic flavor and aroma for a longer time (McHugh and Senesi, 2000).

The films on minimally processed apples "Royal Gala" showed an average decrease of 38% in respiratory rates and more than 50% in ethylene production compared to the control, with alginate being the most effective fruit puree (Fontes et al., 2008).

Studies using mango puree to produce edible films showed that the coating reduced weight loss and slowed the ripening of mangoes in fresh and minimally processed mangoes; they had a longer shelf life under these experimental conditions (Sothornvit and Rodsamran, 2008). According to Azeredo et al. (2010), edible films made from fruit puree can combine the mechanical properties to increase shelf life along with the color and aroma from the pigments and volatile compounds of fruits.

8.4 ALTERNATIVES FOR MANAGEMENT OF MICROORGANISMS

The Food and Drug Administration (FDA, 2011), under the ruling 21 C.F.R. 173.315, has approved the use of sodium hypochlorite, chlorine dioxide, hydrogen peroxide, peracetic acid, and ozone as plant protection agents in the processing of fresh fruits and vegetables.

The use of chlorine, between 150 and 200 mg L^{-1}, was sufficient to control fungi and bacteria in minimally processed cabbage (Fantuzzi, 1999; Silva, 2000) and collards (Carnelossi, 2000). In minimally processed carrots, the chlorine concentration (between 150 and 200 μL L^{-1}) is recommended based on type of processing (slices, cubes, sticks, or grated) (Silva, 2003). Lower residual chlorine concentrations, ranging from 5 to 20 mg L^{-1}, were effective in controlling microorganisms in minimally processed pineapple and papaya (Sarzi, 2002), mango (Donadon et al., 2007), melon (Bastos et al., 2000), guava (Mattiuz et al., 2003), and star fruit (Teixeira et al., 2007).

However, although chlorine is a potent disinfectant with strong oxidizing properties, studies have shown that it cannot completely oxidize organic materials, leading to the formation of unwanted by-products when in water, such as chloroform ($CHCl_3$) and other trihalomethanes, which are suspected carcinogens.

In alkaline pH, chlorine can also react with organic nitrogen bases to produce chloramine (NH_2Cl), which is a known carcinogen.

Chlorine dioxide (ClO_2) in aqueous solution has been postulated as an alternative to sodium hypochlorite (NaClO) for minimally processed products, with the advantage of avoiding the risks associated with the formation of trihalomethanes. Chlorine dioxide gas (ClO_2) can be produced commercially on-site by the reduction of sodium chlorate or by the oxidation of sodium chloride and chlorine gas (Cl_2) using expensive equipment which also requires trained personnel to operate and maintain (Andrade and Pinto, 2008). In this context, when the chlorine dioxide was used in the sanitization of minimally processed products, the formation of the trihalomethanes was negligible in minimally processed carrots (Klaiber et al., 2005), lettuce (López-Gálvez et al., 2009), and other commercially prepared salads (COT, 2006).

Hydrogen peroxide (oxygenated water) is characterized by containing a pair of oxygen atoms (–O–O), which are highly oxidative with the release of O_2 in aqueous solutions and thus create antimicrobial activity, mainly for Gram-positive and Gram-negative bacteria (Andrade and Pinto, 2008). It is used in the formulation of sanitizer solutions based on peracetic acid. The O_2 released is corrosive and requires care in handling and strict management of the active ingredient concentration.

Peracetic acid is known for its germicidal properties and therefore has been recommended as a potential substitute for chlorine, with the advantage of having good biodegradability (Andrade and Pinto, 1999). However, for some minimally processed fruits and vegetables, peracetic acid, when acidified in the medium, can induce a loss of selective permeability of cell membranes, causing cell decompartmentalization, with the consequent loss of tissue structure and the formation of undesirable flavor compounds (personal observation).

Ozone (O_3), which is highly volatile and toxic, has to be prepared on-site using expensive equipment that requires trained personnel for its operation. Inhalation of 20 ppm for 1 h or 50 ppm for 30 min can be fatal. However, the use of oxygenated water was found to be effective in reducing microbial populations and in controlling the physiological metabolism of minimally processed celery (Zhang et al., 2005) and cantaloupe (Selma et al., 2008).

Currently, one can say that there is a demand for alternative strategies in controlling microorganisms in postharvest practices. Such strategies consider the use of natural chemical compounds, generally classified as safe substances (GRAS substances), as reviewed by Oms-Oliu et al. (2010). There is a need for more research on more effective treatments with less chemical risk, such as electrolyzed water (Al-Haq and Sugiama, 2004; Abadias et al., 2008; Huang et al., 2008) and pulsating ultraviolet (UV) light (CFSAN-FDA, 2000).

Electrolyzed water is a chlorine-based disinfectant that provides a high efficiency of sanitization with low amount of free chlorine, so it is considered an environmentally friendly method for disinfecting minimally processed fruits and vegetables (Koide et al., 2009). Studies have shown that electrolyzed water destroys the transport of proteins within the cell membranes of fungi and bacteria, leading to the rupture of membranes and increased infiltration of the electrolyzed water. In fungi, the electrolyzed water induces oxidation of the cell walls and disrupts the metabolism of organic compounds; also the more resistant survival structures of fungi can be eliminated, depending on the concentration used (Huang et al., 2008). The electrolyzed water is considered more effective as a bactericide and fungicide than chlorine, because it penetrates more easily in the irregularities of the fruit surface (Al-Haq et al., 2005) and minimally processed vegetables such as celery (Zhang et al., 2005). Cut cabbage in slightly acidic water (pH 6.1 with 20 mg L^{-1} available chlorine) was more effective in controlling microorganisms (aerobic bacteria, molds, and yeasts) than the hypochlorite solution containing 150 mg L^{-1} of available chlorine, which is traditionally recommended (Koide et al., 2009).

Another technology used for surface decontamination of minimally processed fruits and vegetables is short-wave radiation in the UV region (Allende et al., 2006; Krishnamurthy et al., 2007). Radiation emitted between 200 and 280 nm (UV-C) induces the formation of breaks in DNA molecules, preventing reproduction and protein synthesis, resulting in a germicidal effect (Bintsis et al., 2000); this provides the advantage of not generating by-products or waste of chemicals capable of altering the sensory characteristics of the final product (Guerrero-Beltrán and Barbosa-Canovas, 2004). The UV-C can be applied continuously or pulsed. In the continuous model, also known as conventional, the UV-C is applied continuously. In the case of the pulsed model, UV light, stored in a capacitor, is released in intermittent flashes, instantly increasing the energy intensity which makes the pulsed UV more effective and faster to inactivate microorganisms (CFSAN-FDA, 2000). This model presents an additional advantage compared to conventional methods, the flash that occurs every 300 μs creates a large temperature gradient between the inner and outer parts of the cell which causes a disruption of the microorganisms membrane, further expanding the effect of the pulsed UV mechanism (McDonald et al., 2000). Several reviews of the applications of UV in the postharvest of fruit and vegetables can be found in Allende and Artes (2003), Smith et al. (2002), Yaun et al. (2004), and Geveke (2005).

The application of UV-C in minimally processed fruit causes an increase in the concentrations of antioxidants, polyphenols, and flavonoids (Alothman et al., 2009). Thus, according to these authors, in addition to its use for decontaminating the surfaces of minimally processed fruits, this new technology can be used to increase the beneficial compounds for consumer health.

In the future, it is expected that systems using electrolyzed water and pulsed UV will be used to guarantee the safety of minimally processed foods than the currently used chlorine-based compounds, but without changing the sensory properties and nutritional quality of these products.

GLOSSARY OF SPECIES

Products	Species (Binomial Name)
Apple	*Malus domestica* Borkh.
Banana	*Musa* spp.
Broccoli	*Brassica oleracea* L. var. italica group
Cabbage	*Brassica oleracea* L. var. capitata
Carrot	*Daucus carota* L.
Celery	*Apium graveolens* L.
Chicory, endive, escarole	*Cichorium endivia* L.
Collard	*Brassica oleracea* L. var. acephala
Garlic	*Allium sativum* L.
Guava	*Psidium guajava* L.
Kiwi	*Actinidia deliciosa*
Lettuce	*Lactuca sativa* L.
Mango	*Mangifera indica* L.
Melon	*Cucumis melo* L.
Onion	*Allium cepa* L.
Papaya	*Carica papaya* L.
Pineapple	*Ananas comosus* L. Merr.
Potato	*Solanum tuberosum* L.
Spinach	*Spinacia oleraceai* L.
Star fruit	*Averrhoa carambola* L.
Tomato	*Lycopersicon esculentum* Mill.
Zucchini	*Cucurbita pepo*

REFERENCES

Abadias, M., Usall, J., Oliveira, M., Alegre, I., Viñas, I. 2008. Efficacy of neutral electrolyzed water (NEW) for reducing microbial contamination on minimally-processed vegetables. *International Journal of Food Microbiology* 123: 151–158.

Abe, K., Watada, A. E. 1991. Ethylene absorbent to maintain quality of lightly processed fruits and vegetables. *Journal of Food Science* 56: 1493–1496.

Abeles, F. B., Morgan, P. W., Saltveit Jr., M. E. 1992. *Ethylene in Plant Biology*. San Diego, CA: Academic Press, 414 pp.

Ahvenainen, R. 1996. New approaches in improving the shelf life of minimally processed fruit and vegetables. *Trends Food Science and Technology* 7: 179–187.

Al-Haq, M. I., Sugiyama, J. 2004. Application of electrolyzed water in food processing. *ASAE/CSAE Annual Meeting*, August 1–4, Ottawa, Ontario, Canada. (ASAE paper No., 04–6178)

Al-Haq, M. I., Sugiyama, J., Isobe, S. 2005. Applications of electrolyzed water in agriculture and food industries. *Food Science and Technology Research* 11: 135–150.

Allen, L., Nelson, A. I., Steinberg, M. P., Mcgill, J. N. 1963. Edible corn-carbohydrate food coatings, I. Development and physical testing of a starch-algin coating. *Food Technology* 17: 1437–1441.

Allende, A., Artes, F. 2003. UV-C radiation as a novel technique for keeping quality of fresh processed "Lollo Rosso" lettuce. *Food Research International* 36: 739–746.

Allende, A., McEvoy, J. L., Luo, Y., Artes, F., Wang, C. Y. 2006. Effectiveness of two-sided UV-treatments in inhibiting natural microflora and extending the shelf-life of minimally processed Red Oak leaf lettuce. *Food Microbiology* 23: 241–249.

Alothman, M., Bhat, R., Karim, A. A. 2009. UV radiation-induced changes of antioxidant capacity of fresh-cut tropical fruits. *Innovative Food Science and Emerging Technologies* 10: 512–516.

Andrade, N. J., Pinto, C. L. O. 1999. *Higienização na Indústria de Alimentos*. Viçosa, Brazil: Centro de Produções Técnicas, 96 pp.

Andrade, N. J., Pinto, C. L. O. 2008. Higienização na Indústria de Alimentos e Segurança alimentar. In: Bastos, M. S. R. (Org.). *Ferramentas da Ciência e Tecnologia para a Segurança dos Alimentos*. Embrapa Agroindústria Tropical, Fortaleza, Brazil: Banco do Nordeste do Brasil, pp. 41–66.

Arruda, M. C., Jacomino, A. P., Spoto, M. H. F., Gallo, C. R. 2004. Conservação de melão rendilhado minimamente processado sob atmosfera modificada ativa. *Revista Brasileira de Ciência e Tecnologia de Alimentos* 24: 53–58.

Avena-Bustillos, R. J., Cisneros-Zavallos, L. A., Krochta, J. M., Saltveit M. E. 1993. Optimization of edible coatings on minimally processed carrots using response surface methodology. *Transactions of the ASAE (American Society of Agricultural Engineers)* 36: 801–805.

Azeredo, H. M. C., Mattoso, L. H. C., Avena-Bustillos, R. J., Ceotto Filho, G., Munford, M. L., Wood, D., McHugh, T. H. 2010. Nanocellulose reinforced chitosan composite films as affected by nanofiller loading and plasticizer content. *Journal of Food Science* 75: N1–N7.

Bai, J. H., Saftner, R. A., Watada, A. E., Lee, Y. S. 2001. Modified atmosphere maintains quality of fresh cut Cantaloupe (*Cucumis melo* L.). *Journal of Food Science* 66: 1207–1211.

Ballantyne, A., Stark, R., Selman, J. 1988. Modified atmosphere packaging of shredded lettuce. *International Journal of Food Science and Technology* 23: 267–274.

Barth, M. M., Kerbel, E. L., Broussard, S., Schimidt, S. J. 1993. Modified atmosphere packaging protects market quality in broccoli spears under ambient temperature storage. *Journal of Food Science* 58: 1070–1072.

Bastos, M. S. R., Souza Filho, M. S. M., Alves, R. E., Filgueiras, H. A. C., Borges, M. F. 2000. Processamento mínimo de abacaxi e melão. In: *Encontro nacional sobre processamento mínimo de frutas e hortaliças, 2, 2000*. Anais… Viçosa, Brazil: Universidade Federal de Viçosa, pp. 89–94.

Bintsis, T., Litopoulou-Tzanetaki, E., Robinson, R. 2000. Existing and potential applications of ultraviolet light in the food industry. A critical review. *Journal of Science Food and Agriculture* 80: 1–9.

Bolin, H. R., Huxsoll, C. C. 1991. Control of minimally processed carrot (*Daucus carota*) surface discoloration caused by abrasion peeling. *Journal of Food Science* 56: 416–418.

Brecht, J. K. 1995. Physiology of lightly processed fruits and vegetables. *HortScience* 30: 18–22.

Burns, J. K. 1995. Lightly processed fruits and vegetables: Introduction to the colloquium. *HortScience* 30: 14–17.

Cameron, A. C., Talasila, P. C., Joles, D. W. 1995. Predicting film permeability needs for modified-atmosphere packaging of lightly processed fruits and vegetables. *HortScience* 30: 25–34.

Cantwell, M. 1992. Postharvest handling systems: Minimally processed fruits and vegetables. In: Kader, A. A. (Ed.) *Postharvest Technology of Horticultural Crops*, 2nd edn. Davis, CA: Division of Horticultural and Natural Resources, University of California, pp. 273–281.

Carnelossi, M. A. G. 2000. *Fisiologia pós-colheita de folhas de couve (Brassica oleracea cv. acephala) minimamente processadas*. 81 f. Tese (Doutorado em Biologia Vegetal) – Universidade Federal de Viçosa. Viçosa, Brazil: UFV, Viçosa.

Cassaro, K. P. 1999. *Conservação de bulbos de cebola minimamente processados*. 78 f. Trabalho (Graduação em Agronomia)—Universidade Estadual Paulista, Faculdade de Ciências Agrárias e Veterinárias, Jaboticabal.

CFSAN—FDA. 2000. *Kinetics of Microbial Inactivation for Alternative Food Processing Technologies: Pulsed Light Technology*. Atlanta, GA: Center for Food Safety and Applied Nutrition, Food and Drug Administration. Disponível em. www.cfsan.fda.gov/comm/ift-puls.html (accessed April 19, 2006).

Chitarra, M. I. F. 1999. Alterações bioquímicas do tecido vegetal com o processamento mínimo. In: *SEMINÁRIO SOBRE HORTALIÇAS MINIMAMENTE PROCESSADAS*, 1999, Piracicaba. Palestra… Piracicaba: Escola Superior de Agricultura "Luiz de Queiróz" (USP), 9 pp.

COT, Committee on Toxicity of Chemicals in Food, Consumer Products and the Environment, 2006. COT statement on a commercial survey investigating the occurrence of disinfectants and disinfection by-products in prepared salads. http://cot.food.gov.uk/pdfs/cotstatementwashaids200614.pdf (accessed February 2, 2012).

Couture, R., Cantwell, M. I., Ke, D., Saltveit, Jr., M. E. 1993. Physiological attributes related to quality attributes and storage life of minimally processed lettuce. *HortScience* 28: 723–725.

Cuq, B., Gontard, N., Guilbert, S. 1995. Edible films and coatings as active layers. In: Rooney, M. L. (Ed.) *Active Food Packaging*. London, U.K.: Blackie Academic & Professional, pp. 111–142.

Cutter, C. N. 2006. Opportunities for bio-based packaging technologies to improve the quality and safety of fresh and further processed muscle foods. *Meat Science* 74: 131–142.

Davis, G., Song, J. H. 2006. Biodegradable packaging based on raw materials from crops and their impact on waste management. *Industrial Crops and Products* 23: 147–161.

Donadon, J. R., Souza, B. S., Durigan, J. F. 2007. Processamento mínimo de manga. In: Moretti, C. L. (Ed.) *Manual de Processamento Mínimo de Frutas e Hortaliças*. Rio de Janeiro, Brasília: Embrapa/SEBRAE, pp. 273–282.

Durango, A. M., Soares, N. F. F., Andrade, N. J. 2006. Microbiological evaluation of an edible antimicrobial coating on minimally processed carrots. *Food Control* 17: 336–341.

Durigan, J. F., Alves, R. E., Silva, E. O. 2007. *Técnicas de processamento mínimo de frutas e hortaliças e suas relações com o mercado*. Fortaleza, Brazil: Instituto Frutal, 110 pp. (Coleção Cursos Frutal).

Durigan, J. F., Sargent, S. A. 1999. Uso do melão Cantaloupe na produção de produtos minimamente processados. *Alimentos e Nutrição* 10: 69–77.

Fantuzzi, E. 1999. Atividade microbiana em repolho (*Brassica oleraceae* cv. capitata) mini-mamente processado. Dissertação (Mestrado em Ciência e Tecnologia de Alimentos)—Universidade Federal de Viçosa, Viçosa, Brazil: UFV, Viçosa.

FDA. 2011. Secondary direct food additives permitted in food for human consumption. Code of federal regulations. Title21—Foods and Drugs. Vol. 3. Revised in April-2011. Part 173. Section 173.315: http//www.accessdata.fda.gov/ (accessed February 2, 2012).

Feitosa, T., Bruno, L. M., Borges, M., De, F. 2008. Segurança Microbiológica dos Alimentos. In: Bastos, M. S. R. (Org.). *Ferramentas da Ciência e Tecnologia para a Segurança dos Alimentos*. Fortaleza, Brazil: Embrapa Agroindústria Tropical: Banco do Nordeste do Brasil, pp. 21–39.

Fontes, L. C. B., Sarmento, S. B. S., Spoto, M. H. F., Dias, C. T., Dos, S. 2008. Conservação de maçã minimamente processada com o uso de películas comestíveis. *Ciência e Tecnologia de Alimentos* 28: 872–880.

Freire Junior, M. 2008. Uso de revestimentos comestíveis em produtos minimamente processados. In: *V Encontro Nacional sobre Processamento Mínimo de Frutas e Hortaliças: Palestras, Mini-cursos e Resumos*, Universidade Federal de Lavras, Lavras, Brazil, pp. 38–42.

Garcia, M., Martino, M., Zaritzky, N. 2000. Lipid addition to improve barrier properties of edible starch-based films and coatings. *Journal of Food Science* 65: 941–947.

Geraldine, R. M., Soares, N. F. F., Botrel, D. A., Gonçalves, L. A. 2008. Characterization and effect of edible coatings on minimally processed garlic quality. *Carbohydrate Polymers* 72: 403–409.

Geveke, D. J. 2005. UV inactivation of bacteria in apple cider. *Journal of Food Protection* 68: 1739–1742.

Gorny, J. R. 2001. *A Summary of CA and MA Requirements and Recommendations for Fresh-Cut (Minimally Processed) Fruits and Vegetables*. Postharvest Horticulture Series No. 22A, University of California, Davis, CA. pp. 95–145.

Guerrero-Beltrán, J. A., Barbosa-Cánovas, G. V. 2004. Review: Advantages and limitations on processing foods by UV light. *Food Science and Technology International* 10: 137–147.

Guilbert, S., Biquet, B. 1996. Edible films and coatings. In: Bureau, G., Multon, J. L. (Eds.) *Food Packaging Technology*. New York: Wiley-VCH Inc., Vol. 1, pp. 315–353.

Howard, L. R., Griffin, L. E. 1993. Lignin formation and surface discoloration of minimally processed carrot sticks. *Journal of Food Science* 58: 1065–1067.

Howard, L. R., Griffin, L. E., Lee, Y. 1994. Steam treatment of minimally processed carrots sticks to control surface discoloration. *Journal of Food Science* 59: 356–358.

Huang, Y. R., Hung, Y. C., Hsu, S. Y., Huang, Y. W., Hwang, D. F. 2008. Application of electrolyzed water in the food industry. *Food Control* 19: 329–345.

Hurst, W. C. 1995. Sanitation of lightly processed fruits and vegetables. American Society for Horticultural Science ASHS. *Annual meeting* N°90, Nashville TN, ETATS-UNIS (27/06/1993). vol. 30, n° 1, pp. 13–40 (26 ref.), pp. 22–24.

Hyodo, H., Tanaka, K., Yoshisaka, J. 1985. Induction of 1-amino-cyclopropane-1-carboxylic acid syntase in wounded mesocarp tissue of winter squash fruit, and the effects of ethylene. *Plant Cell and Physiology* 26: 161–167.

IFPA. 1999. *Fresh-Cut Produce Handling Guidelines*, 3rd edn. Newark, NJ: Produce Marketing Association, 39 pp.

Kader, A. A. 1995. Regulation of fruit physiology by controlled/modified atmospheres. *Acta Horticulturae* 398: 139–146.

Kaya, S., Maskan, A. 2003. Water vapour permeability of pestil (a fruit leather) made from boiled grape juice with starch. *Journal of Food Engineering* 57: 295–299.

Ke, D., Saltveit, Jr., M. E. 1989. Wound-induced ethylene production, phenolic metabolism and susceptibility to russet spotting in iceberg lettuce. *Physiologia Plantarum* 76: 412–418.

Kim, D. M., Smith, N. L., Lee, Y. C. 1993. Quality of minimally processed apple slices from selected cultivars. *Journal of Food Science* 58: 1115–1117.

Kim, D. M., Smith, N. L., Lee, Y. C. 1994. Effect of heat treatment on firmness of apples and apple slices. *Journal of Food Processing* 18: 1–8.

King, A. H. 1983. Brow seaweed extracts (Alginates). In: Glicksman, M. E. D. (Ed.) *Food Hydrocolloids*. Boca Raton, FL: CRC Press, Vol. 2, pp. 115–190.

Klaiber, R. G., Baur, S., Wolf, G., Hammes, W. P., Carle, R. 2005. Quality of minimally processed carrots as affected by warm water washing and chlorination. *Innovative Food Science and Emerging Technology* 6: 351–362.

Koide, S., Takeda, J. I., Shi, J., Shono, H., Atungulu, G. G. 2009. Disinfection efficacy of slightly acidic electrolyzed water on fresh cut cabbage. *Food Control* 20: 294–297.

Krishnamurthy, K., Demirci, A., Irudayaraj, J. M. 2007. Inactivation of *Staphylococcus aureus* in milk using flow-through pulsed UV-light treatment system. *Journal of Food Science* 72: 233–239.

Lee, T. H., Mcglasson, W. B., Edwards, R. A. 1970. Physiology of disks of irradiated tomato fruit. I. Influence of cutting and infiltration on respiration, ethylene production and ripening. *Radiation Botany* 10: 521–529.

López-Gálvez, F., Allende, A., Selma, M. V., Gil, M. I. 2009. Prevention of *Escherichia coli* cross-contamination by different commercial sanitizers during washing of fresh-cut lettuce. *International Journal of Food Microbiology* 133: 167–171.

López-Malo, A., Palou, E., Welti, J., Corte, P., Argaiz, A. 1994. Shelf-stable high moisture papaya minimally processed by combined methods. *Food Research International* 27: 545–553.

Luengo, R. F. A., Lana, M. M. 1997. *Processamento Mínimo De Hortaliças*. Fortaleza, Brasília: Embrapa Hortaliças, 3 pp. (Comunicado Técnico da Embrapa Hortaliças, 2).

Mancini, F., McHugh T. H. 2000. Fruit-alginate interactions in novel restructured products. *Nahrung* 44: 152–157.

Mattiuz, B. H., Durigan, J. F., Rossi Jr., O. D. 2003. Processamento mínimo em goiabas 'Paluma' e 'Pedro Sato': Avaliação química, sensorial e microbiológica. *Ciência e Tecnologia de Alimentos* 23: 409–413.

Mattiuz, B. H., Durigan, J. F., Sarzi, B. 2001. Aspectos fisiológicos de goiabas 'Pedro Sato' e 'Paluma' submetidas ao processamento mínimo. In: *CONGRESSO BRASILEIRO DE FISIOLOGIA VEGETAL*, 8, 2001, Ilhéus. Anais… Ilhéus, Brazil: Sociedade Brasileira de Fisiologia, 2001. CD-ROM.

McDonald, K. F., Curry, R. D., Clevenger, T. E., Unklesbay, K., Eisenstrack, A., Golden, J., Morgan, R. D. 2000. A comparison of pulsed and continuous ultraviolet light sources for the decontamination of surfaces. *IEEE Transactions on Plasma Science* 28: 1581–1587.

McGlasson, W. B., Partt, H. K. 1964. Effects of wounding on respiration and ethylene production by cantaloupe fruit tissue. *Plant Physiology* 39: 128–132.

McHugh, T. H, Huxsoll, C. C, Krochta, J. M. 1996. Permeability properties of fruit puree edible films. *Journal of Food Science* 61: 88–91.

McHugh, T. H., Senesi, E. 2000. Apple wraps: A novel method to improve the quality and extend the shelf life of fresh-cut apples. *Journal of Food Science* 65: 480–485.

Moleyar, V., Narasimham, P. 1994. Modified atmosphere packaging of vegetables: An appraisal. *Journal of Food Science and Technology* 31: 267–278.

Nguyen-The, C., Carlin, F. 1994. The microbiology of minimally processed fresh fruits and vegetables. *Critical Reviews in Food Science and Nutrition* 34: 371–401.

Nicoli, M. C., Anese, M., Severini, C. 1994. Combined effects in preventing enzymatic browning reactions in minimally processed fruit. *Journal of Food Quality* 17: 221–229.

O'Connor-Shaw, R. E., Roberts, R., Ford, A. L., Nottingham, S. M. 1994. Shelf life of minimally processed honeydew, kiwifruit, papaya, pineapple and cantaloupe. *Journal of Food Science* 59: 1202–1206.

Oms-Oliu, G., Rojas-Graú, M. A., González, L. A., Varela, P., Soliva-Fortuny, R., Hernando, M. I. H., Munuera, I. P., Fiszman, S., Martín-Belloso, O. 2010. Recent approaches using chemical treatments to preserve quality of fresh-cut fruit: A review. *Postharvest Biology and Technology* 57: 139–148.

Phillips, C. A. 1996. Review: Modified atmosphere packaging and its effects on the microbiological quality and safety of produce. *International Journal of Food Science and Technology* 31: 463–479.

Pirovani, M. E., Piagentini, A. M., Guemes, D. R., Pentima, J. H. 1998. Quality of minimally processed lettuce as influenced by packaging and chemical treatment. *Journal of Food Quality* 21: 475–484.

Prado, M. E. T., Chitarra, A. B., Bonnas, D. S., Pinheiro, C. M., Mattos, L. M. 2003. Armazenamento de abacaxi 'Smoth Cayenne' minimamente processado osb refrigeração e atmosfera modificada. *Revista Brasileira de Fruticultura* 25: 97–70.

Priepke, P. E., Wei, L. S., Nelson, A. I. 1976. Refrigerated storage of prepackaged salad vegetables. *Journal of Food Science* 41: 379–382.

Raybaudi-Massilia, R. M., Mosqueda-Melgar, J., Martín-Belloso, O. 2008. Edible alginate-based coating as carrier of antimicrobials to improve shelf-life and safety of fresh-cut melon. *International Journal of Food Microbiology* 121: 313–327.

Rodriguez, M. T., Gonzáles, M. P., Linares, J. M. 1987. Degradation of chlorophyll and chlorophyllase activity in senescing barley leaves. *Journal of Plant Physiology* 129: 369–374.

Rojas-Graü, M. A., Soliva-Fortuny. R., Martin-Belloso, O. 2009. Edible coatings to incorporate active ingredients to freshcut fruits: A review. *Trends in Food Science and Technology* 20: 438–447.

Rojas-Graü, M. A., Tapia, M. S., Martin-Belloso, O. 2008. Using polysaccharide-based edible coatings to maintain quality of fresh-cut Fuji apples. *LWT—Food Science and Technology* 41: 139–147.

Rolle, R., Chism, G. W. 1987. Physiological consequences of minimally processed fruits and vegetables. *Journal of Food Quality* 43: 274–276.

Sabater, B., Rodriguez, M. T. 1978. Control of chlorophyll degradation in detached leaves of barley and oat through effect of kinetin on chlorophyllase levels. *Physiologia Plantarum* 43: 274–276.

Saltveit, M. E. 1999. Effect of ethylene on quality of fresh fruits and vegetables. *Postharvest Biology and Technology* 15: 279–292.

Sarzi, B. 2002. *Conservação de abacaxi e mamão minimamente processados: Associação entre o preparo, a embalagem e a temperatura de armazenamento.* 2002. 100 f. Dissertação (Mestrado em Agronomia)—Faculdade de Ciências Agrárias e Veterinárias, Universidade Estadual Paulista, Jaboticabal, Brazil.

Sarzi, B., Durigan, J. F., Lima, M. A., Mattiuz, B. 2001a. Comportamento respiratório de mamão minimamente processado quando armazenado sob diferentes temperaturas. In: *CONGRESSO BRASILEIRO DE FISIOLOGIA VEGETAL*, 8, Ilhéus. Anais... Ilhéus, Brazil: Sociedade Brasileira de Fisiologia, CD-ROM.

Sarzi, B., Durigan, J. F., Teixeira, G. H. A., Donadon, J. R. 2001b. Efeito da temperatura e do tipo de corte na conservação de abacaxi minimamente processado (PMP). In: *CONGRESSO BRASILEIRO DE FISIOLOGIA VEGETAL*, 8, Ilhéus. Anais... Ilhéus, Brazil: Sociedade Brasileira de Fisiologia, CD-ROM.

Selma, M. V., Ibáñez, A. M., Cantwell, M., Suslow, T. 2008. Reduction by gaseous ozone of *Salmonella* and microbial flora associated with fresh-cut cantaloupe. *Food Microbiology* 25: 558–565.

Senesi, E., McHugh, T. H. 2002. Film e coperture eduli com matrici a base di fruta. *Industrie Alimentari* 41: 289–94.

Silva, E. O. 2000. *Fisiologia pós-colheita de repolho (Brassica oleracea var. capitata) minimamente processado*. Viçosa, MG, 2000. 79 f. Tese (Doutorado em Biologia Vegetal)—Universidade Federal de Viçosa Viçosa, Brazil.

Silva, V. A. 2003. *Fisiologia de cenoura minimamente processada*. Viçosa, 78f. Dissertação (Mestrado em Biologia Vegetal)—Universidade Federal de Viçosa, Viçosa, MG.

Singh, B., Wang, D. J., Salunkhe, D. K., Rahman, A. R. 1972b. Controlled atmosphere storage of lettuce. 2. Effects on biochemical composition of leaves. *Journal of Food Science* 37: 52–55.

Singh, B., Yang, C. C., Salunkhe, D. K., Rahman, A. R. 1972a. Controlled atmosphere storage of lettuce. 1. Effects on quality and respiration rate lettuce heads. *Journal of Food Science* 37: 48–51.

Smith, W. L., Lagunas-Solar, M. C., Cullor, J. S. 2002. Use of pulsed ultraviolet laser light for the cold pasteurization of bovine milk. *Journal of Food Protection* 50: 108–111.

Soliva-Fortuny, R. C., Martín-Bellos, O. 2003. New advances in extending the shelf-life of fresh-cut fruits: a review. *Trends in Food Science and Technology* 14: 341–353.

Sothornvit, R., Rodsamran, P. 2008. Effect of a mango film on quality of whole and minimally processed mangoes. *Postharvest Biology and Technology* 47: 407–415.

Teixeira, G. H. A., Durigan, J. F., Alves, R. E., O'hare, T. J. 2007. Use of modified atmosphere to extend shelf life of fresh-cut carambola (*Averrhoa carambola* L. cv. Fwang Tung). *Postharvest Biology and Technology* 44: 80–85.

Teles, C. S. 2001. *Avaliação física, química e sensorial de couve (Brassica oleracea L. var acephala) minimamente processada, armazenada sob atmosfera modificada*. Viçosa, MG, 104 f. Dissertação (Mestrado em Ciência e Tecnologia de Alimentos)— Universidade Federal de Viçosa, Viçosa, Brazil.

Vanetti, M. C. D. 2004. Segurança microbiológica em produtos minimamente processados. In: *ENCONTRO NACIONAL SOBRE PROCESSAMENTO MÍNIMO DE FRUTAS E HORTALIÇAS*, 3, 2004, Viçosa. Palestras, Resumos e Oficinas. Viçosa, Brazil: CEE, pp. 30–32.

Vilas-Boas, E. V. B., Vilas-Boas, B. M., Giannoni, J. A., Rosane, J. M. 2007. Tendências na área de processamento mínimo de frutas e hortaliças: avanços tecnológicos. In: *II Simpósio Brasileiro de Pós-Colheita: Frutas, Hortaliças e Flores: Palestras e Resumos*, Universidade Federal de Viçosa, Viçosa, Brazil, pp. 109–119.

Watada, A. E., Abe, K., Yamuchi, N. 1990. Physiological activities of partially processed fruits and vegetables. *Food Technology* 44: 116–122.

Watada, A. E., Qi, L. 1999. Quality of fresh-cut produce. *Postharvest Biology and Technology* 15: 201–205.

Wiley, R. C. 1997. *Minimally Processed Refrigerated Fruits and Vegetables*. New York: Chapman and Hall, 357 pp.

Wills, R. H., Mcglasson, B., Graham, D., Joyce, D. 1998. *Postharvest—An Introduction to the Physiology and Handling of Fruits, Vegetables and Ornamentals*. New York: CABI International, 262 pp.

Yamauchi, N., Watada, A. E. 1991. Regulated chlorophyll degradation in spinach leaves during storage. *Journal of American Society of Horticultural Science* 116: 58–62.

Yang, S. F., Hoffman, N. E. 1984. Ethylene biosynthesis and its regulation in higher plants. *Annual Review of Plant Physiology* 35: 155–189.

Yaun, B. R., Summer, S. S., Eifert, J. D., Marcy, J. E. 2004. Inhibition of pathogens on fresh produce by ultraviolet energy. *International Journal of Food Microbiology* 90: 1–8.

Zagory, D. 2000. What modified atmosphere packaging can and can't do for you. *16th Annual Postharvest Conference and Trade Show*, March 14–15, 2000, Yakima, WA.

Zhang, L., Lu, Z., Yu, Z., Gao, X. 2005. Preservation of fresh-cut celery by treatment of ozonated water. *Food Control* 16: 279–283.

9 Enzyme Maceration

Sueli Rodrigues

CONTENTS

9.1 INTRODUCTION

9.1.1 PLANT CELL WALL

The plant cell wall serves a variety of functions. Along with protecting the intracellular contents, the structure bestows rigidity to the plant, provides a porous medium for the circulation and distribution of water, minerals, and other nutrients, and houses specialized molecules that regulate growth and protect the plant from disease. Growing pant cells are surrounded by a polysaccharide-rich primary wall. This wall is part of the apoplast, which is the free diffusional space outside the plasma membrane. Apoplast itself is largely self-contiguous and contains everything that is located between the plasma membrane and the cuticle. The primary wall and middle lamella account for most of the apoplast in growing tissue. The symplast, the inner side of the plasma membrane in which water and low-molecular-weight solutes can freely diffuse, is another unique feature of plant tissues.

The main chemical components of the primary plant cell wall include cellulose, a complex carbohydrate made up of several thousand glucose molecules linked end to end. In addition, the cell wall contains two groups of branched polysaccharides, the pectins and cross-linking glycans. Organized into a network with the cellulose microfibrils, the cross-linking glycans increase the tensile strength of the cellulose, whereas the coextensive network of pectins provides the cell wall with the ability to resist compression. In addition to these networks, a small amount of protein can be found in all plant primary cell walls. Some of this protein is thought to increase mechanical strength, and part of it consists of enzymes, which initiate reactions that form, remodel, or breakdown the structural networks of the wall. Such changes in the cell wall directed by enzymes are particularly important for fruit to ripen and leaves to fall in autumn.

The secondary plant cell wall, which is often deposited inside the primary cell wall as a cell matures, sometimes has a composition nearly identical to that of the earlier-developed wall. More commonly, however, additional substances, especially lignin, are found in the secondary wall. Lignin is the general name for a group of polymers of aromatic alcohols that are hard and impart considerable strength to the structure of the secondary wall. Lignin is what provides the favorable characteristics of wood to the fiber cells of woody tissues and is also common in the secondary walls of xylem vessels, which are central in providing structural support to plants. Lignin also makes plant cell walls less vulnerable to attack by fungi or bacteria, as do cutin, suberin, and other waxy materials that are sometimes found in plant cell walls.

The middle lamella, a specialized region associated with the cell walls of plants, is rich in pectins. Special conduits called plasmodesmata allow the cells to communicate with one another and share their contents. Primary and secondary walls contain cellulose, hemicellulose (xylan, glucuronoxylan, arabinoxylan, or glucomannan), and pectin in different proportions. The cellulose fibrils are embedded in a network of hemicellulose and lignin. Figure 9.1 presents the structure of vegetable cell wall.

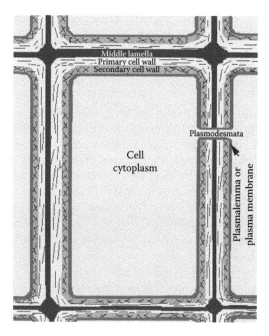

FIGURE 9.1   Scheme of vegetal cell wall.

The ripening of fruits and vegetables is associated with changes in wall structure and composition. Plant-derived beverages often contain significant amounts of wall polysaccharides. In many cases, the presence of cell wall polysaccharides is undesirable and compromises the product quality.

9.2 MAIN COMPONENTS OF CELL WALL

Pectic substance is the generic name used for the compounds that are acted upon by the pectinolytic enzymes. They are high-molecular-weight, negatively charged, acidic, complex glycosidic macromolecules (polysaccharides) that are present in the plant kingdom. They are present as the major components of middle lamella between the cells in the form of calcium pectate and magnesium pectate. Pectic substances account for 0.5%–4.0% of the fresh weight of plant material (Jayani et al., 2005). Cell wall composition is presented in Figure 9.2.

9.2.1 PECTIN

Pectic substances are responsible for the consistency, turbidity, and fruit juice appearance. They are also responsible for the increase in juice viscosity, which affects the juice concentration and filtration (Alkorta et al., 1998; Fernández-Gonzáles et al., 2004; Uenojo and Pastore, 2007).

Pectic substances mainly consist of galacturonans and rhamnogalacturonans in which the C-6 carbon of galactate is oxidized to a carboxyl group, the arabinans and the arabinogalactans (Whitaker,1990). These substances are a group of complex colloidal polymeric materials, composed largely of a backbone of anhydrogalacturonic acid units (Cho et al., 2001; Codner, 2001). The carboxyl groups of galacturonic acid are partially esterified by methyl groups and partially or completely neutralized by sodium, potassium, or ammonium ions (Jayani et al., 2005). The primary chain consists of a-D-galacturonate units linked α (1, 4), with 2%–4% of L-rhamnose units linked β (1, 2) and β (1, 4) to the galacturonate units. The side chains of arabinan,

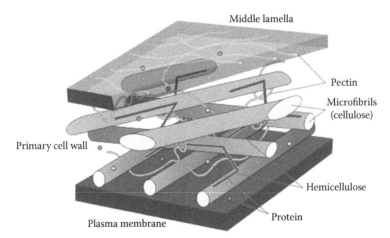

FIGURE 9.2   Detailed structure of a vegetal cell wall.

galactan, arabinogalactan, xylose, or fucose are connected to the main chain through their C1 and C2 atoms (Blanco et al., 1999). The relative molecular masses of pectic substances range from 25 to 360 kDa (Sakai et al., 1993). Pectic substances are classified into four main types as follows:

1. *Protopectin* is the water insoluble pectic substance present in intact tissue. Protopectin on restricted hydrolysis yields pectin or pectic acids. Pectic substances do not have a defined molecular weight.
2. *Pectic acid* is the soluble polymer of galacturonans that contains negligible amount of methoxyl groups. Normal or acid salts of pectic acid are called pectates.
3. *Pectinic acid* is the polygalacturonan chain that contains >0% and <75% methylated galacturonate units. Normal or acid salts of pectinic acid are called pectinates.
4. *Pectin (polymethyl galacturonate)* is the polymeric material in which, at least, 75% of the carboxyl groups of the galacturonate units are esterified with methanol.

9.2.2 CELLULOSE

Cellulose is a homopolymer composed of (1, 4)-linked β-D-glycoside residues. The linear chains of parallel alignment are tightly linked by hydrogen bonds to form microfibrils (Zykwinska et al., 2005). That is, in native cellulose (cellulose I), the cellulose polymer chains are stacked together during biosynthesis in polymer bundles known as fibrils or microfibrils (Larsson and Westlund, 2005; Liu et al., 2006).

9.2.3 HEMICELLULOSE

Hemicellulose contains many different sugar monomers (cellulose contains only anhydrous glucose). In hemicelluloses, besides glucose, xylose, mannose, galactose,

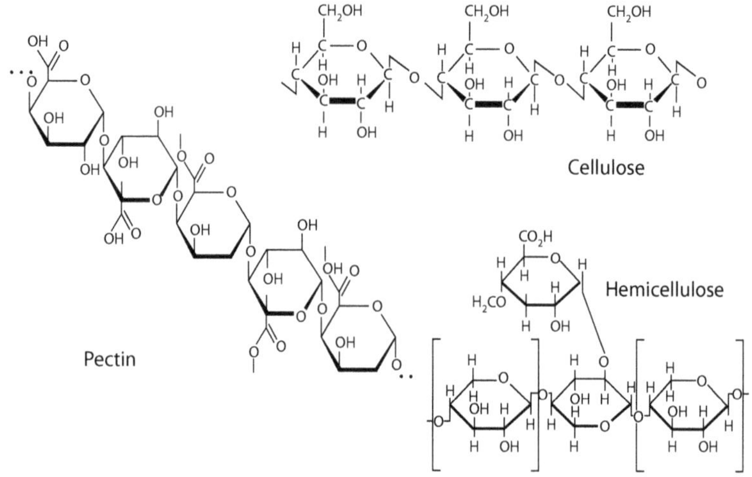

FIGURE 9.3   Cell wall polysaccharides.

rhamnose, and arabinose can be included as sugar monomers. Hemicelluloses contain most of the D-pentose sugars and occasionally small amounts of L-sugars as well. Xylose is always the sugar monomer present in the largest amount, but mannuronic acid and galacturonic acid also tend to be present. Hemicellulose consists of shorter chains of sugar units than cellulose and is a branched polymer. Figure 9.3 shows the structure of pectin, cellulose, and hemicellulose.

9.3 ENZYME MACERATION

Enzyme maceration is a technology largely applied in fruit processing to breakdown the cell wall polysaccharide chains to increase the juice extraction yield and improve the release of functional components present in the fruit such as anthocyanin, vitamins, polyphenols, aroma compounds, and so on. The cloudiness of fruit juices and wines is mainly due to pectins, which are usually associated with other plant polymers and cells debris. The enzyme treatment destroys the cell wall structure. The most enzymes applied in this technology attack the cell wall pectins. However, the commercial enzymes used in fruit juice processing are in fact a pool of several enzymes and usually contain, besides pectinases, cellulases and hemicellulases, among others.

The main enzymes used in the fruit maceration are depicted in Table 9.1. These enzymes can be applied alone or in preparations combining two or more enzymes taking advantage of the synergistic effect of the enzyme pool. Nowadays, most of the enzymes applied in fruit maceration are from microbial origin. *Aspergillus japonicus, Aspergillus niger, Mucor flavus, Aspergillus awamori, Penicillium italicum, Bacillus macerans, Erwinia carotovora, Saccharomyces pastorianus, Saccharomyces cerevisiae*, and *Kluyveromyces marxianus* are some strains that are able to produce macerating enzymes. The degradation mechanism and the chain

TABLE 9.1
Enzymes Applied in Fruit Juice Processing

Enzyme	Function
Polygalacturonase	Random digestion of pectins
Pectin methyl esterase	Removal of methyl esters and release of methanol turning the pectin digestible by polygalacturonase
Pectin lyase	Cleaves pectin into oligosaccharides without esterase action
Hemicellulases	Hydrolysis of cell wall polysaccharides
Endoglucanases	Hydrolysis of cellulose
Xylanases	Hydrolysis of xylan and arabinoxylan
Cellulases	Hydrolysis of cell wall polysaccharides
Polygalacturonase	Hydrolysis of polygalacturonan in the cell walls
Rhamnogalaturonase	Degrade ramified hairy regions of pectin

Source: Adapted from Bhat (2002); Jayani, R.S. et al., *Process Biochem.*, 40, 2931, 2005; Uenojo and Pastore (2007).

TABLE 9.2

Commercial Enzymes for Fruit Maceration

Brand Name	Enzyme Preparation	Manufacturer
Carezyme	Cellulase from *A. niger*	Novozymes
Celluclast	Cellulase from *Trichoderma reesei* ATCC 16921	Novozymes
Everlase	Protease from *Bacillus* sp.	Novozymes
Flavourzyme	Protease from *Aspergillus oryzae*	Novozymes
Glucanex	Lysing enzymes from *Trichoderma harzianum*	Novozymes
Novozyme 188	Cellobiase from *A. niger*	Novozymes
Pectinex[a]	Pectinase	Novozymes
Pentopan Mono BG	Xylanase from *Thermomyces lanuginosus*	Novozymes
Viscozyme	Multienzyme complex containing a wide range of carbohydrases, including arabinase, cellulase, β-glucanase, hemicellulase, and xylanase	Novozymes
Lallzyme C-MAX™	Cinnamyl esterase-free pectinase	Lallemand
Lallzyme MMX™	Beta-glucanase and pectinase blend sourced from *Trichoderma* sp. and *A. niger*	Lallemand
Lallzyme C™	High polygalacturonase and pectin lyase with controlled levels of pectin esterase	Lallemand
Lallzyme Cuvée Blanc	Specific blend of pectinases and glycosidase	Lallemand
Pectolyase	Polygalacturonase from *A. japonicus*	Duchefa Biochemie B.V.
Onozuka R-10	Cellulase from *Trichoderma viride*	PhytoTechnology Laboratories
Macerozyme R10	Pectinase, cellulase, and hemicellulase	PhytoTechnology Laboratories

[a] There are a number of preparations of Pectinex (Pectinex MASH, Pectinex Ultra, Pectinex mash XXL, etc.).

cleavage can be endo, exo, or random. Trans-elimination is the prevalent mechanism in pectin lyases (Jayani et al., 2005). Table 9.2 presents some commercial enzymes for fruit juice and wine processing.

The choice of the right enzyme preparation is usually the key of the processing development and must take into account the fruit tissue composition and the desired final product. Enzymes usually present a narrow range of pH and temperature for stability and activity. Macerating preparations usually contain more than one enzyme from different microbial sources, and these enzymes can present different optimal pH and temperatures. For instance, polygalacturonase from *Fusarium oxysporum* presents optimum activity temperature at 69°C and pH 11 (Pietro and Roncero, 1996). The optimal activity of pectin lyase from *Penicillium italicum* is obtained at 40°C and pH from 6.0 to 7.0 (Alana et al., 1990). The fruit pH and its composition also affect the final results in a process, and in some case, pH correction might affect the final product quality. The use of high temperatures in the maceration step might also be undesirable because it may impart nutritional losses.

9.4 CASE STUDIES

9.4.1 Enzymatic Prepress of Elderberry and Blackcurrant Juice

The industrial processing of elderberries mainly takes place in Northern Europe where the berries are processed for juice, juice concentrate, wine, and jelly manufacture (Jensen et al., 2001; Landbo et al., 2007). Elderberries have received increased attention due to their high contents of anthocyanins that are widely used as a color ingredients in various beverages and in home wine sets and may also provide nutritional benefits as antioxidant phytonutrients (Netzel et al., 2005). In industrial berry juice production, prepress addition of pectinases is prevalent as this enzyme addition increases the juice yield by breaking down the pectin in the plant cell walls and in the middle lamellae between the plant cells. Apart from promoting juice extraction, studies on black currant juice have demonstrated that enzymatic prepress treatment with multicomponent pectinase preparations enhanced the release of anthocyanins and other phenolics into the juice (Bagger-Jørgensen and Meyer, 2004; Landbo and Meyer, 2004). Although enzyme-catalyzed breakdown of the plant cell wall matrix and middle lamella may first increase the immediate turbidity in the juice (Grassin and Fauquembergue, 1996), studies with black currant juice have shown that forced enzymatic prepress treatment, that is, increased enzyme dosage and prolonged reaction time, may also improve the juice clarity (Landbo and Meyer, 2004). Immediate turbidity in fruit juices is generally assumed to be mainly due to the presence of pectin and other fractions of fruit cell wall material (Grassin and Fauquembergue, 1996).

Increased pectinolytic enzyme dose, longer maceration time, and elevated reaction temperature all had significantly positive effects on the juice yield. Increased enzyme dose and maceration temperature also increased the yields of anthocyanins in the elderberry juice, while none of the reaction parameters affected the juice turbidity. With the optimal treatment with a pectinolytic enzyme preparation (Pectinex BE 3 L), produced by a cloned *Aspergillus* strain, a maximal juice yield of 77% w/w of the berry mash, an anthocyanin yield of 2380 mg/kg fresh berry mash, and a turbidity level of 128 formazin nephelometric units (FNU) were obtained. Enzymatic prepress treatment generally decreased turbidity levels by 30% as compared to pressing without prior enzymatic treatment. A comparison of the responses obtained after the optimal enzymatic treatment with five different pectinolytic enzyme preparations showed that the *A. niger* preparation Pectinex BE Color gave slightly better juice and phenol yields and lower turbidity levels than the other enzyme preparations tested. In conclusion, the results demonstrated that juice yields and phenolic yields in elderberry juice could be improved with enzyme treatment and that the optimal reaction conditions for obtaining the best juice yield, highest phenolics, and lowest turbidity levels could be rationally identified via statistical factor level optimization (Landbo et al., 2007).

9.4.2 Pesticide Removal from Tomatoes

Pesticides used during agricultural production have increasingly caused concerns because of their adverse effects on human health as they leave residues in varying extents on agricultural produce such as vegetables and fruits. Pesticide residues in plants may be unavoidable even when good agricultural practices are used. On the other hand, it was observed that pesticide residues in plant produce are reduced by processing or some household preparation stages such as washing, peeling, cooking, etc. (Dikshit et al., 2003).

Romeh et al. (2009) evaluated the effects of the use of enzymes on the residual levels of profenofos, imidacloprid, and penconazole in tomato fruits and products. According to them, tomato fruits can be safely harvested for human consumption or for processing purposes 3 days after the spray time of imidacloprid and penco-nazole and 7 days after in the case of profenofos. Pesticide residues were greatly decreased in tomato juice under cold or hot break compared with that taken from unwashed tomato fruits. A sharp decline in profenofos level was noted after treat-ment by Pectinex ultra SP-L and Benzyme M during tomato crushing showing that enzyme treatment can be applied to remove pesticides from fruits.

9.4.3   ENZYME MACERATION OF APPLE JUICE

The effects of different commercial enzymatic mash treatments on yield, turbidity, color, and polyphenolic and sediment of procyanidin content of cloudy apple juice were studied by Oszmiánski et al. (2009). Addition of pectolytic enzymes to mash treatment had positive effect on the production of cloud apple juices by improving polyphenolic contents, especially procyanidins and juice yields (68.3% in control samples to 77% after Pectinex Yield Mash). As a summary of the effect of enzymatic mash treat-ment, polyphenol contents in cloudy apple juices significantly increased after Pectinex Yield Mash, Pectinex Smash® XXL, and Pectinex XXL maceration were applied, but no effect was observed after Pectinex Ultra-SPL I Panzym XXL use compared to the control samples. The content of polymeric procyanidins represented 50%–70% of total polyphenols, but in apples, polymeric procyanidins were significantly lower in juices than in fruits and also affected by enzymatic treatment (Pectinex AFP L-4 and Panzym Yield Mash) compared to the control samples. The enzymatic treatment decreased procyanidin content in most sediment with the exception of Pectinex Smash XXL and Pectinex AFP L-4. Generally in samples that were treated by pectinase, radical scavenging activity of cloudy apple juices was increased compared to the untreated reference samples. The highest radical scavenging activity was associated with Pectinex Yield Mash, Pectinex Smash XXL, and Pectinex XXL enzyme and the lowest activity with Pectinex Ultra SP-L and Pectinex APFL-4. However, in the case of enzymatic mash treatment, cloudy apple juices showed instability of turbidity and low viscosity. These results must be ascribed to the much higher hydrolysis of pectin by enzymatic preparation, which is responsible for viscosity. During 6 months of stor-age at 4°C, small changes in analyzed parameters of apple juices were observed.

9.4.4   CLARIFICATION OF CHERRY JUICE

Industrial juice clarification is accomplished by a combination of enzymatic depecti-nization, gelatin–silica sol, and/or bentonite treatment. The gelatin–silica sol treatment step is particularly slow and mischievous and requires comprehensive downstream processing to obtain clarified juice. Pinelo et al. (2010) studied the enzymatic clari-fication of industrially pressed cherry juice using pectinase (Pectinex Smash) and protease (Enzeco), both enzyme preparations derived from *Aspergillus* spp. The effects of the enzyme clarification treatments were assessed immediately after the particular clarification treatment (immediate turbidity) and during 14 days of cold storage (turbidity development). The protease treatment resulted in significant reduction of immediate turbidity, but had low clarification impact during the subsequent cold storage. In contrast, pectinase addition exerted a weak effect on immediate turbidity reduction, but effectively decreased the turbidity development during storage. Conventionally, immediate turbidity is presumed to be caused by pectin, while turbidity development during cold storage (haze formation) is assumed to be due to protein–phenol interactions. The results suggest that proteins play a decisive role in the formation of immediate turbidity in cherry juice, and point to that pectin may contribute to turbidity development during cold storage of cherry juice.

9.4.5 Bilberry Mash Treatment

Szajdek et al. (2009) studied the effect of different bilberry mash treatment methods, that is, heat treatment, enzymatic treatment, and combined treatment (enzymatic maceration preceded by heat treatment), on total phenolics, anthocyanin profile, and ascorbic acid, as well as on DPPH and OH radical scavenging activity of juices. Three fractions of juice containing different groups of phenolic compounds were evaluated. Enzymatic maceration of fruit mash was carried out using three enzymatic preparations Pektopol PT-400, Pectinex BE Color, and Gammapect LC Color. Juices obtained by combined treatment of fruit mash (enzymatic maceration preceded by heat treatment) were the richest source of total phenolics, the concentration of which ranged from 2304 to 4418 mg/L in these juices. These juices also showed the strongest DPPH and OH radical scavenging activity. Juices produced from fruit mash subjected to enzymatic maceration contained the highest amount of anthocyanins. The juices differed considerably with regard to their anthocyanin profile, determined by HPLC. Ascorbic acid content was at a comparable level in all juices and varied between 13.5 and 16.9 mg 100 g. Fraction I, containing mostly anthocyanins, showed the highest DPPH and OH radical scavenging activity, while fraction III, dominated by phenolic acids, was the least active.

9.4.6 Peach Juice Clarification

Santin et al. (2008) studied the peach juice clarification using enzymatic treatment. The enzymatic hydrolysis was performed to reduce the juice viscosity and pulp content and consequently to increase juice extraction. Two commercial pectinase preparations (AFP L3 and ULTRA SP WOP from Novozymes) as well as a cellulase (Onozuka—R10) were tested. The hydrolysis using Pectinex AFP L3 (Novozymes) at 25°C for 60 min presented the best results of pulp (48%) and viscosity (68%) reduction. Process using Pectinex AFPL-3 at 25°C for 60 min was the most effective condition to achieve the highest reductions of pulp content (48%) and viscosity (68%) of the juice. Physical and chemical analyses showed that the enzymatic treatment is effective for reduction of peach pulp viscosity, pulp content, and turbidity and does not influence other juice parameters such as pH, total acidity, vitamin C, and soluble solids.

9.4.7 White Pitaya Maceration

Raw white pitaya (*Hylocereus undatus*) juice is very cloudy and viscous. These physical properties lead to the use of enzymatic treatment to clarify the juice. White pitaya juice was treated with pectinase enzyme (Pectinex Ultra SP-L) at different maceration times (20–100 min), reaction temperatures (30°C–50°C), and enzyme concentrations (0.01%–0.1%). The use of Pectinex Ultra SP-L at 0.06% enzyme concentration at 49°C and 40 min maceration time resulted in a stable juice with no sedimentation and haze formation in the final product. At this procession, conditions resulted in minimum viscosity (3.47 cP), minimum absorbance value at 660 nm (0.0065), maximum juice yield (80.3%), and maximum L^* value (89.63).

9.4.8 Other Applications

Besides fruit juice processing, maceration enzymes might be also used in other applications: enzyme cocktail, used for the production of animal feeds to reduce the food viscosity, which increases absorption of nutrients, liberates nutrients, either by hydrolysis of nonbiodegradable fibers or by liberating nutrients blocked by these fibers, and reduces the amount of feces (Hoondal et al., 2002). Another application is in citrus peel oil extraction. Lemon oil can be extracted using pectinases (Scott, 1978).

Nowadays, the main limitation for the application of enzymes in fruit processing is costs. The costs of enzymatic treatment in fruit juice processing and enzyme storage, which usually requires cold storage, are relevant variables. Some recent works have been published on enzyme immobilization for fruit juice processing with promising results.

REFERENCES

Alana, A., Alkorta, I., Dominguez, J.B., Llama, M.J., and Serra, J.L. 1990. Pectin lyase activity in a *Penicillium italicum* strain. *Applied and Environmental Microbiology* 56: 3755–3759.

Alkorta, I., Garbisu, C., Llama, M.J., and Serra, J.L. 1998. Industrial applications of pectic enzymes—A review. *Process Biochemistry* 33: 21–28.

Bagger-Jørgensen, R. and Meyer, A.S. 2004. Effects of different enzymatic pre-press maceration treatments on the release of phenols into blackcurrant juice. *European Food Research and Technology* 219: 620–629.

Bhat, M.K. 2000. Cellulases and related enzymes in biotechnology. *Biotechnology Advances* 18: 355–383.

Blanco, P., Sieiro, C., and Villa, T.G. 1999. Production of pectic enzymes in yeasts. *FEMS Microbiology Letters* 175: 1–9.

Cho, S.W., Lee, S., and Shin, W. 2001. The X-ray structure of *Aspergillus aculeatus* polygalacturonases and a modeled structure of the polygalacturonase–octagalacturonate complex. *Journal of Molecular Biology* 314: 863–878.

Codner, R.C. 2001. Pectinolytic and cellulolytic enzymes in the microbial modification of plant tissues. *Journal of Applied Bacteriology* 84: 147–160.

Dikshit, A.K., Pachaury, D.C., and Jindal, T. 2003. Maximum residue limit and risk assessment of etacyfluthrin and imidacloprid on tomato. *Bulletin of Environmental Contamination and Toxicology* 70: 1143–1150

Fernández-González, M., Úbeda, J.F., Vasudevan, T.G., Otero, R.R.C., and Biones, A.I. 2004. Evaluation of polygalacturonase activity in *Saccharomyces cerevisiae* wine strains. *FEMS Microbiology Letters* 237: 261–266.

Grassin, C. and Fauquembergue, P. 1996. Fruit juices. In T. Godfrey and S. West (Eds.), *Industrial Enzymology*, 2nd edn. London, U.K.: MacMillan Press, pp. 225–264.

Hoondal, G.S., Tewari, R.P., Tewari, R., Dahiya, N., and Beg, Q.K. 2002. Microbial alkaline pectinases and their industrial applications—A review. *Applied Microbial and Biotechnology* 59: 409–418.

Jayani, R.S., Saxena, S., and Gupta, R. 2005. Microbial pectinolytic enzymes: A review. *Process Biochemistry* 40: 2931–2944.

Jensen, K., Christensen, L., Hansen, P., Jørgensen, U., and Kaack, K. 2001. Olfactory and quantitative analysis of volatiles in elderberry (*Sambucus nigra* L) juice processed from seven cultivars. *Journal of the Science of Food and Agriculture* 81: 237–244.

Landbo, A.K., Kaack, K., and Meyer, A.S. 2007. Statistically designed two step response surface optimization of enzymatic prepress treatment to increase juice yield and lower turbidity of elderberry juice. *Innovative Food Science and Emerging Technologies* 8: 135–142

Landbo, A.K. and Meyer, A.S. 2004. Effects of different enzymatic maceration treatments on enhancement of anthocyanins and other phenolics in black currant juice. *Innovative Food Science and Emerging Technologies* 5: 503–513.

Larsson, P.T. and Westlund, P.O. 2005. Line shapes in CP/MAS 13C NMR spectra of cellulose I. *Spectrochimica Acta Part A* 62: 539–546.

Liu, C.-F., Ren, J.-L., Xu, F., Kiu, J.-J., Sun, J.-X., and Sun, R.-C. 2006. Isolation and characterization of cellulose obtained from ultrasonic irradiated sugarcane bagasse. *Journal of Agriculture and Food Chemistry* 54: 5742–5748.

Netzel, M., Strass, G., Herbst, M., Dietrich, H., Bitsch, R., Bitsch, I. et al., 2005. The excretion and biological antioxidant activity of elderberry antioxidants in healthy humans. *Food Research International* 38: 905–910.

Oszmiánski, J., Wojdylo, A., and Kolniak, J. 2009. Effect of enzymatic mash treatment and storage on phenolic composition, antioxidant activity, and turbidity of cloudy apple juice. *Journal of Agriculture and Food Chemistry* 57: 7078–7085.

Pietro, A.D. and Roncero, M.I.G. 1996. Purification and characterization of an exo-polygalacturonase from the tomato vascular wilt pathogen. *Fusarium oxysporum* f.sp. lycopersici. *FEMS Microbiology Letters* 145: 295–298.

Pinelo, M., Zeuner, B., and Meyer, A.S. 2010. Juice clarification by protease and pectinase treatments indicates new roles of pectin and protein in cherry juice turbidity. *Food and Bioproducts Processing* 88: 259–265

Romeh, A.A., Mekky, T.M., Ramadan, R.A., and Hendawi, M.Y. 2009. Dissipation of profenofos, imidacloprid and penconazole in tomato fruits and products. *Bulletin of Environmental Contamination and Toxicology* 83: 812–817

Sakai, T., Sakamoto, T., Hallaert, J., and Vandamme, E.J. 1993. Pectin, pectinase and protopectinase: Production, properties and applications. *Advances in Applied Microbiology* 39: 231–294.

Santin, M.M., Treicherl, H., Vladuga, E., Cabral, L.M.C., and Di Luccio, M. 2008. Evaluation of enzymatic treatment of peach juice using response surface methodology. *Journal of the Science of Food and Agriculture* 88: 507–512.

Scott, D. 1978. Enzymes, industrial. In: M. Grayson and D. Ekorth (Eds.), *Kirk-Othmer Encyclopedia of Chemical Technology*, NY: Wiley, pp. 173–224.

Szajdek, A., Borowska, E.J., and Czaplicki, S. 2009. Effect of bilberry mash treatment on the content of come biologically active compounds and the antioxidant activity of juices. *Acta Alimentaria* 38: 281–289.

Uenojo, M. and Pastore, G.M. 2007. Pectinases: aplicações industriais e perspectivas. *Quimica Nova* 30: 388–394.

Whitaker, J.R. 1990. Microbial pectinolytic enzymes. In: W.M. Fogarty and C.T. Kelly (Eds.), *Microbial Enzymes and Biotechnology*, 2nd edn. London, U.K.: Elsevier Science Ltd.

Zykwinska, A.W., Ralet, M.C.J., Garnier, C.D., and Thibault, J.F.J. 2005. Evidence for vitro binding of pectin side chains to cellulose. *Plant Physiology* 139: 397–407.

10 Fruit and Fruit Juices as Vehicles for Probiotic Microorganisms and Prebiotic Oligosaccharides
Advances and Perspectives

Sueli Rodrigues

CONTENTS

10.1 INTRODUCTION

10.1.1 PROBIOTICS

According to the currently adopted definition, probiotics are "Live microorganisms, which when administered in adequate amounts confer a health benefit on the host" (FAO/WHO, 2001). Nowadays, probiotics represent probably the archetypal functional food and are defined as alive microbial supplement, which beneficially affect the host by improving its intestinal microbial balance (Kalliomaki et al., 2001; Brown and Valiere, 2004) and inhibiting pathogens and toxin-producing bacteria (Metchnikoff, 1907). Multiple reports have described their health benefits on gastrointestinal infections, antimicrobial activity, improvement in lactose metabolism, reduction in serum cholesterol, immune system stimulation, antimutagenic properties, anticarcinogenic properties, antidiarrheal properties, improvement in inflammatory bowel disease, and suppression of *Helicobacter pylori* infection by addition of selected strains to food products (Gomes and Malcata, 1999; Agerholm-Larsen et al., 2000; Gotcheva et al., 2002; Nomoto, 2005; Shah, 2007). Lactic acid bacteria (LAB) and bifidobacteria, the most studied and widely employed bacteria within the probiotic field, are normal components of the intestinal microbiota and have a long tradition of safe application within the food industry.

The most frequently used bacteria in probiotic products include the *Lactobacillus* and *Bifidobacterium* species. *Lactobacillus acidophilus* strains showed good activity in the human digestive tract, relieving chronic constipation (Rettger, 1935). More recently, the following intestinal LAB species with alleged health beneficial properties have been introduced as probiotics, including *Lactobacillus rhamnosus*, *Lactobacillus casei*, and *Lactobacillus johnsonii* (Tannock, 2003). Lactobacilli strains have been largely employed in several commercial products. The main employed strains are *L. acidophilus*, *L. amylovorus*, *L. bulgaricus*, *L. casei*, *L. fermentum*, *L. gasseri*, *L. johnsonii*, *L. lactis*, *L. paracasei*, *L. plantarum*, *L. reuteri*, *L. rhamnosus*, and *L. salivarius*. Recently, the use of *Bifidobacterium* species has increased in the market. The main commercial strains are *B. adolescentis*, *B. animalis*, *B. bifidum*, *B. brevis*, *B. infantis*, *B. lactis*, and *B. longum*.

According to Espinoza and Navarro (2010), probiotics have been added to yogurt and other fermented dairy products. However, dairy probiotic products may represent inconveniences due to their lactose and cholesterol content. In addition, there is an increasing demand for vegetarian probiotic products (Ray and Sivakumar, 2009). This fact has led to the development of probiotic products from various food matrices including fruits (Prado et al., 2008) and vegetables (Yoon et al., 2006).

Technological advances have made it possible to alter some structural characteristics of fruit and vegetable matrices by modifying food components in a controlled way (Betoret et al., 2003). This could make them ideal substrates for the culture of probiotics since they already contain beneficial nutrients such as minerals, vitamins, dietary fibers, and antioxidants (Yoon et al., 2004), while lacking the dairy allergens that might prevent consumption by certain segments of the population (Luckow and Delahunty, 2004). There is a genuine interest in the development of fruit juice–based functional beverages with probiotics because they have taste profiles that are appealing to all age groups and because they are perceived as healthy and refreshing foods (Tuorila and Cardello, 2002; Yoon et al., 2004; Sheehan et al., 2007).

According to Siró et al. (2008), probiotic dairy products are the principal functional food found in the market. These products accounted for sales of around U.S.$1.35 billion in 1999 (Hilliam, 2000; Ouwehand, 2007) and about 56% of functional foods total U.S.$31.1 billion global sales in 2004 (Benkouider, 2005). The main markets of dairy probiotics are Scandinavia, the Netherlands, Switzerland, Croatia, and Estonia. Greece, France, and Spain can be considered as developing markets (Mäkinen-Aakula, 2006). Germany, France, the United Kingdom, and the Netherlands account for around two-thirds of all sales of functional dairy products in Europe (Hilliam, 2000). Such products have shown an impressive growth, bringing the market volume in Germany from around U.S.$5 million in 1995 to

U.S.$419 million in 2000, of which U.S.$301 million account for pro-, prebiotic, and other functional yogurts, and around U.S.$118 million for functional dairy drinks (Menrad, 2003). In Central-Eastern Europe, for example, in Czech Republic, Hungary, or Romania, the probiotic market is dominated by international companies such as Unilever or Danone, and most of the national producers are only able to adopt technologies and product ingredients developed in other countries (Banykó, 2007). Estimates target a total of U.S.$19.6 billion on probiotic product sales in 2013, a compound annual growth rate (CAGR) of 4.3% (Granato et al., 2010).

There is, however, an extensive research and development activity concerning probiotics, which resulted in a great number of special new dairy products such as Synbiofir drinking kefir, Synbioghurt drinking yoghurt, HunCult fermented drink, Milli Premium sour cream, Aktivit quark dessert, New Party butter cream, and Probios cheese cream (Szakály, 2007; Siró et al., 2008).

Fruit juice has also been suggested as a novel, appropriate medium for fortification with probiotic cultures because it is already positioned as a healthy food product, and it is consumed frequently and loyally by a large percentage of the world population (Tuorila and Cardello, 2002). However, research has shown that, for example, perceptible off-flavors such as dairy, medicinal, and savory flavors were associated with probiotic orange juices (Luckow and Delahunty, 2004; Luckow et al., 2006), which might be a limitation of the product consumption.

10.1.2 PREBIOTICS

Besides the probiotic products, there are also the prebiotic functional foods, in which the development has increased lately. A prebiotic is a specific ingredient that allows changes, both in the composition and/or activity of the gastrointestinal microflora that confers benefits upon host well-being and health (Roberfroid, 2007). Usually prebiotics are oligosaccharides, but the definition may include noncarbohydrates. The most prevalent forms of prebiotics are nutritionally classified as soluble fiber. Prebiotics are not metabolized in the stomach and reach the large intestine where they are selectively metabolized by bifidobacteria and LAB. Thus, it is assumed that a prebiotic should increase the number and/or activity of the probiotic species in the large intestine. The healthy beneficial effects of prebiotic intake are, in fact, due to the increase of number of probiotic microbiota. Prebiotics may be naturally found in some foods such as raw chicory root, raw Jerusalem artichoke, and raw garlic, or they might be obtained using enzymes such as glycosyltransferases, inulinases, pectinases, and others. Prebiotics are food components that confer a health benefit on the host associated with modulation of the intestinal microbiota (Sako et al., 1999; Monaco et al., 2011).

Oligosaccharides, consisting of a mixture of hexose oligomers with a variable extent of polymerization degree, are food products with interesting nutritional properties. They may be naturally present in food, mostly in fruits, vegetables, or grains, or produced by biosynthesis from natural sugars or polysaccharides and added to food products because of their nutritional properties or organoleptic characteristics. The dietary intake of oligosaccharides is difficult to estimate, but it may reach 3–13 g/day per person (for fructooligosaccharides [FOSs]), depending on the population. The resistance to enzyme reactions occurring in the upper part of the gastrointestinal tract allows oligosaccharides to become "colonic nutrients," as some intestinal bacterial species express specific hydrolases and are able to convert oligo-saccharides into short-chain fatty acids (acetate, lactate, propionate, butyrate) and/or gases by fermentation. Oligosaccharides that selectively promote some interesting bacterial species (lactobacilli, bifidobacteria) and thus equilibrate intestinal micro-biota are now named prebiotics. The pattern of short-chain fatty acid production in the colon as well as the prebiotic effect, if demonstrated, are dynamic processes that vary with the type of oligosaccharide being affected by the degree of polymeriza-tion, the nature of hexose moieties, the duration of the treatment, the initial composi-tion of flora, or the diet in which they are incorporated. Experimental data obtained in vitro and in vivo in animals, and also recent data obtained in human subjects, support the involvement of dietary oligosaccharides in physiological processes in the different intestinal cell types such as mucin production, cell division, immune cells function, and ionic transport, and also outside the gastrointestinal tract such as hormone production and lipid and carbohydrate metabolism (Delzenne, 2003).

Subsidiaries of major European sugar producers play an important role on the market of prebiotic ingredients (e.g., the Südzucker subsidiary Orafti). In addition, a relatively high number of small or specialized producers are offering functional ingredients as well (e.g., in the probiotic field). So far, only a limited number of biotechnology compa-nies have specialized on this field despite the expected growth perspectives (Menrad, 2000). In general, the suppliers of functional ingredients try to prove the efficacy of a specific substance and sell it to a wide range of food industry companies, thereby creat-ing specific formulations. In this sense, innovative suppliers of food ingredients are of high relevance, in particular for product innovations of SME food companies.

The world demand for prebiotics is estimated to be around 167,000 tons and 390 million Euro (Siró et al., 2008). Among them, FOS, inulin, isomalto-oligosac-charides (IMOs), polydextrose, lactulose, and resistant starch are considered the main prebiotic components. Oligosaccharides, such as soy oligosaccharides (SOSs), galac-tooligosaccharides (GOSs), and xylo-oligosaccharides (XOSs), are also marketed in Japan (Ouwehand, 2007). Inulin and oligofructose, nondigestible fermentable fructans, are among the most studied and well-established prebiotics (Gibson, 2006).

10.1.3 Synbiotics

Synbiotics refer to nutritional supplements combining probiotics and prebiotics in a form of synergism (Holzapfel and Schillinger, 2002). The combination of probiotic and prebiotic in a food product is interesting because in addition to directly introduc-ing live beneficial bacteria to the colon, there is an increase in the number of benefi-cial *Bifidobacterium* and *Lactobacillus* species in the intestinal microbiota with the use of a prebiotic (Gibson et al., 2004; Cardarelli et al., 2008).

10.2 FRUIT AND FRUIT JUICES AS VEHICLE OF PROBIOTICS

Traditionally, probiotic foods are dairy-based products such as fermented milk, yogurts, and cheeses. However, due to the increase of consumers with food allergies and some kind of intolerance to dairy products, alternative vehicles for probiotics have been investigated in the last 10 years. Probiotic fruit-based products are dairy-free, soy-free, and vegan. These products are targeted to those who cannot consume dairy-based products or just right for those who like the refreshing taste of juice. The recommended intake of probiotic microorganisms is 7.00 log CFU/mL (Vinderola and Reinheimer, 2000; Pereira et al., 2011).

Although LAB has been considered a difficult microorganism that demands various essential amino acids and vitamins for growing (Salminen and Von Wrigh, 1993), some probiotic strains have the capability to grow in fruit matrices. Researchers have reported that cell viability depends on the strains used, the characteristics of the substrate, the oxygen content, and the final acidity of the product (Shah, 2001).

10.2.1 FRESH FRUITS AS PROBIOTIC VEHICLE

LAB are the main microorganism studied for functional fruit juice. However, few works have been published on fruit as probiotic vehicle probably because LAB cultivation in semisolid substrate as fruit and fruit pieces is difficult due to low microbial mobility and the difficulty of nutrient access. Rößle et al. (2010a) applied a probiotic microorganism (*L. rhamnosus* GG [LGG]) to fresh-cut apple wedges (cultivar Braeburn) and measured entrapment and stability of the microorganism. Instrumental eating quality parameters (Color Lab, texture, soluble solids, titratable acidity, and pH) and sensory acceptability were also monitored to investigate if application of the probiotic significantly influenced eating quality. Apple samples were cut into skin-on wedges and were dipped in an edible buffer solution containing approximately 10^{10} CFU/mL of LGG. LGG was enumerated on each test day (0, 2, 4, 6, 8, and 10) on whole wedges, on wedges flushed with a buffer solution (2% tri-sodium citrate), and on the flush-off liquid itself. All three sample sets contained 10^8 CFU/g over the test period, which is sufficient for a probiotic effect and is comparable to counts of probiotic bacteria in commercially available dairy products. This included the sample set of wedges, which had been flushed with buffer solution indicating good adherence of the bacteria over the test period. Physicochemical properties of the apple wedges containing LGG compared to the control remained stable over the 10 day period. Cryo-scanning electron microscopy and confocal laser scanning microscopy demonstrated good adherence of LGG to the surface of apple wedges. According to the authors, the process for making this product was relatively simple, and the product would retail from the conventional chill counters of supermarket stores. It is likely that its price would be competitive with existing probiotic dairy products.

In another study, the same grout applied LGG to fresh-cut apple wedges containing prebiotics (oligofructose and inulin). An assessment of the quality, sensory, polyphenol, and volatile attributes of symbiotic and probiotic samples was undertaken. Fructan analysis showed that all prebiotics remained stable over the 14 day storage period and an intake of 100 g of apple supplies 2–3 g of prebiotics. All sample sets contained ca. 10^8 CFU/g over the test period, which is sufficient for a probiotic effect. Browning index, firmness, acidity, and dry matter remained stable throughout the 14 days compared to the control while applying prebiotic coatings resulted in an increase in soluble solids. Sensory assessment indicated acceptable quality of probiotic and synbiotic apple wedges. HPLC analysis showed levels of decrease in polyphenolic compounds. No difference was found between O_2 and CO_2 headspace concentration and volatile production of synbiotic samples and samples only treated with probiotic (Rößlea et al., 2010b).

Vacuum and/or atmospheric impregnation techniques were evaluated by Alzamora et al. (2005) as feasible technologies for exploitations of fruit and vegetable tissues as new matrices into which functional ingredients can be successfully incorporated, providing novel functional product categories and new commercial opportunities. Plant tissues are multiphase systems with an intricate internal microstructure formed by cells, intercellular spaces, capillaries, and pores. Penetration of microorganisms through the cell walls therefore would require physicochemical or enzymic degradation and/or alteration of wall structures. In porous fruits and vegetables, not the wall pores but the intercellular spaces may play a major role regarding the microorganism penetration. These intercellular spaces are commonly called "pores." Intercellular air spaces are common in parenchymatous tissue and have been estimated to be 20%–25% of the total volume in apple, 15% in peach, 37%–45% in mushroom, and 1% in potato. For instance, mature cells of apple parenchyma tissues may be 50–500 µm in diameter with interconnecting air spaces ranging from 210 to 350 µm across (Lapsley et al., 1992). Thus, these spaces are large enough, so microorganisms are able to pass through because most bacteria range from 0.2 to 2.0 µm in diameter.

Rodríguez (1998) conducted basic impregnation studies with different microorganisms (*Saccharomyces cerevisiae*, *L. acidophilus*, and *Phoma glomerata*) to evaluate their penetration into a porous fruit tissue. Granny Smith apple was selected as a model of porous fruit, cut into cylinders, and impregnated with sucrose isotonic solution containing the microorganisms. One vacuum pulse of 2 min at five different absolute pressures (75, 125, 225, 325, and 425 mmHg) at 25°C was applied. For each microorganism, a control was prepared by maintaining the apple samples immersed into the solution for 2 min at atmospheric pressure (675 mmHg). When comparing microbial counts of fresh apple and apple treated under atmospheric conditions, it could be observed that the simple soaking renders a significant increase in microbial counts. This highlights the fact that capillary force and superficial adherence are very important factors that cannot be neglected in any modeling approach of immersion and impregnation operations. The lower the absolute pressure of the vacuum pulse applied, the higher the incorporation of microorganisms.

The fortification of apple cylinders with *Bifidobacterium* spp. "Bb12" (Christian Hansen Corp.) by applying vacuum impregnation was investigated by Maguiña et al. (2002). Apple was impregnated with an isotonic sucrose solution containing 7.95 log10 CFU/g of the microorganism at five different vacuum pressure levels (101, 125, 225, 325, and 425 mmHg). Applied vacuum had a significant effect on incorporation of *Bifidobacterium* spp. The greater incorporation of bifidobacteria was attained at absolute pressures 101 and 125 mmHg, which corresponded to the larger values of the volumetric fraction determined experimentally. In all cases, the microorganism was incorporated at count levels higher than 10^7 cells/g, which is enough to be classified as probiotic in most countries.

Fresh-cut apple and papaya cylinders were successfully coated with 2% (w/v) alginate or gellan film-forming solutions containing viable bifidobacteria. Water vapor permeability in alginate (6.31 and 5.52 × 10^{-9} g/Pa s m²) or gellan (3.65 and

4.89×10^{-9} g/Pa s m^2) probiotic coatings of papaya and apple, respectively, was higher than in the corresponding cast films. The gellan coatings and films exhibited better water vapor properties in comparison with the alginate coatings. Counts higher than 10^6 CFU/g of bifidobacteria were maintained for 10 days during refrigerated storage of fresh-cut fruits, demonstrating the feasibility of alginate- and gellan-based edible coatings to carry and support viable probiotics on fresh-cut fruit (Tapia et al., 2007).

With the aim of developing new functional foods, a traditional product, the table olive, was used as a vehicle for incorporating probiotic bacterial species. Survival on table olives of *L. rhamnosus*, *L. paracasei*, *Bifidobacterium bifidum*, and *B. longum* at room temperature was investigated. The results obtained demonstrated that bifidobacteria and one strain of LGG showed a good survival rate, with a recovery of about 10^6 CFU/g after 30 days. The LGG population remained unvaried until the end of the experiment, while a slight decline (to about 10^5 CFU/g) was observed for bifidobacteria. High viability ($>10^7$ CFU g/L) was observed throughout the 3 month experiment for *L. paracasei* IMPC2.1. This strain, selected for its potential probiotic characteristics and for its lengthy survival on olives, was used to validate table olives as a carrier for transporting bacterial cells into the human gastrointestinal tract. *L. paracasei* IMPC2.1 was recovered from fecal samples in four out of five volunteers fed 10–15 olives per day carrying about 10^9–10^{10} viable cells for 10 days (Lavermicocca et al., 2005).

10.2.2 FRUIT JUICE AS PROBIOTIC VEHICLE

Adding probiotics to juices is more complex than formulating in dairy products because the bacteria need protection from the acidic conditions in fruit juice. However, some recent studies have proved that some strains are able to grow and survive at probiotic levels in fruit juice. However, the use of fruit juice as probiotic vehicle has increased lately.

Wang et al. (2009) assessed the feasibility of noni as a raw substrate for the production of probiotic noni juice by LAB (*L. casei* and *L. plantarum*) and bifidobacteria (*B. longum*). Changes in pH, acidity, sugar content, cell survival, and antioxidant properties during fermentation were monitored. All tested strains grew well on noni juice, reaching nearly 10^9 colony-forming units/mL after 48 h fermentation. *L. casei* produced less lactic acid than *B. longum* and *L. plantarum*. After 4 weeks of cold storage at 4°C, *B. longum* and *L. plantarum* survived under low-pH conditions in fermented noni juice. In contrast, *L. casei* exhibited no cell viability after 3 weeks. Moreover, noni juice fermented with *B. longum* had a high antioxidant capacity that did not differ significantly from that of LAB. According to the authors, *B. longum* and *L. plantarum* are optimal probiotics for fermentation with noni juice.

Pereira et al. (2011) optimized the conditions of *L. casei* NRRL B-442 cultivation in cashew apple juice as well as determined the proper inoculum amount and fermentation time. Moreover, the survivability ability of *L. casei* in cashew apple juice during refrigerated storage (4°C) for 42 days was investigated. The optimum conditions for probiotic cashew apple juice production were initial pH 6.4, fermentation temperature of 30°C, inoculation level of 7.48 log CFU/mL (*L. casei*), and 16 h of fermentation process. It was observed that the *L. casei* grew during the refrigerated storage. Viable cell counts were higher than 8.00 log CFU/mL throughout the storage period (42 days). The values of lightness, yellowness, and total color change increased, and the values of redness reduced along the fermentation and refrigerated storage periods. The product also showed good acceptance in a preliminary sensory evaluation. It is important to emphasize that no heat treatment was necessary and there was no contamination evidence. Thus, the *L. casei* overlapped and controlled other microbial growth and juice spoilage, avoiding costs with heating treatment, as well as its adverse effects, as nutritional losses and sensory changes.

Fonteles et al. (2011) optimized the fermentation pH and temperature and also the fermentation time for development of a new probiotic fermented drink with cantaloupe juice. The fermented juice was subjected to a storage study during 42 days at 4°C. The initial pH and temperature influenced the growth of *L. casei* in cantaloupe juice. However, on the microbial viability, only the effect of temperature was significant. Optimum conditions for a satisfactory growth and viability of *L. casei* in cantaloupe juice were as follows: pH of 6.1 and temperature of 31°C. The fermentation time of 8 h was selected as the optimal fermentation time to prepare the probiotic cantaloupe juice. The cell viability was 8.3 log CFU/mL at the end of the fermentation. This level was kept over the 42 days of refrigerated storage. The consistent growth and viability of the probiotic microorganism in cantaloupe juice during fermentation and storage suggested that melon is a suitable vehicle for *L. casei* delivering.

Charernjiratrakul et al. (2007) reported that *L. plantarum* at 30°C and *Pediococcus pentosaceus* grew well in carrot juice with initial pH of 6.4. Yoon et al. (2006) found that *L. plantarum*, *L. casei*, and *L. delbrueckii* grew rapidly on sterilized cabbage juice without nutrient supplementation reaching nearly 8.00 log CFU/mL at 48 h of fermentation at 30°C. Yoon et al. (2005) who evaluated the viability of *L. plantarum* and *L. delbrueckii* in fermented red beet juice (30°C for 72 h) during refrigerated storage for 28 days reported that although the lactic cultures in fermented beet juice gradually lost their viability during cold storage, the viable cell counts in the fermented beet juice still remained at 6.00–8.00 log CFU/mL after 28 days of cold storage at 4°C. It is important to have a significant number of viable LAB in the product for maximum health benefits (Shah, 2001).

Champagne et al. (2008) evaluated *L. rhamnosus* R0011 inoculated in apple–pear–raspberry juice blend during 28 days in storage conditions simulating consumer use. Polyethylene terephthalate bottles of 1 L were used. The bottles were opened every 7 days, and samples of 250 mL were taken. These authors concluded that consumers can expect good viability of *L. rhamnosus* R0011 over 28 days of storage in a refrigerator, even if the bottles have been opened and cells are exposed to oxygen.

Mousavi et al. (2010) evaluated the production of probiotic pomegranate juice through fermentation by four strains of LAB: *L. plantarum*, *L. delbrueckii*, *L. paracasei*, and *L. acidophilus*. The microbial population of *L. paracasei* and *L. acidophilus* decreased approximately three logarithmic cycles during the first week of cold storage and lost their viability after 2 weeks. Viable cells of *L. delbrueckii* and *L. acidophilus* remained at a reasonable level within 2 weeks (~5 log CFU mL^{-1}). But for all studied strains, viable cell counts decreased dramatically to zero (nondetectable viable cells) after.

10.2.3 Prebiotic Oligosaccharides

Food supplementation with prebiotics has been strongly linked to modulation of the intestinal microbiota by shifting the bacterial balance toward benefic microorganisms, such as bifidobacteria and lactobacilli species (Gibson et al., 1995). Supplementation of inulin in binary cultures led to significantly shorter generation times compared to pure cultures, while no valuable effect was observed on the counts of pure cultures. Inulin supplementation to pure cultures lowered the generation time, with particular concern to *S. thermophilus* and *L. acidophilus*. The generation time of all microorganisms decreased in the following order: monocultures, cocultures, and cocktail. It demonstrated a synergism between *S. thermophilus* and the other strains and a bifidogenic effect of inulin. Enumerations of *L. rhamnosus* in cocktail markedly decreased compared to cocultures likely because of greater competition for the same substrates.

GOS is produced by the transgalactosylation of lactose and has been used as a prebiotic supplement in pediatric nutrition. Growing evidence supports the prebiotic supplementation of infant formula (Sako et al., 1999). A recently new kind of prebiotic oligosaccharides are xylose oligomers. Xylose oligomers are chains of xylose molecules linked with β-1,4-bonds with the degree of polymerization greater than 2 (Lau et al., 2011). Xylose oligomers can be produced from xylan either by chemical or enzymatic hydrolysis or by a combination of both (Vazquez et al., 2000; Moure et al., 2006; Akpinar et al., 2007; Yang et al., 2007; Aachary and Prapulla, 2009). Xylose oligomers have been reported to have a prebiotic effect on humans (Voragen, 1998; Moure et al., 2006) and inhibit the growth of pathogens in human intestines. In addition to pharmaceutical applications, xylose oligomers have been used in fortified foods, antiobesity diets, animal feeds, as agricultural ripening agents, and as yield enhancers because of their lower sweetness and high stability over a wide pH range from 2.5 to 8.0 (Vazquez et al., 2000).

Prebiotics have also been reported to have additional biological activities, beyond their influence on the gut microbiota (Quintero et al., 2011). Specifically, it has been suggested that some prebiotics, in particular, GOSs, may be able to inhibit gastrointestinal infections via antiadhesive activities (Saulnier et al., 2009; Shoaf-Sweeney and Hutkins, 2009). This concept is based on the observation that for most pathogens, the initial adherence to host cell surfaces is one of the first steps before colonization and infection. Adherence is most commonly mediated via lectin-like adhesins expressed by the bacteria that recognize and adhere to ligand-like carbohydrates located on the surface of the host epithelial cells. Bacterial variants that are unable to express functional adhesins are unable to adhere and initiate infections, indicating that adherence is required for pathogenesis (Boddicker et al., 2002; Cleary et al., 2004). Thus, agents that inhibit adherence could potentially reduce infections (Zorf and Roth, 1996; Sharon and Ofek, 2002).

Hess et al. (2011) studied if the satiety effect is linked to colonic fermentation. Short-chain FOS (scFOS) is a fermentable fiber that can be added to foods to influence these actions. The primary objective of the study was to determine if scFOS affects satiety and hunger and has an additive effect on food intake. Using a double-blind crossover design, 20 healthy subjects were assigned to consume two separate doses of 0, 5, or 8 g of scFOS. The first dose was mixed into a hot cocoa beverage and served with a breakfast meal of a bagel and cream cheese. Satiety was assessed with visual analogue scales (VASs) at 0, 15, 30, 45, 60, 90, 120, 180, and 240 min. Ad libitum food intake was measured at a lunch meal provided at the test site at 240 min. Subjects then recorded their food intake over the remainder of the 24 h study day. The second dose of scFOS was consumed in the form of solid, chocolate-flavored chews (51–67 total kcal) without additional food or drink, 2 h prior to the subject's dinner meal. Breath hydrogen measures were collected prior to the breakfast test meal (0 min) and the ad libitum lunch (240 min). Gastrointestinal tolerance was evaluated over the course of the 24 h study day using VAS. All treatments were well tolerated. No differences in subjective satiety over the morning, or food intake at lunch, were found. Over the remainder of the day, the high dose of scFOS reduced food intake in women, but increased food intake in men, suggesting a gender difference in the longer-term response. Breath hydrogen increased in a dose-dependent manner. These results indicate that scFOS undergoes fermentation within 240 min; however, acceptable amounts of scFOS did not enhance acute satiety or hunger.

10.2.4 PREBIOTICS IN FRUIT JUICES

Fortification of fruit juices with novel functional ingredients such as prebiotics is a recent development in the direction of creating new functional foods (Renuka et al., 2009).

Although some studies have been published on the benefits of using prebiotic oligosaccharides in food formulations, few studies on their large-scale production have been published. Usually, enzyme processes are expensive because of the high processing costs of enzyme production, purification, and stabilization. Rabelo et al. (2009) produced prebiotic oligosaccharides in cashew apple juice. Cashew apple is the peduncle of the cashew fruit, which is rich in reducing sugars (fructose and glucose), vitamins, minerals, and some amino acids. The synthesis was done by fermenting the bacteria *Leuconostoc citreum* B742, which produces a glycosyl-transferase able to synthesize prebiotic oligosaccharides. Glucosyltransferases are enzymes that catalyze the transfer of glucosyl residues from a donor molecule to a particular acceptor (Rabelo et al., 2006; Rodrigues et al., 2005, 2006). The use of cashew apple juice as substrate for the production of dextransucrase by *L. citreum* B-742 and the use of the crude enzyme for oligosaccharide enzyme synthesis with high yields (80%) showed to be a potential low-cost industrial process.

The prebiotic effect of the fermented cashew apple juice containing oligosaccharides was evaluated through the *L. johnsonii* growth by Vergara et al. (2010). The prebiotic effect of the cashew apple juice fermented with *Lactobacillus mesenteroides* was demonstrated due to the better growth of *L. johnsonii* in fermented cashew apple juice when compared to the growth observed in the culture media containing only glucose and fructose.

Ramnani et al. (2010) studied the prebiotic effect of fruit and vegetable shots containing Jerusalem artichoke inulin. The test and the placebo shots (100 mL) were produced in three groups: two groups were containing Jerusalem artichoke inulin, and the placebo shots did not contain inulin. The shots containing Jerusalem artichoke inulin were two liquid preparations made of fruit and vegetable juice concentrates and purées: one was predominantly made of pear-carrot-sea buckthorn and Jerusalem artichoke juices or purées (PCS), and the other preparation was predominantly made of plum-pear-beetroot and Jerusalem artichoke juices or purées (PPB). Inulin was not extracted from Jerusalem artichoke but present in the Jerusalem artichoke juice concentrate that was used in the formulation. The placebo was a water-based preparation, with added sugar, thickened and flavored with blood orange, carrot, and raspberry extracts and flavors (but no juice or pureés). The study was carried out in a double-blind, randomized, parallel manner, with volunteers consuming the test products for a 3 week period, followed by a 3 week wash-out period. The test shots were delivered in two different flavors: PCS and PPB. The total dose of inulin consumed by the volunteers was 5 g/day. The primary objective of the study was to monitor changes in levels of the following fecal bacterial populations: total bacteria, bacteroides, bifidobacteria, clostridia, *E. rectale/C. coccoides* group, *Lactobacillus/ Enterococcus* spp., *Atopobium* spp., *F. prausnitzii*, and propionibacteria.

   In the study, consumption of both the PCS and PPB shots containing Jerusalem artichoke inulin resulted in a clear and significant increase in bifidobacteria compared with placebo. The prebiotic effectiveness of Jerusalem artichoke inulin observed is well in line with previous feeding studies, where chicory-derived ingredients have been used. Unlike the clear bifidogenic effect, a small increase in *Lactobacillus/ Enterococcus* group was also observed for both the test shots compared with placebo. In conclusion, the study confirms the prebiotic effectiveness of fruit and vegetable shots containing Jerusalem artichoke inulin as observed by selective increase in bifidobacteria populations and a small increase in lactobacilli. The novel combination of a fruit and vegetable shot with the bacterial modulatory capability of Jerusalem artichoke inulin constitutes a new food format to deliver functional benefits consisting of natural ingredients.

   Renuka et al. (2009) studied the fortification of pineapple, mango, and orange juice with FOSs. Results indicated that sucrose, which is usually used as a sweetener in fruit juice beverages, can be partially substituted with FOS without significantly affecting the overall quality. The fruit juice beverages were evaluated for physicochemical and sensory changes during 6 month storage period at ambient (25°C) and refrigeration temperature (4°C). The pH, total soluble solids, titratable acidity, and color did not change significantly during storage. The initial FOS content of pineapple, mango, and orange juice beverages was 3.79, 3.45, and 3.62 g/100 mL, respectively. The FOS content of the fruit juice beverages stored at refrigeration temperature was 2.00–2.39 g/100 mL after 6 months of storage and 2.69–3.32, 1.65–2.08, and 0.38–0.58 g/100 mL at the second, fourth, and sixth months of storage at ambient temperature, respectively. The sensory analysis showed that the beverages were acceptable up to 4 and 6 months storage at ambient and refrigeration temperature, respectively. The study indicates that fruit juice beverages can successfully be fortified with FOS with shelf life of 4 and 6 months at ambient and refrigeration temperature, respectively. There were no undesirable changes in the physicochemical characteristics of the fruit juice beverages fortified with FOS. Constant pH, TSS, TA, and viscosity of fruit juice beverages clearly indicate that there is no spoilage either due to microbial or enzymatic reaction. The study has opened up a new avenue for the preparation of a commonly available, hugely popular healthy beverage.

10.3 FINAL REMARKS

The changes in the present day consumers' life style have led to a vital change in the marketing trends of food sector. It was mainly the advances in understanding the relationship between nutrition and health that resulted in the development of the concept of functional foods, which means a practical and new approach to achieve optimal health status by promoting the state of well-being and possibly reducing the risk of disease (Siró et al., 2008). International recognition of functional foods has resulted in recent development of this field, which resulted in a significant growth in the fruit juice market, which has attracted the attention of fruit growers, fruit juice distributors, and processors to meet the demand. Fruit juice beverages with neutraceuticals are now gaining more importance in the fruit juice market (Renuka et al., 2009). The consumption of foods and beverages containing functional prebiotics and probiotics is the current global consumer trend (Mark-Herbert, 2004; Verbeke, 2005).

Although the study on the use of fruit and fruit juices as vehicle for prebiotic oligosaccharides and probiotic microorganisms has increased lately, the success of using these matrices to delivery function ingredients is still a challenge. Microbial viability is strongly dependent of the food matrix, processing, and storage conditions. Thus, a strain might be suitable for some fruit and unviable for another. Each case must be extensively studied, and the use of the right processing and the right strain is the key for the product success and acceptance. For prebiotic oligosaccharides, the juice composition and the kind of the glycosidic bonds in the carbohydrate chain are the key factors. To avoid the hydrolysis of these oligosaccharides, both the chemical nature of the oligosaccharide and the juice pH might be taken into account in the product development.

REFERENCES

Aachary, A.A. and Prapulla, S.G. 2009. Value addition to corncob: Production and characterization of xylooligosaccharides from alkali pretreated lignin-saccharide complex using *Aspergillus oryzae* MTCC 5154. *Bioresource Technology*, 100: 991–995.

Agerholm-Larsen, L., Raben, A., Haulrik, N., Hansen, A.S., Manders, M., and Astrup, A. 2000. Effect of 8 week intake of probiotic milk products on risk factors for cardiovascular diseases. *European Journal of Clinical Nutrition*, 54: 288–297.

Akpinar, O., Ak, O., Kavas, A., Bakir, U., and Yilmaz, L. 2007. Enzymatic production of xylooligosaccharides from cotton stalks. *Journal of Agriculture and Food Chemistry*, 55: 5544–5551.

Alzamora, S.M., Salvari, D., Tapia, M.S., López-Malo, A., Welti-Chanes, J., and Fito, P. 2005. Novel functional foods from vegetable matrices impregnated with biologically active compounds. *Journal of Food Engineering*, 67: 205–214.

Banykó, J. 2007. Functional food in the Czech Republic. In *Proceedings of the Fourth International FFNet Meeting on Functional Foods*, March 26–27, Budapest, Hungary.

Benkouider, C. 2005. The world's emerging markets. Functional ingredients. Available on: http://newhope360.com/food/world-s-emerging-markets (accessed on June 2011).

Betoret, N., Puente, L., Díaz, M.J., Pagán, M.J., García, M.J., and Gras, M.L. 2003. Development of probiotic-enriched dried fruits by vacuum impregnation. *Journal of Food Engineering*, 56: 273–277.

Boddicker, J.D., Ledeboer, N.A., Jagnow, J., and Clegg, S. 2002. Differential binding to and biofilm formation on, HEp-2 cells by *Salmonella* enterica serovar *Typhimurium* is dependent upon allelic variation in the fimH gene of the fim gene cluster. *Molecular Microbiology*, 45: 1255–1265.

Brown, A.C., and Valiere, A. 2004. Probiotics and medical nutrition therapy. *Nutrition in Clinical Care*, 7: 56–68.

Cardarelli, H.R., Buriti, F.C.A., Castro, I.A., and Saad, S.M.I. 2008. Inulin and oligofructose improve sensory quality and increase the probiotic viable count in potentially symbiotic *petit-suisse* cheese. *LWT—Food Science and Technology*, 41: 1037–1046.

Champagne, C.P., Raymond, Y., and Gagnon, R. 2008. Viability of *Lactobacillus rhamnosus* R0011 in an apple-based fruit juice under simulated storage conditions at the consumer level. *Journal of Food Science*, 73: 221–226.

Charernjiratrakul, W., Kantachote, D., and Vuddhakul, V. 2007. Probioitc lactic acid bacteria for applications in vegetarian food products. *Songklanakarin Journal Science and Technology*, 29: 981–991.

Cleary, J., Lai, L.C., Shaw, R.K., Straatman-Iwanowska, A., Donneberg, M.S., Frankel, G., and Knutton, S. 2004. Enteropathogenic *Escherichia coli* (EPEC) adhesion to intestinal epithelial cells: Role of bundle-forming pili (BFP), EspA filaments and intimin. *Microbiology*, 150: 527–538.

Delzenne, N.M. 2003. Oligosaccharides: State of the art. *Proceedings of the Nutrition Society*, 62: 177–182.

Espinoza, Y.R., and Navarro, Y.G. 2010. Non-dairy probiotic products. *Food Microbiology*, 27: 1–10.

FAO/WHO Report of a Joint FAO/WHO Expert Consultation on Evaluation of Health and Nutritional Properties of Probiotics in Food Including Powder Milk with Live Lactic Acid Bacteria. 2001. In *Health and Nutritional Properties of Probiotics in Food including Powder Milk with Live Lactic Acid Bacteria*, October, Cordoba, Argentina. Food and Agriculture Organization of the United Nations, World Health Organization.

Fonteles, T.V., Costa, M.G.M., de Jesus, A.L.T., and Rodrigues, S. 2011. Optimization of the fermentation of Cantaloupe juice by *Lactobacillus casei* NRRL B-442. *Food and Bioprocess Technology*, in press (DOI 10.1007/s11947-011-0600-0).

Gibson, G.R. 2006. From probiotics to prebiotics and a healthy digestive system. *Journal of Food Science*, 69: M141–M143.

Gibson, G.R., Beatty, E.R., Wang, X., and Cummings, J.H. 1995. Selective stimulation of bifidobacteria in the human colon by oligofructose and inulin. *Gastroenterology*, 108: 975–978.

Gibson, G.R., Probert, H.M., Van Loo, J., Rastall, R.A., and Roberfroid, M.B. 2004. Dietary modulation of the human colonic microbiota: Updating the concept of prebiotics. *Nutrition Research Review*, 17: 259–275.

Gomes, A.M.P. and Malcata, F.X. 1999. *Bifidobacterium* spp. and *Lactobacillus acidophilus*: biological, biochemical, technological and therapeutical properties relevant for use as probiotics. *Trends in Food Science and Technology*, 10: 139–157.

Gotcheva, V., Hristozova, E., Hrostozova, T., Guo, M., Roshkova, Z., and Angelov, A. 2002. Assessment of potential probiotic properties of lactic acid bacteria and yeast strains. *Food Biotechnology*, 16: 211–225.

Granato, D., Branco, G.F., Nazzaro, F., Cruz, A.G., and Faria, J.A.F. 2010. Functional foods and nondairy probiotic food development: Trends, concepts, and products. *Comprehensive Reviews in Food Science and Food Safety*, 9: 1–11.

Hess, J.R., Birkett, A.M., Thomas, W., and Slavin, J.L. 2011. Effects of short-chain fructooligosaccharides on satiety responses in healthy men and women. *Appetite*, 56: 128–134.

Hilliam, M. 2000. Functional food—How big is the market? *The World of Food Ingredients*, 12: 50–52.

Holzapfel, W.H. and Schillinger, V. 2002. Introduction to pre-and probiotics. *Food Research International*, 35: 109–116.

Kalliomaki, M., Salminen, S., Arvilommi, H., Kero, P., Koskinen, P., and Isolauri, E. 2001. Probiotics in primary prevention of atopic disease: A randomized placebo controlled trial. *Lancet*, 357: 1076–1079.

Lapsley, K.G., Escher, F.E., and Hoehn, E. 1992. The cellular structure of selected apple varieties. *Food Structure*, 11: 339–349.

Lau, C.-S., Bunnell, K., Thoma, G.J., Lay, J.O., Gidden, J., and Carrier, D.J. 2011. Separation and purification of xylose oligomers using centrifugal partition chromatography. *Journal of Industrial Microbiology and Biotechnology*, 38: 363–370.

Lavermicocca, P., Valerio, F., Lonigro, S.L., De Angelis, M., Morelli, L., Callegari, M.L., Carlo, G., Rizzello, C.G., and Visconti, A. 2005. Study of adhesion and survival of lactobacilli and *bifidobacteria* on table olives with the aim of formulating a new probiotic food. *Applied and Environmental Microbiology*, 71: 4233–4240.

Luckow, T. and Delahunty, C. 2004. Consumer acceptance of orange juice containing functional ingredients. *Food Research International*, 37: 805–814.

Luckow, T., Sheehan, V., Fitzgerald, G., and Delahunty, C. 2006. Exposure, health information and flavour-masking strategies for improving the sensory quality of probiotic juice. *Appetite*, 47: 315–323.

Maguiña, G., Tapia, M.S., Briceño, A.G., Rodríguez, C., Sánchez, D., Roa, V., and López-Malo, A. 2002. Incorporación of *Bifidobacterium* spp. en una matriz porosa de fruta por el mecanismo hidrodinámico (PF-08). In *Actas del 2_Congreso Español de Ingenien'a de Alimentos*. September 18–20, 2002, Lleida, España: Universitat de Lleida, pp. 1–4.

Mäkinen-Aakula, M. 2006. Trends in functional foods dairy market. In *Proceedings of the Third Functional Food Net Meeting*, September 18–19, Liverpool, U.K.

Mark-Herbert, C. 2004. Innovation of a new product category—Functional foods. *Technovation*, 24: 713–719.

Menrad, K. 2000. Markt und Marketing von funktionellen Lebensmitteln. *Agrarwirtschaft*, 49: 295–302.

Menrad, K. 2003. Market and marketing of functional food in Europe. *Journal of Food Engineering*, 56: 181–188.

Metchnikoff, E. 1907. Essais optimistes. Paris. *The Prolongation of Life: Optimistic Studies*, P.C. Mitchell (transl. and ed.). London, U.K.: Heinemann.

Monaco, M.H., Kashtanov, D.O., Wang, M., Walker, D.C., Rai, D., Jouni, Z.E., Miller, M.J., and Donovan, S.M. 2011. Addition of polydextrose and galactooligosaccharide to formula does not affect bacterial translocation in the neonatal piglet. *Journal of Pediatric Gastroenterol Nutrition*, 52: 210–216.

Moure, A., Gulló, P., Domínguez, H., and Parajó, J.C. 2006. Advances in the manufacture, purification and applications of xylo-oligosaccharides as food additives and nutraceuticals. *Process Biochemistry*, 41: 1913–1923.

Mousavi, Z.E., Mousavi, S.M., Razavi, S.H., Emam-Djomeh, Z., and Kiani, H. 2011. Fermentation of pomegranate juice by probiotic lactic acid bacteria. *World Journal of Microbiology and Biotechnology*, 27: 123–128.

Nomoto, K. 2005. Review prevention of infections by probiotics. *Journal of Bioscience and Bioengineering*, 100: 583–592.

Ouwehand, A. 2007. Success in applying pro- and prebiotics in dairy products. In *Proceedings of the Fourth International FFNet Meeting on Functional Foods*, March 26–27, Budapest, Hungary.

Pereira, A.L.F., Maciel, T.C., and Rodrigues, S. 2011. Probiotic beverage from cashew apple juice fermented with *Lactobacillus casei*. *Food Research International*, 44: 1276–1283.

Prado, F.C., Parada, J.L., Pandey, A., and Soccol, C.R. 2008. Trends in non-dairy probiotic beverages. *Food Research International*, 41(2): 111–123.

Quintero, M., Perez-Munoz, M.E., Jimenez, R., Fangman, T., Rupnow, J., Wittke, A., Russell, M., and Hutkins, R. 2011. Adherence inhibition of *Cronobacter sakazakii* to intestinal epithelial cells by prebiotic oligosaccharides. *Current Microbiology*, 62: 1448–1454.

Rabelo, M.C., Fontes, C.P.M.L., and Rodrigues, S. 2009. Enzyme synthesis of oligosaccharides using cashew apple juice as substrate. *Bioresource Technology*, 100: 5574–5580.

Rabelo, M.C., Honorato, T.L., Gonçalves, L.R.B., Pinto, G.A.S., and Rodrigues, S. 2006. Enzymatic synthesis of prebiotic oligosaccharides. *Applied Biochemistry and Biotechnology*, 133: 31–40.

Ramnani, P., Gaudier, E., Bingham, M., van Bruggen, P., Tuohy, K.M., and Gibson, G.R. 2010. Prebiotic effect of fruit and vegetable shots containing Jerusalem artichoke inulin: A human intervention study. *British Journal of Nutrition*, 104: 233–240.

Ray, R.C., and Sivakumar, P.S. 2009. Traditional and novel fermented foods and beverages from tropical root and tuber crops—A review. *International Journal of Food Science and Technology*, 44: 1073–1087.

Renuka, B., Kulkarni, S.G., Vijayanand, P., and Prapulla, S.G. 2009. Fructooligosaccharide fortification of selected fruit juice beverages: Effect on the quality characteristics. *LWT—Food Science and Technology*, 42: 1031–1033.

Rettger, L.F., 1935. *Lactobacillus acidophilus: Its therapeutic application*. Yale University Press, London.

Roberfroid, M.B. 2007. Prebiotics: The concept revisited. *Journal of Nutrition*, 137: 830S–837S.

Rodrigues, S., Lona, L.M.F., and Franco, T.T. 2005. The effect of maltose on dextran yield and molecular weight distribution. *Bioprocess and Biosystems Engineering*, 28: 9–14.

Rodrigues, S., Lona, L.M.F., and Franco, T.T. 2006. Optimizing panose production by modeling and simulation using factorial design and surface response analysis. *Journal of Food Engineering*, 75: 433–440.

Rodríguez, M.I. 1998. Estudio de la penetración de microorganismos en frutas mediante el modelo Hidrodinámico (HDM). Thesis. Instituto de Ciencia y Tecnología de Alimentos, Universidad Central de Venezuela, Caracas, Venezuela.

Rößle, C., Auty, M.A.E., Brunton, N., Gormley, R.T., and Butler, F. 2010a. Evaluation of fresh-cut apple slices enriched with probiotic bacteria. *Innovative Food Science and Emerging Technologies*, 11: 203–209.

Rößle, C., Bruntona, N., Gormleyb, R.T., Rossc, P.R., and Butlerd, F. 2010b. Development of potentially synbiotic fresh-cut apple slices. *Journal of Functional Foods*, 2: 245–254.

Sako, T., Matsumoto, K., and Tanaka, R. 1999. Recent progress on research and applications of non-digestible galacto-oligosaccharides. *International Dairy Journal*, 9: 69–80.

Salminen, S. and Von Wrigh, A. (Eds.). 1993. *Lactic Acid Bacteria*. Marcel Dekker Inc., New York.

Saulnier, D.M.A., Spinler, J.K., Gibson, G.R., and Versalovic, J. 2009. Mechanisms of probiosis and prebiosis: Considerations for enhanced functional foods. *Current Opinion in Biotechnology*, 20: 135–141.

Shah, N.P. 2001. Functional foods from probiotics and prebiotics. *Food Technology*, 55: 46–53.

Shah, N.P. 2007. Functional cultures and health benefits. *International Dairy Journal*, 17: 1262–1277.

Sharon, N. and Ofek, I. 2002. Fighting infectious diseases with inhibitors of microbial adhesion to host tissues. *Critical Reviews on Food Science and Nutrition*, 42: 267–272.

Sheehan, V.M., Ross, P., and Fitzgerald, G.F. 2007. Assessing the acid tolerance and the technological robustness of probiotic cultures for fortification in fruit juices. *Innovative Food Science and Emerging Technologies*, 8: 279–284.

Shoaf-Sweeney, K. and Hutkins, R.W. 2009. Adherence, anti-adherence, and oligosaccharides: Preventing pathogens from sticking to the host. *Advances in Food Nutrition Research*, 55: 101–161.

Siró, I., Kálpona, E., Kálpona, B., and Lugasi, A. 2008. Functional food. Product development, marketing and consumer acceptance: A review. *Appetite*, 51: 456–467.

Szakály, S. 2007. Development and distribution of functional dairy products in Hungary. In *Proceedings of the Fourth International FFNet Meeting on Functional Foods*, March 26–27, Budapest, Hungary.

Tannock, G.W. 2003. Probiotics: Time for a dose of realism. *Current Issues in Intestinal Microbiology*, 4: 33–42.

Tapia, M.S., Rojas-Graü, M.A., Rodríguez, F.J., Ramírez, J., Carmona, A., and Martin-Belloso, O. 2007. Alginate- and gellan-based edible films for probiotic coatings on fresh-cut fruits. *Journal of Food Science*, 72: 190–196.

Tuorila, H. and Cardello, A.V. 2002. Consumer responses to an off-flavor in juice in the presence of specific health claims. *Food Quality and Preference*, 13: 561–569.

Vazquez, M.J., Alonso, J.L., Dominguez, H., and Parajó, J.C. 2000. Oylooligosaccharides: Manufacture and applications. *Trends in Food Science and Technology*, 11: 387–393.

Verbeke, W. 2005. Consumer acceptance of functional foods: Sociodemographic, cognitive and attitudinal determinants. *Food Quality and Preference*, 16: 45–57.

Vergara, C.M.A., Honorato, T.L., Maia, G.A., and Rodrigues, S. 2010. Prebiotic effect of fermented cashew apple (*Anacardium occidentale L*) juice. *LWT—Food Science and Technology*, 43: 141–145.

Vinderola, C.G. and Reinheimer, J.A. 2000. Enumeration of *Lactobacillus casei* in the presence of *L. acidophilus*. Bifidobacteria and lactic starter bacteria in fermented dairy products. *International Dairy Journal*, 10: 271–275.

Voragen, A.G.J. 1998. Technological aspects of functional food related carbohydrates. *Trends in Food Science and Technology*, 9: 328–335.

Wang, C.Y., Ng, C.C., Su, H., Tzeng, W.S., and Shyu, Y.T. 2009. Probiotic potential of noni juice fermented with lactic acid bacteria and bifidobacteria. *International Journal of Food Science and Nutrition*, 1: 1–9.

Yang, C.H., Yang, S.F., and Liu, W.H. 2007. Production of xylooligosaccharides from xylans by extracellular xylanases from *Thermobifida fusca*. *Journal of Agriculture and Food Chemistry*, 55: 3955–3959.

Yoon, K.Y., Woodams, E.E., and Hang, Y.D. 2004. Probiotication of tomato juice by lactic acid bacteria. *Journal of Microbiology*, 42: 315–318.

Yoon, K.Y., Woodams, E.E., and Hang, Y.D. 2005. Fermentation of beet juice by: Beneficial lactic acid bacteria. *Lebensmittel-Wissenschaft und Technologie* 38: 73–75.

Yoon, K.Y., Woodams, E.E., and Hang, Y.D. 2006. Production of probiotic cabbage juice by lactic acid bacteria. *Bioresource Technology*, 97: 1427–1430.

Zorf, D. and Roth, S. 1996. Oligosaccharides anti-infection agents. *Lancet*, 347: 1017–1021.

11 Freeze Concentration Applications in Fruit Processing

Mercé Raventós, Eduard Hernández,
and Josep Maria Auleda

CONTENTS

11.1 INTRODUCTION

In conventional processes, such as evaporation, higher levels of concentration can be reached compared with freeze concentration or membrane techniques. However, the advantage of the freeze concentration technique is based on the quality of the product obtained due to the low temperatures used in the process which makes it a very suitable technology for processing of fruit juices. There are two basic methods for concentrating solutions by freezing: suspension and film freeze concentration. Suspension freeze concentration system (FCS) already has operating equipment in the food industry, while film FCS, also called layer crystallization, is still at an experimental stage.

The growing demand for fruit juices of high organoleptic and nutritional quality has led to the search for new or improved food processing technologies. Among the techniques for concentration of liquid foodstuffs, freeze concentration is of particular interest due to the low temperatures used in the process.

Freeze concentration is a technology that can be used in the food processing industry to concentrate fruit juices (Rahman et al., 2006). This process allows removal of water from a solution by cooling or freezing it until high-purity ice crystals are formed and separated to leave a concentrated fluid. The nutritional and sensory quality of freeze-concentrated fruit juices is higher than those concentrated conventionally by means of evaporation due to the low processing temperatures that avoid undesirable chemical and biochemical changes and minimize the loss of organoleptic properties. Some authors have summarized the studies referred to the concentration of fruit juices by cryoconcentration; however, the subject is treated in general making comparisons with other methods for concentration of juice (Ramteke et al., 1993) and with emphasis on the description and operation of equipment involved in the process (Deshpande et al., 1984).

This chapter summarizes the most important studies relating to the suspension and film freeze concentration in fruit juices illustrating the different possibilities that freeze concentration has in fruit juices industry, presents trends, and suggests improvements for the future development of this technology. It is noted that most recent publications refer to the film FCS. The technology used to design, build, and maintain layer crystallization equipment is simple; and it can be available to any operator in the food industry. Layer system will be used in future if its results can be improved in terms of ice purity and degree of fluid concentration.

11.2 FREEZE CONCENTRATION METHODS

According to various researchers (Müller and Sekoulov, 1992; Flesland, 1995; Chen et al., 1998; Miyawaki, 2001; Wakisaka et al., 2001), there are two basic methods for ice crystal formation in solutions. The first is known as suspension crystallization (Figure 11.1a) (Huige and Thijssen, 1972; Hartel and Espinel, 1993), consisting of an initial phase of ice nuclei formation (nucleation), also called crystallization, followed by a second phase which involves the growth of ice nuclei in the solution. The second method is the crystallization of water present in the solution in the form of an ice layer on a cold surface (Müller and Sekoulov, 1992; Flesland, 1995) (Figure 11.1b).

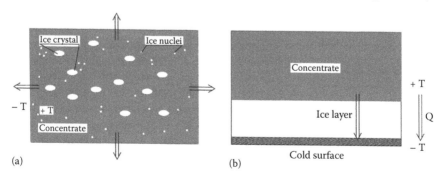

FIGURE 11.1 Two methods for concentration by freezing: (a) suspension and (b) film. The direction of arrows represent heat transfer.

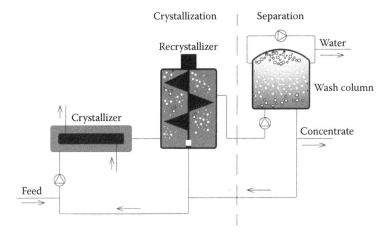

FIGURE 11.2   Schematic suspension FCS.

In industry, FCS consists of three stages: crystallization, growth, and separation of ice crystals, performed with specially designed equipment for each purpose. The system comprises the following equipment (Jansen et al., 2001; Lemmer et al., 2001; Verschuur et al., 2002; van Nistelrooij, 2005): scraped surface heat exchanger (SSHE) (Figure 11.2) to form ice nuclei at high supercooling and low residence times. Ice nuclei are formed on the inner surface of the heat exchanger and then are scraped off by rotating blades. The ice nuclei go to the recrystallizer (Figure 11.2) for ice crystals growth based on the Gibbs–Thomson effect, and then a separation of ice crystals from the concentrate occurs normally in a pressurized wash column (Figure 11.2). In this system, ice crystals of high purity can be attained (Thijssen, 1986).

The second system, film freeze concentration, consists of formation of a single crystal, which grows layer by layer from the solution to be concentrated. The crystal growth (dendrites) tends to be parallel and opposite to the direction of heat transfer (Flesland, 1995). The crystal adheres to the cold surface during the process, facilitating separation of the two phases (Figure 11.3).

This concentration system is based on directional freezing, and the most important crystal form is the dendrite (Flesland, 1995; Chen et al., 1998; Chen and Chen, 2000; Pardo et al., 2002; Gu et al., 2005; Caretta et al., 2006). Heat transfer rates are normally greater than mass transfer rates, due to the high thermal conductivity of ice and low mass diffusion coefficients. Therefore, solute diffusion will be the limiting factor for ice growth, and supercooling (constitutional supercooling) in the tip region will be observed (Rutter and Chalmers, 1953; Ozüm and Kirwan, 1976; Teraoka et al., 2002; Hindmarsh et al., 2005; Ayel et al., 2006). Solute inclusion in ice is difficult to avoid in practical applications, especially for solute concentrations of commercial interest for freeze concentration, which means between 20% and 50% of dissolved solids. The two methods described for the formation of ice in solutions differ in terms of heat extraction, ice growth rate, ice purity, equipment, industrial process, solid–liquid contact surface, and necessary investment (Table 11.1).

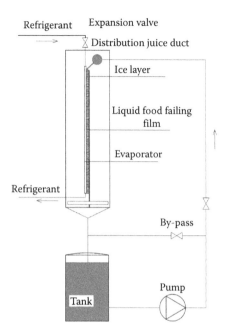

FIGURE 11.3   Schematic film FCS. (From Raventós, M. et al., *J. Food Eng.,* 79, 577, 2007.)

TABLE 11.1
Indicative Data of Film and Suspension Crystallization Systems

	Film Crystallization	Suspension Crystallization
Heat extraction	Through ice layer	Through solution
Ice growth rate (m/s)	$10^{-6}-10^{-7}$	$10^{-7}-10^{-8}$
Ice purity	Low	High
Equipment	No moving parts except pumping equipment	Moving parts in all the equipment are needed
Industrial process	Discontinuous	Continuous
Solid–liquid contact surface	Low	High
Necessary investment (%)	100	150–250

Source: Sánchez, J. et al., *Food Sci. Technol. Int.,* 15(4), 303, 2009.

11.3   SUSPENSION FREEZE CONCENTRATION

Research in FCS has focused on two issues: control of nucleation and growth of ice crystals to obtain large ice crystals, preferably of uniform size, and to separate ice crystals selectively from the concentrate. This system requires separate stages for nucleation and growth of ice crystals since the optimal operating condition requirements for these distinct crystallization phenomena can be significantly different.

The studies using juices have a wide range of aims, including examination of the basis of the process and determination of the organoleptic quality of the juices obtained. The final concentrations attained by this method vary between 45 °Bx and 55 °Bx.

11.3.1 Crystallization

The study of FCS has shown that in fruit juices and sucrose solutions, the most important category of nucleation is secondary rather than primary nucleation. The velocity of nucleation and rate of growth depend on the supercooling of the system, and there is a model based on a power law that can be used to represent the relationship between those variables for sucrose solutions, orange juice, and apple juice. Several studies also found that crystal growth at high supercooling occurs dendritically or with a needle shape, which gives rise to a larger surface area and greater difficulty of separation from the concentrated solution. In highly concentrated sucrose solutions, the nucleation rate is independent of the concentration of the solution, and as the concentration of a sucrose solution increase, the rate of crystal growth declines due to greater viscosity.

Huige and Thijssen (1972) developed the basic knowledge of the FCS process involving supersaturation in a crystallizer with continuous seeding of small crystals that dissolved and promoted the growth of larger crystals already present in the crystallizer (Ostwald ripening mechanism).

Another method of increasing the size of ice crystals is using agglomeration in order to improve the solid–liquid separation. Kobayashi et al. (1996) concluded from storage experiments that agglomeration of ice crystals mainly depends on the seed crystals, the initial crystal size distribution, and the concentration of solute. Extensive agglomeration of ice crystals was observed in glucose solutions with concentrations of 10% and lower, but not in solutions with concentrations of 20% and 30%.

Pronk et al. (2002, 2005) confirms that the Ostwald ripening mechanism in ice suspensions is the most important for crystal size. Agglomeration is observed in some cases but this mechanism plays a minor role. Bayindirli et al. (1993) (Table 11.2) carried out studies for the formulation of a mathematical model to describe the freeze concentration of apple juice, concluding that the kinetic of the cryoconcentration process fits to a sigmoidal curve. A similar behavior of the concentration kinetics can be found in the work of Nonthanum and Tansakul (2008) during the process of cryoconcentration with lime juice. Chiampo and Conti (2002) (Table 11.2) present the results obtained in a freeze concentration pilot plant of Niro Process Technology, using strawberry juice and other kinds of sugar solutions (sucrose solutions at different initial concentrations); it was found that for an optimal functioning of the equipment, the maximum amount of ice in the recrystallizer must not exceed 40% and secondly that the average speed of ice growth is lower in the strawberry juice than in the sugar solutions. The ice productivity of the equipment is lower with strawberry juice than with the sugar solutions with a maximum of $150\,kg/h/m^3$.

Freeze concentration experiments of black currant juice were performed in a pilot plant of Sulzer Chemtech Ltd. (Dette and Jansen, 2010). For the experiments,

TABLE 11.2
Studies That Include Suspension Freeze Concentration of Fruit Juices

Fluid, Author, and Year of Publication	Aim of Study	Equipment	Results
Pineapple juice, Braddock and Marcy (1985)	To examine pilot scale processing parameters and product quality aspects of freeze concentration applied to fresh pineapple juice	Suspension pilot equipment (Model W-8, Grenco Process Technology)	• Flavor of reconstituted freeze concentrated juice was comparable to single strength juice and preferable to evaporator concentrated juice
Orange juice, Braddock and Marcy (1987)	To determine the effects of heat inactivation of enzymes and pulp content on quality of freeze concentrated orange juice	Suspension pilot equipment (Model W-8, Grenco Process Technology B.V.)	• Except for considerable pulp reduction of feedstream juices, there were few differences from normal citrus juice recovery procedures for freeze concentration (the product retained most of the aroma constituents of fresh juice)
Orange juice, Thijssen (1986)	Grenco freeze concentration is compared with evaporation and hyperfiltration	Grenco Process Technology B.V. cryoconcentrator	• Freeze concentration appears as an attractive method: conservation of nutritional and organoleptic characteristics of fresh product, the net added value must be based on the packaged product (production costs make a relatively small contribution of the processing costs to the total costs of the packaged product)
Apple juice, Bayindirli et al. (1993)	To study apple juice cryoconcentration in a single-stage process and a multistage process	Hemispherical porcelain containers	• At the end of the single-stage experiments, only a double concentration could be achieved • With multistage experiments, higher concentration levels could be achieved

TABLE 11.2 (continued)
Studies That Include Suspension Freeze Concentration of Fruit Juices

Fluid, Author, and Year of Publication	Aim of Study	Equipment	Results
Sugar cane juice and sucrose solution, Patil (1993)	To examine freezing points, viscosities, and changes with concentration of raw and clear cane juices	Crystallizer cell with a cooling jacket	• The freezing points of raw juices are lower than those of sucrose solutions • The maximum juice concentration was limited to about 54 °Bx • The decrease in crystal size with increasing concentration and the high viscosities observed would make it impractical to use freezing to increase the Brix value beyond about 50 °Bx
Strawberry juice and sugar solutions, Chiampo and Conti (2002)	To present the results obtained in freeze concentration of fruit juices at a pilot plant	Pilot plan implemented for Niro Process Technology B.V.	• Speed of ice growth in strawberry juice is lower than the speed obtained for sucrose–water solutions • Maximum specific yield, dependent on the concentrated product (150 kg/h/m³)
Lime juice, Nonthanum and Tansakul (2008)	To study the effects of processing conditions on the freeze concentration of lime juice	Batch crystallizer with a cylindrical vessel with a scraper blade	• The processing condition with −18°C cooling medium temperature and 150 rpm rotational speed of the scraper blade was the best among all studied conditions • The loss of the soluble solids with ice crystals during ice separation was relatively high at 35%
Black currant juice, Dette and Jansen (2010)	To study the effects of processing conditions and organoleptic quality on the freeze concentration of black currant in a pilot plant	Pilot plan implemented for Sulzer Chemtech Ltd. For the experiments, a crystallizer (0.2 m²), a growth vessel (0.6 m³), and a wash column are used	• Under steady-state conditions, water production is approximately 10 kg/h • The solid concentration of black currant juice is increased to 39 °Bx from 14.1 °Bx

a crystallizer (0.2 m²), a growth vessel (0.6 m³), and a wash column are used. Under steady-state conditions, water production is approximately 10 kg/h. The solid concentration of black currant juice is increased to 39 °Bx from 14.1 °Bx.

11.3.2　SEPARATION OF ICE CRYSTALS

The separation of ice crystals from concentrated fruit juices can be performed using wash columns, operating either in batch or in continuous mode. Wash columns have been developed to the point where the solute inclusions have been reduced to less than 0.10 kg/m³. Several types of wash columns have been developed and put into industrial application, and they differ in the various crystal bed transport mechanisms (Verdoes et al., 1997). The crystal can be transported by gravity, mechanically (piston/screw), or hydraulic pressure; the last two are known as wash columns with forced transport. For more than two decades, the gravity-type units were the only commercially available ones; problems with scale-up and the effect of back mixing on product purities have limited their commercial application. Forced transport wash columns are smaller and operate with short residence times; crystals obtained are relatively pure and strong enough so that a wash column with forced transport will be preferred over a gravity wash column (Scholz et al., 2002).

The most commonly used wash column for FCSs in the fruit juices industry is the piston-type wash column in countercurrent washing (van Nistelrooij, 2005; Morison and Hartel, 2006).

An important factor is the viscosity of the concentrate because the capacity of ice separators is inversely proportional to the viscosity and directly proportional to the square of the mean diameter of the crystals, as expressed by

$$Q = \frac{\Delta P g d_e^2}{0.2 \mu l} \cdot \frac{\varepsilon^3}{(1-\varepsilon)^2} \tag{11.1}$$

where
　　Q is the draining rate from the crystal bed (cm³/cm²s)
　　ΔP is the pressure difference exerted over the bed by compression or by centrifugal or pressure drop of the filtrate (kg/cm²)
　　d_e is the diameter of the crystals (cm)
　　μ is the viscosity of the liquid (P)
　　l is the thickness of the bed (cm)
　　g is the gravity acceleration (cm/s²)
　　ε is the volume fraction of pores in the bed filled by the liquid phase

11.4　FILM FREEZE CONCENTRATION

A wide variety of liquid foods have been studied in film freeze concentration. In the case of juices, the maximum concentrations attained are around 30 °Bx and with sugar solutions 54 °Bx. There are no references in the papers examined of the use of this sort of equipment in the food industry. There are two film freeze concentration techniques that have been studied in the papers examined: layer freeze concentration and progressive freeze concentration. The difference between these two techniques lies in the equipment used for the formation of ice layer.

Progressive freeze concentration involves the crystallization at the bottom or sides of a vessel or in a pipe. In film FCS, the crystallization occurs on a plate.

11.4.1 Layer Freeze Concentration

In layer freeze concentration, the solution to be concentrated is in contact with a cold surface which consists of a cooled vertical plate on which the fluid descends; ice forms a single layer on the cold surface, and the solution is concentrated continuously throughout the whole procedure. Müller and Sekoulov (1992) suggest that layer freezing process is easier to manage, but when the crystal grows on a cooled surface, this induces a rapid rate of crystallization, and under these conditions, impure ice crystals can be produced. In spite of this moderate ice purity, the disadvantages are compensated by the simplicity of the operation, because there are no moving parts in the equipment and no slurry handling.

Flesland (1995) has been studying the layer crystallization in laminar falling film ($50 < Re < 600$) for freeze concentration applications, using sucrose solutions. Ice growth rates and impurities in the ice or loss of solute are measured for varying conditions. Experimental results show that one step layer crystallization has a lower separation effect than suspension crystallizers with wash columns. The solute inclusions could be reduced by increasing turbulent conditions and by adding more steps and recirculating melt water with high sucrose concentration.

Chen et al. (1998) and Chen and Chen (2000) studied experimentally the solute inclusion in an ice layer on a smooth stainless steel surface under subcooled flow conditions in sucrose and fructose solutions and in orange juice. The average ice growth rate, the solution velocity, the bulk solute/solid concentration, and type of solute have been shown to be the four key factors in determining average distribution coefficient (solute distribution between the ice and the liquid phase). It was concluded that the ice obtained is more pure at low concentrations of soluble solids, low growth rates of ice, and high speeds of the solution on the plate. In solutions with a single component, larger solutes (higher molecular weight) are less likely to be trapped in the ice and produce a more pure ice. No data are provided in the behavior of solutes in solutions with various components, such as fruit juices.

An application at pilot scale of that type of process is presented by Raventós et al. (2007) and Hernández et al. (2009) (Table 11.3), who studied the process of concentration of aqueous solutions of sugars (glucose, fructose, and sucrose) and apple and pear juice in a layer freeze concentrator. In general, a higher degree of concentration was attained in the sucrose solutions in a shorter time in comparison with the results for the glucose and fructose solutions. The best results in terms of purity of the ice were found at low initial concentrations of the sucrose, glucose, and fructose solutions. The kinetic of the cryoconcentration process fits to a linear function in all cases.

TABLE 11.3

Studies That Include Film Freeze Concentration and Other Techniques with Juices

Fluid, Author and Year of Publication	Aim of Study	Equipment	Results
Tomato juice, Liu et al. (1999)	To apply progressive freeze concentration to tomato juice and evaluate its effect on juice quality	Progressive freeze concentrator	• No substantial differences in acidity, vitamin C content, and color quality after reconstitution of freeze concentrated tomato juice compared with unconcentrated juice
Andes berry pulp (*Rubus glaucus Benth*), Ramos et al. (2005)	To apply progressive freeze concentration to Andes berry (*Rubus glaucus Benth*) pulp to study its effects on free volatile composition and concentration	Progressive freeze concentrator Halde (1979) and Liu et al. (1998)	• The color of Andes berry (*Rubus glaucus Benth*) pulp is preserved during the concentration process • This treatment does not change the flavor compounds and even intensifies aroma
Tomato juice, sucrose solution, and coffee extract, Miyawaki et al. (2005)	To apply a tubular ice system to concentrate liquid food. To explore the concentration limit by this method	Tubular ice system (two straight pipes)	• Tubular ice system gave extremely high concentrations with good yields
Sugarcane juice, Rane and Jabade (2005)	To propose a heat pump–based FCS to concentrate sugarcane juice from 20 to 40 °Bx	Heat pump–based FCS (uses layer freezing process)	• Use of heat pump facilitates rejection of a major part of the condenser heat at about 10°C while melting the ice • Bagasse saving of 1338 kg/day or for 1000 kg jaggery can be achieved
Apple juice, Nazir and Farid (2008)	To estimate the erosion rate at different operating conditions with an empirical model based on an analogy between erosion and heat-transfer phenomenon	4 and 5 mm equilateral cylindrical particles made of stainless steel (SS304) fluidized in the inside pipe of a vertical double pipe heat exchanger	• It is possible to carry out freeze concentration of fruit juices in FBHE • The erosion model predicts stable operating limits for the fluidized bed at different bed porosities, particle sizes and concentrations of apple juice

TABLE 11.3 (continued)
Studies That Include Film Freeze Concentration and Other Techniques with Juices

Fluid, Author and Year of Publication	Aim of Study	Equipment	Results
Apple and pear juice, Hernandez et al. (2009)	Different parameters were studied to allow tracking of the process of freeze concentration, such as ice accumulation, variation of the content of soluble solids in the solution and in the ice removed, ice production, and energy consumption	Pilot plant of descending film multiplate freeze concentrator	• Concentrations of 30.2 and 30.8 °Bx with the apple and pear juices, respectively, were obtained • Energy consumption (Wu) has been calculated finding values between 0.26 and 0.30 kWh/kg of juice
Must juice, Hernández et al. (2010)	The process for freeze concentration of white must from Macabeo variety grapes	Pilot plant of descending film multiplate freeze concentrator	• A final concentration of 29.5 °Bx was obtained • An average rate of 1.38 °Bx/h was obtained working at flow rates of 0.8 L/s • The ice production ranged from 1.32 to 1.05 g/m²/s as the concentration of the must increased

Previous works are focused on ice growth rate, solution velocity, concentration, and type of solute, being a dominant factor in the efficiency to remove impurities in the ice. However, it was found that the crystal orientation is another very important factor to eliminate impurities in the ice (Kramer et al., 1971; Okawa et al., 2009). In the study of Okawa et al. (2009), the mechanism of the difference in trap between kinds of crystal orientations was discussed with the existence of a constitutional supercooling.

11.4.2 PROGRESSIVE FREEZE CONCENTRATION

Progressive freeze concentration has been carried out using vertical equipment and tubular equipment.

11.4.2.1 Vertical Progressive Freeze Concentration

A vertical progressive FCS consists of a cylindrical receptacle, a cooling bath, a system for immersing the receptacle in the bath, an agitator for the solution in the ice–solution interface, and an external heating blanket to control the level of ice formed in the receptacle and regulate the growth of the crystal (Figure 11.4).

The variables of the process studied using this type of equipment are the type of solution, rate of immersion or rate of ice growth, rate of agitation, and mechanisms to reduce supercooling. In progressive freeze concentration, the distribution coefficient (K) of solute between the ice and the liquid phase is most important. It is defined as

$$K = \frac{C_S}{C_L} \tag{11.2}$$

where C_S and C_L are the solute concentration in ice and liquid phases, respectively. The value of the partition coefficient K changes between 0 (ideal freeze concentration) and 1 (no concentration).

Miyawaki et al. (1998) showed that the distribution coefficient K is an experimentally observable partition coefficient of solute between ice phase and bulk liquid phase. This was strongly dependent on the advance rate of the ice front and the mass transfer at the ice–liquid interface determined by stirring rate (Liu et al., 1997). The effect of these operating conditions on the partition constant in the progressive freeze-concentration was theoretically analyzed by using a concentration

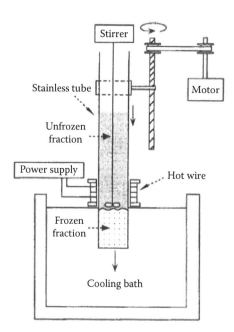

FIGURE 11.4 Apparatus for progressive freeze concentration. (From Miyawaki, O. et al., *J. Food Sci.*, 63, 756, 1998.)

polarization model (Burton et al., 1953). The following equation (Equation 11.3) was developed to describe the partition coefficient:

$$K = \frac{K_0}{[K_0 + (1 - K_0)\exp(-\mu/k)]} \tag{11.3}$$

where
K_0 is limiting partition coefficient at the ice–liquid interface
μ is advance rate of ice front
k is mass transfer coefficient at the interface

K_0 corresponds to the value of the partition coefficient at an infinitesimal advance rate of ice front and/or at infinite mass transfer coefficient because under these conditions $K = K_0$.

Miyawaki et al. (1998) has experimentally determined the mass transfer coefficient, and it has been expressed as a function of stirring speed (N) (Equation 11.4):

$$k = aN^b \tag{11.4}$$

where a and b are constants experimentally determined.

Note that these equations (Equations 11.2 through 11.4) are valid only if K is constant during the concentration process (i.e., quasi-steady state), a condition reached if the volume of solution is sufficiently large.

Progressive freeze concentration has been applied to solutions of glucose, tomato juice, raspberry (*Rubus glaucus*) pulp, and aqueous solutions of sucrose. The results of the studies show that the performance of freeze concentration is strongly influenced by the type of compound present in the solution. Liu et al. (1999) and Halde (1979) found that small solutes are retained in the ice more easily than the larger ones. Moreover, Kawasaki et al. (2006) noted that the solutes of small molecular weight were separated and concentrated more effectively than solutes of larger molecular weight; this corresponded well to the magnitude of the diffusion coefficient of each solute.

On the subject of rate of growth of the ice, Liu et al. (1997) found that a slower rate, on the order of 0.5 cm/h, gives lower values for the distribution coefficient between ice and liquid phases K (0–0.1) in solutions of sucrose. Likewise, Ramos et al. (2005) (Table 11.3), using a similar equipment to that described by Liu et al. (1997), obtained lower purity in the solid phase with an increase in the rate of growth of the ice.

Vigorous agitation of the ice–solution interface prevents the retention of solutes in the ice due to the elimination of the concentration and temperature gradient on the interface. Halde (1979) found that a faster rate agitation in the interface resulted in greater purity of the ice formed with aqueous solutions of glucose. This is similar to the results obtained by Liu et al. (1997) with solutions of glucose, where a faster rate of agitation in the interface (1400 rpm) gave lower values for the distribution coefficient K (0–0.1), and by Liu et al. (1999) with tomato juice, which showed that a faster rate of agitation gave lower concentrations of solids in the ice phase, with the corresponding increase of concentration in the solution. Matsuda et al. (1999) investigated the effects of supersonic radiation on the efficiency (distribution coefficient) of separation and concentration of glucose solutions under various freezing rates. They found that for equal mass concentration, efficiency was greatly improved using supersonic radiation because of the effect of turbulence in the liquid phase by supersonic cavitation. The most recent research by Kawasaki et al. (2006) concluded that concentration efficiency improved with increasing intensity of supersonic irradiation.

With the aim of reducing the likelihood of dendritic growth from the outset of the formation of the solid phase, Liu et al. (1998) proposed a plate with small holes at the bottom of the receptacle for a faster nucleation of the solution trapped in the holes. The result obtained was that in the freeze concentration of a 5% solution of glucose, the use of the plate with holes made it possible to obtain ice with glucose retention 57% lower than in the ice obtained without the use of the plate.

11.4.2.2 Tubular Progressive Freeze Concentration

To increase the productivity of progressive freeze concentration, a tube ice system was proposed by Shirai et al. (1999).

Miyawaki et al. (2005) (Table 11.3) showed that a tubular ice system with a large cooling surface area is an effective method for freeze concentration of tomato juice and sucrose solutions, with an increased productivity and high yield. The system comprises two connected sleeved tubes; the solution circulates inside the tube while the coolant circulates outside. The solid phase is generated on the inner walls of the tubes, and the concentrated solution is recirculated and flows through the annulus that has not yet frozen. In this method, the slower growth rate of ice and the

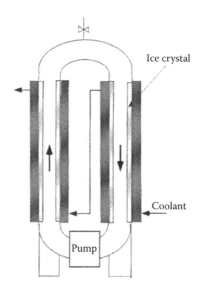

FIGURE 11.5 Tubular ice system for scale-up of progressive freeze concentration. (From Miyawaki, O. et al., *J. Food Eng.*, 69, 107, 2005.)

higher circulation rate gave the lower effective partition constant as was expected theoretically by the concentration polarization model. The effective partition constant was also dependent on the initial solute concentration; the higher concentration gave the higher partition constant of solute. By tubular ice system (Figure 11.5), coffee extract, tomato juice, and sucrose solution were very effectively concentrated to high concentrations with good yields, showing the high potential of progressive freeze concentration for practical applications.

Not all applications included in Tables 11.2 and 11.3 correspond faithfully to one of the two systems described; each application has been included in the system that best represents thought.

11.5   ORGANOLEPTIC QUALITY

Studies performed in the 1980s and early 1990s focused on assessment of the quality of the products obtained as in the studies with pineapple and orange juice carried out by Braddock and Marcy (1985, 1987) (Table 11.2), who compared the aroma and flavor of fresh juice with those of freeze concentrated juice and vacuum evaporation concentrated juice, finding that the freeze concentrated juice preserved its organoleptic qualities better in comparison with the juice concentrated by evaporation.

During the 1990s and the early years of the new century, there were references using fruit juices; some aimed at studying the organoleptic qualities of juices obtained through FCS (Van Weelden, 1994; Lee and Lee, 1998). Lee and Lee (1998) examine changes in quality during the refrigerated storage of clarified pear juice at 10 °Bx obtained by vacuum evaporation, reverse osmosis, and cryoconcentration. The results indicate that after 10 days of storage, there are not significant differences in browning and turbidity. Juices obtained by reverse osmosis and cryoconcentration show a similar sensory quality, and it is superior to that obtained by evaporation.

Concerning the effect of progressive freeze concentration on the characteristics of the fruit juices, Liu et al. (1999) (Table 11.3) showed that freeze concentrated and reconstituted tomato juice showed no material difference in acidity, vitamin C content, or color in comparison with unconcentrated tomato juice. The paper by Ramos et al. (2005) (Table 11.3) assessed the effect of progressive freeze concentration on the retention of compounds of the flavor of raspberry (*Rubus glaucus*) pulp. That paper determined a 20% loss of volatiles during the freeze concentration process. The sensorial analysis, performed by a panel of trained tasters, found that untreated pulp, enzyme-clarified pulp, and freeze concentrated and reconstituted pulp did not show any material difference in appearance, color, taste, aroma, or overall quality.

Freeze concentration technology is very suitable for the concentration of the sensitive juices. The quality of the freeze concentrated juices is excellent, and the taste has to be measurably better than evaporative concentrates, because the aroma, as well as the flavor, remains in the juice.

Ultimately there is no loss of volatiles, making this technique very suitable for the concentration of thermosensitive fluids.

11.6 ECONOMICS IN FREEZE CONCENTRATION

11.6.1 INVESTMENT COSTS

The investment costs of commercially available freeze concentration units with dewatering capacities of 10–15 ton/h are factors 3–4 higher than the costs of evaporation. The investment costs of membrane process are somewhat lower than that of evaporation with aroma recovery equipment.

Currently, freeze concentration is carried out using an SSHE to generate ice crystals which accounts for roughly 30% of the total investment costs in a freeze concentration plant. Fluidized bed freeze concentration is a novel technique that uses a fluidized bed heat exchanger (FBHE) to concentrate liquids through the process of freeze concentration. Ice formed on the cooled surface of the vertical heat exchanger is removed by particles fluidized inside the FBHE (Habib and Farid, 2006). Due to its simplicity in construction, the FBHE is an economically attractive alternative when compared to the SSHE. Experiments show that it is possible to carry out freeze concentration of fruit apple juices in FBHE (Nazir and Farid, 2008) (Table 11.3). A comparison between these two systems for capacities of 100 kW and larger shows that the investment costs of crystallizers with FBHE are about 30%–60% lower than that of SSHE (Figure 11.6). Furthermore, the energy consumption of fluidized bed ice slurry generators is about 5%–21% lower. It can therefore be concluded that the fluidized bed ice slurry generator is an attractive ice crystallizer concerning both investment costs and energy consumption. (Pronk, 2006).

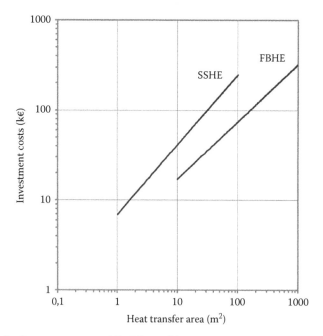

FIGURE 11.6 Investment costs of FBHE and SSHE.

Slurry crystallization (Verschuur et al., 2002; van Nistelrooij, 2005) is perceived as a well-established crystallization process and provides significant cost reductions in crystallizers and recrystallizers. The average achievable crystal size in a slurry crystallizer is smaller than in a conventional process; for certain applications, these smaller crystals can still be perfectly treated in a wash column. Additionally, the specific energy consumption (SEC) is less, which results in a much lower overall production cost for the final user.

A multiplate cryoconcentrator based in a film freeze concentration (Raventós et al., 2007) has the advantages of simplicity and economy in comparison with suspension system, since the concentrate is separated from the ice by gravity and there is no need for washing columns, centrifuges, or presses. In addition, the equipment works at normal air pressure, unlike conventional equipment that requires pressurization.

11.6.2 ENERGY

Several FCSs are in use. The main commercial system is developed by Grenco N.V., the Netherlands, now a subsidiary of Niro Inc. A single stage plant, installed in 1977, had a SEC of 227 kWh/ton water removed (Van Mill and Bouman, 1990). This energy consumption is more than the theoretic energy use of evaporation. A five (multi-) stage plant, installed in 1988, had an SEC of 60 kWh/ton of water removed (Van Mill and Bouman, 1990). This is 50% less than theoretic minimum for evaporation. The use of multistage operated plant reduces the energy requirements considerably. Total energy requirements (refrigeration + energy for drives) for dewatering processes of wine are showed in Table 11.4 (Thijssen and Van Der Malen, 1981).

TABLE 11.4
Energy Requirements for Dewatering Processes

	Electrical (kWh/ton Water Removal)	Thermal (kg Steam/ton Water Removal)
Evaporation 1 effect	4	1100
With jet pump	4	550
With mechanical compression	46	—
Evaporation 2 effect	6	550
With jet pump	6	360
With mechanical compression	38	—
Evaporation 3 effect	8	380
With jet pump	8	270
With mechanical compression	29	—
Evaporation 4 effect	8	290
Evaporation 5 effect	10	230
FC 1 stage condensers at +35°C	110	—
Condensers at +10°C	85	—
FC 4 stage condensers at +35°C	85	—
Condensers at +10°C	60	—
Reverse osmosis	10–50	—

Because freeze concentration and membrane operations utilize primarily electrical energy, the benefit estimate can be done using the "substitution coefficient" (Drioli and Romano, 2001). The substitution coefficient is defined as the ratio between the primary energy (thermal) saved in the new process with respect to the conventional process and the amount of electrical energy consumed, relative to the conventional process:

$$CS = \frac{C_1 - C_2}{E_2 - E_1} \tag{11.5}$$

where
   CS is the substitution coefficient
   C is the consumption of thermal primary energy (MJ or kWh_p)
   E is the consumption of electrical energy (kWh_e)
   1 and 2 are the relative indexes of the conventional and innovating process, respectively

Taking into account that 1 kWh of electrical energy, available at the utilization site, requires to burn, in a power station, 10.5 MJ of primary energy from a combustible source (oil, gas, coal, etc.), the substitution (or process innovation) results are advantageous when CS is greater than 10.5 MJ/kWh (2.9 kWh_p/kWh_e). CS between evaporation and freeze concentration techniques in juices are presented in Table 11.5.

These comparisons suggest that freeze concentration has equivalent or slightly lower energy consumption than a triple effect MVR evaporator, but higher consumption than a reverse osmosis membrane system.

It is more practical to use the relative energy costs in comparing the various concentration processes. The costs of energy depend on the relative costs of steam to

TABLE 11.5

Substitution Coefficients (CS) between Evaporation and Freeze Concentration Techniques in Juices

	FC 1 Stage Condensers at +35°C	FC 1 Stage Condensers at +10°C	FC 4 Stage Condensers at +35°C	FC 4 Stage Condensers at +10°C
Evaporation 1 effect	7.78	10.19	10.19	14.73
with jet pump	3.89	5.09	5.09	7.37
Evaporation 2 effect	3.97	5.22	5.22	7.64
with jet pump	2.60	3.42	3.42	5.00
Evaporation 3 effect	2.79	3.70	3.70	5.48
with jet pump	1.99	2.63	2.63	3.89
Evaporation 4 effect	2.13	2.82	2.82	4.18
Evaporation 5 effect	1.73	2.30	2.30	3.45

electrical energy. FCSs are more attractive as the ratio of steam costs to electrical costs increases. The relative costs of energy for the various concentration processes are given by Thijssen and Van Der Malen (1981). By using the correct ratio, R (costs of 1 ton of steam/costs of 1 kWh), for a specific country and production site, the actual relative costs of energy can be obtained. In the case of Spain, the value of $R = 240–250$, in which case the cost of freeze concentration is comparable to the evaporation of five effects.

Energy can be saved in freeze concentration by reducing the temperature difference between the evaporating and the condensing refrigerant on the vapor compression cycle. The lowering of that temperature difference can be obtained by performing the crystallization in cascade and by using the latent heat of the ice crystals (heat of crystallization) for the condensation of refrigerant. With the cascade array, an important part of the heat of crystallization can be extracted at levels well above the freezing temperature of the final concentrate. An FCS with heat pump was described, and energy of the process was compared with conventional heat evaporation (Rane and Jabade, 2005) (Table 11.3). The FCS, as described when applied to a sugar manufacture process, exhibits significant energy conservation. In FCS using heat pump, it is easy to achieve the equivalent of the temperature difference of 15°C. For this the electrical consumption is 8 kWh for the compressor of the heat pump with 418 MJ (10^5 kcal) cooling capacity. Such a heat pump will produce 1.2×10^3 kg of ice; the COP of such system is therefore 14.3.

11.7 ASPECTS TO CONSIDER IN THE FUTURE DEVELOPMENT OF FREEZE CONCENTRATION

The future development of freeze concentration will be centered on overcoming its drawbacks, on that basis; it might involve the following points:

1. Suspension crystallization attains levels of ice purity that are clearly superior to those attained with film crystallization. For this reason, the future of film crystallization technology will depend on reduction of levels of occlusion in the solid phase.
2. The degree of concentration obtained with film crystallization is still lower than the obtained with the suspension system; an increased in final concentration will need to be attained.
3. Reduction of the level of ice impurity and increase in the degree of concentration of the fluid with the film crystallization system requires progress in the following aspects:
   a. Transport of the fluid to be concentrated in a hydraulic regime with the greatest possible turbulence (on a plate, $Re > 2500$ is recommended, and inside circular conduits, Miyawaki et al. [2005] recommends working with velocities of over 1 m/s).
   b. Application of the characteristic techniques of melt crystallization: Controlled seeding, nucleation mechanically induced by shock waves, ultrasonic vibration, supersonic radiation, or partial melting (sweating) to drain impurities trapped in the ice.

4. One advantage of the film crystallization system is its simplicity, in terms of both the construction and operation of the equipment; nevertheless, in order to optimize operation, a continuous operation system will have to be devised.

5. The design of equipment with the minimum of moving parts in the suspension crystallization system would simplify its operation, and it would make it more competitive.

REFERENCES

Ayel, V., Lottin, O., Faucheux, M., Sallier, D., and Peerhossaini, M. 2006. Crystallization of undercooled aqueous solutions: Experimental study of free dendritic growth in cylindrical geometry. *International Journal of Heat and Mass Transfer* 49: 1876–1884.

Bayindirli, L., Ozilgen, M., and Ungan, S. 1993. Mathematical analysis of freeze concentration of apple juice. *Journal of Food Engineering* 19: 95–107.

Braddock, R.J. and Marcy, J.E. 1985. Freeze concentration of pineapple juice. *Journal of Food Science* 50: 1636–1639.

Braddock, R.J. and Marcy, J.E. 1987. Quality of freeze concentrated orange juice. *Journal of Food Science* 52: 159–162.

Burton, J.A., Prim, R., and Slichter, W.P. 1953. The distribution of solute in crystals grown from the melt. *The Journal of Chemical Physics* 21: 1987–1996.

Caretta, O., Courtot, F., and Davies, T. 2006. Measurement of salt entrapment during the directional solidification of brine under forced mass convection. *Journal of Crystal Growth* 294: 151–155.

Chen, P. and Chen, X.D. 2000. A generalized correlation of solute inclusion in ice formed from aqueous solutions and food liquids on sub-cooled surface. *The Canadian Journal of Chemical Engineering* 78: 312–319.

Chen, P., Chen, X.D., and Free, K.W. 1998. Solute inclusion in ice formed from sucrose solutions on a sub-cooled surface-An experimental study. *Journal of Food Engineering* 38: 1–13.

Chiampo, F. and Conti, R. 2002. Crioconcentrazione di succhi di frutta in un impianto pilota. *Industrie Delle Bevande* 31: 550–554.

Deshpande, S.S., Sathe, S.K., and Salunkhe, D.K. 1984. Freeze concentration of fruit juices. *Food Science and Nutrition* 20: 173–248.

Dette, S.S. and Jansen, H. 2010. Freeze concentration of black currant juice. *Chemical Engineering & Technology* 33(5): 762–766.

Drioli, E. and Romano, M. 2001. Progress and new perspectives on integrated membrane operations for sustainable industrial growth. *Industrial & Engineering Chemistry Research* 40: 1277–1300.

Flesland, O. 1995. Freeze concentration by layer crystallization. *Drying Technology* 13: 1713–1739.

Gu, X., Suzuki, T., and Miyawaki, O. 2005. Limiting partition coefficient in progressive freeze-concentration. *Journal of Food Science* 70: 546–551.

Habib, B. and Farid, M. 2006. Heat transfer and operating conditions for freeze concentration in a liquid–solid fluidized bed heat exchanger. *Chemical Engineering and Processing* 45: 698–710.

Halde, R. 1979. Concentration of impurities by progressive freezing. *Water Research* 14: 575–580.

Hartel, R.W. and Espinel, L.A. 1993. Freeze concentration of skim milk. *Journal of Food Engineering* 20: 101–120.

Hernández, E., Raventós, M., Auleda, J.M., and Ibarz, A. 2009. Concentration of apple and pear juices in a multi-plate freeze concentrator. *Innovative Food Science and Emerging Technologies* 10: 348–355.

Hernández, E., Raventós, M., Auleda, J.M., and Ibarz, A. 2010. Freeze concentration of must in a pilot plant falling film cryoconcentrator. *Innovative Food Science and Emerging Technologies* 11: 130–136.

Hindmarsh, J.P., Russell, A.B., and Chen, X.D. 2005. Measuring dendritic growth in undercooled sucrose solution droplets. *Journal of Crystal Growth* 285: 236–248.

Huige, N.J.J. and Thijssen, H.A.C. 1972. Production of large crystals by continuous ripening in a stirred tank. *Journal of Crystal Growth* 13: 483–487.

Jansen, H., Hernández, M.A., and Martínez, A. 2001. Concentración por congelación de disoluciones acuosas: Un nuevo método para obtener productos innovadores de alta calidad. *CTC Alimentación* 10: 13–15.

Kawasaki, K., Matsuda, A., and Kadota, H. 2006. Freeze concentration of equal molarity solutions with ultrasonic irradiation under constant freezing rate: Effect of solute. *Chemical Engineering Research and Design* 84: 107–112.

Kobayashi, A., Shirai, Y., Nakanishi, K., and Matsuno, R. 1996. A method for making large agglomerated ice crystals for freeze concentration. *Journal of Food Engineering* 27: 1–15.

Kramer, A., Wani, K., and Sulli, J.H. 1971. Freeze concentration by directional freezing. *Journal of Food Science* 36: 320–322.

Lee, Y.C. and Lee, S.W. 1998. Quality changes during storage in Korean cloudy pear juice concentrated by different methods. *Food Sciences and Biotechnology* 7: 127–130.

Lemmer, S., Klomp, R., Ruemekorf, R., and Scholz, R. 2001 Preconcentration of wastewater through the Niro freeze concentration process. *Chemical Engineering and Technology* 24: 485–488.

Liu, L., Miyawaki, O., and Hayakawa, K. 1999. Progressive freeze-concentration of tomato juice. *Food Science and Technology Research* 5: 108–112.

Liu, L., Miyawaki, O., and Nakamura, K. 1997. Progressive freeze-concentration of model liquid food. *Food Science Technology International* 3: 348–352.

Liu, L., Tomoyuki, F., Hayakawa, K., and Miyawaki, O. 1998. Prevention of initial supercooling in progressive freeze-concentration. *Bioscience, Biotechnology, and Biochemistry* 62: 2467–2469.

Matsuda, A., Kawasaki, K., and Kadota, H. 1999. Freeze concentration with supersonic radiation under constant freezing rate—Effect of kind and concentration of solutes. *Journal of Chemical Engineering* 32: 569–572.

Miyawaki, O. 2001. Analysis and control of ice crystal structure in frozen food and their application to food processing. *Food Science and Technology Research* 7: 1–7.

Miyawaki, O., Liu, L., and Nakamura, K. 1998. Effective partition constant of solute between ice and liquid phases in progressive freeze-concentration. *Journal of Food Science* 63: 756–758.

Miyawaki, O., Liu, L., Shirai, Y., Sakashita, S., and Kagitani, K. 2005. Tubular ice system for scale-up of progressive freeze-concentration. *Journal of Food Engineering* 69: 107–113.

Morison, K.R. and Hartel, R.W. 2006. Evaporation and freeze concentration. In: Heldman D.R. and Lund D.B. (eds.), *Handbook of Food Engineering*, 2nd Edn. CRC Press, Boca Raton, FL. pp. 496–550.

Müller, M. and Sekoulov, I. 1992. Waste water reuse by freeze concentration with a falling film reactor. *Water Science and Technology* 26: 1475–1482.

Nazir, M. and Farid, M.M. 2008. Modeling ice removal in fluidized-bed freeze concentration of apple juice. *AIChE Journal* 54: 2999–3006.

Nonthanum, P. and Tansakul, A. 2008. Freeze concentration of lime juice. *Maejo International Journal on Science and Technology* 1: 27–37.

Okawa, S., Ito T., and Saito, A. 2009. Effect of crystal orientation on freeze concentration of solutions. *International Journal of Refrigeration* 32: 246–252.

Ozüm, B. and Kirwan, D.J. 1976. Impurities in ice crystals grown from stirred solutions. Analysis and design of crystallization processes. *AIChE Symposium Series* 153: 1–6.

Pardo, J.M., Suess, F., and Niranjan, K. 2002. An investigation into the relationship between freezing rate and mean ice crystal size for coffee extracts. *Trans IchemE* 80: 176–182.

Patil, A.G. 1993. Freeze concentration: An attractive alternative. *International Sugar Journal* 95: 349–355.

Pronk, P. 2006. Fluidized bed heat exchangers to prevent fouling in ice slurry systems and industrial crystallizers. PhD thesis, Delft University of Technology, Delft, the Netherlands.

Pronk, P., Infante, C.A., and Witkamp, G.J. 2002. Effects of long-term ice slurry storage on crystal size distribution. In: Melinder, A. (ed.), *Proceedings of 5th Workshop on Ice Slurries of the IIR*, May 30–31, Stockholm, Sweden. pp. 151–160.

Pronk, P., Infante, C.A., and Witkamp, G.J. 2005. A dynamic model of Ostwald ripening in ice suspensions. *Journal of Crystal Growth* 275: 355–361.

Rahman, M.S., Ahmed, A., and Chen, X.D. 2006. Freezing-melting process and desalination: I. Review of the state-of-the-art. *Separation and Purification Reviews* 35: 59–96.

Ramos, F.A., Delgado, J.L., Bautista, E., Morales, A.L., and Duque, C. 2005. Changes in volatiles with the application of progressive freeze-concentration to Andes berry (*Rubus glaucus Benth*). *Journal of Food Engineering* 69: 291–297.

Ramteke, R.S., Singh, N.I., Rekha, M.N., and Eipeson, W.E. 1993. Methods for concentration of fruit juices: A critical evaluation. *Journal of Food Science and Technology* 30: 391–402.

Rane, M.V. and Jabade, S.K. 2005. Freeze concentration of sugarcane juice in a jiggery making process. *Applied Thermal Engineering* 25: 2122–2137.

Raventós, M., Hernández, E., Auleda, J.M., and Ibarz, A. 2007. Concentration of aqueous sugar solutions in a multi-plate cryoconcentrator. *Journal of Food Engineering* 79: 577–585.

Rutter, J.W. and Chalmers, B. 1953. A prismatic structure formed during solidification metals. *Canadian Journal of Physics* 31: 15–39.

Sánchez, J., Ruiz, Y., Auleda, J.M., Hernández, E., and Raventós, M. 2009. Review. Freeze concentration in the fruit juices industry. *Food Science and Technology International* 15(4): 303–315.

Scholz, R., Ruemekorf, R., Verdoes, D., and Nienoord, M. 2002. Wash columns—State of the art and further developments. In: Chianese, A. (ed.), *Proceedings of 15th International Symposium on Industrial Crystallization*, September 15–18, Sorrento, Italy, AIDIC Servizi S.r.l., pp. 1425–1430.

Shirai, Y., Wakisaka, M., Miyawaki, O., and Sakashita, S. 1999. Effect of seed on formation of tube ice with high purity for a freeze wastewater treatment system with a bubble-flow circulator. *Water Research* 33: 1325–1329.

Teraoka, Y., Saito, A., and Okawa, S. 2002. Ice crystal growth in supercooled solution. *International Journal of Refrigeration* 25: 218–225.

Thijssen, H.A.C. 1986. The economics and potentials of freeze concentration for fruit juices. *International Federation of Fruit Juice Producers*. XIX Scientific Technical Commission. Symposium Den Haag, May 12–15, The Hague, the Netherlands, pp. 97–103.

Thijssen, H.A.C. and Van Der Malen, B. 1981. Implications on quality of energy savings in the concentration of foods. *Resources and Conservation* 67: 287–299.

Van Mill, P.J.J.M. and Bouman, S. 1990. Freeze concentration of dairy products. *Netherlands Milk and Dairy Journal* 44: 21–31.

Van Nistelrooij, M. 2005. Bridging the cost barrier to freeze concentration. *Food and Beverage Asia*. April/May.

Van Weelden, G. 1994. Freeze concentration: The alternative for single strength juices. *Fruit Processing* 4: 140–143.

Verdoes, D., Arkenbout, G.J., Bruinsma, O.S.L., Koutsoukos, P.G., and Ulrich J. 1997. Improved procedures for separating crystals from the melt. *Applied Thermal Engineering* 17: 879–888.

Verschuur, R.J., Scholz, R., Van Nistelrooj, M., and Scheurs, B. 2002. Innovations in freeze concentration technology. *15th International Symposium on Industrial Crystallization*, September, Sorrento, Italy.

Wakisaka, M., Shirai, Y., and Sakashita, S. 2001. Ice crystallization in a pilot-scale freeze wastewater treatment system. *Chemical Engineering and Processing* 40: 201–208.

12 Refrigeration and Cold Chain Effect on Fruit Shelf Life

José Maria Correia da Costa and Edmar Clemente

CONTENTS

12.1 INTRODUCTION

A great challenge of modern agriculture is to extend the postharvest life of fruits of temperate climate in order to reduce losses and increase the product lifetime on the market. Many forms of storage have been researched as to different temperatures allied to biofilms, biodegradable coatings, controlled atmosphere (CA), and modified atmosphere (MA). The development of new techniques is very important for producers because they can minimize postharvest losses and add value to their product that will have longer trading period. This chapter describes properties and results of research in which storage techniques related to temperature were used to preserve fruits.

12.2 IMPORTANT PHYSICAL AND PROCESS PROPERTIES IN REFRIGERATION

The thermal properties of vegetables and fruits are related to their ability to transfer heat, being essential in the analysis of heat transfer that occurs in thermal processes, such as refrigeration, freezing, and heating, as well as in optimizing the performance of heat transfer equipment, and the knowledge of these properties is essential for the development of food and agricultural science (Castro, 2004).

Thermal conductivity and diffusivity and specific heat of fresh fruits are important properties in the search for adequate knowledge on the operation needs and conditions of cooling equipment. Product temperature and properties are strictly related to the cooling process and to their effect on the accuracy of results (Dússan Sarria and Honório, 2005).

The properties of the cooling media are very important (whether air, water, or other), especially the thermal conductivity and specific heat. On the other hand, transport conditions are also very important in heat transfer calculations (Teruel, 2000).

12.2.1 SPECIFIC MASS

Specific mass or density (ρ) can be defined by the relationship between mass (m) and volume (v) of a biological material. There are three types of density: bulk density, which is equivalent to the mass of each individual product packed in a certain volume, including the empty space inside the package; apparent density, which corresponds to the product total weight divided by the total volume, also including the emptiness; and real density, the ratio between total mass and volume of the product not including the empty space (Castro, 2004).

12.2.2 THERMAL CONDUCTIVITY

The heat transfer is the shifting of energy from one region to another as a result of temperature differences between them, and a form of transfer is conduction. The heat conduction is the transfer of heat associated with the motion of the substance particles without significant displacement or flow of these particles. The method of transferring heat depends on a coefficient known as thermal conductivity. This refers to the amount of heat (flux) by time unit through an area or a thickness, both unitary, with a temperature differential between the faces (Dússan Sarria, 2003). The thermal conductivity is related to the water content, so it increases along with increasing water content. For fruits and vegetables, Equation 12.1 presents thermal conductivity (κ), expressed in W/m°C, in function on water content (U), expressed as humidity percentage (Dússan Sarria, 2003):

$$\kappa = 0.00493U + 0.148 \tag{12.1}$$

12.2.3 THERMAL DIFFUSIVITY

The thermal diffusivity indicates how heat diffuses through a material, being more important in thermal control than conductivity because it expresses how quickly a material adjusts itself to the surrounding temperature. The determination of this property is very important for the food products, and it is essential in prediction of heat transfer processes, such as refrigeration (Venâncio et al., 2006). This property is affected by temperature, composition, water content, homogeneity, and physical structure of the material; and these variables determine how fast heat propagates and how it diffuses through the material.

The diffusivity increases linearly along with increasing water content due to the linear increase of thermal conductivity when water content is greater than 30%. As noted by Riedel (1969), food thermal diffusivity is heavily dependent on water content (U) higher than 40%, such as for guava. This property is defined by Equation 12.2:

$$\alpha = 0.088 \times 10^{-6} + \left[(\alpha_w - 0.088 \times 10^{-6}) \times \frac{U}{100} \right] \tag{12.2}$$

where
α is given in m^2/s
α_w is the thermal diffusivity of water at the product temperature

At 20°C, it is accepted as 0.148×10^{-6} m^2/s, according to Ashrae (1993).

Thermal diffusivity is associated with heat diffusion inside the product during changes of temperature along the time; therefore, a high value of thermal diffusivity means a rapid transfer of heat into the product and a little time for the heat getting out of the product (Dússan Sarria and Honório, 2004).

12.2.4 SPECIFIC HEAT

Specific heat corresponds to the energy required to change the temperature of a unit of product in 1°, based strictly on the amount of energy required and not on the rate at which this temperature change occurs (Fontana et al., 1999). According to Ashrae (1993), the specific heat is directly proportional to water content in fruits and vegetables, as shown in Siebel Equation 12.3 for calculating the specific heat (C_p) of materials according to water content (U):

$$C_p = 0.0335U + 0.837 \; (>0°C) \tag{12.3}$$

where
C_p is given in kJ/kg°C
U is in %

In solid materials from vegetables and fruits, thermal properties such as thermal conductivity, thermal diffusivity, and specific heat are related to type, temperature, and water content of the material. In fresh fruits and vegetables with high water content, the values of thermal conductivity, thermal diffusivity, and specific heat are strongly influenced by water content. In vegetables and fruits, thermal conductivity is much more dependent on the cellular structure, density, and water content than on temperature.

12.2.5 CONVECTIVE COEFFICIENT OF HEAT TRANSFER

The convective coefficient of heat transfer is the rate of heat transfer for each degree of temperature difference across the solid-fluid interface per area unit of solid material surface (Castro, 2004).

The surface coefficient of heat transfer (hc) is not a thermal property of foods, but it is necessary to design equipment for food treatments in which heat transfer by convection is involved (Ashrae, 1993; Pirozzi and Amendola, 2005). The forced air refrigeration of vegetables depends on air velocity and on thermodynamic conditions of the surrounding air, which affects directly the convective coefficient of heat transfer (Ashrae, 1993). The type of packaging, dimensions, opening area, and arrangement in the refrigeration also influence the values of this parameter (Dússan-Sarria, 2003; Thompson et al., 1998), as well as fruit characteristics such as temperature, water content, specific heat, and shape (Thompson et al., 1998).

The convective coefficient of heat transfer ($W/m^2{}^\circ C$) can be obtained from Equation 12.4 (Dincer, 1995):

$$h_c = \frac{(3.2\kappa.R.a)}{(10.3\alpha - a.R^2)} \tag{12.4}$$

where
R is the average radius of fruits (m)
a refers to refrigeration coefficient of (s^{-1})
α is the thermal diffusivity (m^2/s)

Dússan-Sarria (2003) states that the coefficient can reach values between 20 and 35 $W/m^2{}^\circ C$ for air flows from 1 to 3 L/s per kg of product.

12.3 REFRIGERATED STORAGE

The most important factor affecting lifetime of vegetable and fruits is the temperature, which directly influences the rates of chemical and enzymatic reactions. The conservation of vegetables by refrigeration is based on the partial or total suppression of the main agents responsible for the changes in this food group: microbial activity and growth, tissue metabolic activities after harvest, and chemical and enzymatic reactions (Ordóñez, 2005).

Vegetable products are characterized by the susceptibility to undesirable changes in temperature and humidity in the surrounding environment. Sometimes these changes cannot be noticed immediately, but they will be observed throughout the commercial chain through changes of taste, odor, firmness, and other product characteristics (Chitarra and Chitarra, 2005). So the implementation of the "cold chain," according to particular characteristics of each cultivar, ensures the quality maintenance of fruits and vegetables until these products reach the consumer (Ferreira Neto et al., 2006).

The use of low temperatures in fruit storage refers to the concept of thermal load, since this term represents the removal of heat generated by the stored product to reduce its temperature to the desired level (Ferreira Neto et al., 2006). The heat always flows from a hot object to a cold one; in refrigeration, the product is cooled by removing its heat and not by transmitting cold to it (Chitarra and Chitarra, 2005). The storage temperature is closely related to respiratory intensity because it can be reduced by low temperatures, therefore reducing the speed of biochemical reactions, including those related to senescence (Jacomino et al., 2008).

Losses of fruits due to early ripening are so far an obstacle to be overcome. The "cold chain" is defined as the set of systems that ensures the maintenance of product quality from harvest to consumption. Key elements in this chain are fast cooling systems and chambers; refrigerated trucks for land transportation; refrigerated containers for air, rail, and sea transportation; refrigerated displays in supermarkets; and domestic and industrial refrigerators (Teruel et al., 2003).

The storage temperature of vegetables is the most relevant environmental factor because it regulates the rates of all plant physiological and biochemical processes within a physiologically acceptable range, thus controlling senescence and increasing life span (Brackmann et al., 2001). It is important to remark that although cold storage decreases fruit respiration and metabolism, maintaining their quality for a longer period, the delay of all metabolism reactions does not occur (Awad, 1993).

12.4 RAPID REFRIGERATION

Cold storage of fruits should be started immediately after harvest; any delay into the ideal temperature may cause the reduction of time in which the product can remain stored. Refrigeration should be continuous and maintained throughout from producer to consumer.

The time between harvest and cooling will account for the deterioration of fruit products (Hardenburg et al., 1990; Sun and Brosnam, 1999). Strawberries, if not refrigerated, must be sold at no more than 3 days, and if only refrigerated, they can stand for 1 week (Tanabe and Cortez, 1998). To preserve the quality and prolong shelf life of fruits is essential a rapid cooling to temperatures close to storage (Spagnol and Sigrist, 1992). The sooner a plant product reaches its optimal storage temperature, the higher the lifetime and the smaller the losses during marketing.

According to Teruel et al. (2001), the efficiency of this process is characterized by the relationship between time and temperature. To reduce mass loss, it is necessary to reduce the temperature of freshly harvested fruit in the shortest time possible, because the higher the product temperature, the higher is its water vapor pressure (Castro et al., 2000).

As Dússan-Sarria (2003) reports, the amount of heat removed from the product is dependent on the temperature difference between product and environment, on the cooling rate, on the amount of product to be cooled in a specific time, and on the product specific heat. During this stage, the heat transfer from surface to inside the vegetable is made primarily by conduction, when different points of a material are at different temperatures.

Ferri and Rombaldi (2004) compared the delayed cooling of plums for 10, 20, or 30 h and concluded that the longer the delay in cooling, the worse the quality of fruits, as low firmness, extremely mature fruits, and occurrence of rot.

As implied, the process of rapid cooling means a fast removal of heat from freshly harvested product before heading it to the storage stage or transport over long distances.

Precooling is the first step of handling the temperature with the primary purpose of rapid removal of field heat from freshly harvested products (Chitarra and Chitarra, 2005). It is the first step of the "cold chain" and should be applied before the final product storage, in order to rapidly reduce the metabolic processes of respiration and decay (Cantillano, 1986). It is a separate operation of cold storage and requires elaborate equipment and facilities (Bleinroth et al., 1992).

Quite perishable products such as beans, broccoli, cauliflower, fresh corn, tomatoes, leafy vegetables, artichokes, cabbage, carrots, peas, and radishes should be cooled immediately after harvest, preferably still in the field. Among the major fruits that must be cooled immediately after harvest are avocado, strawberry, peach, nectarine, plum, and tropical and subtropical fruits such as guava, mango, papaya, and pineapple stand out (Vissotto et al., 1999).

Metabolic activity generates heat, and the product heat is controlled by environmental temperature. An indicator of field heat is the difference of temperature from freshly harvested product and to its optimum temperature for storage. Rapid cooling is the fast reduction of temperature of freshly harvested product to the optimum temperature for storage, being cooled usually by 7/8 or 88% of the difference in temperature (Gast and Flowers, 1991). A product, even quickly refrigerated, will be well preserved if its heat content is maintained until the final consumption, so if for some reason the "cold chain" is broken, the shelf life and quality will decrease (Dússan-Sarria and Honório, 2005).

Brackmann et al. (2001) evaluated the effect of retarding the cooling time on the quality of apple "Gala." They observed that cooling has a significant effect on maintaining the fruit firmness after 7 days at 20°C and concluded that it is possible delaying the fruit storage for 28 h after harvest, provided it is made a rapid cooling of the material.

There are different methods of cooling, and the right choice depends on the following factors: type of product to be cooled (cultivar, maturity, shape, size, and adaptability to the method), type of equipment available, packaging, physical infrastructure available, manpower required, economic factors, among others (Gast and Flores, 1991; Vissotto et al., 1999).

The main four methods for fruit, depending on the cooling medium, are forced air cooling, hydrocooling, by icing, and by vacuum (Afonso, 2005; Cortez et al., 2002; Teruel et al., 2003).

12.5   MODIFIED ATMOSPHERE

The method of MA storage consists of creating a new condition different from that found in the natural environment. Synthetic films are usually utilized to modify the fruit atmosphere, reducing the oxygen content in the air and increasing the carbon dioxide levels along the time. The reduced oxygen retards the ripening process by decreasing respiratory metabolism and ethylene production, which accounts for changes in flavor, aroma, and texture. However, a percentage of O_2 in the environment must remain, to keep the breathing process and so the metabolic reactions, for preserving the maximum quality of the product (Cortez et al., 2002).

Fruits produce ethylene during their metabolism, a plant hormone with different physiological effects, as acceleration of plant respiration, ripening, and senescence.

Ethylene production increases during fruit ripening, and the control of ethylene levels is utilized along with other methods to prolong shelf life. Ethylene absorbers have been widely used in research to prolong lifetime of climacteric fruits. Removal of ethylene is made by an oxidant ($KMnO_4$) in form of sachets, with high permeability to ethylene, or encapsulated in the structure of films for packaging (Azeredo, 2004).

In MA, after storage, there is no control of the gases surrounding the food, unlike to what happens in a CA, where concentrations of oxygen, carbon dioxide, and even ethylene are monitored and regulated. The use of MA and CA can be used both for preserving vegetables that ripen after harvest and for those with rapid deterioration (Fellows, 2006).

Modifications in the storage microatmosphere can be attained by active or passive means. In passive modification, the atmosphere is created by fruit breathing in the package, until it reaches equilibrium. In the case of active modification, the atmosphere is created by inflating the package-free space with a predetermined gas mixture or by material contained in sachets or incorporated directly into packaging, capable of promoting changes in gas composition. In both cases, the MA is established and maintained by a dynamic balance between respiration and permeation (Azeredo et al., 2000).

When a plastic film packaging is correctly designed, the gas composition inside interferes with the metabolic activity of the packed product. In this condition, there is a reduction in metabolism, and therefore a delay in maturation is achieved.

According to Kader et al. (1989), the selection of a plastic film to attain a favorable MA should be based on the respiratory rate and the optimal concentrations of O_2 and CO_2 to the fruit. For the majority of fruits, an appropriate film should be more permeable to CO_2 than to O_2. It is necessary to maintain a minimum concentration of O_2 in the package so that aerobic respiration continues normally because sharp reductions in the concentration of O_2 can lead to anaerobic condition.

According to Yamashita et al. (2001) and Pinto et al. (2006), the use of plastic films associated with refrigerated environment preserves the integrity of the fruit and allows for better maintenance of sensory attributes.

Plastic films of polyethylene or polyvinyl chloride (PVC) have been widely used, due to their convenience, relatively low cost, and great efficiency, especially when associated with cold storage to prevent loss of fruits. Tropical fruits can have shelf life prolonged by reducing of respiration rate, ethylene production, and consequently ripening decrease in MA (Awad, 1993; Chitarra and Chitarra, 2005).

Carnauba wax applied on the product surface is considered a method of preservation by MA because it modifies the permeability of O_2 and CO_2. The use of waxes as surface coverage of fruits reduces water losses by 30%–50% and slows wrinkling, providing improvement in appearance and modifying its internal atmosphere (Ribeiro et al., 2005).

12.6   FORCED AIR COOLING

In forced air system, the cooling medium is the air, which is forced through the products for reducing the material cooling time (Chitarra and Chitarra, 2005; Cortez et al., 2002). There are several settings in this system, and the most common is to arrange the product in a storage chamber and to head the air flow though the product.

This arrangement can be made with the use of canvas with an air forcer causing a difference of pressure (Cortez et al., 2002).

The rapid cooling system in the chamber should be designed so that the thermal load corresponding to the storage and the field heat are removed, being necessary to promote an efficient circulation of air associated with an appropriate temperature (Leal and Cortez, 1998). Figures 12.1 and 12.2 show the air flowing through the boxes and the fruits.

Most storage chambers do not have refrigeration capacity, or enough air flow to implement this method. Thus, this technology is typically a separate operation and requires equipment with more cooling capacity (Louzada et al., 2003). The air is forced into the products by means of fans, allowing an efficient contact between the cooling medium and the freshly harvested produce, and heat transfer occurs by forced convective process (Vissotto et al., 1999). Once the product is cooled, the air

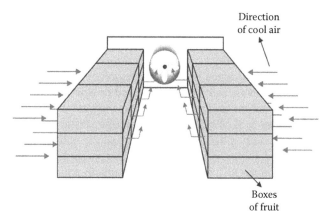

FIGURE 12.1   Air flowing through the boxes in the refrigerating tunnel.

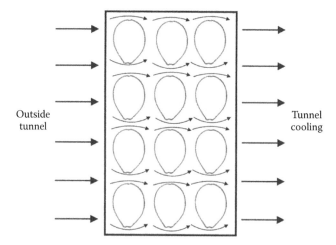

FIGURE 12.2   Air flowing through the fruit inside the package.

flow should be substantially reduced or completely eliminated, because continuous flow can cause severe mass loss for the products (Antoniali, 2000).

The cooling time depends on the box dimensions, opening area, and arrangement; on the product characteristics, such as initial and final temperatures, specific heat, shape, and height of fruit bed arranged inside the package; and on the characteristics of cooling air: temperature, velocity, relative humidity (RH), and thermal properties. The air movement is always in the direction of the product, for avoiding water condensation on the product (Leal and Cortez, 1998; Teruel et al., 2002b).

Teruel et al. (2004) carried out a study with vegetables of different sizes and found that the cooling time varies in proportion to the volume so that larger fruits were cooled for longer times compared with the fruits of smaller sizes.

In slow cooling, performed on conventional chambers, the air is simply moved over the product and not directed to the product, and the heat transfer is made by conduction. The air follows preferential paths that offer less resistance, thus not entering inside the package (Vissotto et al., 1999). In forced air cooling, products are cooled more efficiently, because air is headed to pass within the packaging.

The process takes place as follows: the pallets containing the fruit packages are placed side by side forming a tunnel (Dussan-Sarria and Honorius, 2005). The fan works as an exhaust fan sucking the cool air that is forced transversely through the fruits placed on the cooling tunnel, which is closed at the front and top with canvas to create a negative pressure (Cortez et al., 2002; Teruel et al., 2003).

The described system enables increased interaction between products and cooling medium, as air is forced inside the packages containing the products (Leal and Cortez, 1998). This system can be mounted inside a conventional refrigeration chamber. Thus, the air passing through the boxes inside the tunnel is the cold air coming out of the evaporator (Pinto, 2005). In conventional chambers, the air circulates freely through the chamber at relatively low speeds (less than 1 m/s) without following headed paths and not being forced to pass across the products. In forced air systems, air is conducted through the volume of boxes at higher speeds (between 1 and 5 m/s) in which the heat transfer process is more intense and cooling time is smaller (Teruel et al., 2001).

The packages have a large influence on the process. They should have openings and should be distributed in the way of facilitating the air movement through them, thus promoting an effective exchange of heat between the product and the environment (Cortez et al., 2002; Sanino et al., 2003). The openings should have at least 5% of the effective area for the passage of cool air, because the uniformity of cooling is ensured by the amount of opening areas and by an adequate arrangement (Leal and Cortez, 1998).

Tabolt and Chau (1991) carried out experiments with strawberries, cooling them from 28°C to 5°C. They observed that the time of 7/8 of cooling varied from 160 to 52 min and was dependent on the rate of air circulating in the opening area of packages. Teruel et al. (2002a) studied the effect of two types of packaging on bananas under forced air cooling. The plastic package had 40% of effective opening, and cardboard package had 3.2%. The results confirmed that packages with larger opening showed cooling time 45% lesser than those with smaller opening.

It is important to have openings not obstructed by internal accessories such as trays, bottles, papers, etc. If these accessories are essential to the preservation of the product, the air flow from the fan should be appropriately increased to compensate their effects (Cortez et al., 2002). Fruit cooling also depends on its location in the arrangement; those near the flow of cool air are refrigerated in shorter time than those in a more distant position from the flow (Teruel et al., 2001, 2003).

Thompson et al. (1998) found that cherries packed and cooled from 30°C to 10°C by forced air at a flow rate of 0.002 m³/s/kg of product have a difference of 25 min in cooling time between the cherries in the base of pallets and those on the surface of the pallets. So not always a high flow of cold air means a good cooling process; an evenly well-distributed cold air throughout the storage is needed.

The greater the surface area exposed, the greater the convective coefficient of heat transfer at acceptable rates of air flow. The evaporation rate on the product surface also affects significantly the cooling rate. In the beginning, the evaporative cooling occurs when air is not saturated, and it is inflated through a wet surface. The heat is removed as water evaporation occurs on the product surface, resulting in temperature decrease (Vizzotto et al., 1999).

In the forced air cooling, the rate of air is indicated to remain between 0.001 and 0.006 m³/s/kg of product, which corresponds to 1–6 L/s/kg of product to be cooled (Arifin and Chau, 1988; Teruel et al., 2008). Flows above the recommended may cause an increase in mass transfer, resulting in weight loss of products. Currently, the systems operate according to the air flow predetermined in the project, which depends on the amount of product to be cooled (Teruel et al., 2008).

In the use of forced air cooling for plums, with an air flow of 0.001 m³/s/kg of product, the time to temperature reaches 7/8 of cooling was 4 h; with a double flow rate, the time was reduced by 40% (Thompson et al., 1998).

Compared to the rapid cooling with water, forced air system has the advantage of avoiding contamination with fruit rot spores commonly present in the water, as often occurs when no care is given to water quality (Brackmann et al., 2001). In contrast, the main disadvantages are more additional product handling; difficulty in the use of secondary packaging for protecting the product; uneven cooling (some products reaches ideal temperature before others); and cooling time slower than that by water cooling or vacuum, which may cause excessive loss of water in some products (Cortez et al., 2002; Louzada et al., 2003).

One way to reduce costs of forced air cooling systems is to use them strictly in rapid cooling of products for storing them in refrigeration chambers and later transporting or selling. Some of the major factors that affect cooling costs are hours of operation, medium temperature, and types of packaging (Teruel et al., 2002b).

The use of forced air cooling system can result in products of good quality, in fewer losses for the merchandiser, in increased time for trading with lesser wasting, and in higher profits (Ashrae, 1994).

12.7 CURRENT RESEARCH ON FRUIT REFRIGERATION

12.7.1 ORANGE

Orange trees are plants with perennial leaves which belong to family Rutaceae and genus *Citrus*. Sweet oranges (*Citrus sinensis* L. Osbeck) are the ones with biggest commercial interest, both for fresh consumption and for industrialization (Koller, 2006).

Most part of world's orange production is handled as juice. Even so, the amounts of fresh fruits for home consumption are considerable. Thus, the quality of these fruits must be compatible to the consumer's demands, making the use of postharvest technologies a need (IAPAR, 1992; Figueiredo, 1999; Ormond et al., 2002; Bender, 2006).

The loss of fresh products, including fruits, may go up to 50%, depending on the product and the season. Among the several causes that originate these losses, it is found the nonutilization of refrigerated storage after harvest, which keeps the products adequately cooled until consumption (Carraro and Mancuso, 1994).

The reduction in temperature, to limits that do not bring any damage by cooling to the product, delays the fruits' deterioration, postponing the development of microorganisms and reducing respiration and transpiration processes, which are responsible for quality loss (Volpe et al., 2002).

Cooled storage time changes for each cultivar depending on the applied temperature. According to Chitarra and Chitarra (2005), the great conditions for orange cooled storage are between 3°C and 9°C and from 85% to 90% of relative humidity. In addition, the fruits can be stored during 3–8 weeks, depending on the cultivar and on the climatic conditions.

The results for total soluble solid (TSS), total titratable acidity (TTA), juice percentage, total sugar (TS), and reduced sugar (RS) for "Folha Murcha" cultivar are shown in Tables 12.1 and 12.2 at 25°C and 7°C, respectively. TSS content varied from 9.00 °Bx to 9.65 °Bx at 25°C and from 8.98 °Bx to 9.70 °Bx at 7°C. However, there was no significant difference between the temperature, not even during the storage time (Tables 12.1 and 12.2) (Araújo and Vasconcelos, 1974; Vasconcelos et al., 1976; Bronzi, 1984).

These values differ from those found by Stenzel et al. (2005), who obtained values from 11.29 °Bx to 13.12 °Bx in fruits of the same cultivar. The low content of soluble solids, in the current work, may be assigned to the harvest time of the fruits, which had not completed its ripening, and therefore, they were not appropriate for harvesting.

According to Tables 12.1 and 12.2, it could be observed that titratable acidity, at 25°C, showed significant difference only on the 45th day of storage, when compared

TABLE 12.1

Physicochemical Properties of Oranges (Cultivar "Folha Murcha") Stored at 25°C

	Time (Days)				
	0	**15**	**30**	**45**	**60**
TSS	9.28 a	9.25 a	9.00 a	9.50 a	9.65 a
TTA	1.86 ab	1.79 ab	1.84 ab	2.16 a	1.75 ab
Percent juice	45.64 a	46.62 a	43.49 a	43.74 a	43.24 a
RS	5.36 b	8.46 a	9.08 a	8.85 a	8.54 a
TS	26.74 a	14.97 a	12.10 a	16.34 a	15.60 a

Values followed by the same letter do not differ at Tukey's test ($P > 0.05$).

TABLE 12.2
Physicochemical Properties of Oranges
(Cultivar "Folha Murcha") Stored at 7°C

	Time (Days)				
	0	15	30	45	60
TSS	9.28 a	8.98 a	9.40 a	9.70 a	9.53 a
TTA	1.86 ab	1.81 ab	1.62 b	1.83 ab	2.02 ab
Percent juice	45.64 a	47.23 a	44.38 a	39.41 a	45.27 a
RS	5.36 b	9.28 a	8.43 a	8.65 a	9.16 a
TS	26.74 a	13.10 a	13.63 a	14.38 a	15.39 a

Values followed by the same letter do not differ at Tukey's test ($P > 0.05$).

to the others. On the other hand, at 7°C, the statistical difference happened on the 30th day of storage, with variation from 1.61 to 2.02 mg of citric acid 100 mL^{-1}. Duenhas et al. (2002) found similar values when analyzing oranges of cultivar "Valência," obtaining variation from 1.88 to 2.00 mg of citric acid 100 mL^{-1}.

Regarding TS and RS, a significant difference could be seen, at both temperatures, when comparing the first day of harvest to the others. This result shows that although orange is not a climacteric fruit, alterations in some chemical characteristics may occur after harvest.

Regarding mass loss, a crescent raise could be seen along the storage time, at both temperatures, as it can be observed in Figure 12.3. At 25°C, the mass loss was higher

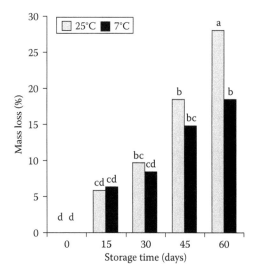

FIGURE 12.3 Mass loss of oranges "Folha Murcha" during 60 days of storage. *Note*: Values followed by the same letter on the line do not differ at Tukey's test ($P > 0.05$).

than 20% at the end of the 60 days. Assman et al. (2006) also observed mass loss during the storage of oranges (cultivar "Pêra"). Storing the fruits for 21 days, they obtained a reduction of 37.6% when compared to the first day of test. According to the same author, this fact is possibly due to the rise in the metabolic activity of the fruit close to the senescence, in addition to a probable rise in ethylene level by its autocatalysis.

It was verified that the higher mass loss occurred at room temperature, showing the positive influence of refrigeration on postharvest quality of the fruits.

The content of vitamin C is an important attribute of quality of citric fruits in general (Kluge et al., 2005, 2007). An oscillation was observed in vitamin C content during storage, both at 25°C and at 7°C (Figure 12.4). The same oscillation was verified by Andrade et al. (2002), who analyzed oranges at different ripening state. The same author observed that greener fruits had lower vitamin C content when compared to riper fruits.

Quantification of phenolic compounds in fruit juices has the finality to evaluate the potential of darkening during or after the processing. It also evaluates the possibility of interference of these compounds on the taste, due to the astringent characteristics of some of them (Fernandes et al., 2007). Figure 12.5 shows the results for total phenolic compounds. At both temperatures, the phenolic compounds' content increases, going from 0.878 to 5.17 mg 100 mL^{-1} at 25°C and 4.90 mg 100 mL^{-1} at 7°C, at 45 days of storage. These results demonstrate that the content of phenolic compounds increases after the orange harvest, not depending on the storage temperature, emphasizing again that these alterations may occur in the fruit, even though it is not climacteric.

According to Figure 12.6, which expresses the carotenoids' content, it is possible to observe that at room temperature (25°C), the values were close to each other and

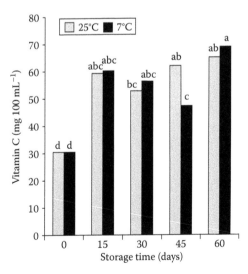

FIGURE 12.4 Vitamin C content found in oranges "Folha Murcha" during 60 days of storage. *Note:* Values followed by the same letter on the line do not differ at Tukey's test ($P > 0.05$).

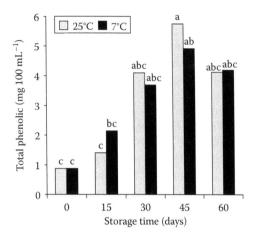

FIGURE 12.5 Total phenolic compounds' content found in oranges "Folha Murcha" during 60 days of storage. *Note:* Values followed by the same letter on the line do not differ at Tukey's test (*P* > 0.05).

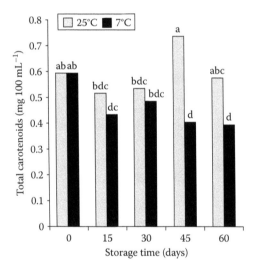

FIGURE 12.6 Total carotenoids' content found in oranges "Folha Murcha" during 60 days of storage. *Note:* Values followed by the same letter on the line do not differ at Tukey's test (*P* > 0.05).

did not vary significantly, when comparing the first and the last days of storage. On the other hand, at 7°C, carotenoids' content was reduced from 0.59 mg 100 mL^{-1} on the first day to 0.39 mg 100 mL^{-1}, and after 60 days of storage, the decrease was significant. This fact can be explained by the low storage temperatures that, according to Chitarra and Chitarra (2005), reduce respiration and delay processes such as pigment synthesis. This content is similar to the ones cited by the same authors, who found values 0.525 and 0.550 mg 100 mL^{-1} for the cultivars "Hamlim" and "Bahia," respectively.

It was observed that a temperature of 7°C is more indicated for postharvest conservation of oranges than 25°C. Oranges from cultivar "Folha Murcha" can be stored for 60 days without losing its chemical characteristics and quality.

12.7.2 STRAWBERRY

Strawberry (*Fragaria x ananassa Duch*) is appreciated by the whole world because of its nutritional qualities, flavor, and peculiar appearance, being consumed "in natura" or industrially processed (Reichert and Madail, 2003). The strawberry is classified as a nonclimacteric fruit; however, it presents respiratory activity, which leads to fast deterioration when maintained at room temperature (Scherer et al., 2003; Santos et al., 2007).

Low temperatures and application of eatable films can aid in the increase of postharvest conservation of fruits (Oliveira et al., 2007; Park et al., 2005; Santos et al., 2007; Tanada-Palmu and Grosso, 2005). Eatable coatings regulate the gaseous exchanges of the product with the external environment and the water loss that results in strawberry mass loss, and also control the loss of the volatiles responsible for fruit "flavor" (taste and aroma) (Chitarra and Chitarra, 2005).

Postharvest coatings may transport nutritional ingredients such as antioxidants, antimicrobials, and flavorings, and may improve the mechanical integrity and the food handling characteristics (Krochta and Mulder-Johnston, 1997). The main limiting factors for life of strawberry are the development of fungi (Siro et al., 2006). Biological control has been an effective method of postharvest control of diseases, using microorganisms such as bacteria and fungi, which act through various mechanisms such as antibiosis, production of lytic enzymes, parasitism, induced resistance, and competition for nutrients and space (Janisciewiez and Korsten, 2002).

Saccharomyces cerevisiae is a yeast with the potential for controlling diseases of plants. It presents the ability to synthesize antibiotic compounds, to compete for space and nutrients in the phylloplane of many plant species, and to have the cell wall elicitors (Piccinin et al., 2005). In strawberries var. Camarosa, treatment with preparations of *S. cerevisiae* reduced the incidence of decay caused by *Botrytis cinerea*, comparable to treatment with fungicides (Gouvêa, 2007). The application of microbial antagonists should be part of an integrated management of postharvest diseases, checking compatibility with other postharvest practices commercial (Baños, 2006).

Postharvest phytopathogen control through the induction of resistance, by using natural processes, is being studied with promising results (Benato et al., 2001). The incorporation of innocuous bioactive compounds, by aspersion or biofilms, has been showing promising perspectives to the use of yeasts in the biocontrol of mycotoxigens and deteriorating fungi (Coelho et al., 2003). Ten species of yeasts were identified in kefir, as reported by Latorre-Garcia et al. (2007), and the main species were *Issatchenkia orientalis*, *S. unisporus*, *S. exiguus*, and *S. humaticus*. The species found in most types of kefir grains of milk collected in Argentina were *Lactococcus lactis* ssp. lactis, *Lactobacillus kefir*, *Lactobacillus plantarum*, *Acetobacter*, and *Saccharomyces*, and the *Lactobacillus* sp. and *L. kefiranofaciens kefirgranum*, usually described in the kefir grains in milk, were not detected (Garrote et al., 2001).

Coating made with cassava starch presents positive aspects; once it is not sticky or poisonous, it is brilliant and transparent, and it may be ingested together with the protected fruit or may be removed with water, besides being a commercial product of low cost (Cereda et al., 1995). The present investigation had as main objective to evaluate biodegradable coatings in organic strawberries, "Camarosa" cultivar, aiming at maintaining the quality and the lengthening of the fruit useful life through the evaluation of aspects such as coloration, mass loss, incidence of rottenness, and total anthocyanin contents during the storage period under refrigeration.

A number of 190 fruits were selected at random for each treatment: strawberries without coating or control (CF); coating of cassava starch 2.0% (CSF); kefir liquid 15% (KL); kefir grains 15% (KG); association of CSF to KL resulting in SKL (in a proportion of 50% each), kefiraride (whey kefir-milk) 15% (KRD); and association of grains and milk kefir 15% (GKL).

CSF coating was obtained by heating the cassava starch suspension in distilled water under rotation/shaking. Kefir grains were cultivated with 30 g of brown sugar and 30 g of grains in 1.0 L distilled water, which was changed every 12 h and maintained at room temperature. One day before its use, the proportion was duplicated. For performing KL treatment, 300 mL of kefir liquid was used, and then the solution was sieved and filled with distilled water to reach the concentration of 15%. KG was obtained from 300 g of kefir grains, added to 1.5 L distilled water, and maintained under heating at 50°C and under light rotation for 30 min. A mixer was used for 2 min to disintegrate the grains. After reaching room temperature, a solution with concentration of 15% was obtained. SKL coating was obtained by associating CSF to KL (in a proportion of 50% each) in a mixer and maintained under rotation for 2 min.

Strawberries were immersed in CSF, KL, KG, and SKL suspensions and kept for 1 min under light movement. They were removed and placed apart on a nylon screen, so as to drain the coating excess liquid. Fruit natural drying occurred in about 3 h.

Following, strawberries were divided into portions of 10 fruits each and conditioned into transparent plastic boxes type PET (tereftalato of polyethylene) with a lid without perforations (17 cm × 9.5 cm × 4.5 cm), and each packing constituted an experimental unit. The experimental outlining was totally at random with five treatments and six replications in each treatment. Fruits were stored in a refrigerated chamber at 10°C ± 2°C, with RH between 60% and 80%. Evaluations were carried out periodically in the first, third, sixth, and ninth day of storage under refrigeration; then mass loss and color were analyzed. The incidence of rottenness and the anthocyanin content were checked just on the ninth day of storage.

Strawberries belonging to "Camarosa" cultivar, produced in organic system and independent of the coatings applied, showed useful life of 9 days if maintained under refrigerated storage. A better result was obtained by Mali and Grossmann (2003) for strawberries from Dover cultivar, coated with yam cassava starch and glycerol, refrigerated at 4°C, once their quality was preserved for 14 days. The latter cultivar has shown characteristics of longer useful life and was submitted to lower storage temperature if compared to the temperature used in the present study.

Results obtained when assaying mass loss and color evolution of strawberries without coating (CF), or coated with different films, produced in organic system, and stored under refrigeration, are shown in Table 12.3.

TABLE 12.3

Evolution of Mass Loss and Color in Organic Strawberries, Belonging to "Camarosa Cultivar," without and with Coatings, Stored at 10°C ± 2°C and 60%–80% RH, for 9 Days

Treatment	Mass/Pulp Loss (%)			Coloration (Levels 1–4)			
	3 Days	6 Days	9 Days	0 Days	3 Days	6 Days	9 Days
CF	2.10	4.36	8.45	3.00	3.18	3.35	3.60
CFS	2.81	5.27	9.14	2.88	3.10	3.23	3.36
SKL	2.20	4.06	6.86	2.95	3.37	3.40	3.47
KL	3.21	5.04	9.25	2.90	3.08	3.26	3.47
KG	2.56	4.68	8.97	2.95	3.38	3.43	3.57

It may be observed that there were no statistical differences in mass loss (%), between the treatments and the control ($P > 0.05$), but there was a significant difference in the averages of storage time ($P < 0.05$). Mass loss, in strawberries preserved with coatings, at the end of the storage time (9 days), varied from 6.86% (SKL) to 9.25% (KL). Such values were superior to the advisable maximum value for strawberries, maximum of 6% of water loss to avoid depreciation of fruit appearance (Cantillano, 2003). The reduction of fruit mass in all treatments presented increasing values, depending on the storage period.

In an experiment carried out by Tanada-Palmu and Grosso (2005), with organic strawberries belonging to "Osso Grande" cultivar, the mass loss was lower than 10% in fruits coated with association of gluten, bee wax, and stearic and palmitic acids, and also in the treatment with PVC films, on the 16th day of storage under refrigeration (7°C–10°C and 60%–80% RH), whereas the fruits of the control group and the ones coated only with gluten presented over 50% of mass loss. Such a great difference between fruit mass losses with and without coatings was not observed in the current study, and that may be attributed to the use of plastic packing in the fruits belonging to the control group, thus providing an MA in its interior and consequently avoiding the mass loss that occurs mainly due to the water steam loss that occurs from fruits to the environment.

Strawberry cultivars known as "Selva" and "Diamante" were studied by Pelayo et al. (2003) and showed useful life of 9 days when stored at 5°C and a useful life of 11 days when 20 kPa CO_2 was added, at the same temperature.

The average of the fruit initial coloration was close to 3.0 (around 95% or plus, with red color), thus showing the fast evolution of the red coloration during the application of treatments and the assembling of the parcels, since the fruits were picked with level of maturation 1 (75% of red color) (see Table 12.3).

At 3 days of storage, all of the fruits presented more than 95% of the epidermis with red color (level 3). These fruits presented significant differences ($P < 0.05$), through Scott-Knott test, a test for grouping averages. The applications of KG and SKL were the products that presented the greatest evolution of red color. On the sixth day of conservation, there were no statistical differences between the treatments applied ($P > 0.05$). However, on the ninth day, it was observed that the application of CSF was the treatment that less evidenced the red coloration. Color values varied from 3.36% (CSF) to 3.60% (CF).

Divergent results for color were verified by Ribeiro et al. (2007), who have not found any significant difference in fruits coated with different starch, carrageen, or quitosan compounds. Del-Valle et al. (2005) also observed that the color of strawberries was not affected by the edible coating made from cactus mucilage. Henrique and Cereda (1999) used cassava starch for strawberry conservation "IAC Campinas" and verified that the application at 3.0% was the one that resulted in better appearance and red color.

The evolution in the incidence of rottenness in "Camarosa" organic strawberries is shown in Table 12.4.

At 3 days of storage, the treatment with KL was the treatment that presented the lowest losses due to rottenness in the strawberries (3.33%). The highest incidence values, at 9 days of refrigerated storage, were observed in the fruits coated with CSF and with KL, an index over 70%. The lowest total rottenness incidence found was 36.18%, which occurred in SKL fruits. Lower values of rottenness incidences, an average of 20%, were found by Malgarim et al. (2006) and Zaicovski et al. (2006) in "Camarosa" strawberries, with a modified temperature or environment, stored under refrigeration, after 9 days at 0°C and for other 3 days at 8°C. According to Baldwin et al. (2006), a significant reduction of diseases was observed in stored strawberry immersed in pectin oligomers, with polymerization degree between 8 and 24 for 10 s. That removed the ethylene production, thus favoring the fruit defense response. Lower values were also reported by Tanada-Palmu and Grosso (2005), who found 40% of rottenness in strawberries coated with gluten after 16 days of storage under refrigeration and 30% in fruits coated with compounds such as gluten, bee wax, and stearic and palmitic acids. Fruits were doubly coated (first in gluten and soon afterward in the formulation

TABLE 12.4

Evolution of Rottenness Incidence in "Camarosa" Organic Strawberries, without and with Coatings, Stored at 10°C ± 2°C and 60%–70% RH, for 9 Days ($n = 6$)

	Incidence of Rottenness (%)			
Treatments	3 Days	6 Days	9 Days	Total
CF	0.00	10.00	48.21	58.21
CSF	0.00	5.00	65.09	70.09
SKL	0.00	3.33	32.85	36.18
KL	3.33	10.56	63.39	73.95
KG	0.00	8.33	50.66	58.99

CF, control; CSF, cassava starch 2.0%; SKL, association of cassava starch 2.0% and liquid of kefir water 15%; KL, liquid of kefir water 15%; KG, grains of kefir water.

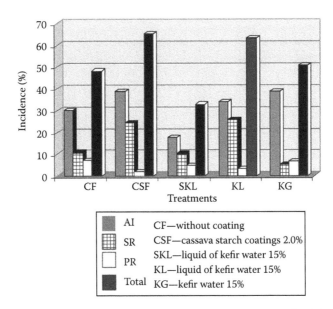

FIGURE 12.7 AI, SR, and PR in organic strawberries, "Camarosa" cultivar, stored at 10°C ± 2°C and RH 60%–80%, after 9 days.

of bee wax and acids). Brackmann et al. (1999) found high incidence of rottenness in "Tangi" cultivar strawberry (values over 40%), stored for 5 days at 20°C, with more than 40% rottenness, even with 20 kPa of CO_2.

In Figure 12.7, the incidence of anthracnose (AI), soft rottenness (SR), and peduncle rottenness (PR) in organic strawberries, "Camarosa" cultivar, without coating (CF) and in the ones coated with CSF, KL, KG, and SKL, stored under refrigeration (10°C and RH 60%–80%) can be observed.

It may be observed that the coatings applied on organic strawberries have reduced the incidence of PR at 9 days of storage under refrigeration. Fruits coated with KG showed a reduction in the incidence of SR if compared to the control fruits. According to Dias et al. (2007), temperatures lower than 8°C–10°C inhibit the development of fungi and sporangia production. However, according to Fortes (2003), one of the most efficient controls of postharvest SR is to maintain the fruits under a temperature below 6°C. The temperature of 10°C, used in the present experiment, was not enough to inhibit the development of SR in the fruits, consequently having 10% of discards in the control group at 9 days of storage.

SR has fungi such as *Rhizopus stolonifer Ehr.* and *R. nigricans Ehr.*, as causal agent (Fortes, 2003). It was observed that fruits carry fungi structures on their surface and that constitute its inoculum. Thus, the infection occurs after harvest, and fungi may be quickly disseminated when in contact with the juice that drips from infected fruit to healthy fruits conditioned in packing. Therefore, it is really important to avoid fungi dissemination (Dias et al., 2005).

Fresh fruit (FF) had low content of all types of microorganisms, with strong development during storage, mainly in yeast, which reached the highest values (10^5), and values were reduced with application of coatings tested (Figure 12.8).

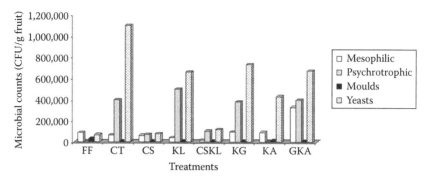

FIGURE 12.8 Incidence of microorganism in organic strawberry without and with coatings, stored at 10°C ± 2°C and 60%–80% RH, for 9 days.

In coatings based on water kefir and milk, although lower than the control, the values were higher than other treatments, possibly by the presence of yeast in these coatings. Reis et al. (2008) found 1.08×10^3 values of the count of fungi and yeasts in strawberries. The Oso Grande cv., newly harvested, along with evolution of the storage period, reaches 4.25×10^3 during the 9 days of storage and intensive reduction in the use of chlorate products. Psychrotrophics were reduced by coating with starch and this combined with chitosan and the water kefir in the control and other treatments. The incidence of mold was quite low in relation to other microorganisms. These results agree with Mali and Grossmann (2003) who found the reduction of mesophilic microorganisms and psychrotrophic mold/yeast strawberries in "Dover" covered with films based on starch from yam throughout the storage period. According with Siro et al. (2006), extending the useful life of strawberry "Camarosa" by means of passive MA rich in oxygen and atmosphere did not favor the increase of acid-resistant pathogens—*Escherichia coli*, *Listeria monocytogenes*, and *Salmonella* sp.—but the restricted development and survival of these are available in the fruit stored at 7°C. Ribeiro et al. (2007) studied coatings based on polysaccharides such as starch (2.0%), carrageenan (0.3%), and chitosan (1%) and observed that the lower rate of development of microorganisms occurred in strawberries coated with chitosan. The chitosan, as reported by Park et al. (2005), controlled the development of *Rhizopus* sp. and *Cladosporium* sp., and when combined with potassium sorbate controlled the aerobic microorganisms and coliforms during storage of strawberries coated with polysaccharide.

Table 12.5 shows the mean values of chemical analyses for treatments carried out with organic strawberries "Camarosa" cultivar, kept under refrigeration (10°C ± 2°C and RH 60%–80%).

The variation between the average pH of the fruit was very small for most treatments, between 3.40 and 3.46, similar to the pH values obtained in fresh fruit. Fruits GKA had average pH of 3.50, significantly superior to other fruits. These values were slightly lower than those found by Cordenunsi et al. (2003) for cultivars "Oso Grande," "Dover," "Campineiro," and "Mazi," which have variation of pH between 3.5 and 3.8, even after refrigerated storage.

For the content of TSSs, the fruits coated with GKA had the highest value, above 8.51 °Bx, differing significantly from other treatments, which ranged between

TABLE 12.5
Values of Chemical Characteristics in Organic Strawberries, "Camarosa" Cultivar, to Fresh Fruit and Fruits Stored (10°C ± 2°C and RH 60%–80%) for 9 Days ($n = 3$)

T	pH	TSS (°Bx)	TTA (g Citric Acid 100 g⁻¹)	Ascorbic Acid (mg 100 g⁻¹)	Anthocyanins (mg 100 g⁻¹)
FF	3.46 a	7.16 a	1.06 a	45.17 b	13.30 a
CF	3.46 a	6.85 a	1.22 b	39.55 a	21.02 b
CS	3.46 a	7.53 a	1.27 c	44.00 b	20.74 b
KL	3.42 a	7.27 a	1.24 b	41.96 a	20.34 b
KG	3.40 a	7.81 a	1.24 b	44.87 b	21.19 b
CSKL	3.44 a	7.47 a	1.22 b	41.37 a	20.94 b
KA	3.43 a	7.86 a	1.24 b	39.30 a	21.19 b
GKA	3.50 b	8.51 b	1.27 c	41.60 a	24.50 c

Values followed by the same letter do not differ in Tukey's test ($P > 0.05$).

T, treatments; FF, fresh fruit; CF, control fruit; CF, cassava starch; CSKL, association of cassava starch 2.0% and kefir liquid 15%; KL, kefir liquid 15%; KG, kefir grains 15%; KA, kefiraride; GKA, association grains kefir and kefiraride; PGN, pelargonidin-3-glycoside; TSS, total soluble solids; TAA, total titratable acidity; n, number of replications.

6.85 °Bx and 7.86 °Bx, values close to the fresh fruit of 7.16 °Bx. Similar initial average, 7.35 °Bx, was found in strawberry "Camarosa" (Malgarim et al., 2006).

Fresh fruit, in this work, showed values of total acidity of 1.06 g of citric acid 100 g⁻¹ of pulp; fruit stored differed statistically, with increases in all treatments and control groups, differing significantly between them to 9 days. Lower values were obtained by Malgarim et al. (2006) in strawberries "Camarosa" fresh 0.6 g citric acid 100 g⁻¹ pulp due to the conditions of cultivation.

The average titratable acidity of fruits coated GKA and CS was 1.27 g citric acid 100 g⁻¹ of pulp, statistically superior to other treatments in the fruits of the ninth day of storage, which showed between 1.22 and 1.24 g 100 g⁻¹ citric acid pulp. It was found that the association with starch did not favor increasing the acidity in fruits coated. Changes were observed in levels of acidity unaccompanied by pH, which can be explained by the buffer capacity of the pulp of the strawberry, due to the high content of citric acid, and through the release of ions to the water, acting as a buffer and opposing to changes in pH (Lehninger, 1976).

The average content of ascorbic acid original was 45.17 mg 100 g⁻¹ pulp. Most of the fruits showed significant reduction in the contents of the storage period, except the fruits coated with CS, and KG did not differ significantly from the fresh fruit. Calegaro et al. (2002) also found a reduction in content of ascorbic acid during storage of strawberry "Oso Grande" cv. Most fruits reached around 40 mg ascorbic acid 100 g⁻¹ for 9 days. However, the fruit receiving resveratrol sprinkled to 4000 ppm had significant reduction in levels of ascorbic acid.

TABLE 12.6

Anthocyanin Content in Organic Strawberries, "Camarosa" Cultivar, Stored for 9 Days ($n = 3$)

Treatments	Anthocyanins (mg PGN 100 g^{-1})
Fresh fruit	13.30
CF	21.02
CSF	20.74
CSKL	20.94
KL	20.34
KG	21.19

CF, control fruit; CF, cassava starch; CSKL, association of cassava starch 2.0% and kefir liquid 15%; KL, kefir liquid 15%; KG, kefir grains 15%.

Anthocyanin contents in "Camarosa" organic strawberries, after 9 days storage under refrigeration, varied from 13.30 (fresh fruit) to 21.19 mg of pelargonidin-3-glycoside (PGN) 100 g^{-1} of pulp (KG) in the different treatments made. Severo et al. (2007) found a content around 12.52 mg of PGN 100 g^{-1} of fresh "Camarosa" strawberry pulp, in commercial maturation condition, similar to results found in this investigation for freshly picked fruits (Table 12.6).

An increase in anthocyanin contents was reported by Gil et al. (1997) during refrigerated storage. Anthocyanin initial values were around 12.02 ± 2 mg 100 g^{-1} for "Selva" strawberry, reaching 15.3 ± 1.2 and 14.2 ± 3.9 mg 100 g^{-1} in a period between 5 and 10 days, respectively. However, they found lower values in fruits treated with concentrations of CO_2 in 10%, 20%, and 40%. Silva et al. (2007) determined the total anthocyanin content in five different strawberry cultivars and found a variation between 20 and 60 mg 100 g^{-1}, in which PGN (77%–90%) and cyanidin-3-glycoside (3%–10%) prevailed. The author has also mentioned aspects such as the influence of the different cultivars, the degree of maturity or ripeness, edaphic-climatic factors, and the postharvest storage in their reports.

The application of postharvest coating SKL, as a result of the association of cassava starch at 2.0% (CSF) to the liquid of kefir water at 15% (KL), was the PVC film, which presented the best results regarding the preservation of the physical characteristics and the reduction in the incidence of rottenness in organic strawberries belonging to "Camarosa" cultivar investigation.

12.7.3 BLUEBERRY

Blueberry (*Vaccinium askey* Read) is a fruit that belongs to the family Ericaceae, and it is classified in the subfamily Vaccinioideae, where the genus *Vaccinium* is found (Antunes and Madail, 2005; Strik, 2005, 2007; Antunes and Raseira, 2006; Donadio et al., 1998; Trehane, 2004; Kalt et al., 2007).

The fruit complete coloration, before its ideal harvesting time, can be numbered as one of the major limitations of these cultivars, for it interferes on the fruit's taste, and it is when the fruit has more acid, tending to cleave its epidermis during rainy seasons (Gough, 1994).

The growing interest for the natural antioxidants made from plant extracts, due to its low toxicity when compared to synthetic antioxidants, is an important factor for the increase of blueberry-cultivated areas. Extracts from fruits, vegetables, cereals, and its industrialized subproducts abound with antioxidants, ascorbic acid, tocopherols, carotenoids, and phenolic compounds; and they have shown an effective antioxidant activity in model systems (Manach et al., 2004; Wolfe et al., 2003; Brazelton and Strik, 2007).

Phenolic compounds are produced in the secondary metabolism of plants, and they defend the plant against pest attack. It has been observed, however, that phenolic compounds are capable of reacting with free radicals, creating stable radicals in humans and animals. This power of neutralization that phenolic compounds have against these radical structures is due to its chemical structure which contains at least one aromatic ring with hydroxyl groups (Giada and Mancini-Filho, 2006).

Besides its antioxidant activity, studies have highlighted multiple functions and important mechanisms related to the phenolic compounds' ability to bond themselves to cell receptors and membrane transporters and to influence on gene expression, signaling, and cell adhesion (Manach et al., 2005).

The use of eatable pellicles has been explored to cover fresh fruits and vegetables, in order to minimize humidity loss and reduce respiration rate and also to bestow them a shining and attractive appearance (Azeredo, 2003). The use of pellicles with this purpose is an economic advantage, avoiding the need for storage in CA, which would imply operational costs and use of equipment. The pellicle function depends mainly on the food product and the kind of deterioration that it is due to (Maia et al., 2000).

Blueberries (cultivar Florida M) were stored at different temperatures (2°C, 5°C, and 25°C room temperature) for a period of 15 days. Kefir was applied in some fruits in order to evaluate its influence on the conservation of these fruits, stored at refrigeration. The blueberries' physicochemical analysis results can be observed in Tables 12.7 through 12.12. The fruits were submitted to a treatment before its storage (T1, without covering, and T2, with kefir covering).

The loss of mass at each studied storage temperatures (room, 2°C, and 5°C) increased during the 15 day storage period. The results obtained for fruits with kefir coating showed lower mass loss. Thus, coating with kefir can be used as an alternative method for conservation of blueberries, increasing its shelf live.

At the end of this period of 15 days, the mass losses for the treatments at 2°C, 5°C, and room temperature were 10.92%, 14.05%, and 33.54%, respectively, for the control treatment. The results for fruits treated with kefir were 5.02%, 9.95%, and 13.02%, respectively. The room temperature was inefficient for the fruit conservation due to the appearance of fungi within 6 day storage and the high mass loss.

The results for the TSSs, observed at 2°C and room temperature, during the storage time, showed no statistical differences among the treatments at a level of significance of 5%. At room temperature, the TSS content was between 11.53 °Bx and 15.73 °Bx for the control treatment and between 10.5 °Bx and 11.43 °Bx for the fruits treated with kefir; it

TABLE 12.7

Physicochemical Characteristics of Blueberries Stored at Room Temperature ($n = 3$)

Storage Time	Treatment	Mass Loss (%)	°Bx	pH	ATT (g 100 g⁻¹ of Citric Acid)
0	T1	0	13.97	2.95	1.59
3	T1	5.08	14.33	2.92	1.68
6	T1	9.89	13.13	3.05	2.16
9	T1	17.23	11.37	2.74	3.24
12	T1	24.55	11.53	2.88	5.53
15	T1	33.54	15.70	2.98	6.08
0	T2	0	11.97	2.96	1.62
3	T2	1.86	11.83	2.87	2.02
6	T2	4.15	12.43	2.95	2.69
9	T2	7.34	11.80	2.56	4.46
12	T2	10.34	10.80	2.88	5.72
15	T2	13.02	10.50	2.93	5.88

n, number of repetitions; T1, treatment without covering; T2, treatment with kefir covering.

TABLE 12.8

Physicochemical Characteristics of Blueberries Stored at 2°C Temperature ($n = 3$)

Storage Time	Treatment	Mass Loss (%)	°Bx	pH	ATT (g 100 g⁻¹ of Citric Acid)
0	T1	0	11.63	2.97	1.63
3	T1	4.24	12.63	2.99	1.43
6	T1	5.59	13.67	2.77	1.94
9	T1	8.09	12.23	2.97	1.96
12	T1	9.83	14.40	3.21	2.19
15	T1	10.92	14.17	3.33	1.95
0	T2	0	11.53	2.96	1.58
3	T2	2.19	13.63	2.94	1.56
6	T2	4.05	13.53	2.79	2.11
9	T2	7.2	13.57	2.93	2.08
12	T2	8.35	14.47	3.14	2.82
15	T2	9.95	14.50	3.26	2.13

n, number of repetitions; T1, treatment without covering; T2, treatment with kefir covering.

TABLE 12.9
Physicochemical Characteristics of Blueberries Stored at 5°C Temperature ($n = 3$)

Storage Time	Treatment	Mass Loss (%)	°Bx	pH	ATT (g 100 g^{-1} of Citric Acid)
0	T1	0	14.23	2.99	1.61
3	T1	3.62	14.27	2.94	1.54
6	T1	5.84	14.43	3.11	1.73
9	T1	8.82	13.47	2.95	1.76
12	T1	10.94	14.10	3.13	2.17
15	T1	14.05	13.87	3.30	2.03
0	T2	0	13.37	2.96	1.58
3	T2	1,12	11.73	2.91	1.63
6	T2	1.98	14.00	3.12	1.66
9	T2	3.65	14.10	2.95	1.85
12	T2	4.11	13.43	3.18	2.36
15	T2	5.02	13.67	3.25	1.17

n, number of repetitions; T1, treatment without covering; T2, treatment with kefir covering.

TABLE 12.10
Carotenoids, Phenolic Compounds, and Anthocyanins in Blueberries Stored at Room Temperature ($n = 3$)

Storage Time	Treatment	Carotenoids (mg/mL)	Phenolic Compounds (mg GAE 100 g^{-1})	Anthocyanins (mg 100 g^{-1})
0	T1	2.00	3390	57.53
3	T1	3.21	4160	176.16
6	T1	3.11	2780	59.17
9	T1	3.28	7910	35.17
12	T1	3.42	2960	27.25
15	T1	3.31	4340	28.02
0	T2	5.21	3990	75.41
3	T2	3.15	5580	181.35
6	T2	3.98	4360	49.95
9	T2	3.26	5750	62.04
12	T2	2.99	2870	45.79
15	T2	3.19	4200	36.64

n, number of repetitions; T1, treatment without covering; T2, treatment with kefir covering.

TABLE 12.11

Carotenoids, Phenolic Compounds, and Anthocyanins in Blueberries Stored at 2°C Temperature ($n = 3$)

Storage Time	Treatment	Carotenoids (mg/mL)	Phenolic Compounds (mgGAE 100 g⁻¹)	Anthocyanins (mg 100 g⁻¹)
0	T1	4.07	3050	65.31
3	T1	3.15	4670	187.19
6	T1	2.93	4690	62.75
9	T1	2.70	5270	87.15
12	T1	2.82	5070	69.35
15	T1	4.01	5020	86.65
0	T2	4.59	3730	63.13
3	T2	3.26	4150	175.99
6	T2	2.62	4690	53.73
9	T2	3.96	5380	79.89
12	T2	2.66	5470	70.83
15	T2	3.55	5450	82.90

n, number of repetitions; T1, treatment without covering; T2, treatment with kefir covering.

TABLE 12.12

Carotenoids, Phenolic Compounds, and Anthocyanins in Blueberries Stored at 5°C Temperature ($n = 3$)

Storage Time	Treatment	Carotenoids (mg/mL)	Phenolic Compounds (mgGAE 100 g⁻¹)	Anthocyanins (mg 100 g⁻¹)
0	T1	4.17	3410	63.09
3	T1	2.91	3990	153.81
6	T1	5.09	4760	60.74
9	T1	4.65	6940	107.56
12	T1	2.98	4160	93.31
15	T1	3.51	5010	73.66
0	T2	4.71	3620	68.60
3	T2	2.68	3850	204.33
6	T2	3.18	4650	65.09
9	T2	2.98	5230	83.11
12	T2	4.05	3760	72.28
15	T2	3.43	5170	81.05

n, number of repetitions; T1, treatment without covering; T2, treatment with kefir covering.

showed that kefir influences the fruit conservation. On days 3 and 12, significant differences could be seen among the treatments stored at 5°C, at the level of 5%. The decrease in TSS content that sometimes could be observed during the storage can be explained by the fact that sugar and acids are used as substrates on the respiration process.

The results for pH at the studied temperatures showed statistical differences among the average values, at the level of significance of 5%. These differences could be seen on the third day of storage at room temperature. From the third to the last day of storage, the pH variations for the fruit treated with kefir were always lower. At 2°C, the differences occurred on days 12 and 15. An increase in pH could be seen along the storage time. On day 0, the pH value for the control treatment was 2.97 and 2.96 for the fruits with kefir. At the end of the storage time, the observed values were 3.33 and 3.26, respectively. The pH value media for blueberries stored at 5°C also showed significant differences, at the level of 5%, only on days 12 and 15.

The results obtained for acidity, at room temperature, showed linear growths for both treatments during the 15 day storage. Among these treatments, significant differences could be observed, at the level of 5%, on days 3, 6, and 9. It could be concluded that room temperature is inefficient for acidity because an excessive increase in this blueberry characteristic may cause undesirable organoleptic alterations, spoiling its commercialization. On day 0, the acidity values were 1.59 and 1.62 g $100 g^{-1}$ of citric acid for T1 and T2, respectively. At the end of the storage time, the observed results were 6.08 and 5.88 $100 g^{-1}$ of citric acid for T1 and T2, respectively. The values observed for titratable acidity, at 2°C and 5°C, showed a nonlinear increase.

For the blueberries stored at room temperature, the carotenoids' content varied from 2.00 mg/mL on day 0 to 3.31 mg/mL at the end of the storage time, for the control treatment. The highest value obtained was 3.24 mg/mL on day 12. In the treatment with kefir, the results varied between 2.99 and 5.21 mg/mL. At 2°C, there were no significant differences in the carotenoids' content, at the level of 5%, among the treatments. The highest values for carotenoids, both in T1 and T2, could be seen on day 0, corresponding to 4.07 and 4.59 mg/mL, respectively. For the storage at 5°C, significant differences could be observed in carotenoids' content, at the level of 5%, on days 6, 9, and 12. However, the highest carotenoids' content could be observed on day 6, in the control treatment, which was 4.65 mg/mL; in treatment T2, the highest content was 4.71 mg/mL, on day 0.

Data on carotenoids' content in small fruits are still very limited. Small fruit culture in Brazil has called the attention of producers, merchants, and consumers, especially in the past few years, due to the confirmation of the presence of antioxidants. The results obtained in this project show that this fruit is a source of these antioxidant compounds.

The phenolic compounds' content showed statistical differences among the treatments, at the level of 5%, when they were stored at room temperature on days 3, 6, and 9. For T1 and T2, the highest value obtained was after 9 days of storage, corresponding to 7910 and 5750 mg GAE $100 g^{-1}$ of fruit, respectively. At the temperatures of 2°C and 5°C, and the level of 5%, there were no significant differences among the treatments during the 15 days. The highest values obtained on storage at 5°C were 6940 and 5230 mg GAE $100 g^{-1}$, respectively, for T1 and T2.

The results obtained for anthocyanins showed a variation during the 15 days of storage, which can be explained by a sort of degradation in the blueberry physiological ripening time. A high content of anthocyanins could be seen in these fruits; however, they are very unstable. At every storage temperature, the anthocyanin content showed no statistical differences among the treatments, at the level of significance of 5%. At room temperature, the anthocyanin contents for T1 and T2 were lower than the contents obtained on the harvest day.

The presence of a high amount of these compounds in blueberry fruits confirms its benefits and importance to health. Therefore, due to the blueberry's short postharvest life, studies aiming to increase it are important for a better use in fresh fruit commercialization as well as its use in processing industry.

During the 15 days of storage, the temperatures of 2°C and 5°C were more adequate for blueberry conservation. However, the use of kefir covering combined with a 5°C temperature is the best alternative to conserve the fruit and increase its life on the supermarkets shelves.

12.7.4 BLACKBERRY

Among several fruit species that have good culture and commercialization perspectives, the blackberry tree is one of the most promising ones (Antunes, 2002).

To promote technologic, economic, environmental, and social development toward organic fruit production, the use of different treatments postharvest, which may be associated to its production, is necessary to reach a loss reduction and maintain the product quality for a longer period of time (Campos et al., 2009). Among the postharvest conservation methods, there is the use of revetment and covering in fruits, which aims an increase in its preservation period associated—or not—to the use of low temperatures. Temperature is probably the factor that most affects fruits and vegetables storage time, for it controls every biochemical and physiologic process (Santos et al., 2008).

Manioc starch can be used in fruit conservation after the emulsification process, which occurs at temperatures higher than 70°C, with excess of water. Kefir grains are white or slightly yellowish clay-like grains, with gelatinous appearance, made of proteins and polysaccharides. It also contains bacteria and yeast which are involved in the fermentation process and which are retaken and reused in a further incubation. Ten species of kefir yeast were identified and the major ones are *Issatchenkia orientalis*, *S. unisporus*, *S. exiguous*, and *S. humaticus* (Latorre-Garcia et al., 2007). The use of kefir in water has been spread out by people who use it and tell the others about it. Since these are handicraft products, and they are not well divulged yet, there is little information and research on its application.

The blackberry is a fruit that possesses a short postharvest life, a fragile structure, and high respiratory activities. Therefore, technology is needed to increase its useful life, guarantee longer conservation time and postharvest quality. Studies with blackberry fruits, cultivar Tupy, were made in order to observe its physicochemical quality, covering them with emulsions made of manioc starch, kefir grains, and water and cooling them at 10°C.

The values for blackberries pH, TA, SS, ratio (SS/TA), and anthocyanins with covering under refrigeration (10°C ± 2°C) can be observed in Tables 12.13 through 12.17.

Antunes et al. (2003) report the values obtained in the beginning of their experiment with the cultivars Comanche and Brazos, where they observed media of pH

TABLE 12.13

pH in Blackberry Fruits, Cultivar Tupy, Stored at 10°C

	T1		T2		T3	
Time	2008	2009	2008	2009	2008	2009
0 day	2.83 ± 0.00 e	3.07 ± 0.03 a	2.90 ± 0.00 d	3.02 ± 0.02 b	2.95 ± 0.01 c	3.01 ± 0.01 b
3 days	2.99 ± 0.06 b	3.02 ± 0.02 ab	2.85 ± 0.02 c	3.08 ± 0.01 a	2.87 ± 0.02 c	3.08 ± 0.05 a
6 days	3.08 ± 0.03 b	3.30 ± 0.05 a	3.04 ± 0.08 b	3.27 ± 0.02 a	3.27 ± 0.03 a	3.32 ± 0.03 a
9 days	3.27 ± 0.02 bc	3.33 ± 0.02 ab	3.14 ± 0.10 d	3.33 ± 0.06 ab	3.18 ± 0.03 cd	3.44 ± 0.02 a
12 days	3.53 ± 0.06 a	3.39 ± 0.03 a	3.9 ± 0.05 a	3.38 ± 0.08 a	3.37 ± 0.14 a	3.52 ± 0.06 a

Values followed by the same letter do not differ in Tukey's test ($P > 0.05$).
T1, without covering; T2, manioc starch 2.5%; T3, water kefir grains 20%.

TABLE 12.14

Total Acidity (g Citric Acid 100 g⁻¹) in Blackberry Fruits Stored at 10°C

	T1		T2		T3	
Time	2008	2009	2008	2009	2008	2009
0 day	1.05 c	1.2 ab	0.86 d	1.30 a	0.80 d	1.20 b
±δ[a]	±0.02	±0.06	±0.08	±0.03	±0.02	±0.04
3 days	1.31 a	1.01 b	1.23 a	1.06 b	1.23 a	1.00 b
±δ	±0.07	±0.08	±0.06	±0.01	±0.04	±0.05
6 days	1.20 a	0.83 c	1.16 a	0.97 bc	0.99 b	0.93 bc
±δ	±0.03	±0.05	±0.05	±0.08	±0.09	±0.08
9 days	0.95 abc	0.83 c	0.98 ab	1.05 a	0.69 d	0.87 bc
±δ	±0.12	±0.05	±0.04	±0.04	±0.03	±0.02
12 days	0.88 bc	0.90 bc	0.85 c	1.12 a	0.88 c	1.03 b
±δ	±0.03	±0.07	±0.05	±0.09	±0.16	±0.09

Values followed by the same letter do not differ in Tukey's test ($P > 0.05$).
T1, without covering; T2, manioc starch 2.5%; T3, water kefir grains 20%.
[a] Standard deviation.

3.59 and 3.39. At the end of the storage time (12 days), the values rose to 3.94 and 4.09, respectively. This rise in pH, reported by the authors, could be observed in this study, with every treatment on both crops. The pH is a parameter that measures acidity in fruits and food; therefore, a rise in its value is directly related to a decrease in acidity, which happens due to the ripening of the fruits (Chitarra and Chitarra, 2005).

The results observed in this study for ripe fruits, cultivar Tupy, were similar to those observed by Tosun et al. (2008), where an increase in pH values during storage could be observed.

TABLE 12.15

Soluble Solids (°Bx) in Blackberry Fruits, Cultivar Tupy, Stored at 10°C

	T1		T2		T3	
Time	2008	2009	2008	2009	2008	2009
0 day	10.13 a	8.33 c	9.28 b	8.15 c	10.08 a	8.45 c
$\pm\delta^a$	±0.05	±0.22	±0.05	±0.38	±0.15	±0.17
3 days	8.93 a	7.43 b	9.25 a	7.55 b	9.15 a	7.60 b
±δ	±0.7	±0.22	±0.35	±0.12	±0.29	±0.24
6 days	9.63 a	7.77 c	9.15 b	7.45 c	9.18 ab	7.70 c
±δ	±0.17	±0.15	±0.24	±0.33	±0.13	±0.14
9 days	8.90 a	7.08 b	8.55 a	6.85 b	8.30 a	7.33 b
±δ	±0.22	±0.35	±0.38	±0.42	±0.47	±0.25
12 days	8.43 a	6.68 b	8.50 a	6.08 bc	8.63 a	5.80 c
±δ	±0.29	±0.41	±0.26	±0.26	±0.25	±0.31

Values followed by the same letter do not differ in Tukey's test ($P > 0.05$).
T1, without covering; T2, manioc starch 2.5%; T3, water kefir grains 20%.
[a] Standard deviation.

TABLE 12.16

Ratio SS/AT in Blackberry Fruits, Cultivar Tupy, Stored at 10°C

	T1		T2		T3	
Time	2008	2009	2008	2009	2008	2009
0 day	9.64 c	6.85 de	10.72 b	6.28 e	12.46 a	7.05 d
$\pm\delta^a$	±0.23	±0.38	±0.52	±0.27	±0.40	±0.12
3 days	6.80 a	7.32 a	7.49 a	7.08 a	7.41 a	7.56 a
±δ	±0.84	±0.77	±0.2	±0.13	±0.37	±0.55
6 days	7.94 abc	9.28 a	8.85 bc	7.62 c	9.23 ab	8.31abc
±δ	±0.16	±0.4	±0.49	±0.6	±1.0	±0.73
9 days	9.35 b	8.51 b	8.63 b	6.49 c	11.93 a	8.36 b
±δ	±1.09	±0.96	±0.72	±0.65	±0.7	±0.29
12 days	9.55 ab	7.47 bc	10.10 a	5.47 c	10.05 a	5.63 c
±δ	±0.7	±0.91	±0.91	±0.67	±2.29	±0.33

Values followed by the same letter do not differ in Tukey's test ($P > 0.05$).
T1, without covering; T2, manioc starch 2.5%; T3, water kefir grains 20%.
[a] Standard deviation.

TABLE 12.17
Total Anthocyanins (mg 100 g⁻¹) in Blackberry Fruits, Cultivar Tupy, Stored at 10°C

	T1		T2		T3	
Time	2008	2009	2008	2009	2008	2009
0 day	15.68 c	71.99 a	15.21 c	62.85 b	19.30 c	62.41 b
±δ[a]	±1.73	±5.9	±0.03	±4.61	±4.10	±1.92
3 days	46.56 c	57.41 ab	41.89 c	64.01 a	49.66 bc	57.85ab
±δ	±8.08	±1.50	±3.25	±6.44	±3.01	±1.27
6 days	89.67 b	70.01 bc	81.59 b	48.25 c	119.82 a	54.71 c
±δ	±16.13	±5.25	±13.58	±7.27	±5.31	±4.52
9 days	128.44 a	67.47 b	125.84 a	54.19 b	117.0 a	61.13 b
±δ	±6.35	±6.72	±12.09	±3.64	±13.54	±4.60
12 days	108.97 ab	52.04 c	85.88 b	46.44 c	120.04 a	38.66 c
±δ	±11.82	±10.48	±18.83	±6.81	±16.56	±3.73

Values followed by the same letter do not differ in Tukey's test ($P > 0.05$).

T1, without covering; T2, manioc starch 2.5%; T3, water kefir grains 20%.

[a] Standard deviation.

Sousa (2007) reports that the reduction in TA and SS content during postharvest storage is due to the fact that sugar and acids are used as respiratory substrates, which decreases its stock. The author analyzed the postharvest behavior of blackberries (cultivar Arapalho) harvested in the beginning and in the end of June. He saw that SS values were 13.0 °Bx and 11.0 °Bx, respectively. Meneghel et al. (2008), while studying the application of coverings with sodium alginate, obtained values 9.1 °Bx, 8.4 °Bx, and 8.1 °Bx for SS. Antunes et al. (2003) evidenced an SS content reduction in blackberry fruits, cultivars Brazos and Comanche, during storage at 20°C.

Ratio SS/TA is used to evaluate the sweet/sour balance in fruits. According to Carvalho et al. (1990), when this ratio is between 12 and 18, it indicates an equilibrated sensorial balance. On day 0 of storage, T3 showed values within this ratio rate.

Meneghel et al. (2008) obtained initial value for anthocyanins in blackberry fruits (cultivar Comanche) equal to 53 mg 100 g⁻¹ and observed a linear increase in these contents during the storage time. In this study, an increase in anthocyanin content could also be seen. When compared to the other treatments, T3 showed a higher content at the end of storage time. Several factors may contribute for anthocyanin degradation, such as sugar (mainly, fructose), which accelerate the darkening process and pH that provides higher stability when between 1 and 3.5.

Differences in the behavior of the chemical characteristics analyzed in both crops may occur, and these conditions may be justified by factors that precede the harvest, such as climatic conditions and harvesting moment (Pinheiro et al., 2005).

Antunes et al. (2003) evaluated the use of PVC film in cultivars Brazos and Comanche and observed a loss rate of 14.86%, at 20°C. In the present study, the treatment with the lowest efficiency happened to be T2, and T3 happened to be the one that reduced the mass loss up to the sixth day of storage.

Cia et al. (2007), while studying cultivars Guarani and Caingangue, obtained losses at a rate of 24% and 19%, respectively, in 3 days of storage at room conditions. In blackberry fruits, cultivar Navaho, stored for 7 days, in addition to 2 days stored at 20°C, the rottenness rate was 14% (Perkins-Veazie et al., 1997). The differences occurring between the two crops for mass loss and rottenness rate can be justified either by climatic conditions of each year or by its fragility at postharvest handling, which may cause cleaves, thus increasing loss rate.

Firmness is a factor that directly influences quality and postharvest life. Meneghel et al. (2008) evaluated three types of covering in blackberry fruits stored at 0°C and observed that the fruits firmness did not decrease with time and was not affected by the coverings. Average was 16, 20, and 19 N. This fact could not be observed in the present study, where the blackberry fruits stored at 10°C showed a reduction in its firmness. Souza (2007) could see in his work on blackberries harvested in the beginning and at the end of June that the fruits' firmness showed a small oscillation (3.5 and 4.20 N) up to the ninth day. The fruits (cultivar Tupy) in the present work showed higher firmness than those found by Souza. At the end of the 12th day of storage, T2 showed the lowest firmness resistance. T2 and T3 interfered positively on both crops up to the sixth day.

To reduce rottenness rate, the use of covering with kefir grains is recommended. This covering also provides lower mass loss, chemical quality, and higher fruit firmness, which allows the fruit consumption until the third day of conservation at 10°C.

12.7.5 COCONUT

Coconut water taste is sweet, slightly astringent. The water's physicochemical characteristics are influenced by variety, maturity stage, soil, irrigation, and climatic conditions (Aragão, 2000). Food industry aims to lengthen the period of time when food remains adequate for consumption. When a fruit is harvested, a natural and irreversible process of degradation starts, which means that an interruption in gaseous balance occurs, with a high oxygen influx and a proportional CO_2 loss. Under this condition (high concentration of O_2 and low concentration of CO_2), cells are not renewed and the respiration rate increases leading to the gradual ripening and eventual senescence of the fruit (Assis and Leoni, 2003).

Commercializing coconut water inside the fruit involves problems related to the product's storage and perishability, hampering its consumption in places distant from the producing regions (Aroucha et al., 2006). Among the conservation methods, there is the use of covering and revetment in fruits, in order to increase its period of preservation. Chitosan is a polymer that, due to its particular properties, has been suggested as an option to this emerging necessity, as well as refrigeration, for the demand for cooled food—which preserves the characteristics of "fresh" fruit, even then it is stored for long periods of time—is growing day after day (David and Fernandes, 1998; Paull, 1999; Kwiatkowski et al., 2008).

Two hundred fruits (cultivar Anão Verde) with a 7 month physiological development were evaluated. The fruits were sanitized with a solution made of sodium hypochlorite 2.0% and dried afterward in order to receive a covering with chitosan (deacetylation degree 98.18%, polymer).

Four treatments were carried out: application of chitosan covering with concentrations 0.0%, 0.5%, 1.0%, and 1.5% in aqueous solution with 0.8% of ascorbic acid and 2% of glycerol. The coconuts were immersed in the covering solution for 3 min, being that the control treatment (0.0% of chitosan) was immersed in distilled water. The fruits were stored in a hothouse BOD at 10°C ± 2°C, for a period of 44 days. Each treatment was carried out in triplicate. The sample unit consisted of eight fruits for evaluation of mass and volume loss and three fruits for evaluation of amount of fresh water.

Evaluation of fruit mass was carried out by weighing in digital scales, and mass loss was calculated, obtaining as a result the weight media for eight identified fruits. Coconut water volume was calculated using volumetric cylinders, and the values for three fruits were compared individually. Coconut water pH was evaluated using a digital potentiometer. TTA and TSS contents were determined according to a methodology described by Instituto Adolfo Lutz (IAL, 2008). RS and TS were determined using Lane–Eynon's method. Ascorbic acid content was determined using 2,6-dichlorephenolindophenol-sodium (DCPI). DCPI in basic or neutral medium is blue, while in acid medium, it is pink and its reduced form is colorless. Titration final stage is detected when a colorless solution becomes pink, when the first drop of DCPI is put into the system with the ascorbic acid already consumed. Lipid content quantification was carried out in a direct extraction with Soxhlet device, using petroleum ether as solvent, with a 6 h reflux (IAL, 2008).

Results for physical evaluations of the fruit mass and coconut water volume can be seen in Tables 12.18 and 12.19.

TABLE 12.18
Monitoring of Mass Loss of Green Coconuts (Cultivar Anão Verde), Covered with Chitosan and Stored at Refrigeration (10°C ± 2°C) ($n = 3$)

Time	0.0% Chitosan	0.5% Chitosan	1.0% Chitosan	1.5% Chitosan
4	3.99 a	3.76 ab	2.77 c	3.01 bc
8	1.44 a	1.30 a	0.96 a	1.16 a
12	1.45 a	1.31 a	1.15 a	1.19 a
16	1.91 a	1.45 ab	1.20 b	0.97 b
20	1.78 a	1.28 b	1.05 b	1.24 b
24	1.79 a	1.26 b	1.23 b	1.25 b
28	1.99 a	1.72 a	1.61 a	1.75 a
32	1.92 a	1.86 a	1.61 a	1.74 a
36	2.03 a	1.28 a	1.26 a	1.70 a
40	1.68 a	1.07 a	1.33 a	1.43 a
44	1.90 a	0.87 b	0.94 b	1.17 ab
Total	21.88	17.16	15.11	16.61

Values followed by the same letter do not differ in Tukey's test ($P > 0.05$). n, number of repition. Time = days.

TABLE 12.19

Green Coconut (Cultivar Anão Verde) Water Volume, Covered with Chitosan and Stored at Refrigeration (10°C ± 2°C) ($n = 3$)

Time	0.0% Chitosan	0.5% Chitosan	1.0% Chitosan	1.5% Chitosan
0	230.00 b	300.00 a	230.00 b	300.00 a
4	290.00 a	258.33 a	335.00 a	315.00 a
8	258.33 b	270.00 ab	353.33 ab	403.33 ab
12	231.61 b	350.00 a	348.00 a	311.67 a
16	230.33 b	355.00 a	251.67 ab	353.33 a
20	250.00 b	386.33 a	390.00 a	386.67 a
24	220.00 b	360.00 a	370.00 a	300.00 a
28	273.33 c	313.33 b	370.00 a	380.00 a
32	316.67 b	310.00 a	396.67 a	331.67 b
36	278.33 b	310.00 b	396.67 a	355.00 ab
40	280.00 a	348.33 a	350.00 a	311.67 a
44	290.00 a	311.67 a	283.33 a	306.67 a

Values followed by the same letter do not differ in Tukey's test ($P > 0.05$). n, number of repition. Time = days.

Day 0 was considered as first day of analysis. On this day (1 day after harvest), the fruits were covered with chitosan, but they were stored at refrigeration. The fruit analysis was carried out every 4 days, during the storage time. For mass loss (%), the fruits were analyzed after the fourth day of storage, comparing to the first day.

It could be seen that mass loss during the 44 days of storage differed significantly in Tukey's test ($P > 0.05$), among the treatments. From the fourth day on, this difference could be noticed in the treatments with 1.0% and 1.5% of chitosan, which showed a lower percentage of mass loss in the fruit. This difference was also noticeable on the 16th day of experiment, when this difference could be seen among the treatments with 0.5%, 1.0%, and 1.5% of chitosan, comparing to the control treatment (0.0% chitosan). This difference remained up to the 24th day of storage. After that, the difference appears again on the 44th day, when the control treatment shows higher mass loss, when comparing to the treatments with 0.5% and 1.0% of chitosan covering. At the end of the storage time, it could be seen that the control treatment showed 21.88% of mass loss, while the coconuts with chitosan covering showed losses below 20%.

In the literature, there are few reports about the application of chitosan in postharvest conservation of fruits. The application of chitosan 0.1% in grapes *Vitis vinifera* L., cultivar Jingxiu, in postharvest conservation, stored at 0°C, reduced fruit's weight loss to values below 3.0% (Meng et al., 2008). Chien et al. (2007a) used chitosan covering in citrus (*Murcott tangor*) and noticed a lower mass loss when compared to fruits without covering, during postharvest storage.

Regarding volume, it could be seen that a variation occurred significantly from the 12th day of storage on, but it was after the 8th day that the treatment with chitosan 0.0% started to show lower coconut water volume, and this statistic difference remains up to the 36th day. According to Costa et al. (2006), coconut (cultivar Anão Verde) water volume may vary between 250 and 400 mL.

Chemical determinations (pH, TTA, and TSS) of water from coconuts covered with chitosan showed statistical differences ($P > 0.05$) among the treatments. The values of pH varied from 4.81 to 5.87, which are within the values established by law, according to Normative n.39 from May 2002, which allows a minimum pH value of 4.3. Resende (2007) tells that coconut water pH is usually within 4.0 and 6.0. According to the author, the most acid pH may be due to the presence of organic acid fragments, free amino acids, CO_2 dissolved during tissue respiration, as well as the presence of fatty acid. Green coconut water is reported to be under hydrostatic pressure which may facilitate CO_2 dissolution in water. TTA showed significant difference ($P > 0.05$) among the treatments from the 8th to the 32nd day of experiment. TSS did not show any significant difference among the treatments on days 12 and 32. According to Aroucha et al. (2006), color, taste, and odor characteristics must be specific, and TSS content at 20°C must be at most 7.0 °Bx, and these values were found in these treatments. Resende (2007) evaluated the application of chitosan 1.5% in green coconuts (cultivar Anão Verde) and obtained a decrease in TSS values until the end of storage time (40 days). According to the author, this reduction in TSS may indicate a higher metabolism of consumed sugar by respiration. In the work of Aroucha et al. (2006), treatment of coconuts with paraffin covering provided better quality during conservation since pH and TSS kept above the values obtained for fruits without paraffin. It could be observed that pH values for water of coconuts with chitosan 1.5% covering were higher than those obtained for fruits without chitosan on days 0, 8, 16, 24, 40, and 44. For TSS, the values found were superior to the control ones on days 28, 40, and 44, for all treatments with chitosan covering.

It could be noticed that RS values varied from 3.78 to 5.60 g 100 mL^{-1}. These values showed statistical differences from the 12th day of storage on. The treatments with chitosan covering showed higher RS content. TS content also showed significant difference in Tukey's test ($P > 0.05$). These varied from 4.35 to 6.33 g 100 mL^{-1}. The treatments with chitosan also showed higher values of TS in coconut water. A degradation tendency both for TS and RS can be noticed from the 12th day on, being that there is an increase in RS content from the 24th day on.

Ascorbic acid content in coconut water varied from 15.55 to 85.33 mg 100 mL^{-1}. According to Aragão (2000), ascorbic acid—or Vitamin C—content in coconut water (cultivar Anão Verde) is 57 mg 100 mL^{-1}, being that these fruits were harvested with 7 month development. Significant differences between the treatments with chitosan and the control treatment could be seen from the eighth day of storage on at refrigeration ($P > 0.05$). From the 4th to the 20th day, ascorbic acid content kept regular, but from this day on, it showed a reduction to values below 50 mg 100 mL^{-1}. There was a reduction in ascorbic acid content for every fruit in the experiment, being that the control fruits showed the highest degradation index. Ascorbic acid value represents vitamin C content that even in this work showed low contents, and it has become important as researches establish a relation between the need for its consumption in a diet and human health (Resende, 2007).

The degradation of these compounds occurs due to metabolic activities which continue to be done after harvest. According to Assis et al. (2003), plants use atmospheric oxygen as electron acceptor during phosphorylation and release CO_2. When a fruit is harvested, the gaseous balance is interrupted, a high oxygen influx happens, and there is a considerable loss of CO_2. Under these conditions, cells are not renewed, and the respiration process increases, which causes a metabolic decline leading the fruit into a gradual ripening and/or senescence. Covering in the fruits causes a partial pore obstruction, reducing gaseous exchange and allowing the fruit to be conserved for a longer period of time (Chitarra and Chitarra, 2005).

Chien et al. (2007b) evidenced that the application of covering of chitosan 0.2%, 0.5%, and 1.0% maintains TTA, TSS, and ascorbic acid contents in minimally processed pitayas, when compared to the control at storage. Besides, it also shows a lower moisture loss. According to Aroucha et al. (2006), green coconuts, stored at refrigeration, remain marketable for 16 days. Resende (2007) observed that coconuts with chitosan 1.5%, stored at 12°C ± 2°C, remained conserved for 40 days. The fruits in the present study remained conserved up to 44 days, at 10°C ± 2°C.

Fruits covered with chitosan and kept at refrigeration showed lower mass loss and maintained its chemical quality. Green coconuts (cultivar Anão Verde) stored with chitosan covering presented advantages for chitosan covering when combining chitosan 0.5%—or more concentrated—with the used temperature.

REFERENCES

Afonso, M. R. A. Resfriamento a vácuo de alfaces hidropônicas (*Lactuca sativa* L.) cv. Salad Bowl: avaliação do processo e da vida pós-colheita. Campinas 2005. Tese (Doutor em Engenharia de alimentos)—Faculdade de Engenharia de Alimentos—Universidade Estadual de Campinas.

Andrade, R. S., Diniz, M. C. T., Neves, E. A., and Nóbrega, J. A. 2002. Determinação e distribuição de ácido ascórbico em três frutos tropicais. *Eclética Química*. 27: 1–8.

Antolniali, S. 2000. Resfriamento rápido com ar forçado paraconservação pós-colheita de alface "crespa". Dissertação (Mestrado em Engenharia Agrícola)—Universidade Estadual de Campinas, Faculdade de Engenharia Agrícola. Campinas, Brazil, p. 125.

Antunes, L. E. C. 2002. Amora-preta: nova opção de cultivo no Brasil. *Ciência Rural*. 32: 151–158.

Antunes, L. E. C., Duarte-Filho, J., and Souza, C. M. 2003. Conservação pós-colheita de frutos de amoreira-preta. *Pesquisa Agropecuária Brasileira*. 38: 413–419.

Antunes, L. E. C. and Madail, J. C. M. 2005. Mirtilo: que negócio é este?. *Jornal da Fruta*. 13: 8–9.

Antunes, L. E. C. and Raseira, M. C. B. 2006 Cultivo do mirtilo (*Vaccinium* spp.). Pelotas: Embrapa Clima Temperado. 99 pp. (Embrapa Clima Temperado. Sistema de Produção, 8).

Aragão, W. M. 2000. *A importância do coqueiro-anão verde (Coletâneas Rumos & Debates)*. Embrapa, Petrolina, Brazil.

Araújo, C. M. and Vasconcelos, H. O. 1974. Informações sobre a qualidade da laranja 'Folha Murcha' cultivada no estado do Rio de Janeiro. Arquivos da Universidade Federal Rural do Rio de Janeiro.UFRRJ, Rio de Janeiro, 4, pp. 19–28.

Arifin, B. B. and Chau, K. V. 1998. Cooling of strawberries in cartons with new vent hole designs. *ASHRAE Transactions* 94(1): 1415–1426.

Aroucha, E. M. M., Queiroz, R. F., Nunes, G. H. S., and Tomaz, H. V. Q. 2006. Qualidade pós-colheita do coco anã verde submetida ao recobrimento com parafina, durante armazenamento refrigerado. *Revista de Biologia e Ciências da Terra*. 6: 42–49.

ASHRAE. *Handbook of Fundamentals*. American Society of Heating, Refrigerating and Air Conditioning Engineers, Atlanta, GA, 1993.

ASHRAE. Methods of precooling fruits, vegetables and cut flowers. In: *Refrigeration Systems and Applications Handbook*, Chapter 10. American Society of Heating, Refrigerating and Air Conditioning Engineers, Atlanta, GA, 1994

Assis, O. B. G. and Leoni, A. M. 2003. Filmes Comestíveis de Quitosana. *RevistadeBiotecnologia—Ciência and Desenvolvimento*. 6: 33–38.

Assmann, A. P., Citadin, I., Kalicz, C. A., Locatelli, M. C., and Danner, M. A. 2006. Armazenamento de caqui CV. 'Fuyu' e laranja CV. 'Pêra' em atmosfera modificada sob diferentes temperaturas. Synergismus scyentifica UTFPR, Pato Branco. 1: 133–143.

Awad, M. 1993. *Fisiologia pós-colheita de frutos*. Nobel, São Paulo, Brazil, 114pp.

Azeredo, H. M. C. de. 2003. Películas comestíveis em frutas conservadas por métodos combinados: potencial da aplicação. *Boletim do CEPPA*. 21: 267–278.

Azeredo, H. M. C. de. 2004. *Fundamentos de Estabilidade de Alimentos*. Embrapa Agroindústria Tropical, Fortaleza, Brazil, 195pp.

Azeredo, H. M. C. de., Faria, J. A. F., and Azeredo, A. C. de. 2000. Active packaging for foods. *Ciência Tecnologia de Alimentos*. 20(3): 337–341.

Baldwin, E., Narciso, J., Cameron, R., and Plotto, A. 2006. Effect of pectin oligomers on strawberry fruit decay and ethylene production. *HortScience*. 41: 1044.

Baños, S. B. 2006. El control biológico en la reducción de enfermedades postcosecha em productos hortofrutícolas: uso de microorganismos antagônicos. *Revista Iberoamericana de Tecnología Postcosecha*. 8: 1–6.

Benato, E. A., Cia P., and Souza, N. L. 2001. Manejo de doenças de frutos pós-colheita. Revisão. *Anual de Patologia de Plantas*. 9: 403–440.

Bender, R. J. 2006. Colheita, beneficiamento, embalagem, conservação e comercialização. In: *Citricultura: Laranja: Tecnologia de produção, pós-colheita, industrialização e Comercialização*. Porto Alegre, Brazil.

Bleinroth, E. W., Sigrist, J. M. M., Ardito, E. de F. G. et al. 1992. *Tecnologia de pós-colheita de frutas tropicais*. ITAL, Campinas, Brazil, 203pp.

Brackmann, A., Hunsche, M., and Balem, T. A. 1999. Efeito de filmes de PVC esticável e polietileno no acúmulo de CO_2 e na manutenção da qualidade pós-colheita de morango cultivar Tangi. *Revista Brasileira de Agrociência*. 5: 89–92.

Brackmann, A., Steffens, C. A., and Mello, A. M. 2001. Efeito do pré-resfriamento e temperatura de armazenamento na qualidade de ameixas, cvs. Pluma 7 e Reubennel. *Revista Brasileira de Agrociência* 7(1): 18–21.

Brazelton, D. and Strik, B. C. 2007. Perspective on the U.S. and global blueberry industry. *Journal of American Pomological Society*. 61: 144–147.

Bronzi, L. P. 1984. Estudo da epiderme foliar de duas cultivares de *Citrus sinensis*: Laranja Folha Murcha e Laranja Pêra. In: *Congresso Brasileiro de Fruticultura, Anais... Florianópolis*, Sociedade Brasileira de Fruticultura, 2, pp. 622–631.

Calegaro, J. M., Pezzi, E., and Bender, R. J. 2002. Utilização de atmosfera modificada na conservação de morangos em pós-colheita. *Pesquisa Agropecuária Brasileira*. 37: 1–6.

Campos, R. P., Rodovalho, M. A., and Clemente, E. 2009. Coating on 'Camarosa' organic strawberries stored at low temperature. *Brazilian Journal Food Technology*. 12: 60–67.

Cantillano, R. F. F. 1986. Pré-resfriamento de frutas de clima temperado. Embrapa. Centro Nacional de Pesquisa de Fruteiras de Clima Temperado. Pelotas: 16pp. (Boletim de Pesquisa, 12).

Cantillano, R. F. 2003. Colheita e pós-colheita. In: Santos, A. M., Medeiros, A. R., Morango. *Produção*. Embrapa Clima Temperado (Pelotas, RS) Brasília: Embrapa Informação Tecnológica. (Frutas do Brasil n. 40).

Carraro, F. A. and Mancuso, C. M. 1994. *Manual de exportação de frutas*. Ministério da Agricultura, do Abastecimento e da Reforma Agrária, Brasília.

Castro, L. R. 2004. Análise dos parâmetros relacionados ao resfriamento a ar forçado em embalagens para produtos hortícolas. Universidade Estadual de Campinas, Faculdade de Engenharia Agrícola, Campinas, Brazil, 161pp. (Tese de Doutorado).

Castro, J. V., Park, K. J., and Honório, S. L. 2000. Determinação de curvas de resfriamento de uvas itália em dois sistemas de acondicionamento. *Engenharia Agrícola* 20(1): 34–44.

Cereda, M. P., Bertollini, A. C., Silva, A. P., Oliveira, M. A., and Evangelista, R. M. 1995. Películas de Almidón para la preservación de frutas. In: *Anais do Congresso de Polímeros Biodegradábles*. Avances y perspectivas, Buenos Aires, Argentina.

Chien, P. J., Sheu, F., and Lin, H. R. 2007a. Coating citrus (*Murcott tangor*) fruit with low molecular weight chitosan increases postharvest quality and shelf life. *Food Chemistry*. 100: 1160–1164.

Chien, P. J., Sheu, F., and Lin, H. R. 2007b. Quality assessment of low molecular weight chitosan coating on sliced red pitayas. *Journal of Food Engineering*. 79: 736–740.

Chitarra, M. I. and Chitarrra, A. B. 2005. *Pós-colheita de frutas e hortaliças: fisiologia e manuseio*. 2nd ed. revisada e ampliada. Universidade Federal de Lavras, Lavras, Brazil.

Cia, P., Bron, I. U., Valentini, S. R. T., Pio, R., and Chagas, E. A. 2007. Atmosfera modificada e refrigeração para conservação pós-colheita da amora-preta. *Bioscience Journal*. 23: 11–16.

Coelho, A. R., Hoffmann, F. L., and Hirooka, E. Y. 2003. Biocontrole de doenças pós-colheita de frutas por leveduras: perspective de aplicação e segurança alimentar. *Semina: Ciências Agrárias*. 24: 337–358.

Cordenunsi, B. R., Nascimento, J. R. O., and Lajolo, F M. 2003. Physicochemical changes related to quality of five strawberry fruit cultivars during cool storage. *Food Chemistry*. 83: 167–173.

Cortez, L. A. B., Siveira Jr., V., and Afonso, M. R. A. 2002. *Resfriamento de frutas e hortaliças*. Embrapa Informação Tecnológica, Brasília, DF, Brazil, 428pp.

Costa, J. M. C., Alves, M. C. S., Clemente, E., and Felipe, E. M. V. 2006. Características físico-químicas e minerais de água de coco de frutos da variedade Anã Amarelo em diferentes períodos de maturação. *Acta Scientiarum Agronomy*. 28: 173–177.

David, P. R. B. S. and Fernandes, Z. F. 1998. Conservação de água de coco por refrigeração. *Boletim do CEPPA*. 16: 1–12.

Del-Valle, V., Hernández-Munoz, P., Guarda, A., and Galotto, M. J. 2005. Development of a cactus-mucilage edible coating (*Opuntia fícus* Indica) and its application to extend strawberry (*Fragaria ananassa*) shelf-life. *Food Chemistry*. 91: 751–756.

Dias, M. S. C., Canuto, R. S., Santos, L. O., and Martins, R. N. 2005. *Doenças do morango*. *Informe Agropecuário*. 26: 40–43.

Dias, M. S. C., Costa, H., and Canuto, R. S. 2007. Manejo de doenças do morangueiro. *Informe Agropecuário*. 28: 64–77.

Dincer, I. 1995. Thermal cooling data for figs exposed to air cooling. *International Communications Heat Mass Transfer* 22(4): 559–566.

Donaldo, L. C., Nachtigal, J. C., and Sacramento, C. K. 1998. *Frutas Exóticas*. FCAV-UNESP, Sao Paulo, Brazil.

Duendas, L. H., Villas Boas, R. L., Sousa, C. M. P., Ragozo, R. A., and Bull, L. T. 2002. Fertirrigação com diferentes doses de NPK e seus efeitos sobre a produção e qualidade de frutos de laranja (*Citrus sinensis O.*) 'Valência'. *Revista Brasileira de Fruticultura*. 24: 214–218.

Dussán-sarria, S. S. and Honório, S. L. 2005. Parâmetros de resfriamento rápido do figo (*Ficus carica* L.) cv. Roxo de Valinhos embalado em caixa de exportação. *Revista UDO Agrícola* 5(1): 96–102.

Felllows, P. J. 2006. *Tecnologia do Processamento de Alimentos: princípios e prática*, 2ª edição. Artmed. Porto Alegre, 602pp.

Fernandes, A. G., Maia, G. A., Sousa, P. H. M., Costa, J. M. C., Figueiredo, R. W., and Prado, G. M. 2007. Comparação dos teores em vitamica C, carotenóides totais, antocianinas totais e fenólicos totais do suco tropical de goiaba nas diferentes etapas de produção e influência da armazenagem. *Alimentos e Nutrição*.18: 431–438.

Ferreira Neto, M., Gheyi, H. R., Holanda, J. S. de, Medeiros, J. F. de, and Fernandes, P. D. 2002. Qualidade do fruto verde de coqueiro em função da irrigação com água salina. *Revista Brasileira de Engenharia Agrícola e Ambiental*. 6: 69–75.

Ferri, V. C. and Rombaldi, C. V. 2004. Resfriamento rápido e armazenamento de caquis (*Diospyrus kaki*, L.), cv. Fuyu, em condições de atmosfera refrigerada e modificada. *Revista Brasileira de Fruticultura* 26(1): 36–39.

Figueiredo, J. O. 1999. Cultivares de laranjeiras no Brasil. In: 1° *SIMPÓSIO INTERNACIONAL DE FRUTICULTURA*. *Proceedings*, Botucatu, Brazil, Vol. 1, pp. 87–108.

Fontana, A. J., Varith, J., Ikediala, J., Reyes, J., and Wacker, B. 1999. Thermal properties of selected foods using a dual needle heat-pulse sensor. In: *ASAE/CSAE-SCGR Annual International Meeting*, 1999, Toronto, Ontario, Canada. ASAE, St. Joseph, MI, 10pp.

Fortes, J. F. 2003. Doenças. In. Santos. A. M., Medeiros, A. R., *Morango. Produção. Embrapa Clima Temperado*. Embrapa Informação Tecnológica, Pelotas, Brasília, 81 pp. (Frutas do Brasil n. 40).

Garrote, G. L., Abraham, A. G., and Antoni, G. L. 2001. Chemical and microbiological characterization of kefir grains. *Journal of Dairy Research*. 68: 639–652.

Gast, K. L. B. and Flores, R. 1991. *Precooling Produce: Fruits and Vegetables. Postharvest Management of Commercial Horticultural Crops*. Cooperative Extension Service, Manhattan, NY.

Giada, M. L. R. and Mancini, F. J. 2006. Importância dos compostos fenólicos da dieta na promoção da saúde humana. *Publicativo UEPG Ciências Biológicas e da Saúde*. 12: 7–15.

Gil, M. I., Holcroft, D. M., and Kader, A. A. 1997. Changes in strawberry anthocyanins and other polyphenols in response to carbon dioxide treatments. *Journal of Agricultural and Food Chemistry*. 45: 1662–1667.

Gough, R. E. 1994. *The Highbush Blueberry and its Management*. Haworth Press, Nova York.

Gouvêa, A. 2007. *Controle em campo e pós-colheita de doenças e metabolismo do morangueiro após tratamento com Saccharomyces cerevisiae*. Tese (Doutorado). Departamento de Fitotecnia e Fitossanitarismo, Setor de Ciências Agrárias, Universidade Federal do Paraná, Curitiba, Brazil.

Hardenburg, R. E., Watada, A. E., and Wang, C. Y. *The Commercial Storage of Fruits, Vegetables, and Florist and Nursery Stocks*, Agriculture Handbook No. 66. United States Department of Agriculture, Washington, DC, 1990.

Henrique, C. M. and Cereda, M. P. 1999. Utilização de biofilmes na conservação pós-colheita de morango (*Fragaria ananassa* Duch) cv. IAC Campinas. *Ciência e Tecnologia. Alimentos*. 19: 231–233.

Instituto Adolfo Lutz (IAL). Ministério da Saúde. Agência Nacional de Vigilância Sanitária. 2008. Métodos Físico-Químicos para Análise de Alimentos/Ministério da Saúde, Agência Nacional de Vigilância Sanitária. (Série A: Normas Técnicas e Manuais Técnicos). Brasília.

Instituto Agronômico do Paraná (IAPAR). 1992. *A citricultura no Paraná*. IAPAR, Londrina-PR-Brasil.

Janisciewicz, W. J. and Korsten, L. 2002. Biological control of postharvest diseases of fruits. *Annual Review of Phytopathology*. 40: 411–441.

Jacomino, A. P., Arruda, M. C., Bron, I. U., and Klunge, R. A. 2008. Transformações bioquímicas em produtos hortícolas após a colheita. In: Koblitz, M. G. Bioquímica de alimentos: Teoria e aplicações práticas. Rio de Janeiro, Brazil: Guanabara Koogan, cap. 6, pp. 154–189.

Kader, A. A., Zagory, D., and Kerbel, E. L. 1989. Modifield atmosphere packaging of fruits and vegetables. *Critical reviews in Food and Nutrition*. 28(1): 1–30.

Kalt, W., Joseph, J. A., and Shukitt-Hale, B. 2007. Blueberries and human health: A review of current research. *Journal of the American Pomological Society*. 61: 151–160.

Kluge, R. A. et al. 2005. Temperatura de armazenamento de tangores 'Murcote' minimamente processados. *Revista Brasileira de Fruticultura* 25(3): 533–536.

Kluge, R. A., Jomori, M. L. L., Eedagi, F. K., Jacomino, A. P., and Del Aguila, J. S. 2007. Danos de frio e qualidade de frutas cítricas tratadas termicamente e armazenadas sob refrigeração. *Revista Brasileira de Fruticultura*. 29: 233–238.

Koller, O. C. 2006. Origem e importância econômica da cultura da laranjeira. In: *Citricultura: Laranja: Tecnologia de produção, pós-colheita, industrialização e Comercialização*. Porto Alegre, Brasil.

Krochta, J. M. and Mulder-Johnston, C. 1997. Edible and biodegradable polymer films: challenges and opportunities. *Trends in Food Science and Technology*. 8: 228–237.

Kwiatkowski, A., Clemente, E., Scarcelli, A., and Vida, J. B. 2008. Quality of coconut water 'in natura' belonging to green dwarf fruit variety in different stages of development, in plantation on the northwest area of Paraná, Brazil. *Journal of Food, Agriculture and Environment*. 6: 90–93.

Latorre-García, L., Castillo-Agudo, L., and Polaina, J. 2007. Taxonomical classification of yeasts isolated from kefir base don the sequence of their ribosomal RNA genes. *World Journal of Microbiology and Biotechnology*. 23: 785–791.

Leal, P. A. M. and Cortez, L. A. B. 1998. Métodos de pré-resfriamento de frutas e hortaliças. In: *II Curso de Atualização em Tecnologia de Resfriamento de Frutas e Hortaliças*. Faculdade de Engenharia Agrícola (UNICAMP), Campinas, Brazil, pp. 81–91.

Lehninger, A. L. 1976. *Bioquímica: componentes moleculares das células*. 2nd ed. Edgar Blucher, São Paulo, Brazil.

Louzada, M. I. F., Sestari, I., and Heldwein, A. B. 2003. Pré-resfriamento de maçã (*Malus domestica* Borkh.), cv. Fuji, em função da temperatura e velocidade do ar. *Revista Brasileira de Fruticultura* 25(3): 555–556.

Maia, L. H., Porte, A., and Souza, V. F. de. 2000. Filmes comestíveis: aspectos gerais, proprie-dades de barreira à umidade e o oxigênio. *Boletim do CEPPA, Curitiba*.18: 1–34.

Malgarim, M. B., Tibola, C. S., Zaicovski, C. B., Ferri, V. C., and Silva, P. R. 2006. Modificação da atmosfera e resveratrol na qualidade pós-colheita de morangos 'Camarosa'. *Revista Brasileira de Agrociência*. 12: 67–70.

Mali, S. and Grossmann, M. V. E. 2003. Effects of yam starch films on storability and quality of fresh strawberries (*Fragaria ananassa*). *Journal of Agricultural and Food Chemistry*. 51: 7005–7011.

Manach, C., Mazur, A., and Scalbert, A. 2005. Polyphenols and prevention of cardiovascular diseases. *Current Opinion in Lipidology*. 16: 77–84.

Manach, C., Morand, C., Rémésy, C., and Jiménez, L. 2004. Polyphenols: food sources and bioavailability. *American Journal of Clinical Nutrition*. 79: 727–747.

Meneghel, R. F. A., Benassi, M. T., and Yamashita, F. 2008. Revestimento comestível de algi-nato de sódio para frutos de amora-preta (*Rubus ulmifolius*). *Semina: Ciências Agrárias*. 29: 609–618.

Meng, X. H., Li, B. Q., Liu, J., and Tian, S. P. 2008. Physiological responses and quality attri-butes of table grape fruit to chitosan preharvest spray and postharvest coating during storage. *Food Chemistry*. 106: 501–508.

NETO, J. F. et al. 2006. Avaliação das Câmaras frias usadas para o armazenamento de frutas e hor-taliças no entreposto terminal de São Paulo (CEAGESP). *Engenharia Agrícola* 26(3): 832–839.

Oliveira, C. S., Grden, L., and Ribeiro, C. O. M. 2007. Utilização de filmes comestíveis em ali-mentos. Série em Ciência e Tecnologia de Alimentos: *Desenvolvimentos em Tecnologia de Alimentos*. 01: 52–57.

Ormond, J. G. P., Paula, S. R. L., Faveret-Filho, P., and Rocha, L. T. M. 2002. Agricultura orgânica: quando o passado é futuro. *BNDES Setorial.* 15: 3–34.

Ordóñez, J. A. P. 2005. Tecnologia de Alimentos: Componentes dos Alimentos e Processos. *Porto Alegre: Artmed* 1: 294.

Park, S., Stan, S. D., Daeschel, M. A., and Zhao, Y. 2005. Antifungal coatings on fresh strawberry (*Fragaria x ananassa*) to control mold growth during cold storage. *Journal of Food Science.* 70: 202–207.

Paull, R. E. 1999. Effect of temperature and relative humidity on fresh commodity quality. *Postharvest Biology and Technology.* 15: 263–277.

Pelayo, C., Ebeler, S. E., and Kader, A. A. 2003. Postharvest life and flavor quality of three strawberry cultivars kept at 5°C in air or air + 20 kPa CO_2. *Postharvest Biology Technology.* 27: 71–183.

Perkins-Veazie, P., Collins, J. K., Clark, J. R., and Risse, L. 1997. Air shipment of 'Navaho' blackberry fruit to Europe is feasible. *HortScience.* 32: 129–132.

Piccinin, E., Di Piero, R. M., and Pascholati, S. F. 2005. Efeito de *Saccharomyces cerevisiae* na produtividade de sorgo e na severidade de doenças foliares no campo. *Fitopatologia Brasileira.* 30: 5–9.

Pinheiro, A. M. P., Vilas Boas, E. V. B., and Mesquita, C. T. 2005. Ação do 1-metilciclopropeno (1-MCP) na vida de prateleira da banana 'Maçã'. *Revista Brasileira de Fruticultura.* 27: 25–28.

Pinto, L. K. A., Martins, M. L. L., Resende, E. D., Almeida, R. F., Vitorzi, L., and Pereira, S. M. F. 2006. Influência da atmosfera modificada por filmes plásticos sobre a qualidade do mamão armazenado sob refrigeração. *Ciência Tecnologia de Alimentos* 26(4): 744–748.

Pirozzi, D. C. Z. and Amendola, M. 2005. Modelagem matemática e simulação numérica do resfriamento rápido de morango com ar forçado. *Engenharia Agrícola* 25(1): 222–230.

Reichert, L. J. and Madail J. C. M. 2003. Aspectos econômicos. In: Santos A M, Medeiros A R M. *Morango. Produção. Embrapa Clima Temperado.*Embrapa Informação Tecnológica, Pelotas, Brasília, pp. 1–81 (Frutas do Brasil n. 40).

Reis, K. C., Siqueira, H. H., Alves, A. P., Silva, J. D., and Lima, L. C. O. 2008. Efeito de diferentes sanificantes sobre a qualidade de morango cv. Oso Grande. *Ciência e Agrotecnologia.* 32: 196–202.

Resende, J. M. 2007. *Revestimentos biodegradáveis para conservação do coco 'Anão Verde.* Tese (Doutorado em Engenharia Agrícola) – Universidade Estadual de Campinas, Campinas, Brasil.

Ribeiro, V. G., Assis, J. S., Silva, F. F., Siqueira, P. P. X., and Vilaronga, C. P. P. 2005. Armazenamento de goiabas 'Paluma' sob refrigeração e em condição ambiente, com e sem tratamento com cera de carnaúba. *Revista Brasileira de Fruticultura* 27(2): 203–206.

Ribeiro, C., Vicente A. A, Teixeira J. A., and Miranda, C. 2007. Optimization of edible coating composition to retard strawberry fruit senescence. *Postharvest Biology Technology.* 44: 63–70.

Riedel, L. 1969. Measurement of thermal diffusivity on foodstuffs rich in water. *Kaltetechnik-Klimatisierung* 21(11): 315–316.

Sanino, A., Cortez, L. B., and Teruel, B. M. 2003. Vida de prateleira do tomate (*Lycopersicum esculentum*), variedade "Débora", submetido a diferentes condições de resfriamento. Workshop de tomate: Perspectivas e pesquisas, Campinas, 28 maio de.

Santos, C. A. A., Castro, J. V., Picoli, A. A., and Rolim, G. S. 2008. Uso de quitosana e embalagem plástica na conservação pós-colheita de pêssegos 'douradão'. *Revista Brasileira de Fruticultura.* 30: 88–93.

Santos, L. O., Martins, R. N., Durigan, J. F., and Mattiuz, B. 2007. Técnicas de conservação pós-colheita do morango. *Informe Agropecuário* 28: 84–87.

Scherer, E. E., Verona, L. A., and Signor, G. 2003. Produção agroecológica de morango no Oeste Catarinense. *Agropecuária Catarinense.* 16: 20–24.

Severo, J., Azevedo, M. L., Chim, J., Schreinert, R. S., Silva, J. A., and Rombaldi, C. V. 2007. Avaliação de compostos fenólicos, antocianinas e poder antioxidante em morangos cv. Aromas e Camarosa. In: *IX ENPOS. Pesquisa e responsabilidade ambiental.* Faculdade de Agronomia Eliseu Maciel, Pelotas, Brazil.

Silva, F. L., Escribano-Bailón, M. T., Alonso, J. J. P., Rivas-Gonzalo, J. C., and Santos-Buelga, C. 2007. Anthocyanin pigments in strawberry. *Food Science and Technology.* 40: 374–382.

Siro, I., Devlieghere, F., Jacxsens, L., Uyttendaelle, M., and Debevere, J. 2006. The microbial safety of strawberry and raspberry fruits packaged in high-oxygen and equilibrium-modified atmospheres compared to air storage. *International Journal of Food Science and Technology.* 41: 93–103.

Souza, M. B. 2007. Amora: Qualidade pós-colheita. Folhas de Divulgação AGRO 556. Ministério da Agricultura Pecuária e Abastecimento – MAPA, Sao Paulo, Brasília.

Spagnol, W. A. and Sigrist, J. M. M. 1992. *Pré-resfriamento. Tecnologia pós-colheita de frutas tropicais.* Manual n.9. ITAL, Campinas, Brazil.

Stenzel, N. M. C., Neves, C. S. V. J., Gonsalez, M. G. N., Schols, M. B. S., and Gomes, J. C. 2005. Desenvolvimento vegetativo, produção e qualidade dos frutos de laranjeira 'Folha Murcha' sobre seis porta-enxertos no Norte do Paraná. *Ciência Rural.* 35: 1281–1286.

Strik, B. 2005. Blueberry: an expanding world crop. *Chronica Horticulturae.* 45: 7–12.

Strik, B. C. 2007. Horticultural practices of growing highbush blueberries in the ever-expanding U.S. and global scene. *Journal of the American Pomological Society.* 61: 148–150.

Sun, D. W. and Brosnan, T. 1999. Extension of the vase life of cut daffodil flowers by rapid vacuum cooling. *International Journal of Refrigeration* 22(6): 472–478.

Tabolt, M. T. and Chau, K. V. 1991. Precooling strawberries. Institute of Food and Agricultural Sciences, University of Florida Circular 492, 8pp.

Tanabe, C. S. and Cortez, L. A. B. 1998. Perspectivas da cadeia do frio para frutas e hortaliças no Brasil. MERCOFRIO 98—Feira e Congresso de Ar Condicionado, Refrigeração, Aquecimento e Ventilação do Mercosul.

Tanada-Palmu, P. S. and Grosso, C. R. F. 2005. Effect of edible wheat gluten-based films and coatings on refrigerated strawberry (*Fragaria ananassa*) quality. *Postharvest Biology Technology.* 36: 199–208.

Teruel, M. B. 2000. Estudo teórico—experimental do resfriamento de laranja e banana com ar forçado. Faculdade de Engenharia Mecânica/UNICAMP, Campinas, Brazil, 300pp. (Tese de Doutorado).

Teruel, B., Cortez, L. A., Leal, P., and Lima, A. G. B. 2001. Estudo teórico do resfriamento com ar forçado de frutas de geometrias diferentes. *Ciências e Tecnologia de Alimentos* 21(2): 228–235, maio-ago.

Teruel, B., Cortez, L., and Neves F. L. 2003. Estudo comparativo do resfriamento de laranja valência com ar forçado e com água. *Revista Ciência e Tecnologia de Alimentos* 23(2): 174–178.

Teruel, B., Kieckbusch, T., and Cortez, L. 2004. Cooling parameters for fruits and vegetables of different sizes in a hydrocooling system. *Scientia Agricola* 61(6): 655–658.

Teruel, B., Silveira, P., Marques, F., and Cappelli, N. 2008. Interface homem-máquina para controle de processos de resfriamento com ar forçado visando à economia de energia. *Ciência Rural* 38(3): 705–710.

Thompson, J. F., Mitchell, F. G., Rumsey, T. R., Kasmire, R. F., and Crisosto, C. H. 1998. Commercial Cooling of Fruits, Vegetables and Flowers, Cap. 4 e 5. Universidade da Califórnia—Divisão de agricultura e recursos naturais, Publicação 21567, pp. 33–34.

Tosun, I., Ustun, N. S., and Tekguler, B. 2008. Physical and chemical changes during ripening of blackberry fruits. *Science Agricola.* 65: 87–90.

Trehane, J. 2004. *Blueberries, Granberries and Other Vaccinimus.* Ed. Timber Press, Portland, OR.

Vasconcelos, H. O., Duarte, L. S. N., and Araújo, C. M. 1976. Estudo cromossômico de (*Citrus sinensis* (L.) Osbeck) laranja 'Folha Murcha'. In: *CONGRESSO BRASILEIRO DE FUTICULTURA*, Rio de Janeiro, Anais. Campinas, Sociedade Brasileira de Fruticultura, Vol. 1, pp. 639–643.

Venâncio, T., Engelsberg, M., Azeredo, R. B. V., and Colnago, L. A. 2006. Metodologia de medida da difusividade térmica por RMN-CWPF. *Comunicado Técnico* 72, ISSN 1517-4786, 5pp.

Vissotto, F. Z., Kieckbush, T. G., and Neves, F. L. C. 1999. Pré-resfriamento de frutas e hortaliças com ar-forçado. *Boletim da Sociedade Brasileira de Ciência e Tecnologia de Alimentos* 33(1): 106–114.

Volpe, C. A., Schoffel, E. R., and Barbosa, J. C. 2002. Influência da soma térmica e da chuva durante o desenvolvimento de laranjas-'Valência' e 'Natal' na relação entre sólidos solúveis e acidez e no índice tecnológico do suco. *Revista Brasileira de Fruticultura.* 24: 436–441.

Wolfe, K., Wu, X., and Liu, R. H. 2003. Antioxidant activity of apple peels. *Journal of Agricultural and Food Chemistry.* 53: 609–614.

Yamashita, F., Tonzar, A. M. C., Fernández, J. G., Morya, S., and Benasi, M. T. 2001. Embalagem individual de mangas cv. Tommy Atkins em filme de plástico: efeito sobre a vida de prateleira. *Revista Brasileira de Fruticultura.* 23(2): 288–292.

Zaicovski, C. B., Tibola, C. S., Malgarim, M. B., Ferri, V. C., Pegolaro, C., Dal Cero, J., and Silva, P. R. 2006. Resveratrol na qualidade pós-colheita de morangos 'Camarosa'. *Revista Brasileira de Agrociência.* 12: 443–446.

13 Vacuum Frying of Fruits Applications in Fruit Processing

Rosana G. Moreira

CONTENTS

13.1 INTRODUCTION

Vacuum frying is a process that is carried out under pressures well below atmospheric levels, preferably below 50 Torr (6.65 kPa). It is an effective means of reducing the oil content in fried foods while producing, for example, potato chips with the same texture and color of those fried in atmospheric conditions (Da Silva and Moreira, 2008; Garayo and Moreira, 2002); in addition, this process lowers acrylamide content (Granda et al., 2004) and enhances the product's organoleptic and nutritional qualities (Da Silva and Moreira, 2008; Dueik et al., 2010; Nunes and Moreira, 2009; Shyu and Hwang, 2001).

Several studies have shown that less oil is absorbed during the vacuum frying process using different pretreatment and de-oiling steps (Garayo and Moreira, 2002; Mariscal and Bouchon, 2008; Moreira et al., 2009; Nunes and Moreira, 2009; Pandey and Moreira, 2010; Yagua and Moreira, 2011). Oil absorption at the surface of chips increases during vacuum frying processes because of the higher heat and mass transfer rates and the existence of a pressurization step, thus increasing the final oil content compared to traditional frying for the same working temperature (Troncoso and Pedreschi, 2009).

The pressurization process plays an important role in the oil absorption mechanism. It can increase or decrease oil absorption depending on the amount of surface oil and free water present in the product (Garayo and Moreira, 2002). The amount of surface oil present at the moment of pressurization is a determining aspect for the final oil content of the product; therefore, a de-oiling process must be used to remove surface oil under vacuum after the product is fried (Moreira et al., 2009). In potato chips, about 14% of the total oil content (TOC) is located in the core (internal oil), and the remaining 86% of the oil content is surface oil (Moreira et al., 2009). A de-oiling mechanism (centrifuging system) can remove surface oil before the pressurization step and is able to reduce the TOC in about 80%–90%.

In recent years, vacuum-fried fruits and vegetables have become very popular in southeastern Asia. A couple of companies in Asia have developed such a system for the process of fruits (apple, pineapple, grapes, banana, guava, mango, peach, etc.) and vegetables (sweet potato, potato, pumpkin, carrots, etc.) into chips and fried fishes and shellfishes (octopus and cuttlefish). The trend is expanding to other continents as well. This chapter discusses the different vacuum frying systems, vacuum frying operation, and vacuum-fried product characteristics.

13.2 VACUUM FRYING SYSTEMS

13.2.1 COMPONENTS OF A VACUUM FRYING SYSTEM

A vacuum frying system consists of the following components: (1) the frying vessel, (2) the de-oiling system, (3) the condenser, and (4) the vacuum pump.

Figure 13.1 illustrates a schematic of a laboratory-size vacuum system used at the Food Engineering laboratory of the Biological and Agricultural Engineering Department at Texas A&M University (College Station, TX). The frying vessel has

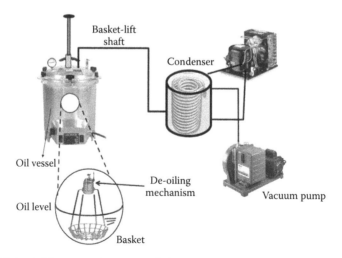

FIGURE 13.1 A laboratory-size vacuum frying system.

a capacity of 6 L and a resistance heater that provides a maximum temperature at 140°C. Inside the vessel, there is a basket and a centrifuge (de-oiling system) with a motor attached to the basket shaft, rotating up to 750 rpm (63 g units). Vacuum is achieved in the vessel by a dual seal vacuum pump (model 1402 Welch Scientific Co., Skokie, IL) with a vacuum capacity of 1.33 kPa (10 Torr). Between the vessel and the pump, a condenser is needed to collect the water vapor coming from the potato chips to avoid water going to the pump. The condenser consists of a refrigeration unit (1/4 HP) connected to a helically coiled heat exchanger (evaporator).

The process consists of loading the products into the fryer basket, closing the lid, and then depressurizing the vessel. When the pressure in the vessel achieves the target vacuum, the basket is submerged into the hot oil. Once the product is fried, the basket is raised, it spins for a period of time to remove the excess surface oil, the vessel is pressurized up to atmospheric pressure, and the fried products are stored in polyethylene bags inside of desiccators for further analysis.

Many companies in Asia and Europe have designed different commercial size vacuum frying systems. Some examples of these units are described in the following text.

13.2.2 BATCH FRYING SYSTEMS

An example of a commercial batch vacuum frying system is the one developed by *I-Tung Machinery Industry* (Taipei, Taiwan). It has a capacity of 50–100 kg of potato chips/h and 20–50 kg of fruits/h and operates at a vacuum of 10.7 Torr (1.42 kPa). *Apple Snack Corporation Ltd.* (Hirosaki-shi, Aomori-ken, Japan) is another company specialized in vacuum frying technology that manufactures vacuum frying equipment for processing of different types of fruits and vegetables.

A pilot-scale batch system developed by BMA (the Netherlands) can be used to test the effect of oil temperature, frying time, vacuum level, and de-oiling process on the quality of the final product.

13.2.3 CONTINUOUS FRYING SYSTEMS

Continuous vacuum frying was a concept developed by Florigo (H&H Industry Systems BMA, the Netherlands) in the early 1970s to produce high-quality french fries. Due to the improvement in the quality of the raw materials and blanching techniques, the use of vacuum fryers almost died out with exception of one or two production companies that still insist in producing a nonblanched product. Today, the BMA automatic continuous vacuum fryers are mainly used to produce fruit chips and very delicate snack products.

The BMA vacuum frying pan is installed in a stainless steel vacuum tube. The raw product is introduced to the fryer through a rotary device. The frying pan is designed to meet the different product specifications. A conveyor takes the finished product out of the fryer toward the outlet system. A closed chamber located at the exit of the vacuum tube prevents air from entering the vacuum zone. A special conveyor system takes the product from one zone to another. The vacuum is created by vacuum pumps. The frying system also contains a frying oil circulation and a filtration system that have been adapted to the special conditions in the vacuum frying installation (BMA, 2010).

The capacity of this unit is about 350 kg/h of potato chips. Certain applications require a two-stage frying process. In that case, the product is prefried in an atmospheric fryer and then subjected to vacuum frying until the final moisture content is reached. For two-stage frying, a unit with an output of up to 1500 kg/h (potato chips) is available.

13.2.4 DE-OILING UNIT

Many vacuum frying units are equipped with centrifuges for de-oiling the product after frying. The centrifuges are installed in a special vacuum dome attached to the vacuum fryer (BMA, Woerden, the Netherlands).

13.3 VACUUM FRYING CHARACTERISTICS

13.3.1 TEMPERATURE CHANGE DURING FRYING

The vacuum frying process can be divided into four periods: depressurization (DP), frying (FR), pressurization (PR), and cooling (CL). Yagua and Moreira (2011) measured the temperature history at the center of the potato chip (PC), temperature of the vacuum vessel headspace (HS), and temperature at the surface of the potato chip (PS), as well as the pressure history (P) during the frying process of potato chips at 120°C (Figure 13.2) using the equipment illustrated in Figure 13.1. During the depressurization period (DP), potato slices are placed in the headspace of the frying vessel, and there is a waiting period until the pressure value goes down to 1.33 kPa to start the second period. In the next step, immersion frying (FR), heat and mass

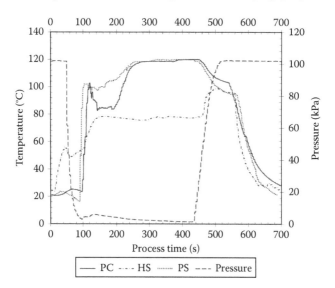

FIGURE 13.2   Temperature history at the center of the potato chip (frying process at 120°C).

transfer phenomena occur; heat is convected from the oil to the surface of the product and then conducted from the surface to the center of the product, water is evaporated from the potato, and a small amount of oil is absorbed by the chip. After frying has been completed, the chips are taken out of the frying medium and held up in the vessel headspace while the vacuum is broken and the system recovers up to atmospheric pressure (PR) which is achieved in about 1 min; an initial cooling of the product also occurs in this stage. Once the system is back to atmospheric pressure, the chips are removed from the frying vessel and allowed to cool down to ambient temperature (CL). As the potato chips cool, during the pressurization and cooling steps, the pressure inside the pores changes, thus creating a differential in pressure, ΔP, between the surface and the center of the product. This pressure difference generates a driving force for the oil at the surface to penetrate the pores. However, during the cooling period ΔP is much smaller than the one in the pressurization period.

The temperature at the center of the potato has a subtle increase followed by a slight decrease during the depressurization period. The increase in temperature takes place when the potato is in the headspace of the vacuum vessel where the temperature is high (80°C–90°C) due to heat irradiating from the oil, while the decrease in temperature is a consequence of lowering the pressure of the system, since the potato chip is trying to reach the saturation temperature at the new pressure. For the remaining time of immersion in oil, the temperature at the center of the chip increased until it reached equilibrium with the frying medium.

During the final two steps of the process, pressurization and cooling, the temperature of the chip starts declining down to the ambient temperature (~22°C). The rate of cooling during the pressurization period is slower than the one after the chips are out of the fryer, considering that during the pressurization step, the chips are suspended in the warm headspace of the vessel.

The surface temperature (TS) history (Figure 13.2) is very close to the one for temperature at the center of the chip (TC) with some major differences at the beginning of the immersion frying period. During the depressurization step, the increase and slight decrease of temperature are also noted. The most important difference between the center and the surface temperature histories is that the latter does not show the temperature spike followed by the evaporation plateau seen in the center temperature history.

The temperature at the headspace (HS) of the vacuum vessel (Figure 13.2) is of great importance in the understanding of the center and surface temperature history behavior.

Before placing the lid on the vessel, the thermocouple reading is ambient temperature, but once the vessel is closed, the temperature increases (around 60°C) due to the radiant heat from the hot frying oil. Temperature in the vessel headspace remains constant during the depressurization step. Once the chips are submerged into the oil, the headspace temperature increases up to 80°C due to the water vapor migrating from the potato chips to the headspace. For the rest of the frying period, this temperature stays fairly constant.

After the frying period is over, the basket holding the hot product is raised up to the headspace causing an instant raise in temperature. The headspace maintains this final temperature until the vessel is open and the lid is placed at ambient temperature.

During the depressurization period, the pressure (P) is lowered from 101.6 ± 0.15 to 1.4 ± 0.06 kPa at a rate of 1.9 ± 0.06 kPa/s (Figure 13.2). Once the desired pressure is achieved, the potato slices are submerged into the oil causing vapor release and a short increase in pressure of about 7–10 kPa. The vacuum pump rapidly accounts for the increase in pressure and efficiently adjusts the pressure back to the lowest level; from this moment, pressure remains constant until the end of the frying.

13.3.2 MOISTURE LOSS OF POTATO CHIPS DURING FRYING

In vacuum frying, the initial warm-up period is very short, it may last 1–5 s (Garayo and Moreira, 2002), and the material reaches the evaporation temperature faster than in other drying processes. Water starts evaporating as soon as the raw material is in contact with the oil. The drying rate is very fast during the first 60–100 s of the process and then slows down as the product reaches equilibrium (Yagua and Moreira, 2011).

13.3.3 OIL UPTAKE OF POTATO CHIPS DURING FRYING

Oil content is one of the main quality parameters for fried products. Some of the parameters affecting the final oil content of fried products are product shape, temperature and frying time, moisture content, porosity, pore-size distribution, and pre- and posttreatments (Bouchon and Aguilera, 2001; Moreira et al., 1997; Saguy and Pinthus, 1995).

Oil uptake is essentially a surface-related phenomenon resulting from the competition between drainage and suction into the porous crust once the potato is removed from the frying oil and starts cooling (Bouchon and Pyle, 2005a,b; Gamble and Rice, 1987; Moreira et al., 1997). Vertical placement and centrifugation are two methods of oil removal in fried products before the cooling period. Vertical placement is used commercially, and mechanical means are employed to place the chips at the desired angle. During centrifugation, the centrifugal force acts perpendicular to the surface of the chips and separates the oil directly from the porous surface (Pandey and Moreira, 2010).

Moreira and Barrufet (1998) proposed that the oil uptake by tortilla chips can be described in terms of capillary forces and described the mechanism of oil uptake using the percolation theory. Kawas and Moreira (2002) found that the pore-size distribution developed during frying of tortilla chips is an important factor for oil absorption during the cooling period since small pores formed in the chips trapped more air during frying, resulting in higher capillary pressure during cooling and, as a consequence, higher internal oil content (IOC).

The TOC of fried products can be separated into two regions: the oil on the core of the product (internal oil) that is absorbed during frying and the surface oil which is absorbed during cooling. Moreira et al. (1997) measured oil content on the surface (surface oil) and the core (internal oil) of tortilla chips to determine oil distribution during frying and cooling.

Several studies have shown that less oil is absorbed during the vacuum frying process using different pretreatment and de-oiling steps (Garayo and Moreira,

2002; Mariscal and Bouchon, 2008; Moreira et al., 2009). In addition, Troncoso and Pedreschi (2009) showed that oil absorption at the surface increases during vacuum frying processes because of the higher heat and mass transfer rates and the existence of a pressurization step, thus increasing the final oil content compared to traditional frying for the same working temperature.

Mir-Bel et al. (2009) noted that oil absorption during the cooling stage is greatly influenced not only by the difference in temperature but also by the vacuum break conditions as the system is restored to atmospheric pressure. The researchers found that the volume of oil absorbed by the product is inversely proportional to the pressurization velocity, meaning that lower velocities favor oil absorption, showing an increase of 70% for potato chips compared to the oil content when the vacuum breaks abruptly.

Yagua and Moreira (2011) measured the TOC, the surface oil content (SOC), and the IOC of the potato chips fried under vacuum at 120°C (Figure 13.3). The final value of oil content is comparable with that obtained in traditional frying since oil is drastically absorbed during the pressurization stage of the process where the pressure difference drives the surface oil into the product.

Oil absorption occurs very rapidly since almost 75% of the TOC is absorbed between the first 70 and 80 s of frying. After this time interval, the TOC remains fairly constant until the frying process is finished. The curves for TOC during vacuum frying at different temperatures show the same shape and behavior; the only difference is that the equilibrium value is reached faster at higher temperatures. This indicates that the final oil content of the chips is a function of time and the remaining moisture within the product, which increases with decreasing temperature (Garayo and Moreira, 2002).

The effect of the pressurization step can be diminished by the use of the de-oiling system. SOC was defined as the difference between TOC and IOC, and it is the oil

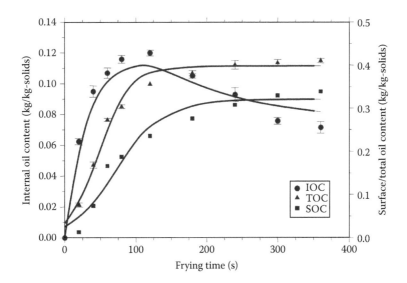

FIGURE 13.3   TOC, SOC, and IOC of the potato chips fried under vacuum at 120°C.

removed by the de-oiling system before the pressurization step. The SOC follows the same behavior as that of the TOC in which the maximum value is reached fast, and then it remains constant for the rest of the process. By evaluating the difference between the TOC and the IOC, the importance of the de-oiling system in vacuum frying processes, discussed by Moreira et al. (2009), is confirmed.

It is interesting to note that the IOC decreases as frying time increases. These results indicate that probably, as the chip is removed from the oil bath, the adhered oil intends to flow into the pore spaces during the pressurization period. The amount of moisture in the product seems to be related to oil absorption, which is higher when more free spaces are available (when the amount of moisture is small). That is, an increase in the surrounding pressure, during the pressurization period, causes the water vapor inside the pore to condense (at constant T and P). As the water vapor condenses, the pressure difference ($P_{surroundings} - P_{pore}$) causes the oil to be absorbed into the pore space. Under these operating conditions, most of the oil will be absorbed during the pressurization process.

Once the moisture is reduced to a critical level (during 70–120 s), oil absorption during the pressurization process decreases (since ($P_{surroundings} - P_{pore}$) is negligible). Air then diffuses faster than the oil into the pore spaces, thus blocking the oil to flow into the product. In this case, most of the oil is absorbed during cooling.

These observations indicate that the pressurization process plays an important role in the oil absorption mechanism. It can increase or decrease oil absorption in the product depending on the amount of surface oil and moisture presented in the product. The lower the vacuum pressure and the higher the frying temperature, the highest is the oil content of the chips during the initial period of frying when moisture is still available in the product. Therefore, for these conditions, the lowest oil content is obtained when the chips are fried as close as to the equilibrium moisture content.

13.4   VACUUM-FRIED FRUITS AND VEGETABLES

Table 13.1 shows the oil content of different products fried under atmospheric and vacuum frying systems. In general, the oil content of vacuum-fried products is lower than the atmospheric fried ones. When de-oiling process is used with vacuum-fried products, the oil content is reduced substantially. Oil content of chips fried under atmospheric conditions can be as low as 33% when some type of device is used to remove the oil content at the product's surface (Moreira et al., 1999). Without de-oiling mechanisms, some products would result in higher oil content. In this table, green beans, mango, and blue potato chips would result in higher oil content if a de-oiling system were not used in the vacuum frying system.

Examples of commercially available vacuum-fried fruits and carrots (Table 13.2) show the effect of using de-oiling process to produce higher-quality products with low oil content.

Another way of reducing the oil absorption at the surface of vacuum-fried products is by pretreating the product before frying. Examples of pretreatments include (1) blanching/drying (blanching slices in hot water at 85°C for 3.5 min and air drying at 60°C until a final moisture content of ~0.6 kg water/kg dry solid) and (2) pretreating slices in sulfite solution (slices soaked in a 3.5 kg/m³ sodium metabisulfite

TABLE 13.1

Oil Content of Different Products Fried in Atmospheric and Vacuum Frying Systems

Product	Oil Content (% w.b.)	
	Atmospheric Frying	Vacuum Frying
Potato chips[a,b]	33.2 ± 0.1[b]	9.8 ± 0.2
Potato chips[c,d]	39.1 ± 0.2	44.3 ± 0.2
Blue potato chips[c,e]	32.9 ± 0.1	34.5 ± 1.1
Green beans[c,f]	28.7 ± 0.4	27.3 ± 0.3
Mango chips[c,g]	31.0 ± 0.9	33.2 ± 0.3
Sweet potato chips[c,h]	39.0 ± 0.2	29.6 ± 0.6
Apple chips[c,i]	18.7 ± 0.1	13.7 ± 0.2

(a) atmosphere, (v) vacuum.

[a] De-oiled.

[b] (v) $T = 120°C$, frying time $= 360$ s ($P = 1.33$ kPa) centrifuged for 40 s at 700 rpm (Moreira et al., 2009).

[c] Not de-oiled.

[d] (a) $T = 140°C$, (v) $T = 115°C$ ($P = 5.37$ kPa) (fried to 2% w.b. final moisture content) (Troncoso et al., 2008).

[e] (a) $T = 165°C$, frying time $= 180$ s; (v) $T = 121°C$, frying time $= 420$ s ($P = 1.33$ kPa) (Da Silva and Moreira, 2008).

[f] (a) $T = 165°C$, frying time $= 300$ s; (v) $T = 121°C$, frying time $= 330$ s ($P = 1.33$ kPa) (Da Silva and Moreira, 2008).

[g] (a) $T = 165°C$, frying time $= 240$ s; (v) $T = 121°C$, frying time $= 180$ s ($P = 1.33$ kPa) (Da Silva and Moreira, 2008).

[h] (a) $T = 165°C$, frying time $= 240$ s; (v) $T = 130°C$, frying time $= 120$ s ($P = 1.33$ kPa) (Da Silva and Moreira, 2008).

[i] (a) $T = 160°C$, frying time $= 900$ s; (v) $T = 115°C$, frying time $= 900$ s ($P = 15$ kPa) (Mariscal and Bouchon, 2008).

solution at 20°C for 3 min and pH adjusted to 3.0) (Troncoso et al., 2009). Potato variety can also affect the oil content of vacuum-fried chips.

Pretreated (air drying) apple slices produce lower oil content chips compared to apple slices fried at atmospheric pressure or vacuum ($P = 15$ kPa) (Mariscal and Bouchon, 2008). After frying for 15 min at $T = 140°C$ (atmospheric) and $T = 95°C$ (vacuum), atmospheric fried slices absorb 21% more oil than vacuum-fried slices and more than three times the amount of oil absorbed by predried vacuum-fried slices. At higher frying temperature ($T = 150°C$ and 160°C [atmospheric] and $T = 105°C$ and 115°C [vacuum]), the difference in oil uptake between predried vacuum-fried and vacuum-fried only slices decreases at the longest frying time, whereas the difference between the latter and atmospheric frying increases. For instance, after frying for 15 min at $T = 160°C$ (atmospheric) and $T = 115°C$ (vacuum), atmospheric fried slices absorb 47% and 55% more oil than vacuum-fried and predried vacuum-fried slices, respectively.

TABLE 13.2

Oil Content of Different Vacuum-Fried Products (Commercial Samples) Fried in Vacuum Systems with a De-Oiling Process (Centrifuge inside the Vacuum Chamber)

Product	Oil Content (% w.b.)
Apple chips	21
Kiwi chips	30
Peach chips	20
Pineapple chips	20
Sweet potato chips	17
Carrot chips	24

Source: Apple & snack Co, Ltd (http://www.net24. ne.jp./~applesnack/e/Lowfat/lowfat.html), Aomori, Japan.

Enzymatic browning of banana slices occurs practically in vacuum-fried banana slices during processing (Apintanapong et al., 2007). Control of enzymatic browning can be achieved by various methods, and chemical treatment is one of them. Banana slices treated with 0.5% cysteine and 1% citric acid before vacuum frying (9.3 kPa) at 90°C produces high-quality product in terms of color, texture, and flavor.

Although oil content is an important quality parameter for fried products, the overall quality characteristics have to include color, texture, and flavor. Vacuum frying is the only frying technology that can be used to fry delicate, high-sugar content products like fruits and vegetables. In general, vacuum-fried snacks retain more of their natural colors and flavors due to the reduced oxidation and lower frying temperature.

Fruits and vegetables are sources of many vitamins and antioxidants. Consumers often find it difficult to eat more fruits and vegetables because they believe they are too expensive, spoil too quickly, or take too long to prepare. There are many highly nutritious vegetables and fruits that could be vacuum fried (Table 13.2). Carotenoids make corn yellow, carrots orange, and tomatoes red. More than 600 carotenoids have been found in plants. About half of the roughly 50 carotenoids in the human diet are absorbed into the blood stream. Lycopene and beta-carotene each constitutes about 30% of plasma carotenoids. Only alpha, beta, and a few other carotenes (not lycopene or lutein) can be converted to vitamin A. Both alpha-carotene and beta-carotene are protective against liver cancer and lung cancer in cell culture and animal studies. Heating frees up carotenoids especially beta-carotene and lycopene. Carotenoids are nearly insoluble in water and are best absorbed when associated with oils. Anthocyanins (flavonoid polyphenolics) are water-soluble glycosides and acyl-glycosides of anthocyanidins, making them susceptible to losses during the frying process. Anthocyanins make cherries and strawberries red and blueberries blue. Anthocyanins have anti-inflammatory effects. The length of time and the method of frying are important factors that affect phytochemical/nutraceutical stability.

Dueik et al. (2010) fried carrot slices under vacuum (6.5 kPa) at temperatures 98°C–118°C and showed that vacuum frying may reduce the oil content of carrot crisps by nearly 50% compared to atmospheric fried chips. Furthermore, they preserve around 90% of trans-α-carotene and 86% trans-β-carotene, which leads to the preservation of the color of raw carrots.

Perez-Tinoco et al. (2008) vacuum-fried (24 kPa) pineapple chips at 106°C–117°C and noted that the total phenolic content and dehydroascorbic acid content increased with increasing frying time and temperature. The chips had an appetizing golden yellow color and low fat content and high residual content of vitamin C.

When frying delicate products like fruits, a pretreatment is needed to provide texture (firm structure) to the fruit slices before they can be processed as chips. Often, osmotic dehydration (OD) is used in combination with vacuum frying to produce high-quality fruit snacks (Da Silva and Moreira, 2008). In recent years, vacuum-fried fruits and vegetables have become very popular in southeastern Asia. Swi-Bea et al. (2004) described the process of transforming raw mango into chips in three steps: (1) pretreatment: blanching of mango slices in hot water (95°C), cooling, freezing (−20°C), and then thawing in a solution containing 30 °Bx maltose syrup and 1% calcium chloride; (2) vacuum frying at 100°C and 70 Torr to produce chips; and (3) de-oiling by centrifugation and flavoring by adding seasoning.

Da Silva and Moreira (2008) observed that mango chips fried under atmospheric conditions were of lower quality than those fried under vacuum (1.33 kPa and 120°C), though texture characteristics of the fried products were not affected by the frying method. Final total carotenoids (mcg/g d.b.) were higher by 19% compared to the atmospheric fried chips. Sensory panelists overwhelmingly preferred ($P < 0.05$) the vacuum-fried mango chips for color, texture, taste, and overall quality.

OD has been used to reduce the initial moisture content, conserve, and retain the initial quality of processed fruits and vegetables (Heng et al., 1990; Torreggiani and Bertolo, 2001). During osmotic processing, water flows from the product into the concentrated osmotic solution, while small amounts of the osmotic solute are transferred from the solution into the product (Dermesonlouoglou et al., 2007).

Because of the high sugar content of the product after OD, vacuum frying is an excellent technology to produce high-quality (in terms of sensory, physical properties, and reduced oil content) deep-fat fried fruit chips. Atmospheric frying alone cannot be used to fry fruits (Da Silva and Moreira, 2008) because the product's texture and color completely deteriorate, resulting in the collapse of the product's structure and overcooked appearance (dark color).

Nunes and Moreira (2009) showed that the high-quality mango chips could be produced with an osmotic solution (maltodextrin) concentration of 65 (w/v) and temperature of 40°C, which resulted in the highest dehydration efficiency index (water loss/sugar gain) and provided a good texture characteristic. The pretreated mango slices when vacuum fried for 2 min at 120°C produced the lowest oil content (22% w.b.) chips. More than 45% of the oil content was reduced by de-oiling the samples by centrifuging at 225 rpm for 25 s during the pressurization step. Lower oil content chips, produced by de-oiling for longer times and higher speeds, did not result in higher-quality product in terms of flavor and texture.

Da Silva and Moreira (2008) compared the changes in product quality attributes (PQA), such as color, texture, phytochemicals, and oil content, and sensory characteristic for different fruits and vegetables (sweet potato, blue potato, mango, and green beans) fried in vacuum and traditional fryers. To increase the solid content of the green beans and mango and to provide a better texture, these products need to be pretreated before frying by OD in sugar solutions. In their study, the sensory panelists overwhelmingly preferred the vacuum-fried products for color, texture, taste, and overall quality. Most of the products retained or accentuated their original colors when fried under vacuum. The atmospheric fried products showed excessive darkening and scorching. Anthocyanin and total carotenoid content were significantly high (20%–50% higher) for the products fried in the vacuum fryer than those in the traditional fryer. The texture characteristics of the products for both frying methods were not significantly different. There were significant differences for lightness (L^*), green-red chromaticity (a^*), and blue-yellow chromaticity (b^*) for most products. In general, products fried under atmospheric conditions are darker and redder in color than the products fried under vacuum. Most fruits retain or accentuate their original colors when fried under vacuum. Vacuum frying method clearly reduced color degradation due to the absence of oxidation during the process. In addition, the technology allows to produce fried products with the good texture even when using lower oil temperature (<100°C). This is not possible when using atmospheric frying that requires higher temperature to produce the right crunchiness.

13.5   ADVANTAGES AND DISADVANTAGES OF THE TECHNOLOGY

Because it is a low temperature process, vacuum frying can be used for heat-sensitive products. It is an excellent technology for producing fruit and vegetable chips with the necessary degree of dehydration without excessive darkening or scorching. Additionally, vacuum-fried products have higher retention of nutritional quality (high residual vitamin C presence of phenolic compounds and antioxidant capacity), color is enhanced (less oxidation), and oil degradation is reduced compared to atmospheric frying.

One major drawback is the residual oil left after frying. The process requires a de-oiling mechanism to remove the excessive oil absorption at the surface of the product during frying. The de-oiling process should be carried out into the vacuum chamber immediately after frying when the oil is still very hot and less viscous. The use of de-oiling systems can produce fruits and vegetables with very low oil content without affecting the organoleptic properties of the products.

Another drawback of vacuum frying systems may be the higher cost of equipment compared to atmospheric frying. Nonetheless, the operating expense of vacuum frying is very similar to that of frying at atmospheric pressure, as the cost of depressurization (vacuum) would be offset by the lower frying temperatures and the longer shelf life of the frying oil.

The commercial application of this technology will therefore depend on the additional value-added it can bring to new food products in comparison with other technologies.

REFERENCES

Apintanapong, M., Cheachumluang, K., Suansawan, P., and Thongprasert, N. 2007. Effect of antibrowning agents on banana slices and vacuum-fried slices. *Journal of Food, Agriculture and Environment, 5*(3, 4), 151–157.

BMA Nederland BV [BMA]. 2010. BMA vacuum frying. http://www.bma-nl.com/Vacuum-frying.1241.0.html?&L= (accessed on April 6, 2011).

Bouchon, P. and Aguilera, J.M. 2001. Microstructural analysis of frying of potatoes. *International Journal of Food Science and Technology, 36*, 669–676.

Bouchon, P. and Pyle, D.L. 2005a. Modeling oil absorption during post-frying cooling. Part I: Model development. *Food and Bioproducts Processing, 83*(C4), 253–260.

Bouchon, P. and Pyle, D.L. 2005b. Modeling oil absorption during post-frying cooling. Part II: Solution of the mathematical model, model testing and simulations. *Food and Bioproducts Processing, 83*(C4), 261–272.

Da Silva, P.F. and Moreira, R.G. 2008. Vacuum frying of high-quality fruit and vegetable-based snacks. *LWT-Food Science and Technology, 41*, 1758–1767.

Dermesonlouoglou, E.K., Grannakourou, M.C., and Taoukis, P. 2007. Stability of dehydrofrozen tomatoes pretreated with alternative osmotic solutes. *Journal of Food Engineering, 78*, 272–280.

Dueik, V., Robert, P., and Bouchon, P. 2010. Vacuum frying reduces oil uptake and improves the quality parameters of carrot crisps. *Food Chemistry, 119*(3), 1143–1149.

Gamble, M.H. and Rice, P. 1987. Effect of pre-fry drying of oil uptake and distribution in potato crisp manufacture. *International Journal of Food Science and Technology, 22*, 535–548.

Garayo, J. and Moreira, R.G. 2002. Vacuum frying of potato chips. *Journal of Food Engineering, 55*, 181–191.

Granda, C., Moreira R.G., and Tichy, S.E. 2004. Reduction of acrylamide formation in potato chips by low-temperature vacuum frying. *Journal of Food Science, 69*(8), 405–411.

Heng, W., Guilbert, S., and Cuq, J.L. 1990. Osmotic dehydration of papaya: Influence of process variables on the quality. *Sciences des Aliments, 10*, 831–848.

Kawas, M.L. and Moreira, R. 2002. Characterization of product quality attributes of tortilla chips during the frying process. *Journal of Food Process Engineering, 47*(1), 97–107.

Mariscal, M. and Bouchon, P. 2008. Comparison between atmospheric and vacuum frying of apple slices. *Food Chemistry, 107*, 1561–1569.

Mir-Bel, J., Oria, R., and Salvador, M.L. 2009. Influence of the vacuum break conditions on oil uptake during potato post-frying cooling. *Journal of Food Engineering, 95*, 416–422.

Moreira, R.G. and Barrufet, M.A. 1998. A new approach to describe oil absorption in fried foods: A simulation study. *Journal of Food Engineering, 31*, 485–498.

Moreira, R.G., Castell-Perez, M.E., and Barrufet, M.A. 1999. *Deep-Fat Frying: Fundamentals and Applications.* Gaithersburg, MD: Aspen Publishers.

Moreira, R.G., Da Silva, P.F., and Gomes, C. 2009. The effect of a de-oiling mechanism on the production of high quality vacuum fried potato chips. *Journal of Food Engineering, 92*, 297–304.

Moreira, R.G., Sun, X., and Chen, Y. 1997. Factors affecting oil uptake in tortilla chips in deep-fat frying. *Journal of Food Engineering, 31*, 485–498.

Nunes, Y. and Moreira, R.G. 2009. Effect of osmotic dehydration and vacuum-frying parameters to produce high-quality mango chips. *Journal of Food Science, 74*, 355–361.

Pandey, A. and Moreira, R. 2011. Batch vacuum frying system analysis for potato chips. *Journal of Food Process Engineering.* doi: 10.1111/j.1745-4530.2011.00635.x.

Perez-Tinoco, M.R., Perez, A., Salgado-Cervantes, M., Reynes, M., and Vaillant, F. 2008. Effect of vacuum frying on main physicochemical and nutritional quality parameters of pineapple chips. *Journal of the Science of Food and Agriculture, 88*, 945–953.

Saguy, I.S. and Pinthus, I.J. 1995. Oil uptake during deep fat frying: Factors and mechanisms. *Food Technology*, *49*, 142–145.

Shyu, S. and Hwang, S. 2001. Effects of processing conditions on the quality of vacuum fried apple chips. *Food Research International*, *34*, 133–142.

Swi-Bea, J., Wu, M.-C., and Wei, Y.-P. 2004. Tropical fruits. In: Barrett, D.M., Somogyi, L.P., and Ramaswamy, H.S., eds. *Processing Fruits: Science and Technology*. 2nd edn. Boca Raton, FL: CRC Press, p. 695.

Torreggiani, D. and Bertolo, G. 2001. Osmotic pre-treatment in fruit processing: Chemical, physical and structural effects. *Journal of Food Engineering*, *49*, 247–253.

Troncoso, E. and Pedreschi, F. 2009. Modeling water loss and oil uptake during vacuum frying of pre-treated potato slices. *LWT-Food Science and Technology*, *42*, 1164–1173.

Yagua, C. and Moreira, R. 2011. Characterization of product quality attributes and thermal properties of potato chips during vacuum frying. *Journal of Food Engineering*, *104*, 272–283.

14 Edible Coatings

Henriette Monteiro Cordeiro de Azeredo

CONTENTS

14.1 INTRODUCTION

Considerable attention has been given to edible food packaging, which is intended to be an integral part of and to be eaten with the food product; thus, they are also inherently biodegradable (Krochta, 2002). The increasing interest and research activity in edible packaging have been motivated by both increasing consumer demand for safe, convenient, and stable foods and also awareness of the negative environmental impacts of nonbiodegradable packaging waste. Although the terms *edible films* and *edible coatings* are sometimes referred to as synonyms, there is a difference in that films are preformed separately and then applied to food surface or sealed into edible pouches, whereas coatings are formed directly onto food surfaces.

Edible films and coatings have similar functions as those of conventional packaging, acting as barriers against water vapor, gases, and flavor compounds and improving structural integrity and mechanical-handling properties of foods. Although edible films and coatings are not meant to entirely replace conventional packaging, the efficiency of food protection can be enhanced by combining primary edible packaging and secondary nonedible packaging. Although the basic functions of edible coatings are similar to those of conventional packaging, their use requires an external packaging because of handling and hygienic reasons (Debeaufort et al., 1998). In this context, edible coatings can reduce conventional packaging requirements and waste, since they are able to improve overall food quality, to extend shelf life, and to improve economic efficiency of packaging materials.

Materials used for edible coatings are usually not as effective as petroleum-derived polymers in terms of mechanical and barrier properties. However, such properties are not especially important for fruits and vegetables, since a moderate gas permeation is required in order to allow the product to maintain respiration rates at a minimum but not negligible level.

According to Guilbert and Biquet (1986), wax coatings have been used to delay dehydration of citrus fruits in China since the twelfth century. However, only in the 1930s have emulsions of oils or waxes in water been developed to be spread on fruit surfaces to improve their appearance, as well as control the ripening process, and to retard water loss (Debeaufort et al., 1998).

Coating fruits is not yet a common process in the food industry. Industrial fruit coating consists mostly in keeping fruits in motion (e.g., by vibration or rolling) and simultaneously applying the coating dispersion so that the fruits are exposed to it. Spray coating is the most commonly used technique for applying food coatings, including a batch tank and spraying nozzles to deposit the coating dispersion on food pieces as they move over a conveyor roller (Debeaufort and Voilley, 2009), which should drive them to a drying step. For coating dispersions with high viscosities, screw or drum coaters are more indicated.

This chapter presents a review of developments in edible coatings and their applications for fresh and minimally processed (fresh-cut) fruits.

14.2 REQUIREMENTS FROM A COATING MATERIAL

The characteristics required from an edible coating depend on the specific requirements of the product to be coated, including the primary degradation modes to which it is most susceptible. Fresh and minimally processed fruits have complex requirements concerning packaging systems, since such products are still metabolically active. The main requirements for a fruit coating are described in the following:

1. Moderately low permeability to oxygen and carbon dioxide in order to slow down respiration and overall metabolic activity, retarding ripening and its related changes. On the other hand, the metabolic activity must not be reduced to a degree that creates anaerobic conditions, which promote physiological disorders and accelerate quality loss (Kester and Fennema, 1986; Debeaufort et al., 1998). Edible coatings for fruits should control the ripening by reducing oxygen penetration in the fruit rather than by decreasing CO_2 and ethylene evaporation rates, that is to say, the CO_2/O_2 permeability ratio (related to selectivity) should be as high as possible. Proteins and polysaccharide coatings present much higher ratios (from 10 to 25) than those of conventional plastic films (lower than 5.73) (Debeaufort et al., 1998). The decreased metabolic activity provided by edible coatings has also been known to retard softening changes (Conforti and Zinck, 2002;

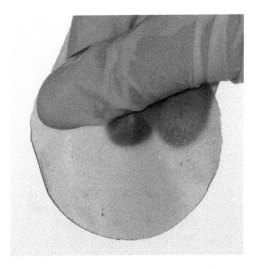

FIGURE 14.1 Sample of an edible film from acerola puree and alginate.

Zhou et al., 2011), which result from the loss of turgor pressure and degradation of cell walls, contributing to a decrease in fruit brittleness and firmness (Zhou et al., 2008). The degradation of cell wall structure has been attributed to activity of enzymes such as pectin methylesterase, cellulase, and polygalacturonase on polysaccharides present in the cell wall (Goulao and Oliveira, 2008).

2. Low water vapor permeability in order to retard desiccation (Garcia and Barret, 2002). In the case of minimally processed fruits, this is especially difficult, since the product surface usually has a very high water activity, which tends to decrease the performance of hydrophilic coatings (Hagenmaier and Shaw, 1992).

3. Sensory inertness or compatibility. Edible coatings were traditionally supposed to be tasteless so would not interfere with the flavor of the product (Contreras-Medellin and Labuza, 1981). Alternatively, they may have sensory properties compatible with those of the food. For instance, fruit purees have been studied as film-forming edible materials (McHugh et al., 1996; Senesi and McHugh, 2002; Rojas-Graü et al., 2006, 2007a; Azeredo et al., 2009) which can be used as edible coatings for fruits due to the presence of film-forming polysaccharides in their compositions. Figure 14.1 presents a photograph of an edible film elaborated with acerola puree and alginate.

14.3 CHEMICAL COMPOSITION OF EDIBLE COATINGS AND THEIR APPLICATIONS TO FRUITS

Edible films and coatings may be classified according to the kind of material from which they are derived. Each chemical class has its inherent properties, advantages, and limitations for being used as films.

14.3.1 Polysaccharide-Based Coatings

Polysaccharides are long-chain biopolymers formed from mono- or disaccharide repeating units linked by glycosidic bonds. They are widely available and usually have low cost. Most polysaccharides are neutral, although some gums are negatively charged. As a consequence of the large number of hydroxyl and other polar groups in their structure, hydrogen bonds play important roles in film formation and characteristics. Negatively charged gums, such as alginate, pectin, and carboxymethyl cellulose (CMC), tend to present some different properties depending on the pH (Han and Gennadios, 2005).

Polysaccharide films are usually formed by disrupting interactions among polymer segments during coacervation and forming new intermolecular hydrophilic and hydrogen bonds upon evaporation of the solvent (Janjarasskul and Krochta, 2010). Because of their hydrophilicity, polysaccharide films provide a good barrier to CO_2 and O_2; hence, they retard respiration and ripening of fruits (Cha and Chinnan, 2004). On the other hand, similarly to other hydrophilic materials, their polarity determines their poor barrier to water vapor (Park and Chinnan, 1995) as well as their sensitivity to moisture, which may affect their functional properties (Janjarasskul and Krochta, 2010).

14.3.1.1 Starches

Starches are polymers of D-glucopyranosyl, consisting of a mixture of the predominantly linear amylose and the highly branched amylopectin (Figure 14.2). Native starch molecules arrange themselves in semicrystalline granules in which amylose and amylopectin are linked by hydrogen bonding. When heat is applied to native starch in presence of water, the granules swell and hydrate, which triggers the "gelatinization" process, characterized by the loss of crystallinity and molecular order, followed by a dramatic increase in viscosity (Kramer, 2009).

Amylose responds to the film-forming capacity of starches, since it forms cohesive and relatively strong films (Han et al., 2006). On the other hand, amylopectin films are brittle and noncontinuous, since its branch-to-branch structure interferes with intermolecular associations, disrupting film formation (Peressini et al., 2003).

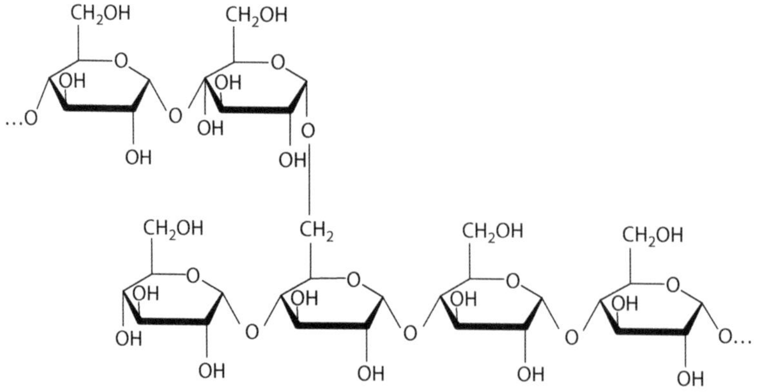

FIGURE 14.2 Chemical structure of a starch fragment.

Indeed, García et al. (1998), comparing quality of strawberries coated with starches from different sources, observed a significant effect of amylose content on color, weight loss, and firmness of coated strawberries.

Application of starch films is limited by two major drawbacks. The films are often very brittle, requiring the presence of plasticizers to improve their flexibility (Peressini et al., 2003; Mali et al., 2004). Moreover, the high hydrophilicity of starch causes its barrier properties to decrease with increasing relative humidity; therefore, starch is not the best option when working with minimally processed fruits (Olivas and Barbosa-Cánovas, 2009).

14.3.1.2 Cellulose and Its Derivatives

Together with starch, cellulose and its derivatives (such as ethers and esters) are the most important raw materials for elaboration of edible films (Peressini et al., 2003). Cellulose, the most abundant natural polymer on earth, is an essentially linear natural polymer of (1→4)-β-D-glucopyranosyl units (Figure 14.3). Its tightly packed polymer chains and highly crystalline structure make it insoluble in water. Water solubility can be conferred by etherification; the water-soluble cellulose ethers, including methyl cellulose (MC), hydroxypropyl cellulose (HPC), hydroxypropyl-methyl cellulose (HPMC), and CMC, have good film-forming properties (Cha and Chinnan, 2004; Janjarasskul and Krochta, 2010). An MC coating has been demonstrated to retard softening and to reduce respiration rates of avocados (Maftoonazad and Ramaswamy, 2005).

14.3.1.3 Anionic Polysaccharides

Pectins are water-soluble anionic heteropolysaccharides composed mainly of (1→4)-α-D-galactopyranosyluronic acid units, in which some carboxyl groups of galacturonic acid are esterified with methanol (Figure 14.4). High-methoxyl pectins (HMPs) have a degree of esterification (DE) above 50%, whereas low-methoxyl

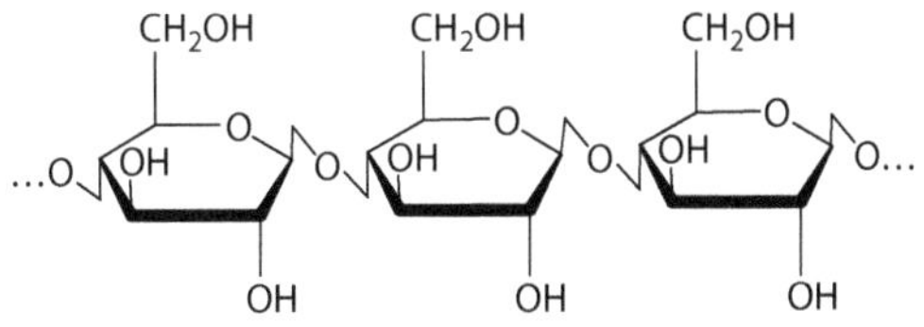

FIGURE 14.3 Chemical structure of cellulose.

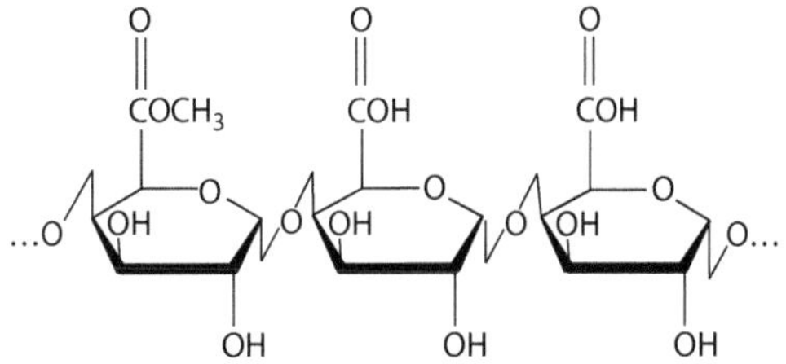

FIGURE 14.4 A fragment of pectin containing esterified and nonesterified carboxyl groups of galacturonic acid.

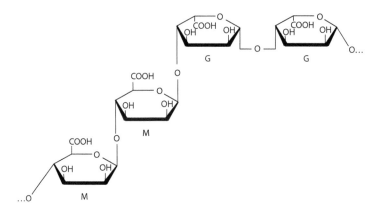

FIGURE 14.5 Basic structure of alginates, containing units of *mannuronic* (M) and *gulu-ronic* (G) acids.

pectin (LMP) has a DE below 50%. The ratio of esterified to nonesterified galact-uronic acid determines the behavior of pectin in food applications, since it affects solubility and gelation properties of pectin (Baldwin et al., 1995). HMP forms gels with sugar and acid, whereas LMP forms gels in the presence of divalent cations such as Ca^{2+}, which links adjacent LMP chains via ionic interactions, forming a tridimensional network (Janjarasskul and Krochta, 2010).

Alginates, which are extracted from brown seaweeds, are salts of alginic acid, a linear copolymer of D-mannuronic and L-guluronic acid monomers (Figure 14.5), containing homogeneous poly-*mannuronic* and poly-*guluronic* acid blocks (M and G blocks, respectively) and MG blocks containing both uronic acids. These highly anionic polysaccharides suffer instantaneous gel formation by reacting with di- or trivalent cations, similarly to LMP. Calcium ions, especially effective, have been applied as gelling agents (Cha and Chinnan, 2004). Calcium ions pull alginate chains together via ionic interactions, after which interchain hydrogen bonding occurs (Kester and Fennema, 1986). Films can be formed either from evaporating water from an alginate gel or by a two-step procedure involving drying of alginate solution followed by treatment with a calcium salt solution to induce cross-linking (Janjarasskul and Krochta, 2010). The strength and permeability of films may be altered by changing calcium concentration and temperature, among other factors (Kester and Fennema, 1986). Apples previously immersed in calcium chloride and coated with alginate presented lower weight loss, softening, and browning when compared to untreated fruits (Olivas et al., 2007).

14.3.1.4 Chitosan

Chitosan, a linear polysaccharide consisting of β (1→4) linked residues of *N*-acetyl-2-amino-2-deoxy-D-glucose (glucosamine) and 2-amino-2-deoxy-D-glucose (*N*-acetyl-glucosamine) (Figure 14.6), produced by partial deacetylation of chitin, presents a cationic character which confers unique properties on this polysaccharide, such as antimicrobial activity and the ability to carry and slow-release functional ingredi-ents (Coma et al., 2002). Chitosan coatings were effective in extending shelf life of

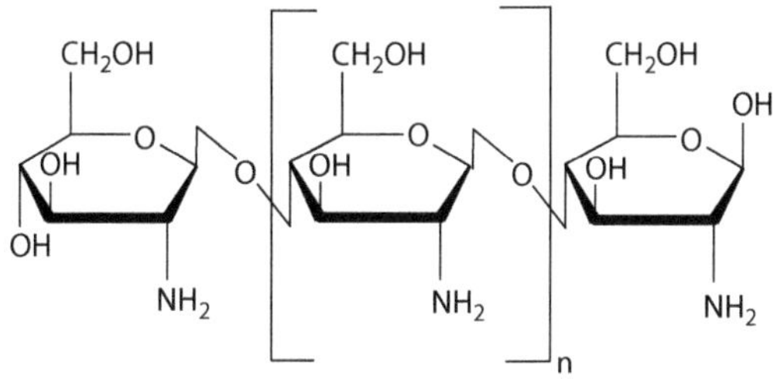

FIGURE 14.6 Chemic structure of chitosan.

several fruits; reducing rates of water loss (Dong et al., 2004; Hernández-Muñoz et al., 2006; Chien et al., 2007; Lin et al., 2011), ascorbic acid loss (Dong et al., 2004; Chien et al., 2007), softening (Ali et al., 2011), and enzymatic browning (Dong et al., 2004; Jiang et al., 2005); and delaying the ripening process by decreasing respiration rates (Ali et al., 2011; Lin et al., 2011). Moreover, they were reported to have retarded microbial growth in fruit surfaces (Hernández-Muñoz et al., 2006; Chien et al., 2007; Campaniello et al., 2008). The polycationic structure of chitosan probably interacts with the predominantly anionic components (lipopolysaccharides, proteins) of microbial cell membranes, especially Gram-negative bacteria (Helander et al., 2001).

14.3.1.5 Some Novel Polysaccharide Coatings

Ali et al. (2010) evaluated the performance of a novel coating based on gum arabic on tomato fruits. When compared to uncoated fruits, tomatoes coated with 10% gum arabic presented significant lower rates of changes in weight, color, firmness, titratable acidity, soluble solids content, ascorbic acid content, and decay percentage. Maqbool et al. (2010) observed that, although gum arabic alone has not presented antifungal effects against *Colletotrichum musae* (the causal agent of anthracnose) in bananas, composite coatings obtained from combining gum arabic with chitosan presented better antifungal effects when compared to chitosan alone. However, the authors did not present an explanation for the observed synergistic effect of gum arabic and chitosan.

Some new edible coatings have been obtained from mucilages, which are heteropolysaccharides obtained from plant stems. Prickly pear cactus mucilage as an edible coating was demonstrated to have extended shelf life of strawberries, retarding their softening (Del-Valle et al., 2005). *Aloe vera* mucilage, composed mainly by polysaccharides, has been used as a coating for grapes (Valverde et al., 2005) and sweet cherries (Martínez-Romero et al., 2006), retarding their respiration and ripening rates, softening, weight loss, and color changes.

14.3.2 PROTEIN-BASED COATINGS

Proteins are linear, random copolymers built from up to 20 different monomers. The main mechanism of formation of protein films involves denaturation of the protein initiated by heat, solvents, or change in pH, followed by association of peptide chains through new intermolecular interactions (Janjarasskul and Krochta, 2010).

While hydroxyl is the only reactive group in polysaccharides, proteins may be involved in several possible interactions and chemical reactions (Hernandez-Izquierdo and Krochta, 2008), such as chemical reactions through covalent (peptide and disulfide) bonds and noncovalent (ionic, hydrogen, and van der Waals) interactions. Moreover, hydrophobic interactions may occur between nonpolar groups of amino acid chains (Kokini et al., 1994).

The most peculiar properties of proteins compared to other film-forming materials are denaturation, electrostatic charges, and amphiphilic character. Protein conformation can be affected by many factors, such as charge density and hydrophilic–hydrophobic balance (Han and Gennadios, 2005). Proteins have good film-forming properties and good adherence to hydrophilic surfaces (Baldwin et al., 1995). Protein-derived films provide good barriers to O_2 and CO_2 but not to water (Cha and Chinnan, 2004). Their barrier and mechanical properties are impaired by moisture owing to their inherent hydrophilic nature (Janjarasskul and Krochta, 2010).

Gelatin is a substantially pure protein food ingredient, obtained by a mild heat treatment of collagen under acidic or alkaline conditions, when collagen is partially denatured, but recovers part of its triple helix structure upon cooling. On dehydration, films are formed with irreversible conformational changes (Badii and Howell, 2006; Dangaran et al., 2009).

Total bovine milk proteins consist of about 80% casein and 20% whey proteins. Edible films and coating from casein, whey proteins, and total milk proteins have been reported. Simple milk protein films provide good barriers to gases because of their complex intermolecular bindings (Chen, 1995). Caseinate films are water soluble, although the use of a buffer solution at the pI of casein results in water-insoluble films (Krochta et al., 1990). Avena-Bustillos and Krochta (1993) observed that barrier properties of casein may be enhanced by cross-linking with calcium ions or by adjusting its pH to the isoelectric point. Coatings from both calcium caseinate (Le Tien et al., 2001) and whey proteins (Le Tien et al., 2001; Pérez-Gago et al., 2003) delayed browning of apples, which was attributed to the oxygen barrier of the coatings and/or to its antioxidative properties (Pérez-Gago et al., 2003). Whey protein coatings enhanced the appearance of plums by creating glossy surfaces and improved their firmness (Reinoso et al., 2008).

Zein, the prolamin (alcohol-soluble) fraction of corn protein, has been used in coating formulations (Cha and Chinnan, 2004). Zein coatings are usually prepared by dissolving zein in aqueous ethyl alcohol (Ghanbarzadeh et al., 2007). Zein is rich in nonpolar amino acids, with low proportions of basic and acidic amino acids. Its consequent water insolubility favors the water vapor barrier of its films (Dangaran et al., 2009).

Wheat gluten is composed mainly of gliadins and glutenins (alkali- and acid-soluble) fractions. Gliadin, the viscous component, constitutes a heterogeneous protein group characterized by single polypeptide chains associated by hydrogen bonding and hydrophobic interactions, having intramolecular disulfide bonds. Glutenins form an extensive network of intermolecular disulfide bonds (Hernández-Muñoz et al., 2003). Because gluten is water insoluble, a complex solvent system with basic or acidic conditions in the presence of alcohol and disulfide bond-reducing agents is required to prepare casting solutions (Cuq et al., 1998). Strawberries with gluten coatings presented lower weight loss and softening when compared to uncoated ones (Tanada-Palmu and Grosso, 2005).

14.3.3   LIPID COATINGS

Unlike other macromolecules, lipids are not biopolymers, being not able to form cohesive, self-supporting films. So, they are either used as coatings or incorporated into biopolymers to form composite films, providing a good water vapor barrier, thanks to their low polarity (Greener and Fennema, 1989).

Lipid-based edible films have variable behavior against moisture transfer, depending on their particular properties. Polarity of lipids depends on the distribution of chemical groups, the length of aliphatic chains, and the presence and degree of unsaturation (Morillon et al., 2002). Unsaturated fatty acids are less efficient to control moisture transfer because of their higher polarity when compared to saturated ones. Indeed, Hagenmaier and Baker (1997) observed that coatings containing stearic or palmitic acids were more efficient to reduce desiccation of oranges than those with oleic acid.

Waxes are esters of long-chain aliphatic acids with long-chain aliphatic alcohols (Rhim and Shellhammer, 2005). They are more resistant to water diffusion than most other edible film materials because of their very low content of polar groups (Kester and Fennema, 1986) and their high content in long-chain fatty alcohols and alkanes (Morillon et al., 2002). There are a variety of naturally occurring waxes, derived from vegetables (e.g., carnauba, candelilla, and sugar cane waxes), minerals (e.g., paraffin and microcrystalline waxes), or animals—including insects (e.g., beeswax, lanolin, and wool grease)—while some other waxes are synthetically produced, such as carbowaxes and polyethylene wax (Rhim and Shellhammer, 2005).

Triglycerides or neutral lipids are esters of fatty acids with glycerol. Mono- and ditriglycerides (partial esters) can also be used as coating materials. Their functional properties, especially water vapor permeability, are dependent on their chemical structures. Long-chain triglycerides are insoluble in water, whereas short-chain molecules are partially water soluble. Unsaturated fatty acids have significantly lower melting points and increased moisture transfer rates than the corresponding saturated ones (Rhim and Shellhammer, 2005). Branching of acyl chain also results in increased water vapor permeability because of the increased mobility of hydrocarbon chains and less efficient lateral packing of acyl chains (Janjarasskul and Krochta, 2010).

Acetic acid esters of monoglycerides, called acetylated monoglycerides, have been also used as food-coating materials. Their moisture barrier tends to improve with increasing degree of acetylation, possibly because of removal of hydrophilic hydroxyl groups. According to Bourlieu et al. (2008), there are certain inconveniences about their use, such as undesirable aftertaste and the tendency of highly saturated acetylated monoglycerides to crack and flake during storage.

14.3.4   COMPOSITE COATINGS

Multicomponent or composite coatings are usually made to explore the complementary advantages of each component as well as to minimize their disadvantages. Most composite coatings associate a hydrophobic compound (often lipids) and a hydrophilic structural matrix (Guilbert and Biquet, 1986). Since polysaccharides and proteins are less permeable than lipids to gases, their presence in coatings is useful to reduce respiration rates and to retard fruit senescence, while lipids reduce desiccation.

Composite films and coatings can be produced as either bilayers or stable emulsions. In bilayer composite films, the lipid forms a second layer over the polysaccharide or protein layer. On the other hand, in emulsion composite films, the lipid is dispersed and entrapped in the supporting biopolymer matrix (Pérez-Gago and Krochta, 2005).

García et al. (2000) demonstrated that lipid addition to starch films decreased their crystalline-amorphous ratio, which is expected to increase film diffusibility and, consequently, permeability. On the other hand, lipid addition also reduces the hydrophilic–hydrophobic ratio of films, which decreases their water solubility and therefore water vapor permeability.

The optimum lipid concentration in the preparation of composite edible coatings should be determined by considering both the effect of decreasing the moisture transfer rate and the physical strength of the films, since excessive levels of some lipid materials can result in the coating becoming brittle (Shellhammer and Krochta, 1997).

A gluten coating reduced softening and weight loss of strawberries, especially when lipids (beeswax, stearic, and palmitic acids) were incorporated. However, the lipid addition impaired the acceptance of the strawberries in terms of appearance and flavor (Tanada-Palmu and Grosso, 2005). Similarly, Vargas et al. (2006) observed that, although the addition of a lipid component (oleic acid) has improved the water vapor resistance of chitosan-coated strawberries, it has decreased their acceptance.

14.4 COATING/FRUIT SURFACE SUITABILITY

Film formation as a coating depends on two types of interaction: cohesion and adhesion. Cohesion is related to attractive forces between molecules of film components, influencing the mechanical strength of films (Guilbert et al., 1996; Sothornvit and Krochta, 2005). The presence of ingredients incompatible with the main biopolymer in the film-forming dispersion causes the cohesion and the film strength to decrease (Han and Gennadios, 2005). Excessive cohesive forces result in brittleness, which may be overcome by addition of plasticizers. Adhesion, by its turn, is related to attractive forces between film and substrate; it is important for film casting and coating processes (Sothornvit and Krochta, 2005). A poor adhesion results in incomplete coating or easy peel-off of a film from the surface (Han and Gennadios, 2005).

For the optimization of a coating solution, the control of adhesion and cohesion coefficients is very important since the former promotes spreading of the liquid whereas the later promotes its contraction (Ribeiro et al., 2007).

The coating procedure involves wetting the fruit surface with a coating solution. An adequate coating solution spreads spontaneously on the fruit surface (Mittal, 1997). However, it is hardly possible to have a coating solution/dispersion perfectly suited to the surface in terms of polarity and surface tension. Then, efforts should be made in order to find the best formulation possible in terms of compatibility with the surface. The interfacial tension between the coating solution and the fruit surface must be estimated, based on the surface tension of both the surface and the coating solution and the contact angle between both.

Finding a suited coating solution can be a challenge. The lower the surface (or interfacial) tension of a solution, the higher its surface (or interfacial) activity (Gaonkar, 1991). Most solutions are water based, and water has a high surface tension (72.8 dyn cm^{-1}), whereas most solid surfaces have lower surface tension (Nussinovitch, 2009). Since an adequate coating involves compatibility between the coating solution and the solid surface, the surface tension of the coating solution must be reduced in order to adjust it to the lower surface tension of the surface, thus lowering the interfacial tension and enhancing adhesion.

Surface-active agents, such as emulsifiers and other amphiphilic chemicals in the film-forming solution, reduce the surface tension of the coating solution, thus decreasing the difference between the solid surface energy and the surface tension of the coating solution and ultimately increasing the work of adhesion (Han and Gennadios, 2005). The surfactant Tween 80 was added to a chitosan coating solution in order to reduce its surface tension and enhance its wettability, improving its adhesion to apple skin (Choi et al., 2002).

Another strategy which can be used to improve compatibility of hydrophilic coatings with fruit surfaces is by means of chemical modification of biopolymers, by introducing hydrophobic groups into their structure. Hydrophobization has been successfully used by Vu et al. (2011) to enhance compatibility of a chitosan coating to strawberry surface.

14.5   ACTIVE COATINGS FOR FRUITS

Besides their function as inert barriers to extend fruit stability, edible coatings can also interact with the coated fruit and/or the surrounding environment in a desirable way, thus constituting an active coating. Active properties of coatings may be related to release of compounds (e.g., antimicrobials and antibrowning agents) which can retard fruit degradation or to absorption/scavenging of undesirable compounds (e.g., ethylene) which might accelerate degradation. Edible coatings have been presented as excellent way to carry additives since they are able to maintain effective concentrations of the additives on the fruit surfaces, where they are mostly needed, reducing the impact of such chemicals on overall acceptability of the fruit (Oms-Oliu et al., 2010).

Incorporating antimicrobial compounds into edible coatings help preserve the quality of fresh-cut fruits, which are more perishable than the corresponding whole fruits, because of the skin removal, favoring microorganisms to invade and grow on product. Since antimicrobials are mostly needed on the product surface, their application on a coating helps minimize antimicrobial usage (Vodjani and Torres, 1990). Antimicrobials most commonly used include organic acids and their salts. Incorporation of malic and lactic acid into soy protein coatings retarded microbial growth on fresh-cut melon (Eswaranandam et al., 2006). Alginate-based edible coatings incorporated with essential oils of cinnamon, palmarosa, and lemongrass have been reported to be more effective to extend the microbiological stability of melons when compared to coatings without the essential oils (Raybaudi-Massilia et al., 2008). On the other hand, the acceptance of melons was decreased by addition of cinnamon oil.

The incorporation of the sulfur-containing amino acids *N*-acetylcysteine and gluta-thione into edible coatings has been shown to reduce microbial growth (Oms-Oliu et al., 2008; Rojas-Graü et al., 2008).

Fresh-cut fruits can also be benefitted by the incorporation of antibrowning agents into edible coatings in order to retard color changes caused by oxidation of phenolic compounds by polyphenol oxidase (PPO) in presence of oxygen. The incorporation of ascorbic acid into edible coatings has been reported to retard enzymatic brown-ing of fruits (Brancoli and Barbosa-Cánovas, 2000; McHugh and Senesi, 2000; Pérez-Gago et al., 2006). The thiol-containing amino acids, *N*-acetylcysteine and glutathione, incorporated into edible coatings, besides reducing microbial growth in fruit surfaces, were also effective to inhibit browning of fresh-cut fruits (Pérez-Gago et al., 2006; Rojas-Graü et al., 2007b, 2008; Oms-Oliu et al., 2008), since they react with quinones formed during the initial phase of enzymatic browning to yield colorless products or to reduce *o*-quinones back to *o*-diphenols (Richard et al., 1992). Edible coatings containing sour whey have been demonstrated to be effective in reducing oxidative browning of cut apples, when compared to other coatings based on milk proteins (Shon and Haque, 2007).

14.6   FINAL CONSIDERATIONS

Edible coatings have demonstrated to be effective to increase stability of a vari-ety of fresh and minimally processed fruits. Despite the great benefits from using edible coatings, commercial applications of this technology on a broad range are still very limited. Enhancement of the water resistance of polysaccharide films is still required, which can be obtained by chemical modification of the biopolymer matrix or incorporation of hydrophobic ingredients. Moreover, mechanical and bar-rier properties of biopolymer films still need improvement. Several studies have tried to improve performance of biopolymers by incorporation of nanostructures such as nanocellulose (Azeredo et al., 2009, 2010) and nanoclays (Mangiacapra et al., 2006; Olabarrieta et al., 2006; Petersson and Oksman, 2006). However, the use of nano-structures in edible coatings still requires careful investigation about their safety, since their size may allow them to penetrate into cells and eventually remain in the human body. Considering evidences supporting that nanoparticles can exhibit differ-ent properties from the corresponding bulk materials, the need for accurate informa-tion on their effects to human health is imperative.

REFERENCES

Ali, A., Maqbool, M., Ramachandran, S., and Alderson, P.G. 2010. Gum arabic as a novel edible coating for enhancing shelf-life and improving postharvest quality of tomato (*Solanum lycopersicum* L.) fruit. *Postharvest Biology and Technology* 58: 42–47.
Ali, A., Muhammad, M.T., Sijam, K., and Siddiqui, Y. 2011. Effect of chitosan coatings on the physicochemical characteristics of Eksotika II papaya (*Carica papaya* L.) fruit during cold storage. *Food Chemistry* 124: 620–626.
Avena-Bustillos, R.J. and Krochta, J.M. 1993. Water vapor permeability of caseinate-based edible films as affected by pH, calcium crosslinking and lipid content. *Journal of Food Science* 58(4): 904–917.

Azeredo, H.M.C., Mattoso, L.H.C., Avena-Bustillos, R.J., Ceotto Filho, G., Munford, M.L., Wood, D., and McHugh, T.H. 2010. Nanocellulose reinforced chitosan composite films as affected by nanofiller loading and plasticizer content. *Journal of Food Science* 75(1): N1–N7.

Azeredo, H.M.C., Mattoso, L.H.C., Wood, D., Williams, T.G., Avena-Bustillos, R.J., and McHugh, T.H. 2009. Nanocomposite edible films from mango puree reinforced with cellulose nanofibers. *Journal of Food Science* 74(5): N31–N35.

Badii, F. and Howell, N.K. 2006. Fish gelatin: Structure, gelling properties and interaction with egg albumen proteins. *Food Hydrocolloids* 20: 630–640.

Baldwin, E.A., Nisperos-Carriedo, M.O., and Baker, R.A. 1995. Edible coatings for lightly processed fruits and vegetables. *HortScience* 30(1): 35–38.

Bourlieu, C., Guillard, V., Vallès-Pamiès, B., and Gontard, N. 2008. Edible moisture barriers for food product stabilization. In *Food Materials Science*, eds. J.M. Aguilera and P.J. Lillford, pp. 547–575. New York: Springer.

Brancoli, N. and Barbosa-Cánovas, G.V. 2000. Quality changes during refrigerated storage of packaged apple slices treated with polysaccharide films. In *Innovations in Food Processing*, eds. G.V. Barbosa-Cánovas and G.W. Gould, pp. 243–254. Lancaster, PA: Technomic.

Campaniello, D., Bevilacqua, A., Sinigaglia, M., and Corbo, M.R. 2008. Chitosan: Antimicrobial activity and potential applications for preserving minimally processed strawberries. *Food Microbiology* 25: 992–1000.

Cha, D.S. and Chinnan, M.S. 2004. Biopolymer-based antimicrobial packaging—A review. *Critical Reviews in Food Science and Nutrition* 44(4): 223–237.

Chen, H. 1995. Functional properties and applications of edible films made of milk proteins. *Journal of Dairy Science* 78: 2563–2583.

Chien, P.J., Sheu, F., and Yang, F.H. 2007. Effects of edible chitosan coating on quality and shelf life of sliced mango fruit. *Journal of Food Engineering* 78: 225–229.

Choi, W.Y., Park, H.J., Ahn, D.J., Lee, J., and Lee, C.Y. 2002. Wettability of chitosan coating solution on 'Fuji' apple skin. *Journal of Food Science* 67: 2668–2672.

Coma, V., Martial-Gros, A., Garreau, S., Copinet, A., Salin, F., and Deschamps, A. 2002. Edible antimicrobial films based on chitosan matrix. *Journal of Food Science* 67: 1162–1169.

Conforti, F.D. and Zinck, J.B. 2002. Hydrocolloid-lipid coating affect on weight loss, pectin content, and textural quality of green bell peppers. *Journal of Food Science* 67(4): 1360–1363.

Contreras-Medellin, R. and Labuza, T.P. 1981. Prediction of moisture protection requirements for foods. *Cereal Food World* 26(7): 335–343.

Cuq, B., Gontard, N., and Guilbert, S. 1998. Proteins as agricultural polymers for packaging production. *Cereal Chemistry* 75(1): 1–9.

Dangaran, K., Tomasula, P.M., and Qi, P. 2009. Structure and formation of protein-based edible films and coatings. In *Edible Films and Coatings for Food Applications*, eds. M.E. Embuscado and K.C. Huber, pp. 25–56. New York: Springer.

Debeaufort, F., Quezada-Gallo, J.A., and Voilley, A. 1998. Edible films and coatings: Tomorrow's packagings: A review. *Critical Reviews in Food Science* 38: 299–313.

Debeaufort, F. and Voilley, A. 2009. Lipid-based edible films and coatings. In *Edible Films and Coatings for Food Applications*, eds. M.E. Embuscado and K.C. Huber, pp. 135–168. New York: Springer.

Del-Valle, V., Hernández-Muñoz, P., Guarda, A., and Galotto, M.J. 2005. Development of a cactus-mucilage edible coating (*Opuntia ficus indica*) and its application to extend strawberry (*Fragaria ananassa*) shelf-life. *Food Chemistry* 91: 751–756.

Dong, H., Cheng, L., Tan, J., Zheng, K., and Jiang, Y. 2004. Effects of chitosan coating on quality and shelf life of peeled litchi fruit. *Journal of Food Engineering* 64: 355–358.

Eswaranandam, S., Hettiarachchy, N.S., and Meullenet, J.F. 2006. Effect of malic and lactic acid incorporated soy protein coatings on the sensory attributes of whole apple and fresh-cut cantaloupe. *Journal of Food Science* 71: S307–S313.

Gaonkar, A.G. 1991. Surface and interfacial activities and emulsion of some food hydrocolloids. *Food Hydrocolloids* 5: 329–337.

Garcia, E. and Barret, D.M. 2002. Preservative treatments for fresh cut fruits and vegetables. In *Fresh-Cut Fruits and Vegetables*, ed. O. Lamikanra, pp. 267–304. Boca Raton, FL: CRC Press.

García, M.A., Martino, M.N., and Zaritky, N.E. 1998. Starch-based coatings: Effect on refrigerated strawberry (*Fragaria ananassa*) quality. *Journal of the Science of Food and Agriculture* 76(3): 411–420.

García, M.A., Martino, M.N., and Zaritky, N.E. 2000. Lipid addition to improve barrier properties of edible starch-based films and coatings. *Journal of Food Science* 65(6): 941–947.

Ghanbarzadeh, B., Musavi, M., Oromiehie, A.R., Rezayi, K., Razmi, E., and Milani, J. 2007. Effect of plasticizing sugars on water vapour permeability, surface energy and microstructure properties of zein films. *LWT-Food Science and Technology* 40: 1191–1197.

Goulao, L.F. and Oliveira, C.M. 2008. Cell wall modifications during fruit ripening: When a fruit is not the fruit. *Trends in Food Science and Technology* 19(1): 4–25.

Greener, I.K. and Fennema, O. 1989. Barrier properties and surface characteristics of edible, bilayer films. *Journal of Food Science* 54(6): 1393–1399.

Guilbert, S. and Biquet, B. 1986. Technology and application of edible protective films. In *Food Packaging and Preservation*, ed. M. Mathlouthi, pp. 371–394. London, U.K.: Elsevier.

Guilbert, S., Gontard, N., and Gorris, L.G.M. 1996. Prolongation of the shelf life of perishable food products using biodegradable films and coatings. *LWT-Food Science and Technology* 29: 10–17.

Hagenmaier, R.D. and Baker, R.A. 1997. Edible coatings from morpholine-free wax microemulsions. *Journal of Agricultural and Food Chemistry* 45: 349–352.

Hagenmaier, R.D. and Shaw, P.E. 1992. Gas permeability of fruit coating waxes. *Journal of the American Society for Horticultural Science* 117: 105–109.

Han, J.H. and Gennadios, A. 2005. Edible films and coatings: A review. In *Innovations in Food Packaging*, ed. J.H. Han, pp. 239–262. London, U.K.: Elsevier.

Han, J.H., Seo, G.H., Park, I.M., Kim, G.N., and Lee, D.S. 2006. Physical and mechanical properties of pea starch edible films containing beeswax emulsions. *Journal of Food Science* 71(6): E290–E296.

Helander, I.M., Nurmiaho-Lassila, E.L., Ahvenainen, R., Rhoades, J., and Roller, S. 2001. Chitosan disrupts the barrier properties of the outer membrane of Gram-negative bacteria. *International Journal of Food Microbiology* 71(2–3): 235–244.

Hernandez-Izquierdo, V.M. and Krochta, J.M. 2008. Thermoplastic processing of proteins for film formation—A review. *Journal of Food Science* 73: 30–39.

Hernández-Muñoz, P., Almenar, E., Ocio, M.J., and Gavara, R. 2006. Effect of calcium dips and chitosan coatings on postharvest life of strawberries (*Fragaria ananassa*). *Postharvest Biology and Technology* 39: 247–253.

Hernández-Muñoz, P., Kanavouras, A., Ng, P.K.W., and Gavara, R. 2003. Development and characterization of biodegradable films made from wheat gluten protein fractions. *Journal of Agricultural and Food Chemistry* 51(26): 7647–7654.

Janjarasskul, T. and Krochta, J.M. 2010. Edible packaging materials. *Annual Review of Food Science and Technology* 1: 415–448.

Jiang, Y., Li, J., and Jiang, W. 2005. Effects of chitosan coating on shelf life of cold-stored litchi fruit at ambient temperature. *LWT-Food Science and Technology* 38: 757–761.

Kester, J.J. and Fennema, O.R. 1986. Edible films and coatings: A review. *Food Technology* 40: 47–59.

Kokini, J.L., Cocero, A.M., Madeka, H., and de Graaf, E. 1994. The development of state diagrams for cereal proteins. *Trends in Food Science and Technology* 5: 281–288.

Kramer, M.E. 2009. Structure and function of starch-based edible films and coatings. In *Edible Films and Coatings for Food Applications*, eds. M.E. Embuscado and K.C. Huber, pp. 113–134. New York: Springer.

Krochta, J.M. 2002. Protein as raw materials for films and coatings: Definitions, current status, and opportunities. In *Protein-Based Films and Coatings*, ed. A. Gennadios, pp. 1–41. Boca Raton, FL: CRC Press.

Krochta, J.M., Pavlath, A.E., and Goodman, N. 1990. Edible films from casein-lipid emulsions for lightly-processed fruits and vegetables. In *Engineering and Food*, Vol. 2, eds. W.E.L. Spiess, and H. Schubert, pp. 329–340. New York: Elsevier.

Le Tien, C., Vachon, C., Mateescu, M.A., and Lacroix, M. 2001. Milk protein coatings prevent oxidative browning of apples and potatoes. *Journal of Food Science* 66(4): 512–516.

Lin, B., Du, Y., Liang, X., Wang, X., Wang, X., and Yang, J. 2011. Effect of chitosan coating on respiratory behavior and quality of stored litchi under ambient temperature. *Journal of Food Engineering* 102: 94–99.

Maftoonazad, N. and Ramaswamy, H.S. 2005. Postharvest shelf-life extension of avocados using methyl cellulose-based coating. *LWT—Food Science and Technology* 38: 617–624.

Mali, S., Grossmann, M.V.E., García, M.A., Martino, M.M., and Zaritzky, N.E. 2004. Barrier, mechanical and optical properties of plasticized yam starch films. *Carbohydrate Polymers* 56: 129–135.

Mangiacapra, P., Gorrasi, G., Sorrentino, A., and Vittoria, V. 2006. Biodegradable nanocomposites obtained by ball milling of pectin and montmorillonites. *Carbohydrate Polymers* 64: 516–523.

Maqbool, M., Ali, A., Ramachandran, S., Smith, D.R., and Alderson, P.G. 2010. Control of postharvest anthracnose of banana using a new edible composite coating. *Crop Protection* 29: 1136–1141.

Martínez-Romero, D., Albuquerque, N., Valverde, J.M., Guillén, F., Castillo, S., Valero, D., and Serrano, M. 2006. Postharvest sweet cherry quality and safety maintenance by *Aloe vera* treatment: A new edible coating. *Postharvest Biology and Technology* 39: 93–100.

McHugh, T.H., Huxsoll, C.C., and Krochta, J.M. 1996. Permeability properties of fruit puree edible films. *Journal of Food Science* 61(1): 88–91.

McHugh, T.H. and Senesi, E. 2000. Apple wraps: A novel method to improve the quality and extend the shelf life of fresh-cut apples. *Journal of Food Science* 65: 480–485.

Mittal, K.L. 1997. The role of the interface in adhesion phenomena. *Polymer Engineering and Science* 17: 467–473.

Morillon, V., Debeaufort, F., Blond, G., Capelle, M., and Voilley, A. 2002. Factors affecting the moisture permeability of lipid-based edible films: A review. *Critical Reviews in Food Science and Nutrition* 42(1): 67–89.

Nussinovitch, A. 2009. Biopolymer films and composite coatings. In *Modern Biopolymer Science*, eds. S. Kasapis, I.T. Norton, and J.B. Ubbink, pp. 295–326. London, U.K.: Academic Press.

Olabarrieta, I., Gallstedt, M., Ispizua, I., Sarasua, J.R., and Hedenqvist, M.S. 2006. Properties of aged montmorillonite-wheat gluten composite films. *Journal of Agricultural and Food Chemistry* 54: 1283–1288.

Olivas, G.I. and Barbosa-Cánovas, G. 2009. Edible films and coatings for fruits and vegetables. In *Edible Films and Coatings for Food Applications*, eds. M.E. Embuscado and K.C. Huber, pp. 211–244. New York: Springer.

Olivas, G.I., Mattinson, D.S., and Barbosa-Cánovas, G. 2007. Alginate coatings for preservation of minimally processed 'Gala' apples. *Postharvest Biology and Technology* 45: 89–96.

Oms-Oliu, G., Rojas-Graü, M.A., González, L.A., Varela, P., Soliva-Fortuny, R., Hernando, M.I.H., Munuera, I.P., Fiszman, S., and Martín-Belloso, O. 2010. Recent approaches using chemical treatments to preserve quality of fresh-cut fruit: A review. *Postharvest Biology and Technology* 57: 139–148.

Oms-Oliu, G., Soliva-Fortuny, R., and Martín-Belloso, O. 2008. Edible coatings with anti-browning agents to maintain sensory quality and antioxidant properties of fresh-cut pears. *Postharvest Biology and Technology* 50: 87–94.

Park, H.J. and Chinnan, M.S. 1995. Gas and water vapor barrier properties of edible films from protein and cellulosic materials. *Journal of Food Engineering* 25: 497–507.

Peressini, D., Bravin, B., Lapasin, R., Rizzotti, C., and Sensidoni, A. 2003. Starch–methylcellulose based edible films: Rheological properties of film-forming dispersions. *Journal of Food Engineering* 59(1): 25–32.

Pérez-Gago, M.B. and Krochta, J.M. 2005. Emulsion and bi-layer edible films. In *Innovations in Food Packaging*, ed. J.H. Han, pp. 384–402. London, U.K.: Academic Press.

Pérez-Gago, M.B., Serra, M., Alonso, M., Mateos, M., and del Rio, M.A. 2003. Effect of solid content and lipid content of whey protein isolate-beeswax edible coatings on color change of fresh cut apples. *Journal of Food Science* 68(7): 2186–2191.

Pérez-Gago, M.B., Serra, M., and del Rio, M.A. 2006. Color change of fresh-cut apples coated with whey protein concentrate-based edible coatings. *Postharvest Biology and Technology* 39: 84–92.

Petersson, L. and Oksman, K. 2006. Biopolymer based nanocomposites: Comparing layered silicates and microcrystalline cellulose as nanoreinforcement. *Composites Science and Technology* 66: 2187–2196.

Raybaudi-Massilia, R.M., Mosqueda-Melgar, J., and Martín-Belloso, O. 2008. Edible alginate-based coating as carrier of antimicrobials to improve shelf-life and safety of fresh-cut melon. *International Journal of Food Microbiology* 121: 313–327.

Reinoso, E., Mittal, G.S., and Lim, L.T. 2008. Influence of whey protein composite coatings on plum (*Prunus domestica* L.) fruit quality. *Food Bioprocessing and Technology* 1: 314–325.

Rhim, J.W. and Shellhammer, T.H. 2005. Lipid-based edible films and coatings. In *Innovations in Food Packaging*, ed. J.H. Han, pp. 362–383. London, U.K.: Academic Press.

Ribeiro, C., Vicente, A.A., Teixeira, J.A., and Miranda, C. 2007. Optimization of edible coating composition to retard strawberry fruit senescence. *Postharvest Biology and Technology* 44(1): 63–70.

Richard, F.C., Goupy, P.M., and Nicolas, J.J. 1992. Cysteine as an inhibitor of enzymatic browning. 2. Kinetic studies. *Journal of Agricultural and Food Chemistry* 40: 2108–2114.

Rojas-Graü, M.A., Avena-Bustillos, R.J., Friedman, M., Henika, P.R., Martín-Belloso, O., and McHugh, T.H. 2006. Mechanical, barrier, and antimicrobial properties of apple puree edible films containing plant essential oils. *Journal of Agricultural and Food Chemistry* 54: 9262–9267.

Rojas-Graü, M.A., Avena-Bustillos, R.J., Olsen, C., Friedman, M., Henika, P.R., Martín-Belloso, O., Pan, Z., and McHugh, T.H. 2007a. Effects of plant essential oil compounds on mechanical, barrier and antimicrobial properties of alginate-apple puree edible films. *Journal of Food Engineering* 81: 634–641.

Rojas-Graü, M.A., Tapia, M.S., and Martín-Belloso, O. 2008. Using polysaccharide-based edible coatings to maintain quality of fresh-cut Fuji apples. *LWT-Food Science and Technology* 41: 139–147.

Rojas-Graü, M.A., Tapia, M.S., Rodríguez, F.J., Carmona, A.J., and Martín-Belloso, O. 2007b. Alginate and gellan based edible coatings as support of antibrowning agents applied on fresh-cut Fuji apple. *Food Hydrocolloids* 21: 118–127.

Senesi, E. and McHugh, T.H. 2002. Film e coperture eduli com matrici a base di fruta. *Industrie Alimentari* 41(420): 1289–1294.

Shellhammer, T.H. and Krochta, J.M. 1997. Whey protein emulsion film performance as affected by lipid type and amount. *Journal of Food Science* 62: 390–394.

Shon, J. and Haque, Z.U. 2007. Efficacy of sour whey as a shelf-life enhancer: Use in antioxidative edible coatings of cut vegetables and fruit. *Journal of Food Quality* 30(5): 581–593.

Sothornvit, R. and Krochta, J.M. 2005. Plasticizers in edible films and coatings. In *Innovations in Food Packaging*, ed. J.H. Han, pp. 403–433. London, U.K.: Academic Press.

Tanada-Palmu, P.S. and Grosso, C.R.F. 2005. Effect of edible wheat gluten-based films and coatings on refrigerated strawberry (*Fragaria ananassa*) quality. *Postharvest Biology and Technology* 36: 199–208.

Valverde, J.M., Valero, D., Martínez-Romero, D., Guillén, F., Castillo, S., and Serrano, M. 2005. Novel edible coating based on *Aloe vera* gel to maintain table grape quality and safety. *Journal of Agricultural and Food Chemistry* 53: 7807–7813.

Vargas, M., Albors, A., Chiralt, A., and González-Martínez, C. 2006. Quality of cold-stored strawberries as affected by chitosan-oleic acid edible coatings. *Postharvest Biology and Technology* 41: 164–171.

Vodjani, F. and Torres, J.A. 1990. Potassium sorbate permeability of methylcellulose and hydroxypropyl methylcellulose coatings: Effect of fatty acids. *Journal of Food Science* 55(3): 841–846.

Vu, K.D., Hollingsworth, R.G., Leroux, E., Salmieri, S., and Lacroix, M. 2011. Development of edible bioactive coating based on modified chitosan for increasing the shelf life of strawberries. *Food Research International*, 44(1): 198–203.

Zhou, R., Li, Y., Yan, L., and Xie, J. 2011. Effect of edible coatings on enzymes, cell-membrane integrity, and cell-wall constituents in relation to brittleness and firmness of Huanghua pears (*Pyrus pyrifolia* Nakai, cv. Huanghua) during storage. *Food Chemistry* 124: 569–575.

Zhou, R., Mo, Y., Li, Y., Zhao, Y., Zhang, G., and Hu, Y. 2008. Quality and internal characteristics of Huanghua pears (*Pyrus pyrifolia* Nakai, cv. Huanghua) treated with different kinds of coatings during storage. *Postharvest Biology and Technology* 49(1): 171–179.

15 Thermal Treatment Effects in Fruit Juices

Fátima A. Miller and Cristina L.M. Silva

CONTENTS

15.1 INTRODUCTION

As a response to consumers' demand to attain new and more differentiated food products of high quality and with guarantee of safe standards, considerable research efforts have been made in the recent years. It is generally recognized that fresh fruits are important components for a well-balanced diet due to their nutritional value and also for presenting differences in color, shape, taste, aroma, and texture. Therefore, fruit juices with the same health benefits are developed and manufactured, providing a more convenient way of consumption. However, the perishable nature of these products, which stated their fast consumption, and also the number of outbreaks associated with nonpasteurized juices identified the importance of an accurate thermal treatment.

A thermal process is a time–temperature schedule that can be classified according to the intensity of the heat treatment. When the temperature is below 100°C, the process is called pasteurization; when it is equal to 100°C, it is designated as canning; and when it is higher than 100°C, it is named sterilization. In all its forms of application, thermal processing persists as the most widely used method of food preservation.

Pasteurization is designed mainly for the inactivation of: (1) pathogenic vegetative bacteria, such as, *Escherichia coli*, *Listeria*, and *Salmonella*; (2) spoilage microorganisms, such as yeasts and moulds; and (3) natural toxins and enzymes. Typical pasteurization conditions for fruit juices are 65°C for 30 min, 77°C for 1 min, or 88°C for 15 s (Ramaswamy et al., 1992). After this mild treatment, the juice is not sterile and, therefore, other preservation techniques are required for product preservation (Lund, 1975). Exceptions are fruit juices with a pH lower than 4.5 that, after treatment, are stable at ambient conditions. In these high-acid foods, the product acidity eliminates the possible development of *Clostridium botulinum* spores, and consequently there is no toxin production.

Therefore, the severity of the process depends on

- The juice's physical and chemical properties
- The juice's heat penetration characteristics
- The target microorganism's or enzyme's heat resistance
- The storage conditions following the thermal treatment

The fruit juices can be heat treated by conventional heating methods, such as canning or hot filling. However, these conventional operations have the tendency to induce permanent changes to the nutritional and sensory attributes of foods. Therefore, advances have been made in the improvement of those food processing operations with the intention of ensuring high-quality food retention levels, while extending the products' shelf-life. The aim is to substantially reduce nutrient damage by reducing heating times and optimization of heating temperatures.

Recent developments have also been made in alternative technologies with the potential to produce safe and better quality foods. Some of these alternative techniques included ohmic and microwave heating and radio-frequency processing. Their industrial application depends not only on their quality and safety capabilities but also on their potential to offer economically competitive advantages.

Juice processing must follow some regulatory requirements and national or international directives that recommend and enforce performance standards or methods. The intention of establishing these regulations is to ensure fruit juice safety and to minimize quality losses.

15.2 THERMAL PROCESSING

Fruit juices are available in the market as ready to drink, concentrated, nectar, syrup, or powders. Although they are differently prepared, all of them are subjected to thermal treatment.

As a liquid product, fruit juices are easily thermally treated using heat exchangers (85°C–92°C, 10–15 s). There are several types of heat exchangers, and the choice for one of them is related to the amount of solid particles presented in the fruit juice.

When the juice is clear and has little solid particles, the most used system is the plate heat exchangers (see Figure 15.2). It consists of a series of thin plates clamped together on a frame and separated by spacing gaskets. The spaces form channels in which the heating fluid and the fruit juice flow, exchanging heat through alternate plates.

When the juice has higher proportion of solid particles, the shell and tube heat exchanger or the scraped surface heat exchanger may be used. These two systems are tubular heat exchangers that mainly consist of one single tube or more, enclosed within a larger tube. The juice flows through the smaller tubes, with the heating medium flowing over the tubes within the larger tube. Although the shell and tube heat exchanger consists of a number of parallel tubes enclosed by a cylindrical shell, the scraped surface heat exchanger is a double-tube equipment with the incorporation of internal blades attached to a moving shaft or frame. This type of heat exchanger is particularly useful for high-viscosity products, since the heat transfer is increased by the produced turbulence.

The heating process may occur before or after the product is packaged. Conventional methods are hot filling and canning.

15.2.1 HOT FILLING

The juice is heated in a heat exchanger to 85°C–95°C and then filled into glass, metal, or certain types of plastic containers. After sealing, the container is inverted for several minutes before cooling, thus pasteurizing the container and destroying any spoilage microorganism that might be present on the cap (Bates et al., 2001). This heating method is a batch process adequate for highly acidic juices such as the ones of apple and grape and also for viscous fruit products like pulps and concentrates. A shelf life from 9 to 12 months is obtained with this procedure (Ramaswamy and Abbatemarco, 1996). The heat exchangers mostly used in this procedure are the scraped surface heat exchangers and the plate heat exchangers.

15.2.2 CANNING

This process consists of three independent stages: (1) hermetically sealing the juice in the container; (2) heat "sterilizing" the sealed unit; and (3) cooling it to ambient temperature for subsequent storage (Nicolaides and Wareing, 2002).

Care must be taken with the headspace present in the top of the container. This space is needed for product expansion during heat processing, but the air must be removed by a steam exhaust treatment. If the air is left there, product surface discoloration and rancidity can occur due to oxidation of the contents.

This process can use as containers not only cans but also glass containers and some heat-resistant plastic films. However, when glass containers are processed (heated or cooled), care must be taken to avoid cracks due to thermal shock. When cans are used as containers, fruit juices are preserved using plain tinplate cans, because minimization of the oxidation process can be achieved by the reducing effect of the tin (Arthey and Ashurst, 1996).

In the traditional canning process, a simple boiling water bath is used for heating. This leads to lower product quality, because of the slow heating and cooling rates.

15.3 AIM OF THERMAL PROCESSING OF JUICE

The major goal of thermal treatment of juice is to prevent enzymatic and microbial deterioration. Thus, the higher the temperature, the greater the microbial load reduction, the enzyme inactivation, and the product shelf life achieved. However, some changes at sensorial and nutritional levels also occur. Therefore, the selection of process conditions must be a compromise that ensures adequate microbial and enzyme inactivation but prevents excessive losses of quality attributes.

15.3.1 CHANGES IN ENZYMATIC ACTIVITY DUE TO HEAT

One of the main objectives of heat treatment of fruit juices is to inactivate enzymes, whose activities result in undesirable changes in sensory quality attributes (loss of color and texture and production of off-flavors and off-odors) and nutritive value (e.g., vitamins) of the product during storage.

Enzymes are proteins that catalyze chemical reactions in which their biological activity is determined by the three-dimensional configuration of the molecule. When enzymes are exposed to high temperatures, protein denaturation can occur. Since protein denaturation is associated with conformational changes, loss of activity or changes in the functionality of the enzymes can happen.

The enzymes that need to be inactivated in fruit juices have not been completely established. However, enzymes such as pectin methylesterase (PME), polyphenoloxidase (PPO), and peroxidase (POD) are generally present and are capable of causing undesirable changes. Among them, PME is dominant and the most heat resistant in several fruits (Awuah et al., 2007).

POD is conventionally used to monitor and evaluate the heat treatment extent, since it is considered one of the most heat-stable enzymes, occurring in a significant number of fruits. Although there is no evidence that POD is directly responsible for the higher quality deterioration in most fruits, when they are heated to a temperature high enough to inactivate POD, it is generally recognized that other undesirable enzymes are inactivated too (Goekmen et al., 2005; Baysal and Demirdoven, 2007).

However, other enzymes are also used to assess heat-processing adequacy. In the case of PPO, due to the changes in appearance and organoleptic properties that it causes, its inactivation is highly desirable. Moreover, recently it was shown that PPO and POD have similar thermostability in grapes, as both lost more than 90% of their relative activity after 5 min of incubation at 78°C and 75°C, respectively (Fortea et al., 2009).

Catalase was the first enzyme used as an indicator for adequate heat treatment. However, it was reported that catalase is inactivated in about 50%–70% of the time required to inactivate POD, which led to its amendment.

Although some of these enzymes might be more important in terms of their quality changing effect, POD is mostly chosen because it is extremely easy to quantify and it is one of the most heat resistant.

Overall, further studies are required to find, for each fruit juice, the critical quality factor that should be retained after heat processing and correlate it with a specific enzyme. Only this can ensure an optimal definition of thermal treatment conditions.

15.3.2 CHANGES IN MICROBIAL VIABILITY DUE TO HEAT

Another major goal of the thermal processing of fruit juices is the reduction of the number of viable microorganisms present, contributing to safer final products. Each and every fruit harbors its own specific and characteristic microflora, which depends on the raw material, processing, and preservation and storage conditions.

Thermal processes are applied to food products usually based on the most heat-resistant microorganism that could present a health hazard. However, the changes in nutritional and quality attributes due to this time–temperature treatment must be known. The thermal resistance of some food quality components is illustrated in Table 15.1.

As can be seen, bacterial spores are far more heat resistant than vegetative cells, since they have higher z-values. In contrast, z-values for nutritional and quality

TABLE 15.1
Kinetic Parameters of Thermal Resistance of Various Food Components

Component	$D_{121°C}$ (min)	z-value (°C)
Vegetative cells	0.002–0.02	5–7
Spores	0.1–5.0	7–12
Enzymes	1–10	7–56
Vitamins	100–1000	25–31
Color, texture, and flavor	5–500	25–45

Source: Lund, D., Heat processing, in *Principles of Food Science: Physical Principles of Food Preservation*, ed. O.R. Fennema, Marcel Dekker, New York, 1975.

D-value is the decimal reduction time and is defined as the time required to inactivate 90% of the population.

z-value is the thermal resistance factor and is defined as the increase in temperature necessary to reduce D-value by 10-fold.

attributes degradation are even higher than those for microorganisms or some heat-labile enzymes, which means that chemical reactions are much less temperature sensitive. This is particularly important when high-temperature short-time (HTST) processing is applied, since there will be a product quality improvement while ensuring its safety. However, special care must be taken with some thermoresistant enzymes and the heat penetration rate to the food product.

When a thermal process is applied, the microbial heat resistance is influenced not only by temperature but also by several other factors. The inherent resistance of the bacterium, which is influenced by the age and growth phase of the culture, pH, water activity, and the composition and consistency of the juice are some issues that must be taken into consideration.

As fruit juices have acidic characteristics (pH < 4.5), the fruit beverage industry usually applies a hot-fill-hold pasteurization process, where keeping the product at 86°C–96°C for approximately 2 min is enough to destroy vegetative mesophilic spoilage microbes (Chang and Kang, 2004). Spoilage incidents still occur involving spore-forming bacteria such as *Clostridium* spp., which, despite being capable of surviving in the juice after the thermal processing, cannot germinate or grow as a result of acidic conditions. However, several works reported spoilage incidents caused by a thermoacidophilic nonpathogenic spore-forming bacterium in a diverse range of fruit juices that were either hot-filled, pasteurized, canned, or ultra heat-treated (Chang and Kang, 2004; Gouws et al., 2005; Walker and Phillips, 2008; Durak et al., 2010; Steyn et al., 2011). Therefore, this bacterium, namely, *Alicyclobacillus acidoterrestris*, has been suggested as the microbial target to be used in the design of adequate pasteurization processes of acidic food products (Silva et al., 1999; Silva and Gibbs, 2001, 2004; Ceviz et al., 2009).

15.4   INFLUENCE OF HEAT ON FRUIT JUICE QUALITY

Independent of the extent of the thermal process, heat can promote reactions that could affect the overall quality of foods. This quality loss may affect several types of quality attributes in fruit juices, as represented in Figure 15.1. In this section, the nutritional and organoleptic degradations are described.

15.4.1   Effect of Heat on Nutritional Attributes

High temperatures have a considerable effect on food nutrients. Nutrients are substances needed for growth, metabolism, and other body functions. Nutritional compounds are divided into two groups: macronutrients and micronutrients. Macronutrients are nutrients required in large amounts that provide calories or energy. There are three broad classes of macronutrients: proteins, carbohydrates, and fats. Micronutrients are nutrients that are needed for life in small quantities, but which the organism itself cannot produce. In this group are included various minerals, vitamins, and antioxidants. Some of these nutrients are susceptible to degradation during heating since they are, in varying degrees, water-soluble, thermally unstable, or sensitive to oxidation.

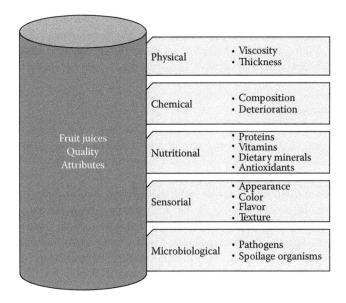

FIGURE 15.1 Overall fruit juice quality dependency.

15.4.1.1 Proteins

Thermal processing can affect protein digestion and consequently limit their absorption (van Boekel et al., 2010). The heat treatment effects on proteins digestibility can be divided into (1) those responsible for altering the secondary, tertiary, and quaternary structure of proteins and (2) those that alter the primary structure (Awuah et al., 2007).

Breaking the secondary, tertiary, and quaternary structures unfolds the proteins—denaturation process—and improves their bioavailability since peptide bonds become readily accessible to digestive enzymes. Modifications of primary protein structures, on the other hand, may lower digestibility and produce proteins that are not biologically available.

Schilling et al. (2008) evaluated the potential protein changes after thermal processing of apple juice. In general, no differences were observed among the proteins before and after the heat treatment, which is probably due to the formation of protein aggregates by covalent bonds that are less digestible than the unprocessed protein.

15.4.1.2 Vitamins

Vitamins are the major sensitive food component to be affected by heat. Beyond temperature, vitamin degradation during heat treatment is dependent on other factors such as oxygen, light, and water solubility (Awuah et al., 2007). In addition, vitamin degradation depends on pH and may be catalyzed by the presence of chemicals, metals, other vitamins and enzymes (Lewis and Heppell, 2000).

The most abundant type of vitamin found in fruit juices is the vitamin C. Vitamin C comprises ascorbic acid (AA) and its oxidized form, dehydroascorbic acid. Fresh fruits mostly contain AA, which is a very water-soluble molecule and also unstable during thermal treatment. Due to its lability, the losses of AA are frequently used as an indicator of food quality.

Other heat-sensitive vitamins are the fat-soluble vitamins A (which in the presence of oxygen is quickly degraded), D, E, and ß-carotene and water-soluble vitamins B1 (thiamine), B2 (riboflavin, which in the presence of an acid environment is quickly degraded), nicotinic acid, pantothenic acid, and biotin C (Ryley and Kajda, 1994).

15.4.1.3 Dietary Minerals

Dietary minerals have no energy value, but they are crucial for normal functioning of the organism. However, if consumed in higher amounts they can become toxic. According to the USDA (2009), the main elements in fruit juices are K, P, Ca, Mg, Na, Fe, Se, Zn, Cu, and Mn.

The degree of thermal processing influence on the mineral content of the product depends on the treatment severity. Some studies have been made on different fruit juices (Gutzeit et al., 2008; Peña et al., 2011). It was reported that some minerals had increased (Mn), some remained equal (Ca, Mg, Cu), and others had decreased (Zn). The authors believe that the increase in some mineral content was probably due to hydrolysis reactions caused by the high temperature.

15.4.1.4 Antioxidants

Antioxidants comprise a broad range of substances that interfere with the propagation of free radicals, preventing oxidative food damage. A high concentration of antioxidants can also result in more stable juices during storage. Other reported beneficial properties of these compounds are in health-protective action, such as antibacterial, anticarcinogenic, and vasodilator actions (Naczk and Shahidi, 2003).

Depending on their structure, antioxidants differ in their polarity and heat stability, which means that heat has different effects on a particular antioxidant. Several studies were carried out to observe the impact of heat on antioxidants. The antioxidant capacity in fruit juices usually decreases significantly after heat processing (Elez-Martínez and Martín-Belloso, 2007; Odriozola-Serrano et al., 2008). The depletion in antioxidant capacity found in the heat-treated juices was due to a substantial loss in total anthocyanins and vitamin C during pasteurization treatments. However, the total antioxidant activity depends on the synergistic and redox interactions between all the different molecules present in the fruit juice (Pellegrini et al., 2003; van Boekel et al., 2010).

In addition to vitamin C and anthocyanins, other microconstituents contribute to the antioxidant capacity in fruit juices, such as vitamin E, carotenoids, and polyphenols.

Polyphenols constitute a substantial part of the total antioxidants in various fruits. These compounds may be classified into different groups: phenolic acids, flavonoids, stilbenes, and lignans. They are highly reactive and may act as substrates for several enzymes, including PPO, POD, glycosidases, and esterases. Generally, polyphenol content in fruits is highly variable and is strongly influenced by the maturity stage, growing areas, variety, extent and storage conditions, and processing procedures.

The effect of food processing on polyphenols activity may differ greatly and is dependent on their concentration, chemical structure, oxidation state, localization in the cell, possible interactions with other food components, and also the type of heat process applied (van Boekel et al., 2010). Therefore, the content and functionality of polyphenols may decrease, increase, or show simple minor changes after thermal processing of fruit juice.

The most studied polyphenols are phenolic acids and flavonoids. Within the phenolic acids, the coumaric, ferulic, and ellagic acids are usually mainly abundant in fruit juices (Arena et al., 2001; Odriozola-Serrano et al., 2008). Although there were discrepant results in several studies, the concentration of most of these phenolic acids has been enhanced, after thermal processing.

Flavonoids are a widespread group of polyphenols which comprise, among others, anthocyanins, flavonols, flavonones, and flavanols (Macheix et al., 1990). The most predominant in fruit juices are the anthocyanins, thus the focus of many investigations. In this case, all researchers agree that heat treatment greatly affects the rates of anthocyanin degradation in fruit juices (Arena et al., 2001; Odriozola-Serrano et al., 2008; Jiménez et al., 2010). None of these studies reported significant differences in flavonol contents before and after juice heat treatment, meaning that this phenolic compound is not affected by thermal processing.

Peña et al. (2011) gathered information about the reasons that may be behind the changes in polyphenol content: (1) biochemical reactions could have occurred during the heat processing, which led to the formation of new phenolic compounds; (2) the thermal processing might have caused significant effects on cell membranes or in phenolic complexes with other compounds, releasing some free phenolic acids or flavonoids; and (3) the thermal process may inactivate PPO, preventing further loss of phenolic compounds.

Carotenoids are considered potent antioxidants, due to their provitamin A activity and free radical scavengers. Therefore, they are related to several health benefits, namely against cancer and coronary heart diseases.

There are basically two types of carotenoids: the carotenoids with oxygen, such as lutein and zeaxanthin, known as xanthophylls, and the carotenoids without oxygen, such as α-carotene and β-carotene (provitamin A carotenoids) and lycopene, known as carotenes. These two groups have significantly different heat sensitivities, the last one being more heat stable. Zulueta et al. (2007) confirmed that lutein and zeaxanthin are highly susceptible to degradation during thermal treatments due to the presence of oxygen in their chemical structure.

Several studies were conducted to evaluate the effect of thermal treatment on these compounds. The most investigated carotenoids are the β-carotene and lutein due to their prevalence in fruit juices. Dhuique-Mayer et al. (2007) evaluated the degradation behavior of carotenoids in citrus juice after heat processing. Carotenes presented losses ranging from 1% to 18%, while the xanthophylls were faster degraded, with losses between 30% and 60%. These results are in accordance with those of previous studies (Lee and Coates, 2003; Gama and Sylos, 2005) that indicate smaller losses for provitamin A carotenoids during thermal pasteurization of orange juice. The carotenoid depletion may be due to its high susceptibility to oxidation, isomerization, and other chemical changes during processing.

However, other works reported a slight increase in carotenoids after heat processing (Sánchez-Moreno et al., 2005; Peña et al., 2011). These observations may be related to the ability of heat treatments to disrupt molecular linkages. Carotenoids are tightly bound to macromolecules, in particular protein and membrane lipids. Thermal processing can promote the release of carotenoids from the matrix due to denaturation of carotene-proteins, leading to better extractability and consequently higher carotenoids concentration (van Boekel et al., 2010).

15.4.2 Effect of Heat on Organoleptic Attributes

As already mentioned, the exposure to high temperatures, even for short periods, has a large influence on the integrity of fruit juices. Thermal processing can result in sensorial changes of appearance, texture, color, and flavor.

15.4.2.1 Color

Color is a primary perceived attribute of a product and is therefore a critical factor in consumer acceptance. One of the main goals of thermal processing is to inactivate the PPO enzyme, the main cause of enzymatic browning, to ensure better color stability of fruit juices during storage. However, color may also be influenced by pigments resulting from nonenzymatic browning reactions or changes in the content of natural pigments, such as chlorophylls, carotenoids, and anthocyanins.

Enzymatic browning is one of the most important color reactions that affect fruit juices. The PPO enzyme is mainly responsible for catalyzing the oxidation of phenolic compounds to quinones, which undergo further polymerization to insoluble dark brown polymers known as melanins. These compounds are recognized to have some antibacterial, antifungal, anticancer, and antioxidant properties (Villamiel et al., 2006).

The formation of yellow and brown pigments in fruit juices during enzymatic browning reactions is controlled by the levels of phenols, the amount of PPO activity, and the presence of oxygen (Spanos and Wrolstad, 1992).

Since enzymatic browning can affect not only the color but also the flavor and the nutritional value of the fruit juices, it can cause substantial economic losses. Therefore, enzymatic browning control measures must be applied. If the heat treatment would be insufficient to completely inactivate PPO, procedures for preventing its activity must be adopted. This can be achieved by nonenzymatic reduction of quinones, chemical modifications or removal of phenolic substracts of PPO, or by using several inhibitors, such as reducing agents, acidulants, chelating agents, and complexing agents (Villamiel et al., 2006).

Nonenzymatic browning is the most complex reaction in food chemistry due to the large number of food components that are able to participate in the reaction through different pathways, giving rise to a complex mixture of products (Olano and Martínez-Castro, 1996). There are several common nonenzymatic browning reactions in foods: Maillard browning, ascorbic acid browning, lipid browning, caramelization, and metal-polyphenolic browning.

Maillard browning is the dominant nonenzymatic browning reaction and takes place between free amino groups from amino acids, peptides, or proteins and the carbonyl group of a reducing sugar. Thus, this reaction causes protein deterioration and consequently a decrease in nutritional food value or in protein digestibility and amino acid availability. The Maillard reaction may be divided into three stages: (1) early Maillard reactions that are chemically well-defined steps without browning; (2) advanced Maillard reactions that lead to the formation of volatile or soluble substances; and (3) final Maillard reactions leading to insoluble brown polymers (Nielsen et al., 2010).

The intermediate reactions and their relative velocities vary with the type of initial reactants and the conditions of the reactions. Among sugars, pentoses are more reactive than hexoses, and hexoses are more reactive than reducing disaccharides. When free amino acids react with sugars, lysine appears to be the most active among them. In peptides and proteins, the N-terminal amino acid is the most reactive, followed by a nonterminal lysine.

Maillard browning is promoted by high storage temperature (i.e., above normal room temperature), high pH and low water activity. In the case of fruit juices, which present low pH and high water activity, the storage temperature must be controlled.

Ascorbic acid browning consists of thermal decomposition of ascorbic acid, under aerobic or anaerobic conditions, in different compounds that provide reactive carbonyls that feed the browning pathway. The aerobic reaction is usually the predominant, where dehydroascorbic acid is formed. This oxidation is sometimes catalyzed by enzymes and metals, which makes this reaction more rapid. Therefore, for ascorbic acid browning control, the fast removal of oxygen from packages is of extreme importance.

Lipid browning occurs when oxidized lipids and lipid-soluble compounds decompose to form browning precursors. In this case, free hydroperoxidic radicals and reactive carbonyls produced by decomposition of lipid hydroperoxides are the precursors of the macromolecular brown pigments (Nielsen et al., 2010). Once more, the absence of oxygen plays an important role in lipid browning control.

Due to the electrophilic character of the carbonyl compounds produced during lipid oxidation, interaction with nucleophiles such as the free amino group of amino acids, peptides, or proteins can also take place, producing end products different from those formed during oxidation of pure lipids (Gillatt and Rossell, 1992).

Caramelization is the nonenzymatic browning where dry sugars or heavily concentrated sugar solutions undergo molecular dehydration with subsequent polymerization to create brown pigments. Temperatures of caramelization are typically around 160°C; therefore, this reaction is not likely to occur from fruit juice processing.

Metal-polyphenolic browning is an oxidative mechanism that results from the interaction of polyphenols present in the fruit juice with iron from external sources. The resulting iron complexes are bluish-black pigments.

Chlorophylls are mainly responsible for the green color of unripe fruits, such as apples, pears, banana, or kiwi. The most important forms of chlorophyll are chlorophyll a and chlorophyll b, occurring in an approximate ratio of 3:1. Chlorophyll degradation starts with the conversion of chlorophyll a to chlorophyllide a, by a reaction catalyzed by chlorophyllase. From chlorophyllides, pheophorbides are produced, which are then converted into tetrapyrrole, which can be cleaved by several oxidative enzymes (Costa et al., 2006). This degradation promotes changes in fruit juice color, from bright green to yellow-green.

pH also has a huge effect on the rate of chlorophyll degradation. At neutral pH, chlorophylls are stable, but in slightly acidic solutions, as in the case of fruit juices, the rate of degradation increases a few folds.

There are few studies on the effect of heat treatment on chlorophyll content in fruit, and they are quite contradictory. Although some studies noted a delay in chlorophyll degradation (Luaces et al., 2005), others observed an enhancement of chlorophyll breakdown after thermal processing (Jing et al., 2009). This is probably related to the thermal stability of the enzymes responsible for the chlorophyll degradation that are present in the fruit juice.

Carotenoids are the major group of coloring substances in nature, responsible for the red, orange, and yellow colors of fruits and vegetables. Thus, the characteristic color of most varieties of fruit juices is mainly due to carotenoid pigments. Carotenoid degradation by heat treatments usually affects not only fruit juice color but also its nutritive value and flavor. The main cause of carotenoid degradation is isomerization and oxidation. In general, carotenoids undergo isomerization with thermal processing, which leads to a final juice with less color intensity. If in the presence of oxygen, carotenoids may be directly oxidized or, through the hydroperoxides, formed during lipid autoxidation. Browning of the fruit juice will be the result of the occurrence of the oxidation process. The retention of this pigment can be improved by the addition of reducing agents, such as potassium metabisulfite, and by the exclusion of oxygen.

Anthocyanins are polyphenolic pigments, responsible for the red, blue, and purple colors of many fruits. Anthocyanin pigments are known to be degraded during thermal processing, which can have a dramatic impact on juice color quality. Therefore, chemical stability studies of anthocyanins have been conducted extensively. The results confirmed that anthocyanin stability is a function of the processing temperature and also intrinsic properties of the product and the process such as pH, storage temperature, chemical structure and concentration of anthocyanins present, light, oxygen, the presence of enzymes, proteins, and metallic ions. Co-pigmentation can also improve anthocyanin's stability as their molecules can react with other natural fruit components, resulting in an enhanced and stabilized color (Talcott et al., 2003). This association results in complexes that absorb more visible light (they are brighter) and light of lower frequency (they look bluer). Most of these co-pigments are flavonoids, although compounds belonging to other groups (e.g., alkaloids, amino acids, nucleotides) can function similarly.

Thermal degradation of anthocyanins induces the loss of the glycosyl moieties of the anthocyanins, by hydrolysis of the glycosidic bond. This results in the formation of different polyphenolic degradation products depending on the severity and nature of the heating process (Patras et al., 2010). The anthocyanin degradation leads to losses in color and eventually may lead to brown products, especially in the presence of oxygen.

In common with other polyphenols, anthocyanins are enzymatically degraded in the presence of PPO. This enzyme can be inactivated by mild heating, and therefore, some authors have reported that the inclusion of a blanching step can have a positive effect on anthocyanin retention and, consequently, on color preservation.

Anthocyanins are more stable in acidic media as the pH of fruit juices. However, anthocyanins are known to display a huge diversity of color variations in the pH range. This can be explained by the ionic nature of anthocyanins that enables the changes in the molecule structure according to the prevailing pH, resulting in different colors and hues at different pH values (von Elbe and Schwartz, 1996).

15.4.2.2 Flavor

Flavor is one of the main food quality attributes that consumers consider of greater importance. Fruit juice flavor is made up of sugars, acids, salts, bitter compounds such as alkaloids or flavonoids, and aroma volatiles and is determined by aroma (odor-active compounds) and taste (Jiang and Song, 2010).

Many of the volatile flavor compounds present in fruit juices are subject to degradation during thermal processing treatments, resulting in significant losses in the natural fruity aromas and the development of cooked aromas due to thermal degradation of the aroma compounds (Su and Wiley, 1998).

Heat processing of low-pH fruit juices can lead to several volatile compound losses (mainly aldehydes and esters), by thermal degradation. During the heating process, glycosidic precursors of various aroma compounds are hydrolyzed to release strong aroma components, which will increase the aroma intensity of the fruit juice (Boylston, 2010).

Thermal processing of fruit juice also results in the formation of off-flavors that are the oxidation products of unsaturated fatty acids or are formed as a result of the Maillard reaction. Furthermore, microbiological contaminations following juice processing can also result in specific off-flavor formation, frequently caused by spoilage microorganisms such as *A. acidoterrestris*.

15.4.2.3 Texture

Heat treatments affect food texture properties, such as viscosity. Viscosity has traditionally been considered an important quality parameter for fruit juices. The consistency and amount of pulp on fruit juices give its viscosity. The ability of a juice to retain its solid portion in suspension is mainly attributed to the total amount of pectic material and structural integrity of the material present (Chou and Kokini, 1987).

An increase in process temperature results in a decrease in juice viscosity. This reduction is mainly due to the degradation of pectin by proteolytic enzymes. Although thermal treatments are known to inactivate the enzymes present, some studies reported residual activity of pectin methyl esterase (Sánchez-Moreno et al., 2005; Wu et al., 2008).

15.5 PROCESS METHODS FOR MINIMIZING QUALITY LOSSES

Food thermal processing induces several biological, physical, chemical, and microbiological changes, leading to sensorial and nutritional modifications that can also damage the product quality, particularly those exposed to high temperature for a relatively long time. Over the years, several processing and packaging techniques have been improved to minimize the impact of heat on products' quality. These techniques arise from variations in conventional methods, by using high-temperature short-time processing methods instead of normal pasteurization, from

using agitating retorts in place of the traditional batch-type still retorts, or by changing cans for special food containers.

Nevertheless, with the intention of further enhancement of fruit juice quality, some novel thermal processing methods have been tested.

15.5.1 CHANGES IN CONVENTIONAL METHODS

As already mentioned, fruit juice quality optimization can be achieved by the appli-
cation of the high-temperature short-time concept using aseptic processing and by the use of agitating retorts or thin profile packages.

15.5.1.1 High-Temperature Short-Time Processing

The HTST process, as the name implies, use higher temperatures and shorter times than conventional thermal processes to achieve pasteurization and sterilization of foods and beverages. This method usually results in minimal degradation of quality attributes, while achieving the same level of sterility as the traditional canning process. The time and temperature conditions depend on several factors, such as size, shape, and type of food.

This concept has been successfully used in thermal processing of liquid foods. The problem is to apply it to solids or to products with large solid particles. Since not all particles flow at the same rate, enough heat must be provided to sterilize the slowest-heating particle. This means that some particles are overprocessed, which leads to a decrease in product quality. Most of developments in HTST processing are related with aseptic systems for their implementation in low-acid particulate food processing.

The standard equipment used for HTST processes are the plate heat exchangers, steam injection, and steam infusion. However, other systems are being developed, since steam is not effective in particle processing.

15.5.1.2 Aseptic Processing

The aseptic process uses the HTST method to achieve food product sterility of enhanced quality. In this process, the juice is heated in a heat exchanger (as the one shown in Figure 15.2) and held for sufficient time in the holding section to complete the required pasteurization treatment. After that, the juice is also cooled inside the system. Finally, the aseptic fruit juice is filled cold and sealed under aseptic conditions (Figure 15.3). A composite package extensively used for aseptic processing is a sterile laminated paper/plastic/foil container, also known as "Tetra Pak." This package acts as a barrier to oxygen, light, moisture, and micro-
organisms and also presents large heat sealability and strength (Ramaswamy and Marcotte, 2005). Another main advantage of aseptic packs is their adaptability to many size containers.

This is a continuous method, with higher energy efficiency due to the fast heat transfer rates. This advantage has turned it into a strong industrial success in the pasteurization of fruit juices. Although the whole system is very expensive, it is cost effective when large amounts of juice are processed. However, this method is

FIGURE 15.2 Plate heat exchanger. (Property of Escola Superior de Biotecnologia, Universidade Católica Portuguesa, Porto.)

FIGURE 15.3 Aseptic filling Tetra Pak machine, in a pilot scale. (Property of Escola Superior de Biotecnologia, Universidade Católica Portuguesa, Porto.)

sometimes not recommended, if heat-resistant enzymes are present in the product. This is due to the fact that increases in processing temperatures will cause faster destruction rates for the microorganisms than for these enzymes.

As already mentioned, oxygen can cause several detrimental changes in the juice quality. As aseptic processing is a closed method, any oxygen present in the system will remain there and cause damages. Therefore, it needs to be removed, before entering in the sterilization step. Nielsen et al. (2010) stated that there are three sources of oxygen in aseptic systems that are not usually a problem in conventional thermal processes: (1) dissolved oxygen, (2) entrapped oxygen, and (3) headspace oxygen. Dissolved oxygen is really in solution within the food matrix and is the only source that participates in the oxidation processes (e.g., ascorbic acid or lipid browning). Entrapped oxygen is present as isolated air bags suspended within the food, which can also hold dissolved oxygen. Dissolved and entrapped oxygen can be effectively reduced by deaeration, and headspace oxygen is easily controlled by flushing.

Although the plate heat exchangers are extensively used for this thermal processing procedure, the tubular heat exchangers and the scraped surface heat exchangers are also applied for this purpose.

15.5.1.3 Agitation Processing

Agitating retorts are sterilizers or autoclaves typically used in the food industry to process canned foods, which can be classified as batch or continuous. The agitation during the thermal processing increases the rate of heat transfer, decreasing the time required to achieve a safe product. Moreover, with shorter heating times, product quality is enhanced.

In batch retorts, the agitation is achieved by a rotating framework inside the retort that holds the baskets, which in turn hold the containers firmly during processing (Carroll, 2011). Therefore, these handling retorts allow high rotational speeds that with the large rate of agitation result in relatively short processing times. Several types of industrial batch retorts are commercially available. Rotary retorts are those in which the cans held in carts may be agitated through axial or end-over-end rotation process. End-over-end agitation refers to rotation of the sealed cans around a circle in a vertical plane, thereby creating a positive movement of the headspace bubble all through the rotating cycle (Figure 15.4). Because the headspace is moving along the length of the can and reverses direction every half a rotation, the end-over-end rotation results in a random and, hence, complete mixing of can contents (Ramaswamy and Marcotte, 2005).

The continuous retort systems consist of at least two cylindrical shells in which processing and cooling takes place in a continuous way. As containers roll via conveyors, the product and headspace bubble move in the container. Various types of continuous retorts are also available and most of them use axial agitation (Figure 15.5). In large thermal processing operations, these retort systems are very efficient, which makes its employment an economic advantage.

Critical factors that need to be controlled in agitating processes include product consistency, headspace, fill of the container, solid–liquid ratio, and type and speed of agitation.

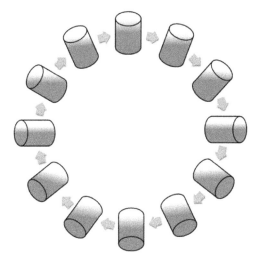

FIGURE 15.4 End-over-end rotation scheme.

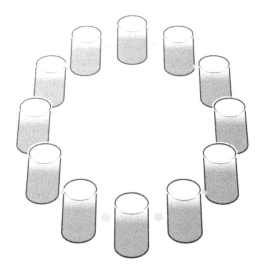

FIGURE 15.5 Axial rotation scheme.

15.5.1.4 Thin Profile Processing

Another way of optimizing quality retention in thermally processed foods is to change the container thickness. Thus, a different container—retort pouch—was developed to use as an alternative to the metal can and glass jar. The retort pouch is a flexible, laminated food package that can withstand thermal processing (Ramaswamy and Marcotte, 2005). It consists of 3- or 4-ply multilayer of polypropylene, aluminum foil, and polyester. To enhance its strength, nylon has also been added as an additional layer. As a result of faster heat penetration to the product center due to the thin profile nature of the container, improvement in product quality is achieved. Its flexible character also promotes an easier distribution, which decreases transportation costs.

Although they act as good obstruction to oxygen, they are not complete barriers, and therefore, the shelf life of the final product can be compromised. Other disadvantages are related to the lack of availability of high-speed filling and sealing machines and the large capital investment that is required.

15.5.2 ALTERNATIVE METHODS

The consumer demand for lightly processed foods that preserve their fresh-like characteristics without compromising their quality and safety has increased rapidly in the last few years. Novel technologies have emerged due to the promising results in the field of fluid foods. Some of them have also been recognized as options to solve some of the problems associated with large particulate products. Therefore, electroheating technologies have been extensively studied and are pointed out as potential alternatives for fruit juice thermal processing. Electroheating methods can be subdivided into either direct electroheating, where electrical current is applied directly to the food (e.g., ohmic heating), or indirect electroheating (e.g., microwave or radiofrequency heating), where the electrical energy is first converted to electromagnetic radiation, which subsequently generates heat within a product (Marra et al., 2009). All these heating methods have the advantage of transferring the energy from its source into the food without heating up the heat transfer surface of the processing equipment.

15.5.2.1 Microwave Heating

Microwave heating is a process that uses electromagnetic waves at frequencies of 915 and 2450 MHz. Both of these frequencies are used in industrial application, while domestic ovens only use the highest one.

In microwave processing, heat is generated within the product by two different mechanisms (dielectric and ionic), the dielectric heating mechanism being the primary one. In this system, the molecules of water present in the food will respond to oscillations in the electromagnetic field. This response is by means of molecular reorientation, resulting in frictional interactions that generate heat. The other heating mechanism is the movement of dissolved ions within the product. This mechanism is involved in the reduction of the penetration depth of the microwave energy, as it implies that the energy distribution can vary within the food.

Several studies were conducted to achieve the effectiveness of this heat treatment on the quality (Cañumir et al., 2002; Picouet et al., 2009; Fratianni et al., 2010) and safety (Koutchma and Ramaswamy, 2000; Picouet et al., 2009) enhancement of fruit juices. In general, all agree that microwave heating can be an alternative to conventional heating, since safe fruit juices with better quality are attained. However, microwave energy has been shown to be ineffective in inactivating browning-related enzymes, proved by the juice oxidation. Therefore, this problem as well as product heating uniformity must be overcome, before the microwave technology can have a successful industrial-scale system in operation.

15.5.2.2 Ohmic Heating

Ohmic heating occurs when an electric current is passed through food, resulting in a temperature rise in the product due to the conversion of the electric energy into heat. The heating process follows Ohm's law, where the conductivity of the food (i.e., the inverse of resistance) establishes the current that passes through the food. Since the electrical conductivity of most foods increases with temperature, ohmic heating becomes very effective. This is especially the case for fruit juices, which contain water and ionic salts in abundance, being the most suitable products for ohmic heating (Palaniappan and Sastry, 1991).

One of the main advantages of ohmic heating is its ability to heat materials rapidly and uniformly, including those containing particulates. By manipulating the ionic contents in formulated products, particulates can be made to heat faster than the liquid (Awuah et al., 2007). Other advantages in the use of this method include the following: there is no need for hot heat transfer surfaces, which reduces fouling problems and thermal damage to a product; the process is ideal for shear-sensitive products, owing to low flow velocities and it can be an HTST process. All these beneficial aspects permit producing high-quality products with minimal structural, nutritional, or sensorial changes. Ohmic heating also has a quiet operation, low maintenance costs, and easier control, which makes this technology, at the industrial level, very promising. Therefore, several studies have been carried out concerning the effects of this technique on specific fruit juice products in comparison with conventional heating (Leizerson and Shimoni, 2005a,b). In these studies, several quality parameters were evaluated, as well as microbial and enzymatic inactivation. Leizerson and Shimoni (2005b) conducted interesting sensory experiments, which indicated that although assessors could distinguish between fresh and pasteurized samples and between pasteurized and ohmic-heated juices, they could not differentiate between fresh and ohmic-heated orange juice. Thus, ohmic heating produced a juice that retained most of the sensory attributes of the initial juice and still reduced microbial and enzymatic activity to required levels.

15.5.2.3 Radio-Frequency Heating

Like microwave heating, radio-frequency heating is a process that uses electromagnetic waves, but applications are restricted to 13.56, 27.12, and 40.68 MHz for industrial, scientific, and medical purposes, respectively. Heat is generated within the product by the same two mechanisms (dielectric and ionic), already explained for microwave heating. In the radio-frequency range, contrary to what happens in microwave heating, dissolved ions are more important for heat generation than the water dipoles in which they are dissolved. Due to the longer wavelengths of this technology compared with those in microwave technique, radio-frequency energy penetrates more deeply into dielectrical materials, without surface overheating or hot/cold spots developing, which are more likely to occur with microwave energy (Marra et al., 2009). As radio-frequency heating is fast, very short treatment times can be developed.

Although the rapid heating and high penetration of radio-frequency energy make this technology an attractive alternative for the processing of liquid foods, studies on the impact of radio-frequency treatment on fruit juice quality and safety are scarce. Geveke and Brunkhorst (2004, 2008) and Geveke et al. (2007) assessed the influence of this method on some quality attributes and microbial inactivation in orange and apple juices. Under their experimental conditions, they demonstrated the potential of radio frequency in the inactivation of *E. coli* and noted no loss in ascorbic acid or enzymatic browning in orange juice.

Notwithstanding these results, further studies are necessary to justify the industrial application of radio frequency for fruit juice heating processes.

15.6   LEGISLATION

Fruit juices are subjected to strict regulatory control and legislation to ensure their safety and quality.

During the 1990s, several outbreaks happened in the United States due to the consumption of fruit juice contaminated with *E. coli*. In response, in 2001 the Food and Drug Administration (FDA) approved new regulations that transmitted food-safety procedures for juice manufacturers. FDA ruled that all juice manufacturers must implement Hazard Analysis and Critical Control Point procedures (HACCP). HACCP is a seven-step system that analyzes how juice is made with a focus on any part of the process that could potentially allow food safety risk. After these identifications, key actions can be taken to reduce or eliminate the hazards. In addition, all juice processors must comply with Good Manufacturing Practices (GMPs).

FDA requires the use of treatments capable of consistently achieving at least a 5 log reduction in the level of the pertinent microorganism in the specific juice (FDA, 2004). As untreated juices do not have, by far, more than 10^5 organisms per gram, applying a 5 log treatment to juice that may contain only 10^1 or 10^2 organisms per gram ensures that the process is adequate to destroy microorganisms of public health significance or to prevent their growth. FDA defines the "pertinent microorganism" as the most resistant microorganism of public health concern that is likely to occur in the juice. By choosing the most resistant pathogen as the target, the treatment will affect all other pathogens that can be present.

In Europe, the Fruit Juice Directive 2001/112/EC (EC, 2001) lays down specifications for fruit juices and fruit-based drinks to guarantee their quality. It specifies several criteria that the product must obey, including which fruits can be used, their minimum content, what ingredients can or cannot be added, and what treatments can be used.

15.7   FINAL REMARKS

Although thermal processing is still the most cost-effective means to reduce microbial populations and enzyme activity, it has several detrimental effects on the quality of fruit juices. To maintain quality attributes, it is important to control microbiological spoilage, enzymatic degradation, and chemical degradation. All these types of food degradation have been reviewed in this chapter as well as the changes carried out in traditional processes to overcome the quality losses. Because a few negative issues persist, alternative methods for conventional heat treatments were exploited. According to the referenced studies, these novel processes seem to be very promising.

Although they might offer faster heating rates, their adoption as a replacement for traditional heat processing would require validation in terms of significant product quality improvement and economic viability. Therefore, a lot of work must be done before the possible implementation of these novel technologies in the food drink industry.

ACKNOWLEDGMENT

Fátima A. Miller gratefully acknowledges her postdoctoral grant (SFRH/BPD/65041/2009) from Fundação para a Ciência e a Tecnologia (FCT).

REFERENCES

Arena, E., Fallico, B., and Maccarone, E. 2001. Evaluation of antioxidant capacity of blood orange juices as influenced by constituents, concentration process and storage. *Food Chemistry* 74: 423–427.

Arthey, D. and Ashurst, P.R. 1996. *Fruit Processing*. London, U.K.: Blackie and Academic & Professional.

Awuah, G.B., Ramaswamy, H.S., and Economides, A. 2007. Thermal processing and quality: Principles and overview. *Chemical Engineering and Processing* 46: 584–602.

Bates, R.P., Morris, J.R., and Crandall, P.G. 2001. *Principles and Practices of Small- and Medium-Scale Fruit Juice Processing*. Rome, Italy: Food and Agriculture Organization of the United Nations.

Baysal, T. and Demirdoven, A. 2007. Lipoxygenase in fruits and vegetables: A review. *Enzyme and Microbial Technology* 40: 491–496.

Boylston, T.D. 2010. Temperate fruit juice flavor. In *Handbook of Fruit and Vegetable Flavors*, ed. Y.E. Hui, pp. 451–462. Hoboken, NJ: John Wiley & Sons, Inc.

Cañumir, J.A., Celis, J.E., de Bruijn, J., and Vidal, L.V. 2002. Pasteurization of apple juice by using microwaves. *Lebensmittel-Wissenschaft und-Technologie* 35: 389–392.

Carroll, R. 2011. Process control of retorts. In *Thermal Processing of Foods*: *Control and Automation*, ed. K.P. Sandeep, Ames, IA: Blackwell Publishing Ltd.

Ceviz, G., Tulek, Y., and Con, A.H. 2009. Thermal resistance of *Alicyclobacillus acidoterrestris* spores in different heating media. *International Journal of Food Science and Technology* 44: 1770–1777.

Chang, S.S. and Kang, D.H., 2004. *Alicyclobacillus* spp. in the fruit juice industry: History, characteristics, and current isolation/detection procedures. *Critical Reviews in Microbiology* 30: 55–74.

Chou, T.D. and Kokini, J.L. 1987. Rheological properties and conformation of tomato paste pectins, citrus and apple pectins. *Journal of Food Science* 52: 1658–1664.

Costa, M.L., Civello, P.M., Chaves, A.R., and Martínez, G.A. 2006. Hot air treatment decreases chlorophyll catabolism during postharvest senescence of broccoli (*Brassica oleracea* L. var. italica) heads. *Journal of the Science of Food and Agriculture* 86: 1125–1131.

Dhuique-Mayer, C., Tbatou, M., Carail, M., Caris-Veyrat, C., Dornier, M., and Amiot, M.J. 2007. Thermal degradation of antioxidant micronutrients in citrus juice: Kinetics and newly formed compounds. *Journal of Agricultural and Food Chemistry* 55: 4209–4216.

Durak, M.Z., Churey, J.J., Danyluk, M.D., and Worobo, R.W. 2010. Identification and haplotype distribution of *Alicyclobacillus* spp. from different juices and beverages. *International Journal of Food Microbiology* 142: 286–291.

EC, 2001. Council directive of 20 December 2001 relating to fruit juices and certain similar products intended for human consumption (2001/112/EC). *Official Journal of the European Communities*, n L 10/58 (12/01/2002).

Elez-Martínez, P. and Martín-Belloso, O. 2007. Effects of high intensity pulsed electric field processing conditions on vitamin C and antioxidant capacity of orange juice and *gazpacho*, a cold vegetable soup. *Food Chemistry* 102(1): 201–209.

Food and Drug Administration (FDA). 2004. *Juice HACCP: Hazards and Controls Guidance, Guidance for Industry,* 1st edn. Division of Plant Product Safety. Available from http://www.fda.gov/Food/GuidanceComplianceRegulatoryInformation/GuidanceDocuments/Juice/ucm072557.htm (accessed date September 15, 2011).

Fortea, M.I., López-Miranda, S., Serrano-Martínez, A., Carreño, J., and Núñez-Delicado, E. 2009. Kinetic characterisation and thermal inactivation study of polyphenol oxidase and peroxidase from table grape (*Crimson Seedless*). *Food Chemistry* 113(4): 1008–1014.

Fratianni, A., Cinquanta, L., and Panfili, G. 2010. Degradation of carotenoids in orange juice during microwave heating. *LWT—Food Science and Technology* 43: 867–871.

Gama, J.J.T. and Sylos, C.M. 2005. Major carotenoid composition of Brazilian Valencia orange juice: Identification and quantification by HPLC. *Food Research International* 38: 899–903.

Geveke, D.J. and Brunkhorst, C. 2004. Inactivation of *Escherichia coli* in apple juice by radio frequency electric fields. *Journal of Food Science* 69(3): FEP 134–FEP 138.

Geveke, D.J. and Brunkhorst, C. 2008. Radio frequency electric fields inactivation of *Escherichia coli* in apple cider. *Journal of Food Engineering* 85(2): 215–221.

Geveke, D.J., Brunkhorst, C., and Fan, X. 2007. Radio frequency electric fields processing of orange juice. *Innovative Food Science and Emerging Technologies* 8(4): 549–554.

Gillatt, P.N. and Rossell, J.B. 1992. The interaction of oxidized lipids with proteins. In *Advances in Applied Lipid Research*, ed. F.B. Padley, pp. 65–118. Greenwich, U.K.: JAI Press.

Goekmen, V., Bahçeci, K.S., Serpen, A., and Acar, J. 2005. Study of lipoxygenase and peroxidase as blanching indicator enzymes in peas: Change of enzyme activity, ascorbic acid and chlorophylls during frozen storage. *LWT—Food Science and Technology* 38(8): 903–908.

Gouws, P.A., Gie, L., Pretorius, A., and Dhansay, N. 2005. Isolation and identification of *Alicyclobacillus acidocaldarius* by 16S rDNA from mango juice and concentrate. *International Journal of Food Science and Technology* 40: 789–792.

Gutzeit, D., Winterhalter, P., and Jerz, G. 2008. Nutritional assessment of processing effects on major and trace element content in sea buckthorn juice (*Hippophaërhamnoides* L. ssp. rhamnoides). *Journal of Food Science* 73(6): H97–H102.

Jiang, Y. and Song, J. 2010. Fruits and fruit flavor: Classification and biological characterization. In *Handbook of Fruit and Vegetable Flavors*, ed. Y.E. Hui, pp. 3–24. Hoboken, NJ: John Wiley & Sons, Inc.

Jiménez, N., Bohuon, P., Lima, J., Dornier, M., Vaillant, F., and Pérez, A.M. 2010. Kinetics of anthocyanin degradation and browning in reconstituted blackberry juice treated at high temperatures (100–180°C). *Journal of Agricultural and Food Chemistry* 58: 2314–2322.

Jing, Y., Mao-run, F., Yu-ying, Z., and Lin-chun, M. 2009. Reduction of chilling injury and ultrastructural damage in cherry tomato fruits after hot water treatment. *Agricultural Sciences in China* 8(3): 304–310.

Koutchma, T. and Ramaswamy, H.S. 2000. Combined effects of microwave heating and hydrogen peroxide on the destruction of *E. coli. Lebensmittel-Wissenschaft und-Technologie* 30: 30–36.

Lee, H.S. and Coates, G.A. 2003. Effect of thermal pasteurization on Valencia orange juice color and pigments. *Lebensmittel-Wissenschaft und-Technologie* 36: 153–156.

Leizerson, S. and Shimoni, E. 2005a. Effect of ultrahigh-temperature continuous ohmic heating treatment on fresh orange juice. *Journal of Agricultural and Food Chemistry* 53: 3519–3524.

Leizerson, S. and Shimoni, E. 2005b. Stability and sensory shelf life of orange juice pasteurized by continuous ohmic heating. *Journal of Agricultural and Food Chemistry* 53: 4012–4018.

Lewis, M. and Heppell, N. 2000. *Continuous Thermal Processing of Foods*. Gaithersburg, MD: Aspen Publications.

Luaces, P., Pérez, A.G., García, J.M., and Sanz, C. 2005. Effects of heat-treatments of olive fruit on pigment composition of virgin olive oil. *Food Chemistry* 90: 169–174.

Lund, D. 1975. Heat processing. In *Principles of Food Science: Physical Principles of Food Preservation*, ed. O.R. Fennema, New York: Marcel Dekker.

Macheix, J.J., Fleuriet, A., and Billot, J. 1990. *Fruit Phenolics*. Boca Raton, FL: CRC Press.

Marra, F., Zhang, L., and Lyng, J.G. 2009. Radio frequency treatment of foods: Review of recent advances. *Journal of Food Engineering* 91: 497–508.

Naczk, M. and Shahidi, F. 2003. Phenolic compounds in plant foods: Chemistry and health benefits. *Nutraceuticals and Food* 8(2): 200–208.

Nicolaides, L. and Wareing, P.W. 2002. Food processing and preservation. In *Crop Post-Harvest: Principles and Practice*, eds. P. Golob, G. Farrell, and J.E. Orchard, pp. 360–372. Gosport, U.K.: Blackwell Publishing.

Nielsen, S., Ismail, B., and Sadler, G.D. 2010. Chemistry of aseptic processed foods. In *Principles of Aseptic Processing and Packaging*, 3rd edn, ed. P.H. Nelson, pp. 75–100. West Lafayette, IN: Purdue University Press.

Odriozola-Serrano, I., Soliva-Fortuny, R., and Martín-Belloso, O. 2008. Phenolic acids, flavonoids, vitamin C and antioxidant capacity of strawberry juices processed by high-intensity pulsed electric fields or heat treatments. *European Food Research and Technology* 228: 239–248.

Olano, A. and Martínez-Castro, I. 1996. Nonenzymatic browning. In *Handbook of Food Analysis*, Vol. 2, ed. M.L. Nollet, pp. 1683–1721. New York: Marcel Dekker.

Palaniappan, S. and Sastry, S.K. 1991. Electrical conductivity of selected juices: Influences of temperature, solids content, applied voltage, and particle size. *Journal of Food Process Engineering* 14(4): 247–260.

Patras, A., Brunton, N.P., O'Donnell, C., and Tiwary, B.K. 2010. Effect of thermal processing on anthocyanin stability in foods: Mechanisms and kinetics of degradation. *Trends in Food Science and Technology* 21: 3–11.

Pellegrini, N., Serafini, M., Colombi, B., Del Rio, D., Salvatore, S., Bianchi, M., and Brighenti, F. 2003. Total antioxidant capacity of plant foods, beverages and oils consumed in Italy assessed by three different in vitro assays. *Journal of Nutrition* 133: 2812–2819.

Peña, M.M.-de la, Salvia-Trujillo, L., Rojas-Graü, M.A., and Martín-Belloso, O. 2011. Impact of high intensity pulsed electric fields or heat treatments on the fatty acid and mineral profiles of a fruit juice-soymilk beverage during storage. *Food Control* 22: 1975–1983.

Picouet, P.A., Landl, A., Abadias, M., Castellari, M., and Viñas, I. 2009. Minimal processing of a Granny Smith apple purée by microwave heating. *Innovative Food Science and Emerging Technologies* 10: 545–550.

Ramaswamy, H.S. and Abbatemarco, C. 1996. Thermal processing of fruits. In *Processing Fruits: Science and Technology*, eds. L.P. Somogyi, H.S. Ramaswamy, and Y.H. Hui, pp. 25–65. Lancaster Basel, PA: Technomic Publishing CO., Inc.

Ramaswamy, H.S., Abdelrahim, K., and Smith, J.P. 1992. Thermal processing and computer modeling. In *Encyclopedia of Food Science and Technology*, ed. Y.H. Hui, p. 2554. New York: John Wiley and Sons, Inc.

Ramaswamy, H. and Marcotte, M. 2005. Food processing: Principles and applications. In *Thermal Processing*, eds. H. Ramaswamy and M. Marcotte, pp. 67–168. Boca Raton, FL: Taylor and Francis.

Ryley, J. and Kajda, P. 1994. Vitamins in thermal processing. *Food Chemistry* 49: 119–129.

Sánchez-Moreno, C., Plaza, L., Elez-Martínez, P., Ancos, B. de, Martín-Belloso, O., and Cano, M.P. 2005. Impact of high pressure and pulsed electric fields on bioactive compounds and antioxidant activity of orange juice in comparison with traditional thermal processing. *Journal of Agricultural and Food Chemistry* 53: 4403–4409.

Schilling, S., Schmid, S., Jäger, H., Ludwig, M., Dietrich, H., Toepfl, S., Knorr, D., Neidhart, S., Schieber, A., and Carle, R. 2008. Comparative study of pulsed electric field and thermal processing of apple juice with particular consideration of juice quality and enzyme deactivation. *Journal of Agricultural and Food Chemistry* 56: 4545–4554.

Silva, F.V.M. and Gibbs, P. 2001. *Alicyclobacillus acidoterrestris* spores in fruit products and design of pasteurization processes. *Trends in Food Science and Technology* 12: 68–74.

Silva, F.V.M. and Gibbs, P. 2004. Target selection in designing pasteurization processes for shelf-stable high-acid fruit products. *Critical Reviews in Food Science and Nutrition* 44: 353–360.

Silva, F.M., Gibbs, P., Vieira, M.C., and Silva, C.L.M. 1999. Thermal inactivation of *Alicyclobacillus acidoterrestris* spores under different temperature, soluble solids and pH conditions for the design of fruit processes. *International Journal of Food Microbiology* 51: 95–103.

Spanos, G.A. and Wrolstad, R.E. 1992. Phenolics of apple, pear, and white grape juice and their changes with processing and storage: A review. *Journal of Agricultural and Food Chemistry* 40: 1478–1487.

Steyn, C.E., Cameron, M., and Witthuhn, R.C. 2011. Occurrence of *Alicyclobacillus* in the fruit processing environment—A review. *International Journal of Food Microbiology* 147: 1–11.

Su, S.K. and Wiley, R.C. 1998. Changes in apple juice flavor compounds during processing. *Journal of Food Science* 63: 688–691.

Talcott, S.T., Brenes, C.H., Pires, D.M., and Del Pozo-Insfran, D. 2003. Phytochemical stability and color retention of copigmented and processed muscadine grape juice. *Journal of Agricultural and Food Chemistry* 51: 957–963.

USDA 2009. USDA national nutrient database for standard reference. Agricultural Research Service. Available from http://www.nal.usda.gov (15/09/2011)

van Boekel, M., Fogliano, V., Pellegrini, N., Stanton, C., Scholz, G., Lalljie, S., Somoza, V., Knorr, D., Jasti, P.R., and Eisenbrand, G. 2010. A review on the beneficial aspects of food processing. *Molecular Nutrition and Food Research* 54(9): 1215–1247.

Villamiel, M., del Castillo, M.D., and Corzo, N. 2006. Browning reactions. In *Food Biochemistry and Food Processing*, ed. Y.H. Hui, pp. 71–102. Oxford, U.K.: Blackwell Publishing.

von Elbe, J.H. and Schwartz, S.J. 1996. Colorants. In *Food Chemistry*, ed. O.R. Fennema, pp. 651–723. New York: Marcel Dekker Inc.

Walker, M. and Phillips, C.A. 2008. *Alicyclobacillus acidoterrestris*: An increasing threat to the fruit juice industry? *International Journal of Food Science and Technology* 43: 250–260.

Wu, J., Gamage, T.V., Vilkhu, K.S., Simons, L.K., and Mawson, R. 2008. Effect of thermosonication on quality improvement of tomato juice. *Innovative Food Science and Emerging Technologies* 9: 186–195.

Zulueta, A., Esteve, M.J., and Frígola, A. 2007. Carotenoids and color of fruit juice and milk beverage mixtures. *Journal of Food Science* 72(9): C457–C463.

16 Effect of Fruit Processing on Product Aroma

Narendra Narain and
Jane de Jesus da Silveira Moreira

CONTENTS

16.1 INTRODUCTION

Fruits supply some of the most important and essential dietary nutrients such as vitamins, minerals, and fiber. The moisture content of fresh fruits is usually quite high, being more than 80%, and hence, these are classified as highly perishable commodities (Orsat et al., 2006). Several processing methods are employed for the conservation of fruits, and some of the principal ones used by fruit processing industry are application of ultraviolet light and radiation, high pressure, ultrasound, membrane separation, high-intensity pulsed electric field, ozone, minimal processing, modified atmosphere packaging, enzyme maceration, freeze concentration, refrigeration and cold chain, vacuum frying, thermal treatment, and edible coatings.

The volatile compounds that are responsible for the aroma and flavor characteristics of fruits are usually labile and thus subject to rearrangements, cyclizations, oxidations, and degradations when submitted to processing procedures involving aeration and temperature increase. Consequently, losses and modifications in the volatile profile are caused during the industrial processing and storage, resulting in products that do not retain the original flavor of the fresh fruit (Facundo et al., 2010).

This chapter reviews initially the different fruit processing methods focusing on product quality, in particular, on fruit aroma, its development during fruit maturation, techniques used in aroma analysis, and aroma retention/loss during processing of fruit products detailing the effects on specific fruits.

16.2 FRUIT PROCESSING

The fruit preservation by drying dates back many centuries when solar drying techniques were used. Usually these procedures did not have any control of processing operations and hence resulted in dried products of poor quality. Later some alternate drying technologies such as freeze, vacuum, osmotic, cabinet, fluidized bed, ohmic, microwave, and a combination of some of these techniques were employed (Bezyma and Kutovoy, 2005; Fernandes et al., 2010).

Except for freeze drying, fruit dehydration resulted in application of heat through conduction, convection, and radiation, or a mixture of these phenomena which lead water to vaporize while these vapors are removed by forced air circulation. The selection of an appropriate drying method depends on several factors such as the type and quality of final fruit product desired, availability of dryer, and drying cost. Heating can also result in negative sensory changes due to the production of undesirable taste and flavor attributes such as overcooked, bitter or burnt, and can also promote changes in color by destroying pigments such as anthocyanins and carotenoids while producing undesirable color compounds by promotion of browning reactions in dehydrated fruit. Energy consumption is another critical parameter in selection of a drying process, and with changing times, it will need to be updated as it will play a bigger role in costing of product. (Raghavan and Orsat, 1998; Raghavan et al., 2005).

Drying of fruits has been principally accomplished by convective drying; however, high temperature of the drying process is an important concern as it results in loss of quality, especially of the aroma constituents in fruits (Nijhuis et al., 1998). The problems associated with conventional convective drying such as loss of color, flavor, and some nutrients in dried fruit product have been reported by several workers (Chua et al., 2000; Mayor and Sereno, 2004). However, among all drying methods employed, freeze drying, vacuum drying, and osmotic dehydration are still considered to be costly for large-scale production of dried fruit products (Khin et al., 2005).

16.2.1   OSMOTIC DEHYDRATION

Fruit subjected to osmosis usually with sugar syrup treatments results in partially dehydrated products, and hence, it is considered as a pretreatment process prior to final drying by other methods. This application has attained prominence in fruit dehydration for production of intermediate moisture foods or as a pretreatment in order to reduce energy consumption or heat damage to fruit (Jayaraman and Gupta, 1992). Application of high hydrostatic pressure to pineapple slices damaged the cell wall structure and resulted in significant changes in the tissue architecture, leaving the cells more permeable. The major advantage of this application is in increased mass transfer rates and relatively better aroma retention of fruit during osmotic dehydration (Rastogi and Niranjan, 2008).

16.2.2   VACUUM DEHYDRATION

The use of vacuum in closed cabinets containing fruits results in the expansion and escape of gas occluded into pores. When the pressure is restored, the pores get occupied by the osmotic solution whereby increasing the available mass transfer surface area. Vacuum pressure (50–100 mbar) is applied to the system for shorter time to achieve the desired result (Sagar and Suresh Kumar, 2010).

16.2.3   PULSED ELECTRIC FIELD APPLICATION

The pulsed electric field treatment of fruit increases the permeability of plant cells, induces cell damage, and results in tissue softening, which in turn decreases the turgor pressure and consequently leads to a reduction in compressive strength. The effective diffusion coefficients of water and solute increase exponentially with electric field strength, and this increase in diffusion coefficient is also attributed to an increase in cell wall permeability, which facilitates the transport of water and solute (Sagar and Suresh Kumar, 2010).

16.2.4   SUPERCRITICAL CO_2 TREATMENT

Of late, this treatment has gained importance for the processing of food and biological materials. The critical point of CO_2 gas is at 304.17 K and 7.38 MPa (Tedjo et al., 2002). The combination of pressure and temperature as process parameters makes it possible to vary the solvent of the medium within certain ranges as desired without having to change the composition of solvent. Although this treatment does not improve the water loss, it retains better the aroma and flavor compounds in a fruit.

16.2.5 ULTRASOUND TREATMENT

Acoustic streaming can affect the thickness of boundary layer which exists between stirred fluid and solid. Cavitation, a phenomenon produced by the sonication, consists in the formation of bubbles in the liquid which can collapse and generate localized pressure fluctuation. This ultimately increases the mass transfer of osmotic treatment. The rate of transfer depends on pressure and frequency of the wave produced by sonication (Raghavan et al., 2005).

16.2.6 MICROWAVE DRYING

Microwave drying commonly employs 2450 MHz of electrical energy; however, this energy may vary between 300 MHz and 300 GHz. These waves are generated by a device known as magnetron inside an oven by increasing the alternating current from domestic power (Orsat et al., 2005). The use of microwave energy for drying has been demonstrated to have moderate low energy consumption (Sagar and Suresh Kumar, 2010) although there is a loss in aroma and degradation of texture of food materials in this processing. The volumetric heating and reduced processing time make microwaves an attractive source of thermal energy. Since microwaves alone cannot complete a drying process, combined techniques such as forced air or vacuum are employed in order to improve the efficiency of the microwave process. Banana slices dried using microwave energy demonstrated that good quality dried products can be obtained by varying power density and duty cycle time (Chou and Chua, 2001). In obtaining juices from several fruits (plums, grapes, and apricots), Cendres et al. (2011) optimized microwave power for best yields, and the juices were brightly colored as it allowed inactivation of endogenous enzymes.

16.2.7 SUPERHEATED STEAM DRYING

Drying is performed with superheated steam in the absence of air in a medium composed entirely of steam. The application of superheated steam raises the temperature above the corresponding saturation temperature at a given pressure, and hence the material gets dried due to the addition of sensible heat. The evaporated water is removed only after the pressure develops beyond certain limit when excess steam is released. The advantage of superheated steam drying is that recycling of drying method is possible, provided additional sensible heat be added. Moreover, this drying is facilitated by converting any conventional convection and conduction dryer so as to use superheated steam (Tatemoto et al., 2007).

Superheated steam flash drying is quite suitable for food products sensitive to high temperature as this process can be performed under vacuum. Pneumatic or flash drying is adjusted according to physical condition used to add heat and removal of water vapor. Low temperature under vacuum is used for certain fruits that might discolor or decompose at high temperature.

16.2.8 HYBRID DRYING

An excellent review of hybrid drying technologies for heat-sensitive foods was published by Chou and Chua (2001). Of late, hybrid drying techniques are being used since the combined technology results in better quality products as compared to individual processes. The number of combinations could be very high, and these would depend on various criteria such as chemical composition and physical parameters of the pulp, the desired product shape and form, quality characteristics of the dried product, etc. Adding a microwave system to a spouted bed system combines the benefits offered by each technology. The microwave action decreases drying time while the fluidization produced by the spouting system improves uniformity in drying, and the product does not get burnt (Feng et al., 1999). Donsi et al. (1998) showed a combination of hot air drying and freeze drying which resulted in high quality of dehydrated fruits and vegetables. Combination drying initially using a conventional drying process followed by a finish microwave or microwave vacuum process has proven to reduce drying time while improving product quality (Erle, 2005; Soysal et al., 2006). Drouzas and Schubert (1996) investigated vacuum-microwave drying of banana slices and reported the product quality in terms of taste, aroma, and rehydration characteristics to be excellent.

16.2.9 FREEZING

Freezing is one of the most common methods used for preserving food as it tends to maintain the original organoleptic attributes of fruit along with the retention of nutrients as close as possible to those of fresh fruit (Menezes et al., 2008; Chassagne-Berces et al., 2010). During freezing, most of the liquid water changes into ice and hence greatly reduces microbial and enzymatic activities. Oxidation and respiration activities also get reduced at low temperatures. However, some physical and chemical changes in food occur due to ice formation, resulting in a subsequent loss of quality (Chassagne-Berces et al., 2010).

16.3 AROMA DEVELOPMENT DURING FRUIT MATURATION

Although flavor includes aroma and taste, aroma is perceived primarily by the nose, while taste receptors exist in the mouth are impacted when the food is chewed (Yamaguchi and Ninomiya, 2000). Fruit aroma characterizes due to the presence of very low concentrations (usually in ppm, ppb) of many volatile organic compounds which are generated during fruit development, maturation, and ripening (Gould, 1983). Despite the fact that hundreds of volatile compounds are often present in fruit, only a few may be odor-active. Generally, citrus fruits such as orange, lemon, and grapefruit are rich in terpenoids while noncitrus fruits such as apple, banana, and pear are characterized by esters and aldehydes.

The fruits are classified into two broad categories—climacteric and nonclimacteric—and each of these characterizes some peculiar behavior on maturation and ripening. Climacteric fruits present a steep rise in respiration activity with maturation and ripening, and these fruits could ripen attached to the plant or after their harvest once they attain a ripened stage which is physiologically adequate for such fruits. The prominent fruits classified in this category are banana, soursop, watermelon, mango, sapoti, etc. However, on the other side, nonclimacteric fruits are those in which respiratory activity is low and almost constant, having a small drop after harvest. The fruits classified as nonclimacteric are lemon, orange, cashew apple, grape, strawberry, etc. (Chitarra and Chitarra, 2005). During ripening of fruit, generally, sweetness increases and astringency decreases. Other biochemical and physiological changes take place which result in development of color, flavor, and texture which softens mainly due to the degradation of pectin by pectic enzymes.

Harvesting firm fruit, which is characterized to immature or mature green stages, has its own advantage as it can withstand better the mechanical damage the fruit is subject to during postharvest handling and processing. Although fruits may have excellent visual quality and acceptance by retailers, it lacks in acceptable volatiles and flavor or aroma attributes in fresh-cut fruits prepared with less ripe fruit (Gorny et al., 1998, 2000; Beaulieu and Lea, 2003; Baldwin et al., 2004; Beaulieu et al., 2004; Soliva-Fortuny et al., 2004). Harvest maturity significantly affects the level of flavor volatiles recovered in fresh-cuts from soft-ripe versus firm-ripe mangoes (Beaulieu and Lea, 2003) and one-fourth, one-half, three-fourth, and full-slip cantaloupe melons (Beaulieu, 2006). Cantaloupe melon fruit harvested at different maturity stages delivers stored cubes differing significantly in quality (Beaulieu and Lea, 2007), flavor, textural attributes, and volatiles (Beaulieu et al., 2004; Beaulieu, 2006). In general, cubes prepared with less mature cantaloupe melon fruit are excessively firm, lack flavor volatiles, and have inferior and less acceptable sensory attributes (Beaulieu et al., 2004).

The pleasant aroma development during ripening of fruits such as pear, peach, and passion fruit results from the beta oxidation of long-chain unsaturated fatty acids, initially breaking the chain of CoA derivative by two carbons each time which later reacts with alcohols that lead to the formation of esters which are considered to be important aroma character of these fruits. During this process, isomerization reactions may also take place, forming *trans–cis* isomers (Rodriguez-Amaya, 2004). The esters of many ripened fruits are also formed by ramifications of amino acids such as leucine. This route of formation of aroma compounds initiates with the loss of amino acids by deamination and decarboxylation reactions. An aldehyde is formed which can be transformed to a corresponding alcohol, and it can react with acetic acid to form isoamyl acetate which characterizes for banana aroma. Alternatively, the aldehyde could be converted to an acid, later esterified with ethanol to produce ethyl-3-methyl butyrate which is a character impact compound for apple aroma. Furthermore, terpenes, especially monoterpenes and sesquiterpenes, which are biosynthesized by isoprenoid way also are important aroma compounds (Rodriguez-Amaya, 2004).

Other compounds which play a major role in fruit aroma are lactones, particularly γ-lactones and δ-lactones. The aroma of configuration (R) in lactones is generally more intense than the (S) configuration (Rodriguez-Amaya, 2004). Studies performed with pear on character and intensity of its aroma revealed the presence of 1,4-decanolide to be the main constituent of this fruit, and this compound was found to be present in 89% in the configuration as (R) and only 11% in the configuration as (S) (Feuerbach et al., 1988).

Passion fruit volatile compounds belong to three main organic classes such as aliphatic, aromatic, and terpenes, followed by C_{13} norterpenoids and monoterpenes. The esters such as ethyl butanoate, ethyl hexanoate, hexyl butanoate, and hexyl hexanoate were related to the formation of yellow passion fruit aroma. These four esters constitute about 95% of total volatiles extracted from simultaneous distillation and extraction (SDE) technique (Hiu and Schreier, 1961). Narain and Bora (1992) monitored volatile compounds such as ethyl butanoate, hexyl butanoate, ethyl hexanoate, ethyl acetate, 2-heptanone, benzaldehyde, and hexanol during postharvest storage of yellow passion fruit and based on the reduction in the concentration of ethyl butanoate and ethyl hexanoate responsible for sweet aroma and on the increase of ethyl acetate during storage of passion fruit concluded that yellow passion fruit retains better aroma only up to 3 days after harvest of the fruit. In general, carbohydrates are the precursors for the formation of esters and alcohols while β-ionone is obtained from carotenoids present in passion fruit.

Although fruit maturity–related changes in flavor volatiles in fresh-cut cantaloupe have been demonstrated, the balance of the nonacetate to acetate esters changed uniformly during storage and independent of initial fruit maturity in fresh-cut cantaloupe, apple, and honeydew. Moreover, several correlations with volatile compounds and classes of compounds were made between several quality and sensory attributes in fresh-cut cantaloupe (Beaulieu and Lancaster, 2007). Whole cantaloupe fruit and cube firmness was positively correlated with increasing acetate ester levels (Beaulieu, 2005), and acetate loss was correlated with firmness loss in stored fresh cuts (Beaulieu, 2006).

Estimation of volatile acids or amines was suggested to be an indicative of off-flavor in stored fruit (Kramer, 1965). Hexanal is a compound typically produced as a result of the activity of lipoxygenase (Theerakulkait and Barrett, 1995). Although aldehyde formation in fresh-cut and damaged tissue is considered normal, it often contributes to important characteristic aroma profile in several fruits. Some off-flavors result from the accumulation of acetaldehyde, ethanol, or ethyl acetate due to fermentative metabolism when the product is exposed to very low oxygen and/or very high carbon dioxide concentrations.

In a recent study (Galvão et al., 2011), the changes in volatile compounds and descriptive odor attributes during maturation of umbu (*Spondias tuberosa*) fruit were monitored. The principal volatile compounds identified in the ripe umbu fruit pulp were 4-methyl-3-penten-2-one, ethyl benzene, 1-penten-3-one, 2-acetyl thiazone, *p*-xylene, limonene, 2,2-dimethyl 4-octenal, 3-hexanol, 2-nonanol, 1-nonanol, 2-pentanol, 2-octanol, 3-methylethyl 2-butanoate, butyl benzoate, 3-allyl cyclohexene, 2-acetyl furan, 2-hexyl furan, β-caryophyllene, and methyl pyrazine (Table 16.1). The surge of terpenic compounds and some esters in the ripe umbu fruits could characterize the importance of these compounds in the overall aroma of umbu fruit. There was a significant increase ($P < 0.05$) in 2-octanol, 1,2-pentanediol, ethyl benzoate, 2-ethyl furan, and furfural in ripe fruits when compared with their concentration in half-ripe fruits. The principal volatile compounds reported by these workers for the characteristic aroma of umbu fruit pulp were β-cis-ocimene, methyl-pyrazine, 2-butyl-thiophene, methyl octanoate, 2-hexyl furan, 2-octanol, (*E*)-2-cyclohexen-1-one, 3-bromo-cyclohexene, 1-heptanol, 2-nonanol, and 1-octanol.

TABLE 16.1
Volatile Compounds in Pulp of Umbu Fruit at Different Stages of Maturity

| Class | Compounds | Retention Index | Peak Area (% Total Area) | |
			Half-Ripe Fruit	Ripe Fruit
Alcohols	2-Pentanol	1097	$1.47^a \pm 0.24$	$0.93^b \pm 0.08$
	2-Methyl-2-pentanol	1103	$0.74^a \pm 0.11$	$0.79^a \pm 0.13$
	2-Methyl-5 bicyclo hexan-2-ol	1152	Nd	0.53 ± 0.08
	3-Hexanol	1192	$1.55^a \pm 0.16$	$1.36^a \pm 0.23$
	2-Butoxy ethanol	1392	$1.13^a \pm 0.08$	$0.76^b \pm 0.14$
	2-Octanol	1406	$0.24^b \pm 0.17$	$0.70^a \pm 0.14$
	2-Nonanol	1513	$3.61^a \pm 0.21$	$3.64^a \pm 0.07$
	1-Nonanol	1655	$2.37^a \pm 0.61$	$2.20^a \pm 0.31$
	1,2-Pentanediol	1782	$0.28^b \pm 0.01$	$0.96^a \pm 0.04$
Aldehydes	(Z) 2-Nonanal	1480	$0.66^a \pm 0.15$	$0.14^b \pm 0.02$
	2,2-Dimethyl-4 octenal	1535	$2.71^b \pm 0.05$	$3.00^a \pm 0.17$
Alkanes	Hexyl-cyclohexane	1264	$0.89^a \pm 0.22$	$0.85^a \pm 0.18$
	4,4-Dimethyl cyclopentane	1677	$0.26^b \pm 0.02$	$0.66^a \pm 0.16$
Aromatics	Toluene	1030	$2.52^a \pm 1.01$	$2.10^a \pm 0.42$
	Ethyl benzene	1112	$8.16^a \pm 1.31$	$7.17^a \pm 2.31$
	p-Xylene	1120	$4.80^a \pm 1.01$	$4.21^a \pm 1.56$
	o-Xylene	1166	$2.18^a \pm 0.12$	$1.65^b \pm 0.15$
	1,2,3-Trimethyl benzene	1317	$0.54^a \pm 0.07$	$0.35^b \pm 0.03$
Esters	Allyl formate	958	$0.51^a \pm 0.15$	$0.41^a \pm 0.11$
	Ethyl butanoate	1021	Nd	$0.70^a \pm 0.01$
	3-Methylethyl-2-butanoate	1150	$1.74^a \pm 0.09$	$1.29^b \pm 0.16$
	Hexyl hexanoate	1582	$0.77^a \pm 0.14$	$0.92^a \pm 0.25$
	3-Methyl-butanoate	1634	2.00 ± 0.18	Nd
	Butyl benzoate	1689	$0.25^b \pm 0.01$	$0.84^a \pm 0.10$
Furans	2-Ethyl furan	948	$0.49^b \pm 0.09$	$0.69^a \pm 0.41$
	2-Hexyl furan	1399	$1.61^a \pm 0.07$	$0.98^b \pm 0.33$
	Furfural	1465	$0.32^b \pm 0.01$	$0.51^a \pm 0.07$
	2-Acetyl furan	1507	$1.50^a \pm 0.32$	$0.97^a \pm 0.01$
Ketones	4-Methyl-2-pentanone	1010	0.67 ± 0.26	Nd
	1-Penten-3-one	1013	$5.71^a \pm 0.12$	$5.05^b \pm 0.01$
	3-Hexanone	1074	$0.76^a \pm 0.02$	$0.69^a \pm 0.04$
	3-Penten-2-one	1117	$1.09^a \pm 0.12$	$0.80^b \pm 0.01$
	4-Methyl-3- penten-2-one	1126	$5.60^b \pm 0.13$	$9.52^a \pm 0.51$
	3-Methyl-cyclopentanone	1203	$1.57^a \pm 0.16$	$1.85^a \pm 0.35$
	α-Ionone	1844	$0.05^a \pm 0.01$	$0.01^a \pm 0.01$
Others	Methyl pyrazine	1257	$1.68^a \pm 0.02$	$1.73^a \pm 0.24$
	2,3-Dimethyl pyrazine	1335	$0.54^a \pm 0.09$	$0.35^b \pm 0.12$

TABLE 16.1 (continued)
Volatile Compounds in Pulp of Umbu Fruit at Different Stages of Maturity

Class	Compounds	Retention Index	Peak Area (% Total Area)	
			Half-Ripe Fruit	Ripe Fruit
Sulfur	4-Methyl-1,2,4-tri-thiazole	1239	$0.54^a \pm 0.11$	$0.55^a \pm 0.28$
compounds	2,4-Dimethyl-thiazole	1271	$1.38^a \pm 0.32$	$0.22^b \pm 0.06$
	5-Methyl thiazole	1287	$0.43^a \pm 0.05$	$0.50^a \pm 0.01$
	2-N-butyl thiophene	1357	$2.55^a \pm 0.42$	$0.07^b \pm 0.05$
	2-Acetyl thiazone	1631	$0.26^b \pm 0.03$	$4.87^a \pm 0.04$
Terpenes	Limonene	1177	Nd	2.09 ± 0.65
	β-Caryophyllene	1564	$2.33^b \pm 0.17$	$3.39^a \pm 0.24$
	α-Caryophyllene	1606	$1.80^a \pm 0.18$	$1.61^a \pm 0.23$

Source: Compounds short-listed from Galvão, M.S. et al., *Food Res. Int.*, 2011.
Means within row having different superscript letters were significantly different at $P < 0.05$.

16.4 ANALYSIS OF AROMA

Aroma consists mostly of volatile compounds, and it may be determined or evaluated with either instrumental or sensorial methods. Sensory evaluation has its own importance as this is directly related to consumer acceptability of fruit and fruit products. On the other hand, instrumental techniques may determine a large number of compounds in a particular fruit or fruit product, but such methods do not give a measure of the contribution of specific compounds for odor or flavor activity which need to be substantiated by complimentary sensory measurement. The most common instrumental techniques used for identification and quantification of volatile compounds are provided by hyphenated instruments such as gas chromatograph coupled with various detectors, in particular, mass spectrometer and gas chromatograph attached to an olfactometer. Moreover, some other equipment such as electronic nose and electronic tongue has also been used lately, but these still have serious limitations.

Aroma determination in fruit and fruit products is a complex task as the volatile compounds usually present in fruit are in extremely low concentrations and these belong to different organic classes and molecular weight. There is also a wide variability in threshold concentration of various compounds, which is defined as the minimum concentration for an odor-producing substance that has a 50% probability to be perceived as a faint odor by a human subject (Sides et al., 2000). Moreover, some compounds have limited chemical stability as these are subjected to photolysis, oxidation, and other reactions. Furthermore, the most important aspect in aroma analysis is the selection of the character impact compounds responsible for the flavor of the fruit from hundreds of organic compounds which may be identified and quantified in a particular fruit.

In general, aroma chemical analysis involves four steps: (1) extraction or capture of volatile compounds, (2) concentration, (3) separation, and (4) identification and quantification. Direct extraction methods involve classical liquid-liquid extraction (LLE), solid-phase extraction (SPE), and supercritical fluid extraction (SFE) procedures; distillation-based methods comprise mainly of vapor distillation and SDE procedures while headspace (HS) technique is based on static headspace (SHS) or dynamic headspace (DHS) capture of volatiles (Augusto et al., 2003). Thus, the chemical characterization of aroma ordinarily demands state-of-the-art techniques for sampling and sample preparation which may require further concentration of volatile extracts before subjecting to separation, detection, and quantitation analysis of volatile compounds (Civille, 1991; Sides et al., 2000). Some of the prominent and currently used sample preparation techniques are described in the following text.

16.4.1 DIRECT EXTRACTION PROCEDURES

The principles of direct extraction methods are simple, and this category includes techniques such as LLE, SPE, or SFE. The volatile constituents present in solid or liquid matrix are extracted by dissolving them in organic solvents. The flavor components are later concentrated by evaporating the solvent usually at low temperatures. The most important factors in the isolation of volatile compounds are selection of solvent and extraction unit (Danisco, 2001). However, of late, SFE has attracted a great deal of attention, and in this procedure, supercritical solvents such as carbon dioxide are used. In this technique, the component mixture is separated based on differences in component volatilities and the differences in interaction of solvent with mixture components. This technique possesses some merit over other direct extraction procedures since critical point for CO_2, which is the most common fluid used in SFE, is usually low at 31.1°C at 73.8 bar, and hence, extraction can be performed at milder and lower temperatures, and there is no need for any further distillation or evaporation of excess solvent as is the case with LLE and SFE.

16.4.2 DISTILLATION-BASED PROCEDURES

Distillation is usually carried out in two different ways: (1) hydrodistillation and (2) steam distillation. In hydrodistillation, the matrix or plant material is mixed or suspended with water in a suitable vessel fitted with a condenser and the mixture is boiled, leading to formation of a condensate which is collected. After the process, an organic, water-insoluble fraction usually an essential oil can be separated from the water. In steam distillation, steam is passed through a vessel containing the matrix–water mixture to yield a similar condensate.

Steam-assisted distillation and hydrodistillation are traditional procedures for isolation of volatile aroma-related compounds from odoriferous samples, such as food and detached parts of plants. Being simple and straightforward procedures, they are still extensively applied for aroma characterization either alone or combined with other sample preparation procedures (Saritas et al., 2001).

Further to the two procedures mentioned earlier, a widespread distillation-based sample preparation for chemical analysis of aromas is SDE, also known as the Likens–Nickerson method. Figure 16.1 shows one of the designs for the Likens–Nickerson SDE apparatus (Nickerson and Likens, 1966; Augusto et al., 2003). An amount of the

FIGURE 16.1 Modified Likens and Nickerson SDE apparatus. 1. Body, 2. Sample flask, 3. Extracting solvent flask, 4. Cold tube, 5. Inlet for purge gas, 6. Cold water inlet and outlet. (Reproduced from Augusto, F. et al., *Trends Anal. Chem.*, 22, 160, 2003.)

matrix sample (fruit pulp, food, etc.,) along with distilled water is contained in flask 2, while flask 3 contains relatively smaller volume of an extracting solvent (denser than water). When the flasks 2 and 3 are heated, SDE proceeds and the water and solvent vapors are conducted to the extractor body, where they condense over the surface of the cold tube 4. The aroma extracts are removed from the matrix by the water vapor and transferred to the organic phase when the liquids condense together on the cold tube.

16.4.3 Headspace Methods

The analysis of the HS can be SHS or DHS. In SHS, the sample is kept in a closed container until it reaches a thermodynamic equilibrium between the volatile liquid or solid and gas phase, at a given temperature. An aliquot of the vapor is collected and injected into gas chromatograph. Due to the partial pressures being significantly lower than the partial pressure of water vapor, the analytes are generally low concentrations (Thomazini and Franco, 2000) while in DHS or purge and trap, a flow of inert gas is bubbled into the sample and the volatile analytes are transferred to a trap containing an adsorbent polymer. Desorption of trapped analytes for subsequent analysis can be performed either with small volumes of adequate solvents (Rankin and Bodyfelt, 1995) or using an online automated thermal desorption (ATD) devices (Valero et al., 1997), the later being suitable for large number of samples analysis. This technique has been used for transfer of volatile components of matrices for the gas phase, and its main advantage is the high detectability since the thermodynamic equilibrium between the sample, the analyte, and the HS is not necessarily required and the detectability of the method is increased by enrichment of the trap with the analytes of interest (Freire et al., 2008).

Both (SHS and DHS) techniques have their limitations and merits. In SHS, the volatiles which are in equilibrium in HS under extraction conditions (defined temperature and time exposure of the matrix) are analyzed, and these are usually highly volatile compounds, while in DHS, a wide distribution of compounds having low and high volatilities is captured, and hence, this technique represents better the volatile profile of fruit pulp (Steinhart et al., 2000). Moreover, in SHS technique, no preconcentration of analyte is involved, and taking into consideration that some compounds are in trace levels or at very low concentrations, these are not captured in the HS (Stevenson et al., 1996). Knudsen et al. (1993) reported that out of 118 references indexed in an extensive literature review on HS analysis of floral fragrances, only nine employed SHS as the sampling or sample preparation procedure. However, a large number of reports describing the use of DHS methods can be found in the literature regarding the chemical characterization of aroma. Thus, this technique is considered to be the most important procedure for fruit aroma analysis (Sides et al., 2000).

16.4.4 SOLID-PHASE MICROEXTRACTION

In the year 1990, Pawliszyn (1999) introduced solid-phase microextraction (SPME) being a fast, simple, and convenient sample preparation method that has become increasingly popular for aroma analysis. This technique does not require any solvent or concentration step for volatile analysis and characterizes for high sensitivity (Kataoka et al., 2000). SPME is based on the sorption of analytes present in the HS of the sample by an extraction phase immobilized on the surface of a fused silica fiber, which can be coated with polymeric liquid or solid phase with a porous material (Augusto and Valente, 2002). Thermodynamic aspects of this technique of sample preparation have been studied, and these show that the extracted amount of analytes by the fiber surface is directly proportional to the analyte concentration in the HS, regardless of the location of the fiber. The thermodynamic theory predicts the effect of temperature, salt, and the polarity of the matrix-coating material to optimize the extraction conditions with a minimum number of experiments (Pawliszyn, 1997). For these reasons, HS-SPME is being presently considered as the best available choice for sample preparation in fruit and fragrance aroma analysis (Sides et al., 2000).

In a study conducted by Augusto et al. (2000), different SPME fibers were compared for the characterization of the aromas of industrialized pulps of Brazilian tropical fruits—cupuassu (*Theobroma grandiflorum*, Spreng.), caja (*Spondias lutea*, L.), siriguela (*Spondias purpurea*, L.), and graviola (*Annona reticulata*, L.). The overall best efficiencies for these aromas were obtained with the usage of Carboxen-PDMS fibers and especially for detection of low-molar mass compounds.

16.4.5 SEPARATION, IDENTIFICATION, AND QUANTIFICATION OF VOLATILE COMPOUNDS

The volatile fraction of a fruit pulp, for whatever be the isolation method of obtaining it, consists of a complex mixture of volatile compounds, which requires a very powerful separation method. High-resolution gas chromatography meets these requirements as this technique is very selective, sensitive, and efficient (Grob and Kaiser, 1994; Franco and Janzantti, 2004). Of late, more stable columns with greater separation resolution efficiencies are commercially available. For analysis of aroma compounds, gas chromatographs are often coupled to detectors, such as thermal conductivity, flame ionization, electron capture, flame photometric, and mass spectrometric detectors. The detector senses the presence of the individual components as they leave the column. The detector output, after suitable amplification, is acquired and processed by a computerized data system which results in a chromatogram of concentration versus time.

The analysis of volatile constituents is directly related to the objective of the study (Reineccius, 2006). Thus, if the objective is only to quantify certain compounds present in fruit, a GC coupled with a flame ionization detector (GC-FID) is good enough (Grob and Kaiser, 2004). However, if the focus of the study is on the qualitative analysis of total volatile compounds, the usage of mass spectrometer (GC-MS) is more appropriate (Reineccius, 2006). The GC-MS is the combination of two powerful analytical tools wherein the GC separates efficiently the components of a complex mixture and the mass spectrometer provides qualitative information of the analytes such as mass spectra to identify them (Rood, 1995; Van Ruth, 2001). Compounds analyzed by mass spectrometry range from acetals, furans, and furanones and other compounds generated from the Maillard browning reactions to nonvolatile bioactive and flavor-active compounds such as phenols, flavonoids, glycosides, saponins, etc. Some recent reviews and books with sections on novel methods of flavor analysis provide more detailed information on these techniques (Pollien et al., 1997; Wilkies et al., 2000; Van Ruth, 2001; Hognadottir and Rouseff, 2003; Barrete et al., 2010; Cullere et al., 2010; Jabalpurwala et al., 2010; Cheong et al., 2011).

Due to the complexity of aroma profiles, rapid analysis using a detector with fast mass spectral acquisition such as time-of-flight mass spectrometry (TOFMS) is recommended. Song et al. (1998) analyzed flavor volatiles in tomato and strawberry fruits and reported that an overlapping eluting compound extracted using SPME coupled to GC-TOFMS was still unresolved. Furthermore, flavor compounds present at extremely low concentrations (ppb, ppt) can still play a significant role in the key aroma profile. The comprehensive two-dimensional gas chromatography (GC-GC) is a powerful technique and is better suited for the analysis of complex mixtures. Using two orthogonal approaches on different combinations of apolar and polar columns, GC-GC enhances the separation efficiency as compared to the conventional one-dimensional GC. When coupled to a fast acquisition detector such as TOFMS, GC-GC-TOFMS has proven to be an established technique, providing better separation capabilities and a full mass spectrum with high-resolution two-dimensional contour plots (Dallüge et al., 2003).

16.4.6 GC-OLFACTOMETRY TECHNIQUES

In flavor analysis and characterization, one of the most difficult tasks is the identification of aroma compounds which contribute to a particular odor whether desirable or not in foods. In this case, a combination technique of gas chromatography attached to olfactometry (GC-O) is commonly employed which contributes in characterizing low threshold odor notes of compounds (Grosch, 1993; Van Ruth and

Roozen, 1994; Marsilli and Miller, 2000; Thomazini and Franco, 2000). The GC-O involves inhaling compounds eluted after gas chromatographic separation by judges, and it reveals not only the important character impact compounds but also the compounds which have negative effect considered as off-odors in foods (Mistry et al., 1997; Reineccius, 2006).

The main olfactometric techniques are divided into three broad categories: (1) successive dilution of the isolate which represents techniques such as CHARM (combined hedonic response measurement) analysis developed by Acree et al. (1984) and AEDA (aroma extract dilution analysis) by Ullrich and Grosch (1987); (2) time-intensity techniques such as OSME, known as smell in Greek (also an abbreviation for Oregon State Method), developed by McDaniel et al. (1990) and FSCM (finger span cross-modality matching) by Etievant et al. (1999); and (3) techniques based on frequency of detection of the odor, such as NIF/SNIF (nasal impact frequency/surface of nasal impact frequency) proposed by Pollien et al. (1997) and OGA (olfactometry global analysis) by Linssen et al. (1993). In principle, what is common in all these techniques is that the compounds eluting from the GC column are directed up to the nose of the judge. However, some differences in their condition setup prevail such as how to isolate the application (original or diluted), data acquisition parameters (descriptors, times, start, end, and maximum intensity in the case of time-intensity technique, software, etc.), and the number of sniffers whether trained or not for aroma note evaluations.

The techniques of CHARM (Acree et al., 1984) and AEDA (Ullrich and Grosch, 1987) involve evaluating a series of dilutions from the original isolate by GC-O wherein two or three trained judges generate the data on odor notes. However, these techniques need a large number of chromatographic runs, which is one of the major disadvantages of these methods (Reineccius, 2006). The odor importance is related to the compound's high degree of dilution, measured in CHARM analysis or flavor dilution factor in AEDA and CHARM techniques. The results are graphically plotted between the CHARM values or the flavor dilution factors versus their Kovats retention indices to obtain aromagrams. The evaluation criterion is based on compound dilutions and threshold value assuming a linear relationship between the intensity of an odor compound and its concentration, and this aspect is usually criticized.

The techniques of frequency and time intensity are based on modern psychophysics and in accordance with the law of Stevens (Silva et al., 2004). McDaniel et al. (1990) developed the OSME technique in which trained judges evaluate the volatile compounds that elute from the chromatographic column and respond to stimuli using an odor intensity scale. Thus, in this technique, the aromagram represents a graph between intensity versus time. Since this technique is not based on successive dilutions, this has the advantage of a shorter time of analysis, as the trained judges need only perform the repetitions of the same isolate, providing an aromagram corresponding to chromatogram (Silva et al., 2004).

A major difficulty in first detecting effluent odors and then describing them simultaneously within time and other chromatographic analytical constraints is encountered by the sniffers who find it difficult to coordinate and still record the data on an intensity scale previously memorized. Pollien et al. (1997) developed a method in which data processing is based on the detection of frequency which is perceived and a sum of individual aromagrams generating a standardized format chromatogram, wherein the normalized height of a compound's peak is called the NIF/SNIF. This technique requires a minimum of 8–10 different untrained sniffers. However, NIF/SNIF does not directly measure the odor intensity, but it increases with concentration and consequently with the odor intensity, and so one can compare intensities of two peaks between different aromagrams. The techniques of frequency of detection of odors, the OSME and AEDA, were compared by Le Guen et al. (2000) and later by Van Ruth and O'Connor (2001), and in both studies, they observed a higher positive correlation between the frequency and OSME techniques. The technique of frequency of detection was considered to be a potential application since it is easier, presents higher repeatability of data reproduction (Van Ruth, 2004), and does not require a larger training for judges (Van Ruth and O'Connor, 2001).

16.5 AROMA RETENTION/LOSS DURING PROCESSING OF FRUIT PRODUCTS

Studies on evaluating aroma losses or retention during fruit processing are quite scarce. Usually the aromas of fresh fruit juice alter during processing or on storage, and the changes occur due to enzymatic hydrolysis, thermal, or other treatments that fruits undergo during processing. In fruit juice concentration, many volatile compounds are lost due to their high volatility, and these compounds are important for the sensorial aroma quality of juices, and hence, these must be recovered and added back to the concentrated juice. These volatile compounds are present at very low concentrations in juices, and their recovery using conventional techniques, such as distillation and partial condensation, is somewhat complex. Until now, only a few studies reporting the effect of processing on the volatile constituents in fruit juices are available, and these are presented on specific fruit basis.

16.5.1 CITRUS JUICE

The effects of the extraction mode and the stage of fruit development on the volatile composition of citrus juices were studied by Barboni et al. (2010). The volatile components were extracted by SPME technique in HS mode coupled to gas chromatography/flame ionization detection for their quantification and GC-MS for their identification. The limonene concentration was higher when the juice was extracted from whole fruits containing peel. The overall concentration of volatiles was lower (20%–40%) when the juice was extracted from peeled fruits compared with juices obtained from entire fruits. The hydrocarbon and oxygenated monoterpene concentrations were higher in juices extracted from whole fruits. The concentration of volatiles correlated directly with the juice extraction process. The essential oils from the flavedo affected adversely on characteristic and fresh juice flavors. With regard to varietal differences, the volatile concentrations were higher in the mandarin juices than in the hybrid and clementine juices. The volatile concentrations increased during ripening until commercial maturation when the concentration of volatile compounds was at its maximum, especially when the juice was obtained from unpeeled fruits.

16.5.2 ORANGE JUICE

The overall flavor changes occurring during the manufacturing of processed orange juice such as juice reconstituted from concentrate were monitored by application of a comparative aroma extract dilution analysis (Buettner and Schieberle, 2001). Ethereal extracts obtained from freshly squeezed and from processed juice showed that the main differences were due to significantly higher flavor dilution factors of the odorants such as acetaldehyde being fresh and pungent and (Z)-hex-3-enal characterizing green note in the fresh juice. However, the flavor dilution factors of several odor-active terpenoid compounds such as limonene, α-pinene, and linalool, as well as 3-isopropyl-2-methoxypyrazine and vanillin, were higher in the reconstituted juice.

The processing and storage effects on key aroma compounds of frozen orange juice concentrate were evaluated by Averbeck and Schieberle (2010). Orange juice reconstituted from concentrate which had been stored at 37°C for 4 weeks revealed an increase in the concentrations of dimethyl sulfide, 2-methoxy-4-vinylphenol, α-terpineol, and 4-hydroxy-2,5-dimethyl-3(2H)-furanone in comparison with the same orange juice before storage. However, lower concentrations of octanal, decanal, (R)-α-pinene, linalool, and (E)-β-damascenone were found after storage, while the concentrations of vanillin and carvone remained almost constant. Similar results were found for same aroma compounds after storage of the orange juice at 20°C for 1 year. Sensory experiments revealed the importance of 2-methoxy-4-vinylphenol and dimethyl sulfide for the typical stale off-flavor of the stored orange juice. Under both storage conditions (37°C for 4 weeks or 20°C for 1 year), the breakthrough odor thresholds of α-terpineol and 4-hydroxy-2,5-dimethyl-3(2H)-furanone were not reached, while the concentrations of dimethyl sulfide and 2-methoxy-4-vinylphenol exceeded their breakthrough odor thresholds, thus confirming the crucial role of these odorants for the off-flavor of stored orange juice from concentrate.

16.5.3 MANDARIN

Mandarins are much more perishable than other citrus fruits, mainly due to their rapid deterioration in sensory attributes after harvest. The biochemical components involved in forming the unique flavor of mandarins and how postharvest storage changes the sensory acceptability were studied by Tietel et al. (2011). This was related to decrease in acidity and in the contents of terpenes and aldehydes, which provide green, piney, and citrus aroma, and increases in ethanol fermentation metabolism products and esters, which result in "overripe" and off-flavors. The aroma of mandarins is derived from a mixture of different aroma volatiles, including alcohols, aldehydes, ketones, terpenes, hydrocarbons, and esters. The vast importance of factors such as its genetic background, maturity stage at harvest, commercial postharvest operation treatments, including curing, degreening, and waxing, and storage conditions were greatly emphasized for the retention of mandarin sensorial aroma quality.

16.5.4 PINEAPPLE JUICE

The changes in flavor attributes of pineapple juice were investigated during industrial processing (Facundo et al., 2009). Two batches of pineapple juice were collected at different processing steps: raw material extraction, finishing, centrifugation, and concentration. The samples were analyzed by qualitative descriptive analysis. Samples collected at the extraction, finishing, and centrifugation steps presented similar flavor profiles among them, which indicated that there were no major changes in the juice quality during these operations. However, the flavor attributes of the processed pineapple juice at different processing steps differed from that of the fresh fruit juice profile because of the presence of higher notes of fermented aroma and taste, indicating that there was no uniformity in the selection of the fruits used for processing. The intensity of these descriptors decreased in the juice collected at the concentration step. The extraction, finishing, and centrifugation steps did not modify substantially the flavor profile of the juice. Major changes in flavor attributes were observed in the concentrated pineapple juice, which showed an increased intensity of undesirable descriptors such as the cooked fruit aroma and flavor and the development of artificial aroma, being strongly characterized by these descriptors that do not represent the fresh pineapple fruit juice flavor. In another work (Pereira et al., 2005) using pervaporation process for clarified pineapple juice, various composite membranes (flat or hollow fiber) were tested, and EPDM membranes showed best performance.

16.5.5 CANTALOUPE MELON

The changes in flavor, enzyme activities, and microorganism survival in cantaloupe melon juice were examined after ultrahigh-pressure treatments (Ma et al., 2010). After the treatment performed at 500 MPa for 20 min, the pressurized juice was tested for sensory attributes, and there was no significant change observed between fresh juice and juice obtained after ultrahigh-pressure treatments. These findings revealed that ultrahigh-pressure treatment is a promising way to process the cantaloupe juice. Freezing process and frozen storage did not significantly reduce either microorganism counts or the enzymatic activities of polyphenoloxidase, peroxidase, and lipoxidase in melon. Liquid nitrogen ultrarapid freezing maintained aroma components of the melon as compared with that of the slow freezing. The aromatic change in frozen melon was mainly related to esters changing continuously in variety and concentration because of enzymolysis or synthetic reactions of esters catalyzed by lipoxygenase. Meanwhile, increasing concentration of linoleic acid and linolenic acid in frozen melon cells promoted the biochemical reactions of unsaturated fatty acids so that enzymolysis of the double bonds in the 9th and 10th carbon of linoleic acid and linolenic acid was accelerated. Consequently, the green notes of frozen storage melon became more and more intense, and the ester fragrance became more and more feeble (Ma et al., 2007).

16.5.6 LITCHI JUICE

The aroma components of clear litchi juice were evaluated during processing, and Table 16.2 presents the several compounds identified and quantified in fresh and processed (enzyme hydrolysis, enzyme inactivation, ultrafiltration, and ultrahigh temperature [UHT] sterilization) litchi juice (Li et al., 2009). The aroma fractions were isolated by SPME process and analyzed by capillary GC-MS. The principal compounds identified in fresh litchi juice were citronellol, geraniol, phenylethyl alcohol, linalool, ethyl alcohol, 1-hexanol, ethyl hexanoate, D-limonene, β-myrcene, 4-carene, γ-terpinene, neral, geranial, hexanal, (E)-2-hexenal, benzaldehyde, and furfural. There was a sharp decrease in total alcohols content in UHT sterilization (82%) and ultrafiltration (63%) processing. Alkenes and aldehydes also decrease to a great extent in a UHT sterilization and ultrafiltration (Table 16.2). The amount of total and seven characteristic aroma compounds (geraniol, citronellol, phenyl-ethyl alcohol, D-limonene, nonanal, geranial, and linalool) increased significantly after enzyme hydrolysis, but decreased during the process of enzyme inactivation, ultrafiltration, and sterilization. A possible addition of aromatic fractions recovered during juice processing helped in obtaining processed litchi juice with an aromatic profile closer to fresh juice. With regard to aroma retention, an ultrafiltration membrane with a larger pore size was found to be better than a membrane with a smaller pore size, and a high-voltage pulsed electric field was better than thermal sterilization.

16.5.7 MANGO JUICE

Heat processing decreased the concentration of esters in mango juice which occurred mainly due to evaporation (El Nemr et al., 1988). A tendency to a greater decrease of ethyl hexanoate after pressing was observed, and it was related to greater affinity of this ester to the pressing cake.

16.5.8 CHERRY JUICE

Four different varieties of fresh cherry fruit were canned and heat processed in a retort (Pierce et al., 1996). The canned cherries were then extracted and analyzed for volatile compounds under similar conditions to the fresh samples. Although the concentration of the chiral compounds changed slightly on processing, the canned cherries contained the same chiral compounds as found in the fresh cherries. Limonene content in canned cherries decreased in total concentration, whereas α-terpineol increased. Moreover, there was a marked decrease in the enantiomeric excess of (R)-(+)-limonene, and small decreases for (R)-(−)-linalool and (R)-(+)-α-terpineol.

TABLE 16.2
Changes in Aroma Compounds (Mean Values, µg/L) in Litchi Juice during Processing

Compound	Raw Juice	Enzyme Hydrolysis	Enzyme Inactivation	Ultrafiltration	UHT Sterilization
Alcohols					
Citronellol	902.01	940.48	591.03	335.37	177.10
Geraniol	969.93	1064.83	721.42	261.88	144.17
Phenylethyl alcohol	326.49	403.24	216.94	50.27	ND
Linalool	42.82	49.58	41.77	21.27	29.87
Ethyl alcohol	377.30	373.14	241.68	112.45	18.77
1-Hexanol	55.24	38.63	21.02	8.26	ND
Benzyl alcohol	10.38	52.04	27.34	ND	ND
α-Terpineol	24.09	28.47	23.59	12.14	9.27
Total	*2891.35*	*3387.28*	*2142.99*	*1070.61*	*522.98*
Esters					
Hexanoic acid, ethyl ester	43.65	41.78	45.38	46.61	40.68
Phthalic acid, isobutyl octyl ester	16.21	17.85	7.89	42.97	29.33
Hexadecanoic acid, ethyl ester	5.29	2.37	0.55	0.62	0.48
Benzoic acid, ethyl ester	ND	3.07	1.20	ND	ND
Total	*48.31*	*45.32*	*13.88*	*51.49*	*36.93*
Ketones					
2,3-Dihydro-3,5-dihydroxy-6-methyl-4H-pyran-4-one	15.43	9.59	5.86	ND	8.03
(Z)-6,10-Dimethyl-5,9-undecadien-2-one	3.43	4.69	3.67	1.25	2.06
Total	*18.86*	*14.28*	*9.35*	*5.0*	*16.23*
Alkenes					
D-Limonene	54.68	62.43	15.50	3.40	1.72
β-Myrcene	76.28	90.73	37.53	12.40	5.04
4-Carene	80.51	92.44	23.15	1.07	ND
(Z)-3,7-Dimethyl-1,3,6-octatriene	16.29	18.63	13.87	5.70	3.50
γ-Terpinene	6.33	4.11	1.37	0.74	ND
Total	*249.09*	*293.34*	*91.42*	*23.31*	*10.26*
Acids					
3-Methyl-2-butenoic acid	1.49	4.37	ND	ND	ND
Total	*26.24*	*52.66*	*35.71*	*3.21*	*8.44*

(*continued*)

TABLE 16.2 (continued)
Changes in Aroma Compounds (Mean Values, µg/L) in Litchi Juice during Processing

Compound	Raw Juice	Enzyme Hydrolysis	Enzyme Inactivation	Ultrafiltration	UHT Sterilization
Aldehydes					
Neral	40.14	84.08	35.84	8.26	5.06
Geranial	70.72	128.90	57.15	14.66	8.48
Hexanal	33.47	17.98	9.91	3.14	ND
(*E*)-2-Hexenal	24.64	43.21	12.66	ND	1.41
Benzaldehyde	11.31	22.82	10.97	1.33	5.14
Nonanal	6.46	17.71	10.81	2.58	3.29
Furfural	ND	ND	6.94	ND	5.48
Total	*232.27*	*317.86*	*146.14*	*29.97*	*35.12*
Others					
Tetrahydro-4-methyl-2-(2-methyl-1-propenyl)-2H-pyran	5.67	10.95	49.68	19.69	33.24
3,6-Dihydro-4-methyl-2-(2-methyl-1-propenyl)-2H-pyran	5.70	8.45	10.52	4.35	8.10
Dimethyl sulfide	ND	ND	ND	ND	38.72
Total	*11.37*	*19.40*	*60.20*	*24.04*	*83.56*

Source: Data abridged and adapted from Li, C. et al., *J. Sci. Food Agric.*, 89, 2405, 2009.
Only mean values of selected compounds shown while total values represent the total concentration of all compounds within their respective organic classes; ND, not detected.

16.5.9 BLACK CURRANT JUICE

The usual processing operations of black currant juice involve crushing, heating, enzyme treatment, pressing, pasteurization, clarification, and filtration. Changes in concentration of aroma impact compounds during black currant juice processing in pilot plant scale were monitored by GC-O and quantified by DHS and GC-MS (Mikkelsen and Poll, 2002). The reduction of aroma compounds during the process varied widely (50%–100%) depending on the processing operations and their conditions. The pressing step and heating treatments had a negative effect on the concentration of aroma compounds. The most important aroma compounds quantitatively were the terpenes 3-carene, terpinolene, and *cis*-ocimene, and the esters ethyl acetate, methyl butanoate, and ethyl butanoate. The relative concentration of the esters such as methyl butanoate, ethyl butanoate, and ethyl hexanoate decreased to less than 10% during the process; and this large reduction was mainly due to heat treatment. The α-pinene content increased until the pressing step, which could be caused by enzymatic release of glucosidically bound α-pinene (Marriott, 1986).

16.5.10 PEARS

The production of butyl and hexyl acetates and the contents of ethanol and acetaldehyde in relation to the activity of some related enzymes (lipoxygenase, pyruvate decarboxylase, alcohol dehydrogenase, and alcohol o-acyltransferase) was studied in "Doyenne du Comice" pears during ripening at 20°C, followed by its long-term cold storage under different conditions: air or controlled atmosphere (CA) in 2 kPa O_2 with various CO_2 partial pressures (0.7, 2, or 5 kPa) (Lara et al., 2003). A decrease in volatile production was found in fruit stored under low O_2 upon return to ambient conditions. Inhibition of volatile biosynthesis by CA storage was due to limited precursor/substrate supply to the related enzymes rather than by enzyme degradation or inactivation. The enzyme activity could be inhibited by the formation of excess product and of oxidation into acetaldehyde along with a shortage of lipid precursors for ester biosynthesis.

16.5.11 APPLES

A study on "Fuji" apples harvested at two different dates, over two consecutive years and stored under different atmospheric conditions (21 kPa O_2 + 0.03 kPa CO_2; 3 kPa O_2 + 2 kPa CO_2; 1 kPa O_2 + 2 kPa CO_2), was performed by Echeverría et al. (2004). After 3, 5, or 7 months of storage plus 1 or 10 days of ripening at 20°C, aroma volatile emission and quality parameters were measured. Highest total aroma emission was recorded in fruits after 5 month storage and 1 day of ripening at 20°C regardless of atmosphere conditions. The compounds contributing mostly to the characteristic aroma of "Fuji" apples were ethyl 2-methylbutanoate, 2-methylbutyl acetate, and hexyl acetate; and their concentrations were higher the first day after removal from storage at 5, 3, and 7 months, respectively.

The integration and optimization of hydrophobic pervaporation were applied to the design and scale-up of pervaporation units for the recovery of natural apple juice aroma (Lipnizki et al., 2002). Both semi-batch and continuous process configurations were considered, and the process conditions were optimized taking the cost consideration of the process. In the case of the continuous process, the cost for the apple juice aroma recovery was between 2.19 and 5.38, while the cost of aroma recovery per kg apple juice was between 0.03 and 0.05. Upon analyzing the optimized processes considering key cost factors associated with pervaporation, membrane lifetime, and membrane cost, it was revealed that the membrane lifetime is more important than the membrane cost and that the continuous process is more sensitive to changes in membrane lifetime and membrane cost. Overall, this study revealed that pervaporation has the potential to become an alternative to conventional processes in recovering and concentrating apple juice aroma compounds. Pereira et al. (2006) also reported that pervaporation presents promising alternative for substituting the conventional aroma recovery techniques such as distillation and partial condensation.

16.5.12 STRAWBERRY

The effect of the postharvest application of methyl jasmonate vapor on total phenolics, antioxidant capacity, anthocyanin profile, and overall aroma composition in strawberry fruits was examined on 5, 7, and 11 days after treatment (Moreno et al., 2010). Several specific key aroma compounds were significantly increased in treated strawberries when compared with untreated samples, and the maximum concentration was registered on the fifth day. However, longer storage time up to 11 days after treatment resulted in a decrease of volatile compounds characterizing for detrimental fruit quality.

16.6 CONCLUSIONS

This chapter discusses the effects of various processing procedures used in obtaining fruit products with the focus being mainly on the aroma characteristics of the fruits. It updates the changes in aroma that occur during fruit development and maturation. The various analytical techniques used for monitoring and determining the aroma qualities of fruits are discussed. Finally, this chapter focuses on evaluating the effects of processing on aroma quality of various fruits and emphasizes the changes regarding loss or retention of aroma inherent quality of fruits.

REFERENCES

Augusto, F., Lopes, A. L., and Zini, C. A. (2003) Sampling and sample preparation for analysis of aromas and fragrances. *Trends in Analytical Chemistry* 22: 160–169.

Augusto, F. and Valente, A. L. P. (2002) Applications of solid-phase microextraction to chemical analysis of live biological samples. *Analytical Chemistry* 21: 428–438.

Augusto, F., Valente, A. L. P., Tada, E. S., and Rivellino, S. R. (2000) Screening of Brazilian fruit aromas using solid-phase microextraction–gas chromatography–mass spectrometry. *Journal of Chromatography A* 873: 117–127.

Averbeck, M. and Schieberle, P. (2010) Influence of different storage conditions on changes in the key aroma compounds of orange juice reconstituted from concentrate. *European Food Research and Technology* 232: 129–142.

Baldwin, E. A., Bai, J., Soliva-Fortuny, R. C., Mattheis, J. P., Stanley, R., Perera, C., and Brecht, J. K. (2004) Effect of pretreatment of intact "Gala" apple with ethanol vapor, heat, or 1-methylcyclopropene on quality and shelf life of fresh-cut slices. *Journal of the American Society for Horticultural Science* 129: 583–593.

Barboni, T., Muselli, A., Luro, F., Desjobert, J.-M., and Costa, J. (2010) Influence of processing steps and fruit maturity on volatile concentrations in juices from clementine, mandarin, and their hybrids. *European Food Research and Technology* 231: 379–386.

Barrete, D. M., Beaulieu, J. C., and Shewfelt, R. (2010) Color, flavor, texture, and nutritional quality of fresh-cut fruits and vegetables: Desirable levels, instrumental and sensory measurement, and the effects of processing. *Critical Reviews in Food Science and Nutrition* 50(5): 369–389.

Beaulieu, J. C. (2005) Within-season volatile and quality differences in stored fresh-cut cantaloupe cultivars. *Journal of Agricultural and Food Chemistry* 53(22): 8679–8687.

Beaulieu, J. C. (2006) Volatile changes in cantaloupe during growth, maturation and in stored fresh-cuts prepared from fruit harvested at various maturities. *Journal of the American Society for Horticultural Science* 131(1): 127–139.

Beaulieu, J. C., Ingram, D. A., Lea, J. M., and Bett-Garber, K. L. (2004) Effect of harvest maturity on the sensory characteristics of fresh-cut cantaloupe. *Journal of Food Science* 69(7): S250–S258.

Beaulieu, J. C. and Lancaster, V. (2007) Correlating volatile compounds, quality parameters, and sensory attributes in stored fresh-cut cantaloupe. *Journal of Agricultural and Food Chemistry* 55: 9503–9513.

Beaulieu, J. C. and Lea, J. M. (2003) Volatile and quality changes in fresh-cut mangoes prepared from firm-ripe and soft-ripe fruit, stored in clamshell containers and passive MAP. *Postharvest Biology and Technology* 30(1): 15–28.

Beaulieu, J. C. and Lea, J. M. (2007) Quality changes in cantaloupe during growth, maturation, and in stored fresh-cuts prepared from fruit harvest at various maturities. *Journal of the American Society for Horticultural Science* 132(5): 720–728.

Bezyma, L. A. and Kutovoy, V. A. (2005) Vacuum drying and hybrid technologies. *Stewart Postharvest Review* 4: 6–13.

Buettner, A. and Schieberle, P. (2001) Application of a comparative aroma extract dilution analysis to monitor changes in orange juice aroma compounds during processing. In *Gas Chromatography-Olfactometry*, ACS Symposium Series Vol. 782, Washington, DC: American Chemical Society, pp. 33–45.

Cendres, A., Chemat, F., Maingonnat, J.-F., and Rnard, C. M. G. C. (2011) An innovative process for extraction of fruit juice using microwave heating. *LWT-Food Science and Technology* 44: 1035–1041.

Chassagne-Berces, S., Fonseca, F., Citeau, M., and Marin, M. (2010) Freezing protocol effect on quality properties of fruit tissue according to the fruit, the variety and the stage of maturity. *Food Science and Technology* 43: 1441–1449.

Cheong, K. W., Tan, C. P., Mirhosseini, H., Chin, S. T., Che Man, Y. B., Hamid, N. S. A., and Basri, M. (2011) Optimization of equilibrium headspace analysis of volatile flavor compounds of Malaysian soursop (*Annona muricata*): Comprehensive two-dimensional gas chromatography time-of-flight mass spectrometry (GCxGC-TOFMS). *Food Chemistry* 125: 1481–1489.

Chitarra, M. I. F. and Chitarra, A. B. (2005) *Pós-colheita de frutas e hortaliças*. Lavras, Brazil: UFLA.

Chou, S. K. and Chua, K. J. (2001) New hybrid drying technologies for heat sensitive foodstuffs. *Food Science and Technology* 12: 359–369.

Chua, K. J., Chou S. K., Ho, J. C., Mujumdar, A. S., and Hawlader, M. N. A. (2000) Cyclic air temperature drying of guava pieces: Effects on moisture and ascorbic contents. *Transactions of the Institution of Chemical Engineers* 78: 72–78.

Civille, G. V. (1991) Food quality: Consumer acceptance and sensory attributes. *Journal of Food Quality* 14: 1–8.

Culleré, L., Ferreira, V., Chevret, B., Venturini, M. E., Sánchez-Gimeno, A. C., and Blanco, D. (2010) Characterisation of aroma active compounds in black truffles (*Tuber melanosporum*) and summer truffles (*Tuber aestivum*) by gas chromatography–olfactometry. *Food Chemistry* 122: 300–306.

Dallüge, J., Beens, J., and Brinkman, U. A. Th. (2003). Comprehensive two-dimensional gas chromatography: A powerful and versatile analytical tool. *Journal of Chromatography A* 1000: 69–108.

Danisco, P. P. 2001. *Comparing Extraction by Traditional Solvents and Supercritical Extraction from an Economic and Environmental Standpoint*, CD-Rom. Versailles, France: ISASF.

Donsi, G., Ferrari, G., Nigro, R., and Maltero, P. D. (1998) Combination of mild dehydration and freeze drying processes to obtain high quality dried vegetables and fruits. *Trans I Chemistry E* 76: 181–187.

Drouzas, A. E. and Schubert, H. (1996) Microwave application in vacuum drying of fruit. *Journal of Food Engineering* 28: 203–209.

Echeverría, G., Fuentes, T., Graell, J., Lara, I., and López, M. L. (2004) Aroma volatile compounds of 'Fuji' apples in relation to harvest date and cold storage technology. A comparison of two seasons. *Postharvest Biology and Technology* 32: 29–44.

El Nemr, S. E., Ismail, I. A., and Askar, A. (1988) Aroma changes in mango juice during processing and storage. *Food Chemistry* 30: 269–275.

Erle, U. (2005) Drying using microwave processing. In: *The Microwave Processing of Foods*. Eds. H. Schubert and M. Regier. Cambridge, MA: Woodhead Publishing, pp. 142–152.

Etievant, P. X., Callement, G., Langlois, D., Issanchou, S., and Coqui-bus, N. (1999) Odor evaluation in gas chromatography—Olfactometry by finger-span method. *Journal of Agricultural and Food Chemistry* 47(4): 1673–1680.

Facundo, H. V. V., Neto, M. A. S., Maia, G. A., Narain, N., and Garruti, D. S. (2010) Changes in flavor quality of pineapple juice during processing. *Journal of Food Processing and Preservation* 34: 508–519.

Feng, H., Tang, J., Mattinson, D. S., and Fellman, J. K. (1999) Microwave and spouted bed drying of frozen blue berries. *Journal of Food Processing and Preservation* 23: 463–479.

Fernandes, F. A. N., Rodrigues, S., Law, C. L., and Mujumdar, A. S. (2010) Drying of exotic tropical fruits: A comprehensive review. *Food and Bioprocess Technology* 4: 163–185.

Feuerbach, M., Frhlich, O., and Schreier, P. (1988) Chirality evaluation of 1,4-decanolide in peach. *Journal of Agricultural Food Chemistry* 36: 1236–1237.

Franco, M. R. B. and Janzantti, N. S. (2004) Avanços na metodologia instrumental da pesquisa do sabor. In: *Aroma e sabor de alimentos: temas atuais*. São Paulo, Brazil: M. R. B. Franco, pp. 17–27.

Freire, M. T. A., Bottoli, C. B. G., Fabris, S., and Reyes, F. G. R. (2008) Contaminantes voláteis provenientes de embalagens plásticas: Desenvolvimento e validação de métodos analíticos. *Química Nova* 31(6): 1522–1532.

Galvão, M. S., Narain, N., Santos, M. S. P., and Nunes, M. L. (2011) Volatile compounds and descriptive odor attributes in umbu (*Spondias tuberosa*) fruits during maturation. *Food Research International* 44(7): 1919–1927.

Gorny, J. R., Cifuentes, R. A., Hess-Pierce, B., and Kader, A. A. (2000) Quality changes in fresh-cut pear slices as affected by cultivar, ripeness stage, fruit size, and storage regime. *Journal of Food Science* 65(3): 541–544.

Gorny, J. R., Hess-Pierce, B., and Kader, A. A. (1998) Effects of fruit ripeness and storage temperature on the deterioration rate of fresh-cut peach and nectarine slices. *HortScience* 33(1): 110–113.

Gould, W. A. (1983) *Food Quality Assurance*. Westport, CT: AVI Publishing Company.

Grob, R. L. and Kaiser, M. A., (eds.) (1994) Environmental Problem Solving Using Gas and Liquid Chromatography, Journal of Chromatography Library Series, vol. 1. Amsterdam, the Netherlands: Elsevier Science Publishers B. V.

Grob, R. L. and Kaiser, M. A. (2004) Qualitative and quantitative analysis by gas chromatography. In: *Modern Practice of Gas Chromatography*. Eds. R. L. Grob and E. F. Berry. New York: John Wiley & Sons, pp. 403–460.

Hiu, D. N. and Schreier, P. J. (1961) The volatile constituents of passion fruit juice. *Journal of Food Science* 26: 557–563.

Hognadottir, A. and Rouseff, R. L. (2003) Identification of aroma active compounds in orange essence oil using gas chromatography–olfactometry and gas chromatography–mass spectrometry. *Journal of Chromatography A* 998: 201–211.

Jabalpurwala, F., Gurbuz, O., and Rouseff, R. (2010) Analysis of grapefruit sulphur volatiles using SPME and pulsed flame photometric detection. *Food Chemistry* 120: 296–303.

Jayaraman, K. S. and Gupta, D. K. (1992) Dehydration of fruit and vegetables—Recent developments in principles and techniques. *Drying Technology* 10: 1–50.

Kataoka, H., Lord, H. L., and Pawliszyn, L. (2000) Applications of solid-phase microextraction in food analysis. *Journal of Chromatography A* 880: 35–62.

Khin, M. M., Zhou, W., and Perera, C. (2005) Development in combined treatment of coating and osmotic dehydration of food—A review. *International Journal of Food Engineering* 1: 1–19.

Knudsen, J. T., Tollsten, L., and Bergström, L. G. (1993) Floral scents—A checklist of volatile compounds isolated by head-space techniques. *Phytochemistry* 33: 253–280.

Kramer, A. (1965) Evaluation of quality of fruits and vegetables. In: *Food Quality*. Eds. G.W. Irving Jr. and S. R. Hoover. Washington, DC: American Association for the Advancement of Science, pp. 9–18.

Lara, I., Miro, R. M., Fuentes, T., Sayez, G., Graell, J., and Lopez, M. L. (2003) Biosynthesis of volatile aroma compounds in pear fruit stored under long-term controlled-atmosphere conditions. *Postharvest Biology and Technology* 29: 29–39.

Le Guen, S., Prost, C., and Demaimay, M. (2000) Critical Comparison of three olfactometric methods for the identification of the most potent odorants in cooked mussels (Mytilus edulis). *Journal of Agriculture and Food Chemistry* 48(4): 1307–1314.

Li, C., Hao, J., Zhong, H., Dang, M., and Xie, B. (2009) Aroma components at various stages of litchi juice processing. *Journal of the Science of Food and Agriculture* 89: 2405–2414.

Linssen, J. P. H., Janssens, G. M., Roozen, J. P., and Posthumus, M. A. (1993) Combined gas chromatography and sniffing pot analysis of volatile compounds of mineral water packed in polyethylene laminated packages. *Food Chemistry* 46: 367–371.

Lipnizki, F., Olsson, J., and Tragardh, G. (2002) Scale-up of pervaporation for the recovery of natural aroma compounds in the food industry Part 2: Optimisation and integration. *Journal of Food Engineering* 54: 197–205.

Ma, Y., Hu, X., Chen, J., Chen, F., Wu, J., Zhao, G., Liao, X., and Wang, Z. (2007) The effect of freezing modes and frozen storage on aroma, enzyme and micro-organism in Hami Melon. *Food Science and Technology International* 13(4): 259–267.

Ma, Y., Hu, X., Chen, J., Zhao, G., Liao, X., Chen, F., Wu, J., and Wang, Z. (2010) Effect of uhp on enzyme, microorganism and flavor in cantaloupe (*Cucumis melo* l.) juice. *Journal of Food Process Engineering* 33: 540–553.

Marriott, R. J. (1986). Biogenesis of blackcurrant (*Ribes nigrum*) aroma. In: *Biogenesis of Aromas*. Eds. T. Parliment and R. Croteau, Washington, DC: American Chemical Society, pp. 184–192.

Marsili, R. T. and Miller, N. (2000) Determination of major aroma impact compounds in fermented cucumbers by solid-phase microextraction-gas chromatography-mass spectrometry-olfactometry detection. *Journal of Chromatography Science* 38: 307–314.

Mayor, L. and Sereno, A. M. (2004) Modelling shrinkage during convective drying of food materials. *Journal of Food Engineering* 61: 373–386.

McDaniel, M. R., Miranda-Lopez, R., Watson, B. T., Micheals, N. J., and Libbey, L. M. (1990) Pinotnoir aroma: a sensory/gas chromatographic approach. In: Flavors and Off-Flavors, ed. G. Charalambous. Amsterdam, the Netherlands: Elsevier.

Menezes, E. M. S., Torres, A. T., and Srur, A. U. S. (2008) Valor nutricional da polpa de açaí (Euterpe oleracea Mart) liofilizada. *Acta Amazônica* 38(2): 311–316.

Mikkelsen, B. B. and Poll, L. (2002) Decomposition and transformation of aroma compounds and anthocyanins during black currant (*Ribes nigrum* L.) Juice processing. *JFS: Sensory and Nutritive Qualities of Food* 67(9): 3447–3455.

Mistry, B. S., Reineccius, T., and Olson, L. (1997) Gas-chromatography—Olfactometry for the determination of key odorants in foods. In: Techniques for Analyzing Food Aroma, ed. R. Marsili. New York: Dekker.

Moreno, F. P., Monagas, M., Blanch, G. P., Bartolome, B., and Castillo, M. L. R. (2010) Enhancement of anthocyanins and selected aroma compounds in strawberry fruits through methyl jasmonate vapor treatment. *European Food Research and Technology* 230: 989–999.

Narain, N. and Bora, P. S. (1992) Posthharvest changes in volatile flavour constituents of yellow passion fruit (passiflora edulis forma flavicarpa). *Journal of Science of Food and Agriculture* 60: 529–530.

Nickerson, G. B. and Likens, S. T. (1966) Gas chromatographic evidence for the occurrence of hop oil components in beer. *Journal of Chromatography A* 21: 1–5.

Nijhuis, H. H., Torringa, H. M., Muresan, S., Yukel, D., Leguijt, C., and Kloek, W. (1998) Approaches to improving the quality of dried fruits and vegetables. *Trends in Food Science and Technology* 9: 13–20.

Orsat, V., Changrue, V., and Raghavan, G. S. V. (2006) Microwave drying of fruits and vegetables. *Stewart Postharvest Review* 6: 4–9.

Orsat, V., Raghavan, V., and Meda, V. (2005) Microwave technology for food processing: An overview. In: *The Microwave Processing of Foods*. Eds. H. Schubert and M. Regier. Karlsruhe, Germany: University of Karlsruhe.

Pawliszyn, J. (1997) *Solid Phase Microextraction—Theory and Practice*. New York: Wiley-VCH.

Pawliszyn, J., Ed. (1999) *Applications of Solid Phase Microextraction*. Cambridge, U.K.: Royal Society of Chemistry.

Pereira, C. C., Ribeiro Jr., C. P., Nobrega, R., and Borges, C. P. (2006) Pervaporative recovery of volatile aroma compounds from fruit juices. *Journal of Membrane Science* 274: 1–23.

Pereira, C. C., Rufino, J. R. M., Habert, A. C., Nobrega, R., Cabral, L. M. C., and Borges, C. P. (2005) Aroma compounds recovery of tropical fruit juice by pervaporation: Membrane material selection and process evaluation. *Journal of Food Engineering* 66: 77–87.

Pierce, K., Mottram, D. S., and Baigrie, B. D. (1996) The effect of processing on the chiral aroma compounds in cherries (*Prunus avium* L.). In: *Chemical Markers for Processed and Stored Foods*, Symposium Series 631. Washington, DC: American Chemical Society, pp. 70–76.

Pollien, P., Ott, A., Montigon, F., Baumgartner, M., Muñoz-box, R., and Chaintreau, A. (1997) Hyphenated headspace-gas chromatography-sniffing technique: Screening of impact odorants and quantitative aromagram comparisons. *Journal of Agricultural and Food Chemistry* 45: 2630–2637.

Raghavan, G. S. V. and Orsat, V. (1998) Electro-technology in drying and processing of biological materials. Keynote presentation at *11th International Drying Symposium (IDS 98)*, August 19–22, Halkididi, Greece, pp. 456–463.

Raghavan, G. S. V., Rennie, T. J., Sunjka, P. S., Orsat, V., Phaphuangwittayakul, W., and Terdtoon, P. (2005) Overview of new techniques for drying biological materials with emphasis on energy aspects. *Brazilian Journal of Chemical Engineering* 22: 195–201.

Rankin, S. A. and Bodyfelt, F. W. (1995) Solvent desorption dynamic headspace method for diacetyl and acetoin in buttermilk. *Journal of Food Science* 60(3): 1205–1207.

Rastogi, N. K. and Niranjan, K. (2008) Enhanced mass transfer during osmotic dehydration of high pressure treated pineapple. *Journal of Food Science* 63: 508–511.

Reineccius, G. (2006) Choosing the correct analytical technique in aroma analysis. In: *Flavour in Food*. Eds. A. Voilley and P. Etiévant. Boca Raton, FL: CRC Press, pp. 81–95.

Rodriguez-Amaya, D. B. (2004) *Aroma e sabor de alimentos*, Livraria Varela, São Paulo, Brazil.

Rood, D. (1995) *A Practical Guide to the Care, Maintenance and Troubleshooting of Capillary Gas Chromatographic Systems*. 2nd edn. Heidelberg, Germany: HuthigVerlag, pp. 323–325.

Sagar, V. R. and Suresh Kumar, P. (2010) Recent advances in drying and dehydration of fruits and vegetables—A review. *Journal Food Science and Technology* 47(1): 15–26.

Saritas, Y., Sonwa, M. M., Iznaguen, H., König, W. A., Muhle, H., and Mues, R. (2001) Volatile constituents in mosses (Musci). *Phytochemistry* 57: 443–457.

Sides, A., Robards, K., and Helliwell, S. (2000) Developments in extraction techniques and their application to analysis of volatiles in foods. *Trends Analytical Chemistry* 19: 322–329.

Soliva-Fortuny, R. C., Alos-Saiz, N., Espachas-Barroso, A., and Martin-Belloso, O. (2004) Influence of maturity at processing on quality attributes of fresh-cut conference pears. *Journal of Food Science* 69(7): 290–294.

Song, J., Fan, L., and Beaudry, R. M. (1998) Application of solid phase microextraction, and gas chromatography/time-of-flight mass spectrometry for rapid analysis of flavor volatiles in tomato and strawberry fruits. *Journal of Agricultural and Food Chemistry* 46: 3721–3726.

Soysal, Y., Oztekin, S., and Eren, O. (2006) Microwave drying of parsle modeling, kinetics and energy aspects. *Biosystems Engineering* 93: 403–413.

Steinhart, H., Stephan, A., and Bücking, M. (2000) Advances in flavor research. *Journal High Resolution Chromatography* 23: 489–496.

Stevenson, R. J., Chen, X. D., and Mills, O. E. (1996) Studies of flavor volatiles with particular reference to dairy protein products. *Food Research International* 29(3–4): 265–290.

Tatemoto, Y., Yano, S., Mawatart, Y., Noda, K., and Komatsu, N. (2007) Drying characteristics of porous material immersed in a bed glass beads fluedized by superheated steam under reduced pressure. *Chemical Engineering Science* 62: 471–480.

Tedjo, W., Eshiaghi, M. N., and Knorr, D. (2002) Impact of non- thermal processing on plant metabolites. *Journal of Food Engineering* 56: 131–134.

Theerakulkait, C. and Barrett, D. M. (1995) Sweet corn germ enzymes affect odor formation. *Journal of Food Science* 60(5): 1034–1040.

Thomazini, M. and Franco, M. R. B. (2000) Metodologia para análise dos constituintes voláteis do sabor. *Boletim da Sociedade Brasileira de Ciência e Tecnologia de Alimentos* 34: 52–59.

Tietel, Z., Plotto, A., Fallik, E., Lewinsohnd, E., and Porata, R. (2011) Taste and aroma of fresh and stored mandarins. *Journal of the Science of Food and Agriculture* 91: 14–23.

Ullrich, F. and Grosch, W. (1987) Identification of the most intense volatile flavour compounds formed during autoxidation of linoleic acid. *Zeitschrift für Lebensmitteluntersuchung und-Forschung A* 184: 277–282.

Valero, E., Miranda, E., Sanz, J., and Martinez-Castro, I. (1997) Automatic thermal desorption in the GC analysis of dairy product volatiles. *Chromatographia* 44: 59–64.

Van Ruth, S. M. (2001) Aroma measurement: Recent developments in isolation and characterization. In: Physics and Chemistry Basis of Biotechnology, eds. M. De Cuyper and J. W. M. Bulte. Amsterdam, the Netherlands: Kluwer Academic Publishers.

Van Ruth, S. M. (2004) Evaluation of two gas chromatography-olfactometry methods: The detection frequency and perceived intensity method. *Journal of Chromatography A* 1054: 33–37.

Van Ruth, S. M. and O'Connor, C. H. (2001) Evaluation of three gas chromatography-olfactometry methods: Comparison of odour intensity-concentration relationships of eight volatile compounds with sensory headspace data. *Food Chemistry* 74: 341–334.

Van Ruth, S. M. and Roozen, J. P. (1994) Gas chromatography/sniffing port analysis and sensory evaluation of commercially dried bell peppers (Capsicum annuum) after rehydration. *Food Chemistry* 51: 165–170.

Wilkes, J. G., Conte, E. D., Kim, Y., Holcomb, M., Sutherland, J. B., and Miller, D. W. (2000). Sample preparation for the analysis of flavors and off-flavors in foods. *Journal of Chromatography A* 800: 3–33.

Yamaguchi, S. and Ninomiya, K. (2000) Umami and food palatability. *Journal of Nutrition* 130: 921S–926S.

17 Sensory Evaluation in Fruit Product Development

Deborah dos Santos Garruti, Heliofábia Virginia de Vasconcelos Facundo, Janice Ribeiro Lima, and Andréa Cardoso de Aquino

CONTENTS

17.1   INTRODUCTION

Let us not forget that the millions of dollars invested in our businesses depend on that small feeling that our products evoke in our customers' mouth

(Platt)

We don't sell products, we sell sensory properties

(Alejandra Muñoz)

Based on these thoughts, we have to agree with Meilgaard et al. (1999) that the primary function of sensory testing is to provide reliable data on which sound decisions may be made. It is an integrated, multidimensional measure with three important advantages: it identifies the presence of notable differences, identifies and quantifies important sensory characteristics in a fast way, and identifies specific problems that cannot be detected by other analytical procedures, as consumer preference, for instance (Nakayama and Wessman, 1979). Comprising a set of techniques for accurate measurement of human responses to foods under minimum potentially biasing effects on consumer perception, sensory analysis attempts to isolate the sensory properties of foods themselves and provides important and useful information to product developers, food scientists, and managers about the sensory characteristics and acceptability of their products (Lawless and Heymann, 1999).

Demands for sensory methodology and technology have grown tremendously around the world, due mainly to the advent of total quality. In addition, the need for understanding people as consumers is something that has been constantly growing and becoming a target of all food industry. Sensory analysis fits into this context as an analytical tool used to translate the link between food products and the consumer, expressing numerical values that can be analyzed and verifying its accuracy through statistical support. Nowadays, most large consumer food companies have departments dedicated to sensory evaluation.

The importance of sensory analysis in the fruit-based food sector is unquestionable, given the variety of applications in food science research, product development, and quality control:

- Improvement of plant varieties and production systems, selection of sources of supply
- Improvement/development of new products and processes
- Product modifications derived from substitution of ingredients and suppliers, changes in processing and packaging, and cost reduction
- Formulation of a product similar to a market leader
- Nutritional enrichment
- Determination of shelf life
- Development of quality standards
- Quality control (raw material and suppliers, processing, end product, packaging)
- Market control (determination of product's acceptability and consumer preferences, determination of market segmentation)

However, a question arises: why do we have to do sensory analysis? Why cannot we monitor those changes by analytical means? The problem is that sensory quality is not an intrinsic property of a food item. It is the result of an interaction between food and the human being. A particular food has its structural, physical, and chemical properties that determine its sensory characteristics, while man carries its culture and food habits. It is also important to consider his psychological condition when he is analyzing the product, which is influenced by his emotional state and physiological and socioeconomic conditions such as age, sex, education, income, and degree of urbanization among others.

In sum, the sensory quality of a product is the way humans perceive them. And human perceptions are the results of complex processes that involve sensory organs and the brain. It now becomes clear that sensory quality must be measured by sensory techniques. Only human sensory data provide information on how consumers perceive or react to food products in real life. Instrumental measurements are useful only when they show good correlation with sensory data (Schiffman, 1996).

However, when man is used as a measuring instrument, strict control of the conditions of tests application and methodology to be used is required, in order to avoid errors of psychological or physiological nature. The principles and practices of sensory evaluation give strict rules for the preparation, coding, and serving of samples under controlled conditions so that the biasing factors are minimized and use techniques drawn from behavioral science that allow numerical data to be collected and statistically treated, establishing lawful relationships between product characteristics and human perception.

17.2 CLASSIFICATION OF SENSORY METHODS

In sensory evaluation, scientific methods are usually classified according to their primary objective (Table 17.1). Two types of methods are generally recognized by the sensory scientists, analytic methods and affective methods, which comprise three classes of tests: discriminative, descriptive, and affective tests. More detailed discussions and explanations on how to conduct, analyze, and interpret each method are given by Amerine et al. (1965), Moskowitz (1983), Stone and Sidel (1993), Lawless and Heymann (1999), Meilgaard et al. (1999).

17.2.1 DISCRIMINATION TESTS

Discrimination or Discriminative tests answer whether any noticeable difference exists among products. It is possible for two or more samples to be physically or chemically different, but this difference may not be perceived by humans. If the difference among samples is very large and obvious, then discriminative tests are not necessary. For example, use these tests if products resulting from a change in ingredients, processing, packaging, or storage show subtle differences and you want to know if they will be perceptible to people.

Discrimination tests are also called difference tests. Meilgaard et al. (1999) subdivide them into overall and directional difference tests.

TABLE 17.1

Classification of Traditional Test Methods in Sensory Evaluation

Analytic		Affective
Laboratory Tests		Consumer Tests
Discrimination	Descriptive	
Are products different in any way?	*How do products differ in specific characteristics?*	*Which product is preferred? How well are products liked? How is the product supposed to be?*
Simple difference	Attribute rating (scales)	*Preference*
Triangle	Time-intensity	Paired preference
Duo-trio	Quantitative descriptive	Ranking preference
Two-out-of-five	analysis	*Acceptance*
A-not-A	Spectrum	Hedonic scaling
Difference from control	Free-choice profiling	Attribute diagnosis (rating)
Similarity		Just-about-right
Paired comparisons		Food action scale (FACT)
n-Alternative forced choice		Purchase intent
Ranking		*Qualitative*
		Focus group
		Focus teams
		One-on-one interviews

Source: Lawless, H.T. and Heymann, H., *Sensory Evaluation of Food: Principles and Practices*, Chapman & Hall, New York, 1999.

1. *Overall difference tests*: Used to check if a significant sensory difference exists between two samples and not in what or how much they are different. Used when no specific attribute(s) can be identified as having been affected. High statistical significant levels do not indicate that the difference is large but only that there is a big chance of a real difference existing. Some of the most used tests in sensory laboratories include the triangle test, duo-trio test, simple difference test, and similarity test, among others.

2. *Directional difference tests*: Will reveal the direction of the difference and which sample has the highest intensity of a particular sensory characteristic. Note that we cannot determine the quantitative measure of those intensities. For instance, we can identify which sample is sweeter, but we do not know if it is a little sweeter or much sweeter. Be aware that a lack of a difference among samples with regard to one attribute does not imply that no overall difference exists. Directional tests can be subdivided according to the number of samples under analysis:

   a. Directional difference between two samples: Paired comparison test or 2-alternative forced choice (2-AFC)

   b. Directional difference among more than two samples: n-alternative forced choice test (n-AFC); ranking test (Friedman analysis); difference-from-control test

17.2.2 DESCRIPTIVE TESTS

The second major class of sensory tests comprises methods that quantify the perceived intensities of the sensory characteristics of a product. These procedures are known as descriptive analyses. All descriptive analysis methods involve the detection and the description of both the qualitative and quantitative sensory aspects of a product. Trained panelists describe the sensory attributes of a sample, often called descriptors. In addition, they rate the intensity of each descriptor to define to what degree it is present in that sample. Meilgaard et al. (1999) explains how two products may contain the same qualitative descriptors, but may differ markedly in the intensity of each, thus resulting in quite different and easily distinctive sensory profiles.

Descriptive analyses are the most sophisticated, comprehensive, and informative sensory evaluation tool. These techniques allow the sensory scientist to obtain complete sensory description of products and help identify underlying ingredient and process variables and other research questions in food product development. The information can be related to consumer acceptance and to instrumental measures by means of statistical techniques such as multivariate regression and correlation (Murray et al., 2001).

Quantitative descriptive analysis or QDA, developed by Stone et al. (1974), is still the most used descriptive method. During several training sessions, the sensory panel is exposed to many possible variations of the product and has the task of generating a set of terms (descriptors) that describe differences among samples. Then, through consensus judges establish definitions for each term and reference standards that should be used to calibrate the intensity scales. However, the actual product evaluation is performed by each judge individually, in booths. Unstructured line scales, anchored with intensity terms, also generated by the panel (e.g., weak–strong) are used, allowing QDA data to be analyzed by both univariate and multivariate statistical techniques: ANOVA of each descriptor, multivariate analysis of variance, principal component analysis (PCA), factor analysis, cluster analysis, and many others. Graphical representation of the data is usually done by radar plots, also known as "cobweb graph" or "star diagram" (Figure 17.1).

Today many product development groups use variations of QDA. The relative simplicity of this technique allows it to be adapted in many different ways; however, any adaptation invalidates the use of the name QDA to describe the procedure.

In free-choice profiling, developed by Williams and Arnold (1984), there is little or no training at all. It allows the panelists to use as many terms as they want to describe the sensory characteristics of a set of samples. The data are analyzed by generalized procrustes analysis (GPA) (Gower, 1975), a multivariate technique that adjusts for the use of different parts of the scale by different panelists and then manipulates the data to combine terms that appear to measure the same characteristic. This technique is very useful in stability studies, where we do not know a priori what sensory characteristics will be developed in the samples, and so we cannot train judges to recognize and measure them. It is also helpful in consumer studies where the objective is to investigate how consumers perceive the products.

Measuring a single descriptor of interest, using scales to express the intensity of a perceived attribute (sweetness, hardness, smoothness, etc.), is also a

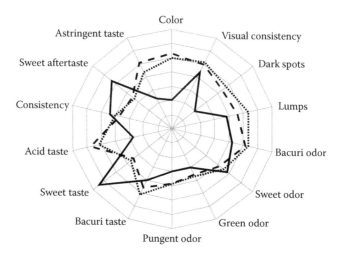

FIGURE 17.1 Star diagram of sensory profile of bacuri nectars by descriptive analysis. —10% pulp without enzyme; —20% pulp with pectinase 1 and cellulase; —20% pulp with pectinase 2 and cellulase.

descriptive technique. The most known scales are category scales, line scales, and magnitude estimation scales, but methods of scaling are under intensive study around the world: cross-modality matching (Stevens and Marks, 1980), labeled magnitude scale (Green et al., 1993, 1996), indirect scaling using Thurstonian model (Baird and Norma, 1978; Frijters et al., 1980; Brockoff and Christensen, 2010), among others.

However, perception of tastes, flavors, and texture in foods is a dynamic phenomenon due to the dynamic nature of processes of breathing, chewing, salivation, swallowing, temperature changes, and tongue movements (Dijksterhuis and Piggott, 2001). By means of conventional scaling methods, panelists can only make a static measurement, which can be a function of an integral of the perception over time, or, more often, a response to the highest intensity perceived. In many cases, this may be the only information required but in other situations it is important to know when the sensation starts, when it reaches the maximum intensity, and how long its duration is. Typical examples are chewing gums and extruded snacks. In the first one, the flavor has to remain as long as possible, and in the other one the flavor needs to "explode" in the mouth and extinguish quickly. Time-intensity (T-I) sensory evaluation provides the opportunity to scale the perceived sensations over time. Today, several commercial sensory analysis software bring T-I scales, but it is possible to develop your own software to collect T-I data. An example is the SCDTI—*Sistema de Coleta de Dados Tempo-Intensidade* (in Portuguese), which means T-I data collecting system, developed at the State University of Campinas, Brazil.

T-I analyses have been widely used in studies of sensory response to sugar-free foods, since they have to mimic all the sucrose sweetness sensations. Bitterness and astringency in several products have also been investigated by time-intensity analysis.

17.2.3 AFFECTIVE TESTS

Nowadays, it is crucial that the industry understands the consumers' needs and their desires and expectations about the products. However, to make the consumer describe is not always easy, as people generally have difficulty in clearly describing what they want. It is very important, then, to have clear goals and use simple methods to match the products' characteristics to the consumer's expectation.

The sensory tests that assess subjective personal responses of customers toward a product are called affective tests. Affective tests measure attitudes such as acceptance and preference. Preference tests determine the customer's preference of a product over the other(s). Acceptance tests quantify the degree of liking or disliking.

Whenever an affective test is conducted, a group of subjects must be selected as a sample of the target population, that is, the population to whom the product is intended. There is no sense in testing the acceptability of a product with people who do not like or do not use that kind of product. In addition, formulating a product for elderly people is different from formulating products for teenagers, for example. In the same respect, developing products for consumers in a highly urbanized area may be different from doing it for people in a rural zone.

Affective tests may be designed as in-house panels (in the lab) or as hall tests (conducted at central locations like fairs, supermarkets, etc.), also called consumer tests. As a general rule, one can use in-house panels for most jobs and then calibrate against consumer tests as soon as possible. Some kind of products require more than that—requiring that the product be tested under its normal conditions of use at the consumer's home (home use tests or home placement tests), where he is an active agent, preparing, serving, proving, and evaluating all aspects of products: package, preparation instructions, sensory attributes, quantity of the portion, and other relevant questions.

Typically, an affective test may involve from 50 to 100 consumers in lab tests, until 300 to 500 in central location and home use tests. The larger size of an affective test arises due to the high variability of individual preferences and thus a need to compensate with increased numbers of people to ensure statistical power and test sensitivity (Lawless and Heymann, 1999).

It is crucial to note that finding no significant preference/acceptance for one sample over another does not mean that there are no perceptible differences among samples. One can equally like both orange and mango juices, but still the orange will be different from mango!

Other usual mistakes in the interpretation of results are, when one sample rates or scores higher preference/acceptance than another, to conclude that "product X was better" than the other. Affective tests do not measure quality. As it was already mentioned, by affective tests we may assess subjective personal responses of a specific group of customers toward a product. Think about a product you like a lot. Let us say it is a fruit candy. Probably it is not the best candy in the world, maybe it is too hard, or too sticky, but it is the one you like best. Then, the right conclusion would be "product X was preferred" or "product X was most accepted" and not "was better" because "better" implies quality judgment.

Common to both preference and acceptance tests is a problem with the univariate analyses of data. There is an implicit assumption that all subjects exhibit the same behavior and that a single value is representative of all subjects. However, individuals' opinions vary or cluster into similar groups, and if they show opposite opinions about the products, mean values will be similar for all products. Although this would suggest that there was no difference in acceptability among the products, this would clearly not be true. One solution, when working with a large number of products (minimum 6), is to use a multivariate statistical analysis called preference mapping (MacFie and Thomson, 1988). The basic data are collected by a larger number of subjects and then individual differences are not averaged, but are built into the model and play an integral role in the fitting algorithm. In the case described earlier, preference mapping would show each individual response and indicate the different opinions of the two groups very clearly.

Greenhoff and MacFie (1994) explain the two distinct ways of dealing with the data, including case studies: internal preference map (MDPREF), which achieves a multidimensional representation of the stimuli, based only on the acceptance/preference data; and external preference map (PREFMAP), which relates product acceptability to a multidimensional representation of stimuli derived from descriptive analysis or instrumental data.

17.2.3.1 Preference Tests

When the objective is to look for the preference of one product or formulation against another, as in product improvement or comparison with a competitive brand, the technique used is the paired preference test, similar to the paired comparison test. Judges receive two coded samples and must choose the sample that is preferred. It is a simple test. Choice is an every-day task for consumers. When the research requires assessing the preference among more than two samples, one can do series of paired preferences, but a ranking preference test is time saving and easier to interpret. This method is also a forced choice, since the participant has to rank several products in either descending or ascending order of preference and is not allowed to have ties in the ranking. The problem with choice tests is that they are not very informative about how well the products were liked by the consumers. If all products are bad, participants will choose the least bad product.

17.2.3.2 Acceptance Tests

When it becomes necessary to determine how well the product is liked by consumers, we will have to collect hedonic or attitude responses from consumers using scales. From relative acceptance scores, one can infer preference; the sample with the higher score is preferred. However, not always a single measure of liking and disliking when a food is tasted in isolation represents the real feeling about it. People's historical preferences may fail to predict their acceptance for certain foods or beverages in an actual tasting (Cardello and Maller, 1982). Context and expectations can affect simple hedonic judgments (Deliza and MacFie, 1996). For this reason, hedonic tests can be complemented by other tests like the ones with appropriateness approach (what judge thinks is a good product) and other behaviorally oriented tests.

17.2.3.2.1 Hedonic Approach

The most popular scale among sensory analysts is the 9-point hedonic scale (Figure 17.2A), developed at the U.S. Army food and Container Institute (Jones et al., 1955; Peryam and Pilgrim, 1957). This method provides a balanced scale for liking with categories labeled with adverbs that represent psychologically equal intervals, with a centered neutral point. Thus, this scale has ruler-like properties, whose equal intervals favor the assignment of numerical values to the responses and the statistical treatment of the data.

However, it has been criticized for a long time for presenting a series of limitations. The main problems are related to the lack of requirements demanded by parametric statistical analyses that are often applied to the data. The 9-point hedonic scale is a category scale, and as such, generates discrete data. It also frequently fails to satisfy the statistical assumptions of normality, homoscedasticity, and additivity required by the Analysis of Variance models (McPherson and Randall, 1985; Pearce et al., 1986; Vie et al., 1991). With a view to overcoming these problems, various authors have proposed alternative methods to generate and statistically analyze sensory data (Miller, 1987; Gay and Mead, 1992; Wilkinson and Yuksel, 1997).

Villanueva et al. (2000) compared the performance of the 9-point hedonic scale and self-adjusting scale (Figure 17.2B), with reference to the discriminative power and statistical assumptions required by the usual ANOVA models, under real consumer test conditions. In a posterior work (Villanueva et al., 2005), the same authors proposed a hybrid hedonic scale (Figure 17.2C). Although the mean values derived from each scale, for each sample, were very similar, the 9-point hedonic scale showed the smallest standard deviation values. However, this scale presented problems with the inequality of sample's variances (lack of homoscedasticity). The self-adjusting scale presented problems with nonnormality of the residuals. For this reason, the significance levels associated with the $F_{samples}$ values, for both scales, are only approximate. The authors

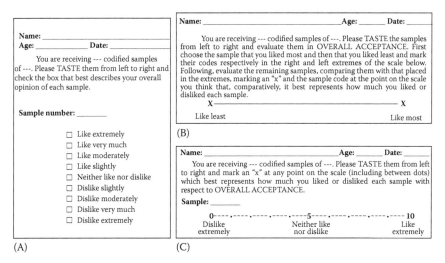

FIGURE 17.2 Example of hedonic scales: (A) 9-point hedonic scale; (B) self-adjusting hedonic scale; (C) hybrid hedonic scale.

suggest that these data should not be analyzed by statistical methods based on assumptions of normality and homoscedasticity but by means of alternative procedures such as generalized linear models (GLM), analysis of categorical data, nonparametric tests, or data transformation for normality. In fact, when proceeding analysis of variance, one should always carry out first a checking of the residuals in order to find out any model inadequacies or violations of the ANOVA model's assumptions. Thus, when the ANOVA model is appropriate, advantages are taken of the greater simplicity and clarity of interpretation that it provides. By its turn, the hybrid hedonic scale data were shown to be adequate in this diagnosis of the ANOVA model. This scale also presented a slightly superior discriminative power than the other two.

Villanueva and Da Silva (2009) studied the performance of these three scales also regarding the generation of the internal preference map (MDPREF). There were strong similarities among the product spaces, but the MDPREF obtained from the hybrid scale showed a slight superiority over the 9-point hedonic and self-adjusting scales with respect to sample segmentation, consumer segmentation, number of significant dimensions, and the proportion of significantly fitted consumers.

However, in those works, authors mentioned that additional experiments must be carried out to confirm the results so far. In our opinion, differences in the scales' performance are not big enough to justify any strong endorsement of the hybrid scale nor any condemnation of the traditional 9-point scale. This is also the opinion of Lawless (2010). In fact, we have found the hybrid scale more difficult to use for both consumers and technicians. If the lab facilities do not have software to collect data, it will be necessary to use a ruler to measure each respondent's evaluation sheet.

Hedonic scaling can also be applied to children and illiterate people. It has used face scales, cartoons characters, or realistic pictures of children or adult faces, representing each one of the verbal points. In many cases, these scales do not perform well. Kroll (1990) showed that verbal descriptors of "good" and "bad," the so-called P&K scale, worked better for children. Below 4–5 years old, acceptance must be inferred from behaviors, such as oral interviews and ad libidum situations or by the amount of ingestion from a standardized sample.

17.2.3.2.2   Appropriateness Approach

As part of a consumer test, researchers often desire to determine the reasons for any preference or rejection by asking additional questions about the sensory attributes. Meilgaard et al. (1999) used category and line scales to assess the intensity of specific attributes and called it attribute diagnosis. For example, the question would be "How intense is the sweetness of this juice?" And the scale would vary from "very week" to "very strong." Differently, just-right scales (Vickers, 1988), also known as just-about-right, assess the intensity of an attribute relative to some mental criterion of the subjects for the product that is under analysis. Example: for the same question given earlier about the sweetness of the juice, the scale would vary from "not at all sweet enough" to "much too sweet," with a middle point corresponding to the ideal sweetness ("just right").

17.2.3.2.3   Behavioral Approach

The food action scale (FACT) developed by Schutz (1965) is an example of a behavioral approach to assessing food acceptability. It is based on consumers' attitudes in relation to the frequency at which they would be willing to consume the product in a given period. Examples of category labels are as follows: "I would eat it every opportunity I had" and "I would eat it only if I were forced to." FACT scale is recommended for testing products with which consumers are not familiar.

Purchase intent scales are very similar to FACT scales, based on consumers' attitudes in relation to their willingness to buy the product if it was for sale. Examples of category labels include "I certainly would not buy it"; "I probably would buy it"; and "I certainly would buy it."

17.2.3.3 Qualitative Affective Tests

Qualitative methods are used to study consumer habits and attitudes that may be useful in predicting the behavior of consumers and to develop the terminology used by them to describe the sensory attributes of the concept or prototype of a product. The main interest of these tests would generate the most varied and possible ideas and reactions on a given product. It is quite useful in product development. In small groups or individual interviews, consumers verbalize their opinions and expectations about the product. The most used techniques so far are focus groups, focus teams, and one-on-one interviews (McQuarrie and McIntyre, 1986; McNeill et al., 2000; Bruseberg and McDonagh-Philp, 2002). In the focus group technique, a group of participants, usually 6–8, sit together for a more or less open-ended discussion about a product or a specific topic. The discussion moderator lets participants introduce themselves and feel comfortable and makes sure that the topics of significance are brought up. To help participants verbalize their needs, interaction among group members is encouraged. The products may or may not be served. The report summarizes what was said and perhaps draws inferences from what was said and left unsaid in the discussion. Generally, the sessions are videotaped.

17.3 PLANNING SENSORY TESTS IN A PRODUCT DEVELOPMENT PROGRAM

Sensory results are only useful when interpreted in the context of hypotheses, background knowledge, and implications for decisions and actions to be taken. Defining the needs of the project is the most important requirement for conducting the right test. Thus, sensory specialists should be full partners with their clients, taking an active role in developing the research program, collaborating on the choosing of the attributes to be analyzed, and setting the experimental designs, which ultimately will be used to answer the questions posed. Only through a process of total involvement they can be in a position to select the most appropriate tests necessary at every point of a research project. They will also be able to contribute interpretations and suggest actions, since they best understand the limitations of tests and what their risks and liabilities may be.

Lawless and Heymann (1999) and Meilgaard et al. (1999) present some guidelines and general steps for choice of techniques (Figure 17.3), reminding us that those rules are generalizations and sometimes the goals, requirements, and resources available in a particular situation will dictate deviations of those principles. Figure 17.3A shows a sensory evaluation flowchart, with an overview of the tasks and decisions in setting up and conducting a sensory test.

Sensory evaluation flowchart

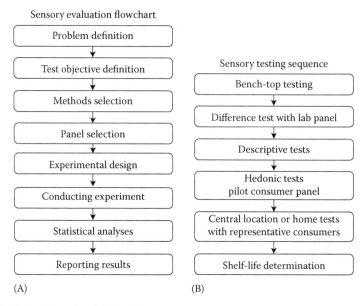

FIGURE 17.3 General guidelines in sensory evaluation: (A) Tasks and decisions in setting up and executing a sensory test; (B) The flow of a sensory evaluation testing program. (From Lawless, H.T. and Heymann, H., *Sensory Evaluation of Food: Principles and Practices*, Chapman & Hall, New York, 1999; Meilgaard, M. et al., *Sensory Evaluation Techniques*, CRC Press, Boca Raton, FL, 1999.)

1. *Problem definition*: First of all, define the problem. We must define what we want to measure. Start analyzing the nature of the project and what is expected from the samples. For example, is it a new product development; product improvement; ingredients, equipment, or process change; or product matching? Are differences desirable or is the accent on proving that no difference exists between the sample and another or former product? Does the product vary only in one or several attributes? Is there any cue like color, consistency, or others that may introduce sensory biases?

2. *Definition of test objective*: Once the objective of the project is clearly stated, the sensory analyst and the project leader can determine the test objective: overall difference or attribute difference? determine a complete sensory profile? and relative preference or acceptability, etc.?

   However, many projects need a sequence of tests for achieving the goal rather than a single test. It is a sequential decision process as in any other problem-solving activity. Figure 17.3B shows the flow of a sensory evaluating testing program during product development/product improvement projects. First, define exactly what sensory characteristics need improvement or need to be evaluated, then determine that the experimental product is indeed different, and finally confirm that the experimental product is liked better than the control. If working with ingredients, equipment, or process changes confirm that no difference exists, and if a difference does exist, determine how consumers view the difference.

a. *Bench-top testing*: This test is mandatory to get familiar with product attributes; choose the ones that will be evaluated and check if sensory variations are obvious or a difference test is needed. All sensory properties should be examined: appearance, aroma, basic tastes, flavor, texture, mouth feelings, aftertastes, and after mouth feelings. Products vary in which attributes? One attribute may influence the analysis of other attributes?

b. *Difference tests*: Although our ultimate interest may lie in whether consumers will like or dislike a new product variation, we must conduct a simple difference test first to see whether any change is perceivable at all. The logic in this sequence is the following: If a screened and experienced discrimination panel cannot tell the difference under carefully controlled conditions in the sensory lab, then a more heterogeneous group of consumers is unlikely to see a difference in their less controlled and more variable world. If no difference is perceived, there can logically be no systematic preference. So a more time-consuming and costly consumer test can sometimes be avoided by conducting a simpler but more sensitive discrimination test first.

c. *Descriptive tests*: If a difference exists, it is usually necessary to know the sensory profile of prototypes or monitoring specific attributes in order to evaluate the effect of process or ingredient variations. In this case, descriptive tests should be applied. This sensory profile will also be useful in determining the specific reasons as to why a product is preferred over the others.

d. *Affective tests*: *Pilot consumer panels.* Consumer sensory evaluation is usually performed toward the end of the product development or reformulation cycle. At this time, the alternative product prototypes have usually been narrowed down to a manageable subset through the use of analytical sensory tests. With in-house panels explore degrees of liking/disliking and identify potential problems for rework. In general, a product tested under true conditions will not give the same result as a product tested by pilot panels. However, testing in pilot panels will help to select samples that should be evaluated in real conditions tests, which are time consuming and involve higher cost and logistics.

e. *Affective tests*: *Central location panels.* Central location evaluation is used to answer the question: Is what we find in pilot panels tests the same as what would be found by real-world consumers? The test also explores degrees of liking/disliking but requires time and expenses and is usually performed by companies that will actually introduce the product in the market. For scientific purposes, answers from pilot consumer panels are found to be sufficient.

f. *Determination of shelf life*: The sensory testing sequence usually ends with a stability study, performed to know how product quality changes during storage and allows establishing shelf life. Shelf life can be defined as the period between manufacture and retail purchase of a food product during which the product is of satisfactory quality (Dethmers, 1979); in other words, the length of time for the product to become unacceptable for sale. A practical use for stability studies is open dating of foods. Open dates are placed on the labels of food products to help consumers in the purchase decision (Labuza, 1982).

Sensory analysis is as important as chemical, physical, and microbiological analyses because it indicates whether the storage products still conform to standards for appearance, aroma, flavor, taste, texture, and functionality. Besides, sensory analysis indicates whether the food is still acceptable to the consumers.

3. *Methods selection*
   a. *Discriminative testing*: The issue here is the kind of difference you want to investigate and the number of samples, as stated earlier. If the objective is to check the overall difference between two samples, triangle test is the most sensitive test, because the probability the panelist has to check the odd sample by chance is one-third while in paired comparison and duo-trio tests it is one-half.

   However, if the samples are complex in any way use duo-trio test, since the task of comparing samples to a standard is easier to panelists. Lawless and Heymann (1999) say that duo-trio test is also more sensitive than the triangle when subjects are familiar with the reference material. For example, if the reference is a product with a long company history and a great deal of ongoing evaluation, deviations from this familiar item may be readily noticed. If the problem relies on a single attribute, forced-choice tests will be more sensitive, since panelists will focus in the intensity of that specific attribute, even if there are other sources of variation.

   b. *Descriptive testing*: When differences in the target attribute are obvious among samples, or you need to measure the difference in more than one attribute, use direct scaling. However, you must follow the descriptive test methodology, since humans are not good absolute measuring instruments and need reference samples of what is a low intensity and what is a high intensity of those attributes in that specific product. Panelists will calibrate themselves according to that frame of references and will be able to become a good relative measuring instrument.

   Choose QDA or one of its free adaptations when a complete sensory profile of the samples is needed, in order to specify the nature of all sensory changes or differences in a set of samples. Sensory profiles can also be correlated to affective data to investigate the reasons for people's likings and disliking about a set of products.

   c. *Affective testing*: Affective tests bring the consumer into the product development. Choose acceptance test to determine how well the product is liked by consumers. For example, how nectar blends from various fruits will be acceptable to consumers who are used to drinking traditional one-fruit nectar. Choose preference tests to determine preference of one product against another. For example, to decide which sweetener is preferable when formulating a sugar-free fruit jam.

d. *Shelf-life tests*: The selection of a particular sensory evaluation procedure for evaluating products in storage is determined by the test purpose. Acceptability assessment by untrained panels is essential for open dating. Discriminative tests with trained panels are useful when one characteristic is more important than others or when some degradation problems are already expected. Descriptive analysis can be used with new products, when there is no information on the behavior of the product under storage.

Depending on the food characteristics, several failure criteria can be used to terminate a shelf-life study (Labuza and Schmidl, 1988). Besides microbiological growth and physical changes, there are many sensory criteria to determine the end of the test (Gacula, 1975):

i.   An increase or decrease in x number of units in a mean panel score. For example, in shelf-life dating of oils, an increase in 2 units of oxidation flavor by a trained panel may determine the end of the test.

ii. Failure time, when a sample reaches an average panel score. This criterion is useful when overall acceptance is used in the stability test. For example, when a 4.5 score is obtained in a 9-point scale for a stored juice.

iii. Just noticeable difference, when the difference between the quality of samples in test and control samples can be detected by trained panels. In this case, a control that can be maintained with only negligible change over time is essential.

iv. Results of a profile descriptive analysis. Comparison between profiles before and after storage may be used to indicate changes in important characteristics.

A question in stability experiments is whether we really need to know the true end of shelf life of a product. Usually, the answer is negative. Most of the time, the producers need the assurance that the product will be acceptable if it is held in the distribution system for a given period of time at certain temperature and humidity. For those situations, stability tests can be planned for a specific time, shorter than that necessary for establishing shelf life. Later, as a complementary data, shelf life can be determined by regression techniques or any other statistical approach.

Another question is whether storing of a food product leads only to deterioration. Actually, some food products require controlled aging to develop characteristic aroma, flavor, and texture. The most known examples are the aging process for wine and cheese. In those products, sensory analysis can also be used for monitoring the product's changes.

4. *Panel selection*: Each kind of test will require different types of panelists.
   a. *Discriminative tests*: It is mandatory that subjects are screened for normal sensory acuity, especially regarding vision, smell, and taste, and for their ability to detect differences among similar products with ingredient or processing variables. Discrimination panels can receive a little training, but they are usually only oriented in the test method.
   b. *Descriptive tests*: Panelists are screened for normal sensory acuity, discriminative ability, and motivation. Descriptive panels are trained, except in free-choice profiling. They are asked to put personal hedonic reactions aside, as their job is only to specify what attributes are present in the product and at what levels of intensity, extent, amount, or duration. It has been a central principle in sensory analysis that you should not rely on consumers for accurate descriptive information. Consumers not only act in a nonanalytical frame of mind but also often have fuzzy concepts about specific attributes, confusing, sour, and bitter, for example. However, in recent years, some researchers are breaking this paradigm, as we are going to see ahead in this chapter.
   c. *Affective tests*: Participants must be chosen carefully to ensure that the results generalize to the population of interest. They can be recruited among users or at least potential users of the new product, who are frequent users of similar products. They possess reasonable expectations and a frame of reference within which they can form an opinion relative to other similar products they have tried. Never use a panelist who has been trained to evaluate that particular product as a consumer, even though he or she uses to consume that product. During training, these panelists have been asked to assume an analytical frame of mind and will not be able anymore to look at that product in an integrative form. Consumers' reactions, on the other hand, are often immediate and based on the integrated pattern, although their attention is sometimes captured by a specific aspect.
5. *Experimental design*
6. *Conducting the experiment*
7. *Statistical analyses*
8. *Reporting results*

These last four steps involve many details and particularities that are not covered in this book, though they must be discussed with the sensory specialists.

17.4   DEVELOPING A TROPICAL FRUIT NECTAR: A CASE STUDY

Next, we see how sensory tests were conducted in a new-product developing project, helping to make important decisions.

Bacuri, a Brazilian tropical fruit, has a pulp with a distinct, strong, acid-sweet, and agreeable flavor (Clement and Venturieri, 1990). However, it shows very high consistency, which hampers the industrial processing steps such as filtration and concentration. In addition, it has not been possible yet to elaborate bacuri nectars within

Brazilian legislation requirements: 30% fruit pulp or at least 20% pulp for fruits with high consistency (Brasil, 2003). Nazaré (2000) and Silva et al. (2007) observed good acceptability only for the nectar formulated with 10%–12% pulp. One way to reduce the pulp consistency is through the technology of enzymatic maceration.

At first, pulp was macerated with a pectinase (P1), and nectars with 20% and 30% macerated pulp were formulated and compared with the control nectar (10% pulp without maceration) in a bench-top testing. The formulation with 30% macerated pulp proved to be too consistent and was discarded. Nectar with 20% macerated pulp was still more consistent than the control and also too acidic.

In the next step, we tried to correct the high acidity adding sugar. We used an in-house panel to make hedonic evaluations, since in this case it mattered to know how sweet people liked the nectar to be. We found that it was necessary to add too much sugar to bring the nectar to a good level of acceptability.

With those results, we turned to test other enzyme preparations, searching for one that could reduce consistency and did not yield so much acid. We tested two pectinases combined (P1 + P2) in two different proportions, and each one of these combined to a cellulase (P1 + C and P2 + C). A difference-from-control test was performed with nectars elaborated with 20% macerated pulps, where the control sample was nectar from P1 macerated pulp, in order to see if those formulations could reduce the consistency and acidity observed for this nectar on the bench-top testing. Only formulations containing cellulase showed reduced consistency and acidity.

A central composite design was used to determine the best combination and concentrations of pectinases and cellulase. Dependent variables were total acidity, consistency, and sugar content, all determined by means of chemical and instrumental analyses. The best performances were observed for P1 + C (1:2) and P2 + C (1:2).

In order to investigate whether the maceration process could modify sensory properties of bacuri nectars, a descriptive profile was determined for products formulated by the optimized enzyme preparations and compared with the control sample (10% nonmacerated pulp). Figure 17.1 presents the star diagram. Both enzyme preparations produced nectars with similar sensory profiles, but very different from control sample. Macerated nectars reached the same consistency of 10% nonmacerated sample but were still more acid. The characteristic aroma of bacuri was enhanced, and panelists perceived stronger green and pungent aromas. However, they were darker in color and showed more dark spots and lumps. Between macerated nectars, enzyme P1 caused less lump formation, but the nectar was more acid and astringent than nectar macerated by P2.

The final step was to evaluate the new products' acceptance by consumers and compare it with the well-established 10% pulp nectar. Hedonic tests were performed in two Brazilian regions: the North region, where bacuri is well known and appreciated, and the Northeastern region, where this fruit is almost unknown. Global acceptance and attribute acceptance were evaluated. Consumers from both regions accepted the macerated samples as much as the control nectar, but in the North region, the scores were higher, as expected. Sensory results indicated that this bacuri nectar has a great potential in the market, even in regions where consumers are not used to this fruit, and the enzymatic maceration studies should continue until an optimized product is made.

17.5 TRENDS IN CONSUMER RESEARCH

In order to put the best possible product on the market or even achieve the "ideal product," it is essential to understand consumer product perception and preferences and relate hedonic responses to sensory product specifications (Worch et al., 2010). New trends in sensory analysis have been related to two great challenges for consumer researchers: (1) get the consumer to describe food properties and (2) understand consumers' reactions (Kleef et al., 2005; Hough, 2010).

17.5.1 Get the Consumer to Describe

In sensory analysis, the classical approach has been to use a trained panel for sensory description of products and consumers only for hedonic evaluations. However, descriptive methods can be expensive and time consuming (both in terms of panelist training and testing time). In addition, it is hard to maintain a trained panel, especially in the industry, where there is a high turnover of employees. In recent years, researchers have been breaking the paradigm and advocating the use of consumers to generate sensory profiling to lead product development. Husson and Pàges (2003) showed that consumers meet the requirements of discrimination, consensus, and reproducibility, Worch et al. (2009) found no significant differences between products profiled by trained or consumer panels. Already known methods like Kelly repertory grid (Kelly, 1955) and more recent methods have been presented like flash profile (FP), check-all-that-apply (CATA), projective mapping techniques, free listing, and ideal profile, among others.

17.5.1.1 Flash Profile

Flash profile (Dairou and Sieffermann, 2002) is adapted from free-choice profiling, where untrained subjects select their own terms to describe and evaluate a set of products simultaneously. The difference is that subjects rank the products on an ordinal scale for each term they individually created, instead of rating. They are asked to focus on the descriptive terms, not on the hedonic terms. Flash profiling (FP) can also be used at the initial stage of a project to create the sensory attributes for the conventional descriptive analyses (Delarue and Sieffermann, 2004) and shows practical feasibility in the evaluation of a large food product set. The individual sensory maps are treated with general procrustes analysis (GPA) to create a consensus configuration. Cluster analysis can also be performed after the GPA on the descriptive terms to assist in the interpretation.

17.5.1.2 Check-All-That-Apply (CATA)

In the CATA method (Lancaster and Foley, 2007), consumers are asked to check all perceived attributes in a specific product, from a list of prechosen terms (Figure 17.4). The actual generation of terms can be performed in many ways: the consumers can choose words to describe the product during the test (like in free-choice profiling), terms can be generated by consumers not testing the product (i.e., a focus group), or terms can be given by a trained panel (descriptive analysis). CATA method requires minimal instruction; it is relatively easy to perform

Check all attributes that describe this sample:

- ☐ Buttery
- ☐ Sweet
- ☐ Milk/dairy flavor
- ☐ Custard/eggy flavor
- ☐ Corn syrup
- ☐ Artificial vanilla
- ☐ Natural vanilla
- ☐ Creamy flavor
- ☐ Soft
- ☐ Hard
- ☐ Gummy
- ☐ Icy
- ☐ Creamy/smooth

FIGURE 17.4   Example of a CATA questionnaire.

and is completed quickly. It is different from scaling since no intensities are given to the attributes. The data are formed by the counts or percentages of consumers that checked each term for each sample. Frequently listed descriptors would be more relevant than those less frequently listed. Data are analyzed by multivariate statistical tools such as correspondence analysis or multiple factor analysis, generating a sensory space similar to PCA. CATA data can also be used for the creation of preference maps correlating hedonic judgment with sensory attributes (Dooley et al., 2010).

17.5.1.3   Free Listing

Free listing (Ares et al., 2010; Hough and Ferraris, 2010) is a variation of CATA and is also called open-ended question. Instead of using a list of terms, consumers are asked to use their own terms to describe the samples. Similar words are grouped, and the matrix of frequencies is also analyzed by correspondence or multiple factor analysis. Cluster analysis can reveal associated descriptors, because they are similar stimuli, for example crispy and crunchy, or because they belong to similar categories in the mind of the consumer, like apple and fruity.

17.5.1.4   Projective Mapping

Projective mapping (Risvik et al., 1994) studies do not involve numerical judgments. They are holistic approaches of characterizing similarities and differences in sensory attributes of products or assessing their preference or liking. Among the techniques that have been gaining popularity are napping (Pagès, 2005), partial napping (Pfeiffer and Gilbert, 2008), free sorting (Abdi et al., 2007), and a combination of them called sorted napping (Lê et al., 2009). Assessors position the products on a two-dimensional surface (e.g., large sheet of paper) according to overall sensory similarities and differences being free to choose the various criteria used to separate the products (Figure 17.5). They are often asked to enhance the map with descriptive terms for each product or drivers for liking and disliking. Multiple factor analysis provides a quick profile showing relationship between products and descriptors, similar to PCA results from conventional profiling.

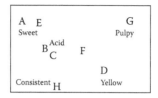

FIGURE 17.5   Example of a projective mapping ballot.

17.5.1.5   Ideal Profiles

While the just-about-right (JAR) scale only asks the deviations from ideal for each attribute and product combination, in the ideal profile method (Punter, 2008; Worch et al., 2009), both perceived and ideal intensities are asked directly (the JAR question "is it just right, too much, or too little" is replaced by the question "how strong is it and what would be the ideal strength"). Worch et al. (2010) compared two different methodologies to analyze JAR and ideal profile data and suggested PLS on dummy variables (Xiong and Meullenet, 2006) for the analysis of JAR data, and the Fishbone method for the analysis of Ideal Profile data.

17.5.2   UNDERSTANDING CONSUMERS

The psychophysical approach of sensory evaluation (sensory scales) is based on the idea that people are rational and can give explicit reasons for their behavior. However, according to Koster (2009), paradigms like uniformity, consistency, objectivity, and conscious choice are fallacies in human behavior. In other words, people are different from one another and you cannot average their behavior; people change their own behavior, evaluations are subjective, and choices are not always rational and conscious. The author considers that the use of scales is efficient but they do not contain all the necessary information.

Sensory science has appropriate psychophysics and marketing techniques to determine consumers' drivers of liking and preference. They allow us not only to hear what people say they like or they do but to see their real behavior, allowing us to obtain a snapshot of people's lives, their experiences, and their relationships. It captures the subject's inner thoughts, feelings and emotions, their values, and rules that guide them, because it is carried out in his or her own environment (Deliza, 2009). Some studies are also reaching, invading, and exploring the domestic environment. To handle this kind of data, some methods have been used like category appraisal, conjoint analysis, experimental design, focus group, free listing, Bayesian networks, laddering empathic design, and information acceleration. Some of these are briefly discussed in the following.

Free listing can also be used to explore behaviors and habits, foods considered appropriate for certain uses or occasions, and feelings related to food consumption. For example, a subject can be asked to list all the things he or she feels while eating a bar of their favorite chocolate. Frequently listed feelings can be useful from a concept development perspective. Associated feelings like, for example, "fattening" associated to "guilty" can help discover why chocolate has a negative image in the minds of some consumers.

Bayesian networks (Craignou and Jouffe, 2008; Craignou and Bezault, 2009), also referred to as belief networks, Bayes nets or causal probabilistic networks are a modern data analysis tool that can handle variability and uncertainty using probability distributions. These techniques can be used for explanation, exploration of information, and prediction of system behaviors and for decision making. Despite their popularity in various fields such as finance, medical diagnosis, robotics, and genetics, their application to food-related problems have only recently emerged. Most current Bayesian network algorithms require discrete variables. Modeling with Bayes nets enables the use of expert knowledge as well as the combination of data from different studies.

Empathic design (Polanyi, 1966; Ulwick, 2002). A multifunctional team is created to observe the actual behavior and environment of consumers. A visual record is made of consumers interacting with their environment. Photographs, videotape, sketches, and notes are tools that make a record of behavior. Data can as well be gathered through responses to questions like "why are you doing that?" Team members have a brainstorming session to transform observations into graphic, visual representations of possible solutions. A nonfunctional, two- or three-dimensional model of a product concept provides a vehicle for further testing among potential consumers.

Information acceleration (Urban et al., 1997). The researcher constructs a virtual buying environment that simulates the information that is available to consumers at the time they make a purchase decision. Respondents are "accelerated" into the future by providing them alternative future environments that are favorable, neutral, or unfavorable toward the new product. In this virtual buying environment, they are allowed to search for information or shop. Measures are taken of respondents' likelihood of purchase, perceptions, and preferences. Based on these measures, a model is developed to forecast sales and simulate strategy alternatives.

ACKNOWLEDGMENTS

The authors are very grateful to Victor Costa Castro Alves and Idila Araujo for figure designs, Maria Aparecida Azevedo Pereira da Silva, for valuable suggestions and ideas for the text, Gustavo Saavedra Pinto, for making available bacuri nectar results.

REFERENCES

Abdi, H., Valentin, D., Chollet, S., and Chrea, C. 2007. Analyzing assessors and products in sorting tasks: DISTATIS, theory and applications. *Food Quality and Preference* 18: 627–640.

Amerine, M.A., Pangborn, R.M., and Roessler, E.B. 1965. *Principles of Sensory Evaluation of Food.* New York: Academic Press.

Ares, G., Giménez, A., Barreiro, C., and Gámbaro, A. 2010. Use of an open-ended question to identify drivers of liking of milk desserts. Comparison with preference mapping techniques. *Food Quality and Preference* 21: 286–294.

Baird, J.C. and Norma, E. 1978. *Fundamentals of Scaling and Psychophysics.* New York: Wiley.

Brasil. 2003. Ministério da Agricultura e do Abastecimento. Secretaria de Defesa Agropecuária. Instrução Normativa n° 12, de 4 de setembro de 2003. *Diário Oficial da República Federativa do Brasil*, Brasília, September 9, 2003.

Brockoff, P.B. and Christensen, R.H.B. 2010. Thurstonian models for sensory discrimination tests as generalized linear models. *Food Quality and Preference* 21: 330–338.

Bruseberg, A. and McDonagh-Philp, D. 2002. Focus groups to support the industrial/product designer: A review based on current literature and designers' feedback. *Applied Ergonomics* 33: 27–38.

Cardello, A.V. and Maller, O. 1982. Relationships between food preference. *Journal of Food Science* 47: 1552.

Clement, C.R. and Venturieri, G.A. 1990. Bacuri e cupuassu. In: *Fruits of Tropical and Subtropical Origin. Composition, Properties and Uses*, eds. S. Nagy, P.E. Shaw, and W.G. Wardowiski, pp. 178–192. Lake Alfred, FL: Florida Department of Citrus.

Craignou, F. and Bezault, M.L. 2009. Identifying drivers of liking for fine fragrances with Bayesian networks. In: *8th Pangborn Sensory Science Symposium*, Florence, Italy. Delegate Manual PL4.3.

Craignou, F. and Jouffe, L. 2008. Studying consumer drivers with Bayesian Networks. In: *9th Sensometrics Meeting*, May 28–30, St Catharine, Ontario, Canada. Oral Presentation.

Dairou, V. and Sieffermann, J.M. 2002. A comparison of 14 jams characterized by conventional profile and a quick original method, the flash profile. *Journal of Food Science* 67: 826–834.

Delarue, J. and Sieffermann, J.M. 2004. Sensory mapping using flash profile. Comparison with a conventional descriptive method for the evaluation of the flavour of fruit dairy products. *Food Quality and Preference* 15: 383–392.

Deliza, R. 2009. Investigating the voice of the consumer in relation to liking. In: *8th Pangborn Sensory Science Symposium*, July, Florence, Italy. Oral presentation.

Deliza, R. and MacFie, H.J.H. 1996. The generation of sensory expectations by external cues and its effects on sensory perceptions and hedonic ratings: A review. *Journal of Sensory Studies* 11: 103–128.

Dethmers, A.E. 1979. Utilizing sensory evaluation to determine product shelf life. *Food Technology* 33: 40–42.

Dijksterhuis, G.B. and Piggott, J.R. 2001. Dynamic methods of sensory analysis. *Trends in Food Science and Technology* 11: 284–290.

Dooley, L., Lee, Y-S., and Meullenet, J.F. 2010. The application of check-all-that-apply (CATA) consumer profiling to preference mapping of vanilla ice cream and its comparison to classical external preference mapping. *Food Quality and Preference* 21: 394–401.

Frijters, J.E.R., Kooistra, A., and Vereijken, P.F.G. 1980. Tables of d' for the triangular methods and the 3-AFC signal detection procedure. *Perception and Psychophysics* 27: 176–178.

Gacula, M.C. 1975. The design of experiments for shelf life study. *Journal of Food Science* 40: 399–403.

Gay, C. and Mead, R. 1992. A statistical appraisal of the problem of sensory measurement. *Journal of Sensory Studies* 7: 205–208.

Gower, J.C. 1975. Generalized Procrustes analysis. *Psychometrika* 40: 35–50.

Green, B.G., Dalton, P., Cowart, B., Shaffer, G.S., Ranking, K., and Higgins, J. 1996. Evaluating the "labeled magnitude scale" for measuring sensations of taste and smell. *Chemical Senses* 21: 323–334.

Green, B.G., Shaffer, G.S., and Gilmore, M.M. 1993. Derivation and evaluation of a semantic scale of oral sensation magnitude with apparent ratio properties. *Chemical Senses* 18: 683–702.

Greenhoff, K. and MacFie, H.J.H. 1994. Preference mapping in practice. In: *Measurement of Food Preferences*, eds. H.J.H. MacFie and D.M.H. Thomson. Glasgow, U.K.: Blackie Academic & Professional.

Hough, G. 2010. Latest trends—Keynote presentation in recent Pangborn and Sensometrics Symposia. In: *6 Ibero-American Sensory Analysis Symposium*, August 19–21, São Paulo, Brazil. Oral presentation.

Hough, G. and Ferraris, D. 2010. Free listing: A method to gain initial insight of a food category. *Food Quality and Preference* 21: 295–301.

Husson, F. and Pagés, J. 2003. Comparison of sensory profiles done by trained and untrained juries: Methodology and results. *Journal of Sensory Studies* 18: 453–464.

Jones, L.V., Peryam, D.R., and Thurstone, L.L. 1955. Development of a scale for measuring soldiers' food preferences. *Food Research* 20: 512–520.

Kelly, G.A. 1955. *The Psychology of Personal Constructs*. New York: Norton.

Kleef, E., Trijp, H.C.M., and Luning, P. 2005. Consumer research in the early stages of new product development: A critical review of methods and techniques. *Food Quality and Preference* 16: 181–201.

Koster, E.P. 2009. Diversity in the determinants of food choice: A psychological perspective. *Food Quality and Preference* 20: 70–82.

Kroll, B.J. 1990. Evaluating rating scales for sensory testing with children. *Food Technology* 44: 78–86.

Labuza, T.P. 1982. *Shelf Life Dating of Foods*. Westport, CT: Food and Nutrition Press.

Labuza, T.P. and Schmidl, M.K. 1988. Use of sensory data in the shelf life testing of foods: Principles and graphical methods for evaluation. *Cereal Foods World* 33: 193–206.

Lancaster, B. and Foley, M. 2007. Determining statistical significance for choose-all-that-apply question responses. *7th Pangborn Science Symposium*, August 12–16, Minneapolis, MN.

Lawless, H.T. 2010. Commentary on "Comparative performance of the nine-point hedonic, hybrid and self-adjusting scales in the generation of internal preference maps." *Food Quality and Preference* 21: 165–166.

Lawless, H.T. and Heymann, H. 1999. *Sensory Evaluation of Food: Principles and Practices*. New York: Chapman & Hall.

Lê, S., Cardoret, M., and Pagès, J. 2009. Combining the best of two worlds, the "sorted napping." *8th Pangborn Sensory Science Symposium*, July 26–30, Florence, Italy. Delegate Manual W5.3.

MacFie, H.J.H. and Thomson, D.M.H. 1988. Preference mapping and multidimensional scaling. In: *Sensory Analysis of Foods*, ed. J.R. Piggot, pp. 381–410. London, U.K.: Elsevier.

McNeill, K.L., Sanders, T.H., and Civille, G.V. 2000. Using focus groups to develop a quantitative consumer questionnaire for peanut butter. *Journal of Sensory Studies* 15: 163–178.

McPherson, R.S. and Randall, E. 1985. Line length measurements as a tool for food preference research. *Ecology of Food and Nutrition* 17: 149–156.

McQuarrie, E.F. and McIntyre, S.H. 1986. Focus groups and the development of new products by technologically driven companies: Some guidelines. *Journal of Product Innovation Management* 1: 40–47.

Meilgaard, M., Civille, G.V., and Carr, B.T. 1999. *Sensory Evaluation Techniques*. Boca Raton, FL: CRC Press.

Miller, A.J. 1987. Adjusting taste scores for variations in use of scales. *Journal of Sensory Studies* 2: 231–242.

Moskowitz, H.R. 1983. *Product Testing and Sensory Evaluation of Foods*. Westport, CT: Food and Nutrition Press.

Murray, J.M., Delahunty, C.M., and Baxter, I.A. 2001. Descriptive sensory analysis: Past, present and future. *Food Research International* 34: 461–471.

Nakayama, M. and Wessman, C. 1979. Application of sensory evaluation to the routine maintenance of product quality. *Food Technology* 33: 38–44.

Nazaré, R.F.R. 2000. Produtos agroindustriais de bacuri, cupuaçu, graviola e açai, desenvolvidos pela Embrapa Amazônia Oriental. *Belém: Embrapa Amazônia Oriental.* Manaus, Brazil: Embrapa Amazônia Oriental.

Pagès, J. 2005. Collection and analysis of perceived product interdistances using multiple factor analysis: Application to the study of 10 white wines from the Loire Valley. *Food Quality and Preference* 16: 642–649.

Pearce, J.H., Korth, B., and Warren, C.B. 1986. Evaluation of three scaling methods for hedonics. *Journal of Sensory Studies* 1: 27–46.

Peryam, D.R. and Pilgrim, F.J. 1957. Hedonic scale method of measuring food preferences. *Food Technology* September 1957: 9–14.

Pfeiffer, J.C. and Gilbert, C.C. 2008. Napping by modality: A happy medium between analytic and holistic approaches. In: *9th Sensometrics Meeting*, July 20–23, St Catharine, Ontario, Canada. Oral presentation.

Polanyi, M. 1966. *The Tacit Dimension.* New York: Doubleday.

Punter, P.H. 2008. Bridging the gap between R&D and marketing: The ideal profile method. In: *Third European Conference on Sensory and Consumer Research*, Hamburg, Germany.

Risvik, E., McEwan, J.A., Colwill, J.S., Rogers, R., and Lyon, D.H. 1994. Projective mapping: A tool for sensory analysis and consumer research. *Food Quality and Preference* 5: 263–269.

Schiffman, H.R. 1996. *Sensation and Perception: An Integrated Approach.* New York: John Wiley & Sons.

Schutz, H.G. 1965. A food action rating scale for measuring food acceptance. *Journal of Food Science* 30: 365–374.

Silva, V.K.L., Figueiredo, R.W., Maia, G.A., Sousa, P.H.M., Figueiredo, E.A.T., and Pinheiro, E.S. 2007. Otimização da formulação de néctar de bacuri (*Platonia insignis* Mart.). In: 7º Simpósio Latino Americano de Ciências de Alimentos, November 4–7, Campinas, SP, Brazil. Book of Abstracts (CD ROM).

Stevens, J.C. and Marks, L.M. 1980. Cross-modality matching functions generated by magnitude estimation. *Perceptions and Psychophysics* 27: 379–389.

Stone, H. and Sidel, J.L. 1993. *Sensory Evaluation Practices.* San Diego, CA: Academic Press.

Stone, H.H., Sidel, J., Oliver, S., Woolsey, A., and Singleton, R.C. 1974. Sensory evaluation by quantitative descriptive analysis. *Food Technology* 28: 24–31.

Ulwick, A.W. 2002. Turn customer input into innovation. *Harvard Business Review* January 1: 92–97.

Urban, G.L., Hauser, J.R., Qualls, W.J., Weinberg, B.D., Bohlmann, J.D., and Chicos, R.A. 1997. Information acceleration: Validation and lessons from the field. *Journal of Marketing Research* 34: 143–153.

Vickers, A. 1988. Sensory specific satiety in lemonade using a just-right scale for sweetness. *Journal of Sensory Studies* 3: 1–8.

Vie, A., Gulli, O., and O'Mhonny, M. 1991. Alternative hedonic measures. *Journal of Food Science* 56: 1–5.

Villanueva, N.D.M. and Da Silva, M.A.A.P. 2009. Comparative performance of the nine-point hedonic, hybrid and self-adjusting scales in the generation of internal preference maps. *Food Quality and Preference* 20: 1–12.

Villanueva, N.D.M., Petenate, A.J., and Da Silva, M.A.A.P. 2000. Performance of three affective methods and diagnosis of the Anova model. *Food Quality and Preference* 11: 363–370.

Villanueva, N.D.M., Petenate, A.J., and Da Silva, M.A.A.P. 2005. Performance of the hybrid hedonic scale as compared to the traditional hedonic, self-adjusting and ranking scales. *Food Quality and Preference* 12: 691–703.

Wilkinson, C. and Yuksel, D. 1997. Modeling differences between panelists in use of measurement scales. *Journal of Sensory Studies* 12: 55–68.

Williams, A.A. and Arnold, G.M. 1984. A new approach to sensory analysis of foods and beverages. In *Progress in Flavour Research. Proceedings of the 4th Weurman Flavour Research Symposium*, ed. J. Adda, pp. 35–50. Amsterdam, the Netherlands: Elsevier.

Worch, T., Dooley, L., Meullenet, J-F., and Punter, P.H. 2010. Comparison of PLS dummy variables and Fishbone method to determine optimal product characteristics from ideal profiles. *Food Quality and Preference* 21: 1077–1087.

Worch, T.W., Le, S., and Punter, P. 2009. How reliable are consumers? Comparison of sensory profiles from consumers and experts. *Food Quality and Preference* 21: 309–318.

Xiong, R. and Meullenet, J.F. 2006. A PLS dummy variable approach to assess the impact of JAR attributes on liking. *Food Quality and Preference* 17: 188–198.

Index

For Product Safety Concerns and Information please contact our EU
representative GPSR@taylorandfrancis.com
Taylor & Francis Verlag GmbH, Kaufingerstraße 24, 80331 München, Germany